Ceramic Engineering

Ceramic Engineering

Edited by **Carl Burt**

NY RESEARCH
P R E S S

New York

Published by NY Research Press,
23 West, 55th Street, Suite 816,
New York, NY 10019, USA
www.nyresearchpress.com

Ceramic Engineering
Edited by Carl Burt

International Standard Book Number: 978-1-63238-479-9 (Hardback)

Printed in the United States of America.

Contents

Preface

This book studies, analyses and upholds the pillars of ceramic engineering in modern times. The various studies that are constantly contributing towards advancing technologies and evolution of this field are examined in detail. Ceramic industry has progressed at a fast pace and found its applications in small households to large industries like aerospace, power and dentistry. Due to their heat and chemical resistant properties ceramics are used widely for preserving and carrying out exothermic reactions. This book presents the complex subject of ceramic engineering in the most comprehensible manner along with the latest researches from across the globe. For someone with an interest and eye for detail, this book covers the most significant topics in this field. It is an essential guide for both academicians and those who wish to pursue this discipline further.

This book is a result of research of several months to collate the most relevant data in the field.

When I was approached with the idea of this book and the proposal to edit it, I was overwhelmed. It gave me an opportunity to reach out to all those who share a common interest with me in this field. I had 3 main parameters for editing this text:

1. Accuracy – The data and information provided in this book should be up-to-date and valuable to the readers.
2. Structure – The data must be presented in a structured format for easy understanding and better grasping of the readers.
3. Universal Approach – This book not only targets students but also experts and innovators in the field, thus my aim was to present topics which are of use to all.

Thus, it took me a couple of months to finish the editing of this book.

I would like to make a special mention of my publisher who considered me worthy of this opportunity and also supported me throughout the editing process. I would also like to thank the editing team at the back-end who extended their help whenever required.

Editor

The use of Waste Materials in Utility Poles, Crossarms, Paver, and Reef Balls Concrete Structures: Advantages and Care

Kleber Franke Portella,[1] **Alex Joukoski,**[1] **João Bosco Moreira do Carmo,**[2] **Camila Freitas,**[2] **Carlos Vicente Gomes Filho,**[3] **and Cinthya Hoppen**[4]

[1] *Instituto de Tecnologia para o Desenvolvimento, CP 19067, 81531-980 Curitiba, PR, Brazil*
[2] *PIPE, UFPR, Centro Politécnico, Jardim das Américas, 81531-980 Curitiba, PR, Brazil*
[3] *PRODETEC, Instituto de Tecnologia para o Desenvolvimento, CP 19067, 81531-980 Curitiba, PR, Brazil*
[4] *PGERHA, UFPR, Centro Politécnico, Jardim das Américas, 81531-980 Curitiba, PR, Brazil*

Correspondence should be addressed to Kleber Franke Portella; kfportella@gmail.com

Academic Editor: Young-Wook Kim

Industrial residues such as sludge from water treatment plants (Swtp) from centrifuged method; electrical porcelain residues (Pw); silica fume (Sf_1 and Sf_2); tire-rubber waste were evaluated in order to be used in concrete structures of electrical energy and environmental sectors, such as utility poles, crossarms, and reef balls technology. The results showed the necessity for evaluating different recycling concentrations in concrete, concomitantly to physicochemical tests allowing to diagnose natural and accelerated aging.

1. Introduction

One of the most recent challenges of modern society is the research of new alternatives of environmentally responsible technologies for the final disposal of residues generated by industrial, domestic, and commercial sectors.

Governmental entities and international communities have been acting together to promote environment protection and pollution reduction through environmental laws concerning well-stablished residue limits and intensive fiscalization.

These actions result in viable application to diverse residues, which can be converted in useful raw materials. An example is the silica fume byproducts, which were initially considered industrial residues, but nowadays are largely used in civil construction due to its greater pozzolanic reaction capacity compared to most of hydraulic cements. Many other different byproducts can be used, like furnace slag, metakaolin, tire-rubber residues, and others [1–10].

This paper emphasizes the importance of the study of recycling conditions and/or residues disposal in civil construction, by using engineering analyses and science materials evaluation. Thus, four different residues (silica fume, sludge from water treatment plant, electrical insulator porcelain and tire-rubber wastes) were tested in concrete core samples; in reef balls and concrete blocks technology for fishing habitat and for creating biomass (animal or plant life); in typical concrete structures of the electrical energy sector, such as crossarms, utility poles, and dams.

2. Materials and Methods

All used materials were submitted to physicochemical analyses and were pretreated prior to concrete samples casting.

2.1. Cement Materials. Filler-modified and sulfate-resistant (type CPII-F 32), high-early strength (type CPV-ARI RS), pozzolan-modified concrete (type CPII-Z 32), and sulfate-resistant pozzolanic (type CPIV-32 RS) Portland cements were used. Each of these was used in order to meet local standards and extend the durability of the structure in aggressive environments such as seawater and coastal regions.

2.2. Artificial and Natural Fine and Coarse Aggregates. Fine and coarse aggregates consisted of washed natural medium sand and crushed basalt stone with maximum nominal size of 4.8 and 19 mm, respectively. They were tested according to the recommendations of Brazilian standards. Synthetic aggregates from waste samples were crushed to both fine and coarse nominal size and also tested.

2.2.1. Sludge from Water Treatment Plant Swtp. The search for economically and environmentally advantageous solutions for the treatment and sludge disposal of WTP remains a challenge, especially for developing countries, living in severe economic constraints and where health problems require emergency solutions. Monthly, approximately 4,000 tons of WTP sludge dry matters are produced throughout the state of Paraná, southern Brazil. In the city of Curitiba (population of 1.9 million people, approximately), capital of Paraná, the potable water supply is provided by Iguaçu, Tarumã, Irai, and Passaúna WTPs, which account for over 50% of all the production of sludge in the state. Passaúna WTP produces, by centrifugation method, about 360 tons/month of sludge. It is the WTP object of study in this work. For this research, collections of hebdomadaires centrifuged sludge were conducted during two months of the year. Afterwards, the final content was homogenized, oven-dried at 110°C, and disaggregated [3].

2.2.2. Electrical Porcelain Waste (Pw). Artificial aggregates of medium and high voltage electrical porcelain waste were obtained by grinding the product in a hammer mill type. The crushed material was classified into different particle sizes after grinding. The fine particulate portion was used to study its potential alkali reactivity by mortar-bar method [11] and the coarse one was separated in four quotas, thus considered: glazed porcelain (as obtained and grounded) with sulfur cement phases (cement waste from the junction of the porcelain to the metallic part of the insulator); glazed porcelain without sulfur; porcelain with sulfur and without its surface glaze and plain porcelain material [12]. The separation of these parts from the raw material is related to the investigation of their potential contributions to the alkali reactivity in concrete.

2.2.3. Silica Fume (Sf). Condensed silica fume (Sf) is a byproduct usually originated from induction arc furnaces in the silicon metal or ferrosilicon alloy industrial processes, where the reduction of quartz to silicon at temperatures up to 2000°C produces SiO vapors, which oxidize and condense in the low-temperature zone to tiny spherical particles consisting of amorphous or noncrystalline silica. The amount of SiO_2 present in this pozzolan is, invariably, close to 80% and is directly related to the existent production process [1].

Currently, Sf was widely used as a supplementary cement material to enhance the strength and durability of concrete. In this research, Sf was also used in order to lower the pH of the resulting concrete to facilitate settlement of marine organisms.

2.2.4. Waste Rubber from Retreading Tires (Tw). The use of rubber waste in concrete is important from the ecological point of view. Population growth and increased use of disposable materials such as packaging, tires, and PET bottles, among others, have caused the accumulation of large quantities of solid waste, which are limiting the capacity of landfills. In 2005, the city of Rio de Janeiro, southwest of Brazil, tires and rubber products accounted for about 0.5% of urban waste and in São Paulo, this quantity is near 3% [13].

Rubber band scroll waste from retreading tires was used without any pretreatment. The composition of the predominant residue was characterized as styrene butadiene rubber by infrared Fourier transform. The average particle size distribution was 4.8 mm.

2.3. Dosage. Ideal concrete mix proportions (by mass for a concrete mixture (w/w)) are listed in Table 1. For each concrete mix, a reference concrete (RC), without addition, was also produced to serve as comparison. To the concrete mixture Tw, two other ratios were studied as a 5 to 15% (w/w) of rubber addition. However, only the 10% (w/w) Tw was considered due to its performance.

2.4. Specimens Casting. For each mixture and material, different types of specimens were casted:

(1) (100×200) mm cylindrical concrete specimens for determining compressive and flexural strength and elastic modulus at 3, 14, and 28 days after casting [14, 15];

(2) (150×300) mm cylindrical concrete specimens for permeability testing and determining specific density, absorption, and porosity of concrete after 28 days of curing;

(3) (300×100) mm cylindrical concrete plates for abrasion resistance of concrete according to ASTM C1138 [16].

Also, six concrete utility poles and crossarms were casted with Pw_1, RC, and Sf_1 and tested for flexural strength after 28 days of concrete curing and also for electrical properties and visual surface inspection during natural ambient exposition. The poles were double-tee cross-section shaped, B type, 11 m long, and with 300 daN of nominal strength, complying with a Brazilian standard [17].

2.5. Physicochemical Characterization of Samples. Cement and natural and artificial aggregates were characterized by physicochemical analyses, previously to their use in the mixtures. Elemental chemical composition and chemical phases were obtained from energy dispersive spectroscopy (EDS), X-ray fluorescence (XRF), and X-ray diffractometry (XRD) methods. PW 2400 Philips fluorescence equipment was used to determine the elemental chemical composition. XRD of specimens were measured using a Philips (X'Pert MPD) diffractometer with Cu-Kα radiation operating at 40 kV and 40 mA. The diffraction patterns were used to identify the structural phases of the specimens. The micrography analyses of fractured concrete surfaces were done using an XL30

TABLE 1: Mix ideal proportions (by mass for a concrete mixture) and properties of fresh concrete with admixtures.

Mixture/sample	Materials code					
	Swtp	Pw	Pw_1*	Sf_1**	Sf_2***	Tw
Cement	1	1	1	1	1	1
Sludge (water treatment plant): Swtp (%)	**8**	0	0	0	0	0
Porcelain waste (Pw) (%)	0	**50**	25	0	0	0
Silica fume (Sf) (%)	0	0	0	**8**	**15**	0
Tire-rubber waste (Tw) (%)	0	0	0	0	0	**10**
Fine aggregate (natural sand)	1.860	1	347	1.837	1.541	1.820
Fine aggregate (artificial)	0.16	1	347	0	0	0
Natural coarse aggregate (19 mm)	2.980	1.497	512	3.306	1.541	2.620
Artificial coarse aggregate (19 mm)	0	1.497	512	0	0	0
Water/cementitious materials ratio	0.51	0.50	0.5	0.46	0.41	0.50
Superplasticizer	0	0	0	1	0.3	0
Slump (mm)	18	25	25	50	50	14
Unit weight (kg/m^3)	2,247	2,219	2,219	2,357	—	2,220
Air content (%)	3.5	0.5	0.5	—	—	2.4

Notes: Portland cement types: filler modified (CPII-F 32); *high-early strength (CPV-ARI RS); **pozzolan modified (CPII-Z 32); ***sulfate resistant (CPIV-32 RS).

model Philips Scanning Electron Microscope (SEM). Gold was applied to the surfaces by sputtering.

Potential alkali reactivity of cement and aggregate was evaluated according to ASTM specification [11].

Nondestructive method, such as electrical half-cell potential with copper-copper sulfate reference electrode, CSE, was used to verify the service life performance of reinforcing steel in concrete in a 3.4% w/w chloride solution and of concrete utility poles submitted to natural aging condition, according to the literature [18–20]. The metallic rebar of the poles was connected as the working electrode. An average potential from thirty measurements was obtained from the testing at both sides of the bottom region of poles, just above the embedment line.

2.6. Environmental Corrosion Stations (ECSs). A marine ECS was built at Caueira beach, in Itaporanga D'Ajuda district, near Aracaju, SE, in northeastern Brazil. It was located at about 2 m above the high tide line of the Atlantic Ocean [19]. Utility poles, crossarms, and concrete material core samples casted with Sf_1 were submitted to natural aging for approximately 500 days. Also, an urban ECS was built at Curitiba, PR, southern Brazil, to test Pw_1 concrete casted in utility poles and crossarms.

Sf_2 concrete admixtures were cast as thick plates (20 × 270 × 300) mm to be previously tested in a marine ECS located at 17 m depth, following a perpendicular line of Praia de Leste beach coast until 30 m deep (25°30′ S). Sample plates were periodically tested by flexural strength and by microstructure concrete surface investigated by Scanning Electron Microscope (SEM) and energy dispersive system analyses (EDS). Afterwards, this composition was cast as reef balls and 1 m^3 concrete blocks technology forms. Both of them were tested in a marine shallow shelf at approximately 30 m of depth for approximately five years [21].

The pH of this resultant concrete mix lowered from 12.3 to 11.4. Concrete with this pH generally needs to age in the ocean for 3–6 months before the pH in the surface region approaches the 8.3 pH of seawater and favour marine organism's settlement [22].

3. Results and Discussion

The physicochemical analyses of cements were in accordance to manufacturer specifications and Brazilian standards.

3.1. Concrete Admixtures

3.1.1. Swtp. The major chemical components obtained by XRF tests from the sludge sample were: 16.55% of silica, 13.07% of alumina, 4.15% of ferrite, 49.79% of volatile materials, and 16.44% of humidity. In natura water treatment sludge was identified as kaolinitic group by XRD as shown in Figure 1.

The result of average compressive strength of 8% (w/w) Swtp concretes at 28 days of curing was 27.6 MPa, being superior to setup limit used for concrete structures as poles for electric energy distribution network. The average flexural strength at 28 days was 3.0 MPa, which is in accordance to the literature data for similar admixtures [1, 20]. Increasing the concentration of sludge from water treatment plant to 10% w/w in concrete, the microstructures presented large porosity, poor compressive strength lower than 15 MPa and the slump test results was 0 mm. This turned the concrete workability to be nonsatisfactory.

The average permeability resulted from 8% (w/w) Swtp was 0.9×10^{-10} cm/s. As the permeability of concrete depends on mix proportions, compaction, curing, and microcracks in the core, and also there is a close relationship between the strength and its durability, the results indicated a good

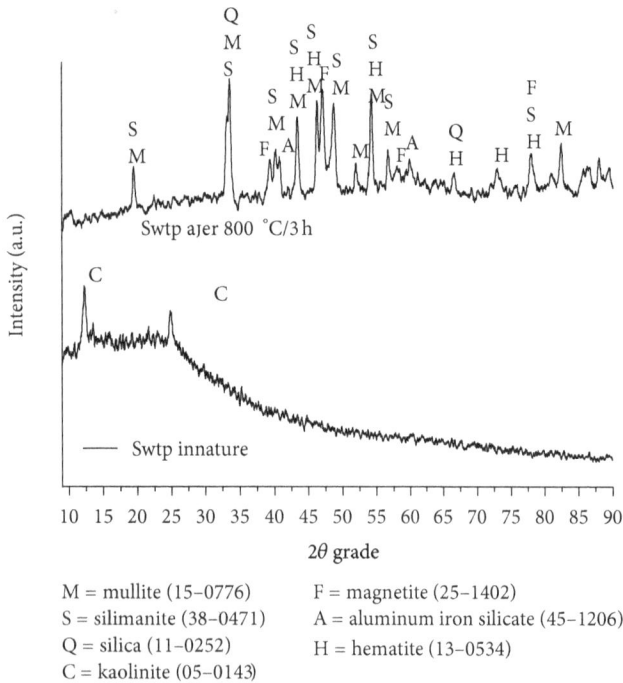

M = mullite (15–0776)   F = magnetite (25–1402)
S = silimanite (38–0471)   A = aluminum iron silicate (45–1206)
Q = silica (11–0252)   H = hematite (13–0534)
C = kaolinite (05–0143)

FIGURE 1: XRD pattern of Swtp phases: "in natura" and after 3 h/800°C treatment.

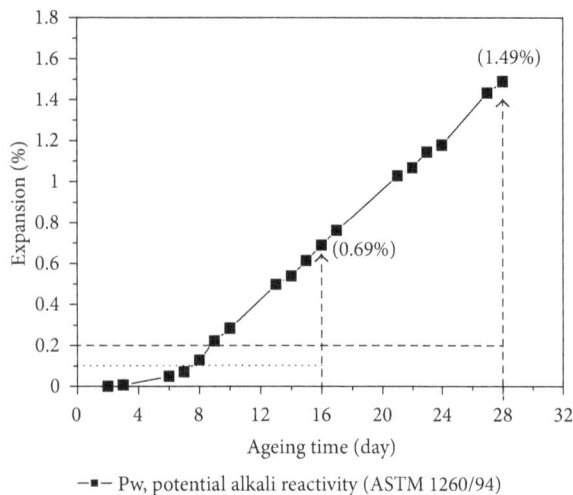

FIGURE 2: Expansion of Pw admixture mortar sample.

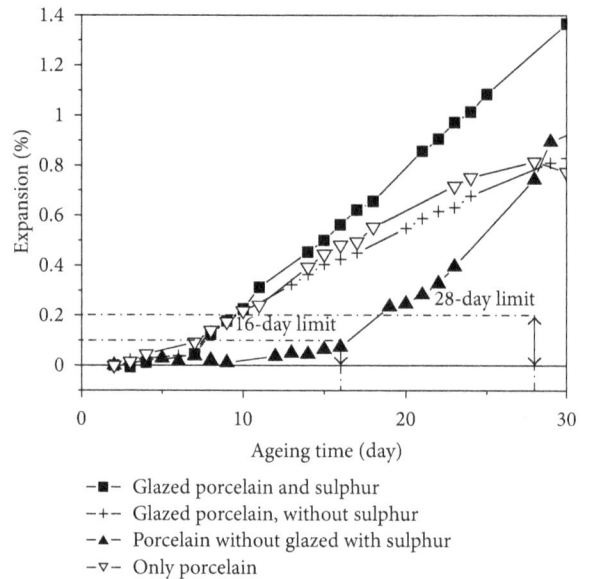

-■- Glazed porcelain and sulphur
-+- Glazed porcelain, without sulphur
-▲- Porcelain without glazed with sulphur
-▽- Only porcelain

FIGURE 3: Expansion of mortar admixtures containing porcelain insulator waste.

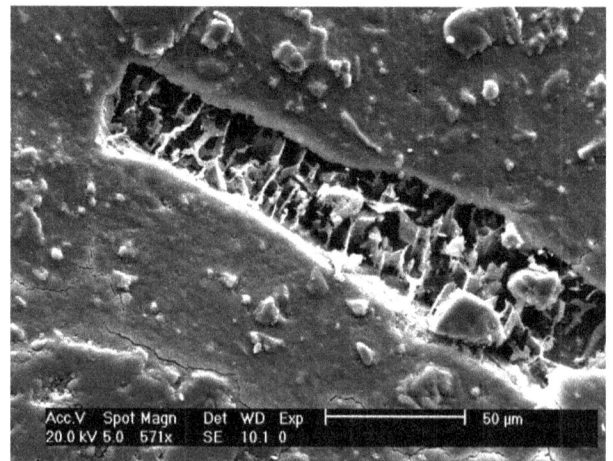

FIGURE 4: SEM of sulphur cement phase micrography in porcelain concrete admixtures.

performance mixture resultant from kaolinitic group and cement phases.

3.1.2. Pw.

Porcelain waste concrete admixture resulted potentially in alkali reactive with CPII-F 32 cement type, as shown in Figure 2.

The expansion tests resultant from different picked-up porcelain material parts and CPII-F 32 cement type showed that porcelain with glazed and sulphur cement phases (as obtained and grounded) is the most damaging to the concrete materials, followed by porcelain parts, porcelain material with glazed but without sulphur cement phase parts, followed by porcelain materials without glazed and with sulphur cement phases. This latest phase presented the lower expansion results, as demonstrated in Figure 3. Sulphur cement phase did not demonstrate larger expansion in its first 16 days agied according to ASTM tests [11]. From 16 to 28 ageing time days, the mortar samples demonstrated positive alkali reaction with an exponential slope expansion results passing to noninnocuous limit, as viewed in Figure 3. Besides, this Pw reinforcing steel samples with sulphur cement phase presented too bulk defects that are capable to enlarge cracking probability risk, as shown in Figure 4, by SEM micrography images.

Pw compressive strength resulted in 30.4 MPa at 28 days of curing, being classified as restrained resistance [14].

Because of the reduced lifespan resulted from the potential reactivity essay with porcelain materials and CPII-F 32

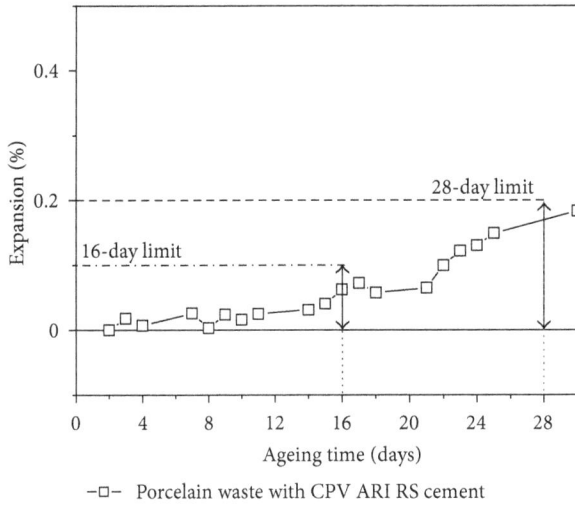

FIGURE 5: Expansion results of Pw mortar casted with high-early strength sulfate-resistant cement (CPV ARI-RS).

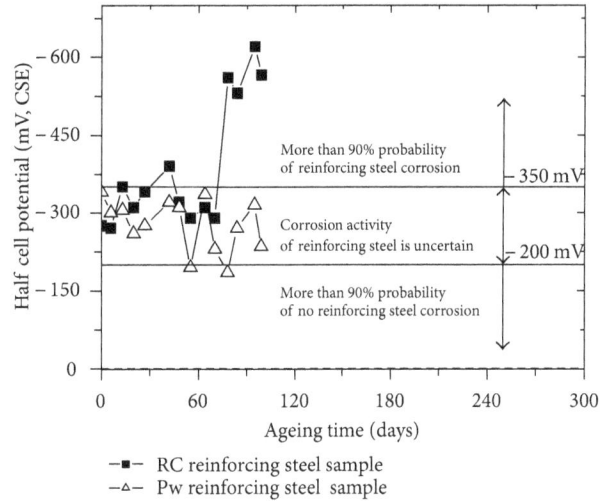

FIGURE 7: Electrical half-cell potential results of RC and Pw reinforcing steel in concrete as a function of aging time in 3.4% NaCl solution.

FIGURE 6: The image shows 3 utility poles and crossarms. Two of them were cast with RC and Pw admixtures (right positions). In detail, is illustrated a CSE system used to measure the seasonal reinforced steel corrosion potential performance.

cement type, the Pw mortar sample was in a second time cast with special sulfate-resistant cement. Portland high-early strength cement (CPV-ARI RS) reduced its expansion limit in 16- and 28-day ageing test to the recommended values (innocuous consideration) as shown in Figure 5. Nevertheless, additional care should be taken because of a positive slope tendence to high delay expansion values.

In Figure 6 is showed an image of utility poles and crossarms under natural ageing in an ECS urban environment located in Curitiba, PR, Brazil. In the right position is a 25% w/w concrete Pw admixture tested with RC utility poles located at the left. In detail is viewed the electrical half-cell potential electrode system for nondestructive test. Even so the utility poles and crossarms were casted using specially cement type (CPV-ARI RS) with 25% w/w porcelain waste in concrete to reduce the probability risk of alkali expansion presented by the 50% w/w Pw one. The rupture of the RC and Pw utility poles by flexural strength was 360 and 440 daN, respectively, being in accordance with the Brazilian specification [15].

Electrical half-cell potential measured during these first five months on urban ambient ECS condition indicated no corrosion activity for RC and Pw structures. As previously reported, both structures are exposed at low aggressive atmosphere.

The electrical results made in RC and Pw reinforcing steel in concrete partially immersed in 3.4% w/w NaCl aqueous solution as function of ageing time are showed in Figure 7. As viewed, Pw reinforcing steel material has been presenting lower corrosion activity performance than RC reinforcing steel.

3.1.3. Sf_1 and Sf_2 Concrete Admixtures. Silica fume admixtures in concrete have been presenting better lifespan performance of concrete structure submitted to salt aggressive environment, as viewed in Figure 8, by electrical half-cell potential rebar measurement results.

Tests made in the reinforcing steel in concrete submitted partially immersed in a 3.4% NaCl aqueous solution have been indicating that Sf_1 samples are having a double lifespan in comparison with RC reinforcing steel, both of them tested at the same laboratory salt aggressive condition.

Visual inspection on Sf_1 and RC utility poles submitted to northeast Brazilian Caueira beach ECS demonstrated that RC structures have been presenting, too, rebar corrosion surfaces with consequently concrete microcracks. Any corrosion

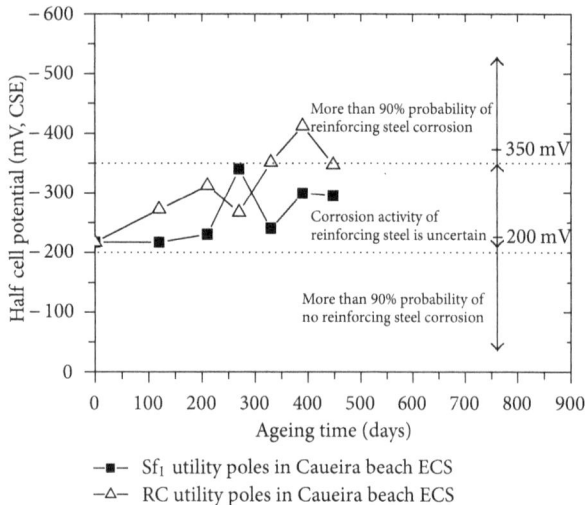

FIGURE 8: Electrical half-cell potential results of RC and Sf_1 utility poles submitted to the natural aging for 480 days in northeast Brazilian Caueira beach ECS.

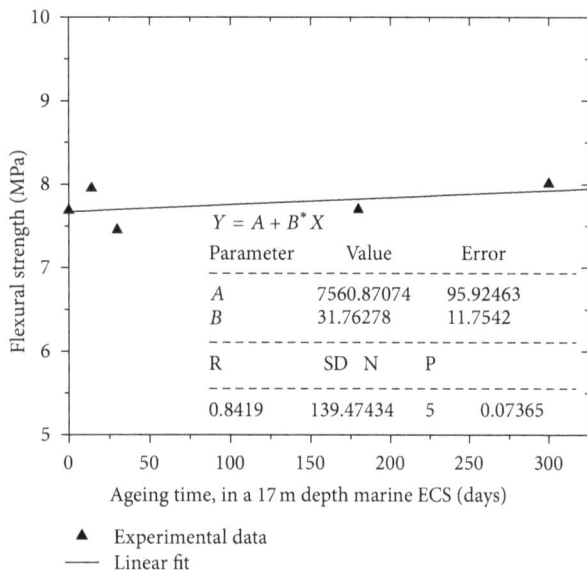

FIGURE 10: Sf_2 concrete admixture and other materials thick plates tested in marine ECS located at 17 m depth, at Praia de Leste beach.

FIGURE 9: Flexural strength of Sf_2 concrete admixture in terms of ageing time in a 17 m depth marine ECS.

FIGURE 11: Sf_2 reef balls technology and block forms of Sf_1 concrete mixture before installation in a marine shallow shelf at Parana state, southern Brazil. The detail shows nowaday jewfish in the block habitat.

surface defects have been viewed on Sf_1 concrete structures exposed at same environmental condition.

Besides the increase of lifespan concrete structures, the silica fume material causes economical positive effects when compared to a reference concrete. The economical differences lowered from 34% in 28 curing days to 24% in 90 curing days, at the same compressive strength results. This phenomenon had been attributed to pozzolan-modified Portland (CPII-Z 32) cement used to cast both samples [5].

Sf_2 mixes casted in thick plates and tested previously in a 17 m depth marine ECS demonstrated good lifespan performance by flexural strength results in the function of ageing time, as well as in terms of biological marine material habitat, as shown in Figures 9 and 10, respectively.

Sf_2 cast as reef balls technology and $1\,m^3$ blocks, as presented in Figure 11, has showed also lifespan good performance in the last 10 years old in a Parana marine shallow shelf at southern Brazil. In Figure 11 is also shown in detail a nowaday jewfish habitat [21].

3.1.4. Tw. Tw concretes cast with up to 20% w/w were tested to be used as paver and curb pieces and as repair materials for hydraulic concrete structures of hydroelectric power plants [4]. Until 15% w/w Tw concrete admixtures, the results showed no serious decreasing in fresh and cured concrete properties relative to RC samples, such as the slump test and compressive strength results. Regarding the 10% w/w Tw concrete admixture, its behaviour was even better than the 15% one.

In Figure 12 are presented paver block manufactured with 10% w/w Tw admixture and, in detail, the surface of paver casted with 20% w/w Tw concrete admixture. As shown,

FIGURE 12: Paver block with 10% w/w Tw concrete mixture. The paver surface view of 20% w/w Tw concrete mixture is shown in detail.

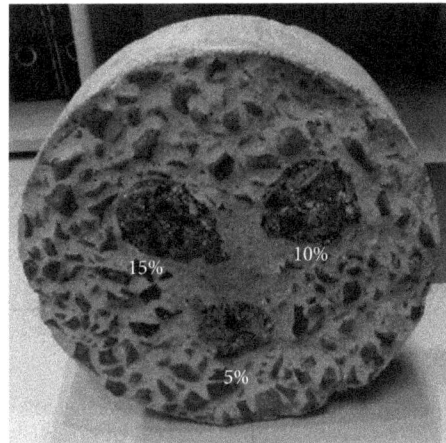

(a)

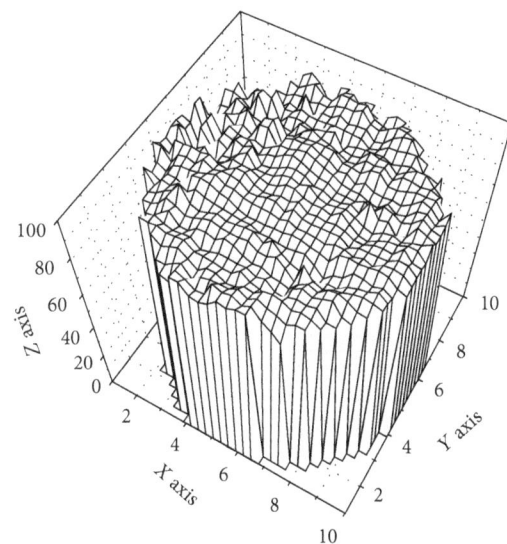

(b)

FIGURE 13: Abrasion-erosion Tw samples image tested and the schematic wearing surface plot results. The wearing surface resulted from three different repair materials concentration: 5% w/w Tw; 10% w/w Tw; 15% w/w Tw.

the 20% w/w Tw composition presents larger quantities of rubber-tired concrete surface efflorescence and porous defect, causing poor visual market aspect and lower resistance.

Compressive strength tests in 10% w/w Tw resulted in 20 MPa, presenting at the same time economic benefits and environmental advantages for materials repair in hydraulic structures [4–21].

Abrasion-erosion test results of Tw samples showed that the wearing was 75% lower than RC samples, meaning that they have better performance. In Figure 13 is showed a sample photograph and the wearing surface plot results after testing with three different repair materials concentration, such as: 5% w/w Tw; 10% w/w Tw; 15% w/w Tw. In all cases Tw materials wearing results are lower than the RC substrate.

4. Conclusion

The Swtp obtained by centrifuged method can be used as concrete admixture up to 10%. The best mechanical behaviour was achieved with 8% w/w Swtp concrete mixture, indicating that it can be used in concrete utility poles and crossarms.

Pw concrete admixture cast with CPII-F 32 cement type had its lifespan reduced due to the large potential reactivity values in all partial chemical phases analysed. These alkali aggregate reactions were reduced after the change of the cement type to a high-early strength and sulfate-resistant one (CPV-ARI RS). This composition could be cast in utility poles and crossarms applied in electrical distribution energy until 25% w/w porcelain waste. Even so is strongly recommended laboratory tests previously cast it in civil structures.

Sf_1 and Sf_2 silica fume mixtures cast as utility poles and crossarms, and reef balls technology and cubic concrete blocks showed good resistance performance when used in high salt Brazilian northeast coastal areas and into 17 m depth marine environment, respectively. Poles, crossarms, reef balls structures and cubic concrete blocks with silica fume

admixtures (Sf_1 and Sf_2) showed good chemical resistance performance in high salt Brazilian Northeast coastal areas and, into 17 m depth marine environment, during 500 days and approximately 5 years of tests exposition, respectively.

Tw composition up to 15% w/w concrete mixture had good mechanical performance during tests, showing that it can be used as paver, curb pieces, and as repair materials for hydraulic structures concrete dams. The 20% w/w Tw composition presented large quantities of rubber-tired concrete surface efflorescence and porous defect, causing poor visual market aspect and low mechanical resistance.

As observed, the worry on the use of recycled materials is not limited to structural stability, but also their durability in concrete structures, meanly submitted to salt aggressive environment.

Acknowledgments

This work was sponsored by Companhia Energética de Sergipe (ENERGIPE), Companhia Paranaense de Energia (COPEL), Instituto de Tecnologia para o Desenvolvimento (LACTEC), Agência Nacional de Energia Elétrica (ANEEL), UFPR/CEM, and Conselho Nacional de Desenvolvimento Científico e Tecnológico (CNPq, Lei 8010/90). The authors wish to thank the (UFPR/PIPE and UFPR/PGERHA), LACTEC/IEP/PRODETEC.

References

[1] P. K. Mehta and P. J. M. Monteiro, *Concreto: Estrutura, Propriedades e Materiais*, PINI, São Paulo, Brazil, 1994.

[2] V. M. John, *Reciclagem de resíduos na construção civil: contribuição à metodologia de pesquisa e desenvolvimento [thesis]*, Universidade de São Paulo, Sao Paulo, Brazil, 2000.

[3] C. Hoppen, K. F. Portella, A. Joukoski, E. M. Trindade, and C. V. Andreóli, "The use of centrifuged sludge from a water treatment plant (WTP) in portland cement concrete matrices for reducingthe environmental impact," *Química Nova*, vol. 29, no. 1, pp. 79–84, 2006.

[4] C. V. G. Filho, *Metodologia para a gestão diferenciada de resíduos da construção civil urbana: estudo de gestão ambiental de rejeitos de pneus [Dissertation]*, PRODETEC, Ribeirão Preto, Brazil, 2007.

[5] J. B. M. do Carmo and K. F. Portella, "Estudo comparativo do desempenho mecânico da sílica ativa e do metacaulim como adições químicas minerais em estruturas de concreto," *Cerâmica*, vol. 54, no. 331, pp. 309–318, 2008.

[6] J. M. R. Dotto, A. G. de Abreu, D. C. C. dal Molin, and I. L. Müller, "Influence of silica fume addition on concretes physical properties and on corrosion behaviour of reinforcement bars," *Cement and Concrete Composites*, vol. 26, no. 1, pp. 31–39, 2004.

[7] A. Magrinho, F. Didelet, and V. Semiao, "Municipal solid waste disposal in Portugal," *Waste Management*, vol. 26, no. 12, pp. 1477–1489, 2002.

[8] J. R. Pan, C. Huang, J. J. Kuo, and S. H. Lin, "Recycling MSWI bottom and fly ash as raw materials for Portland cement," *Waste Management*, vol. 28, no. 7, pp. 1113–1118, 2008.

[9] J. M. L. Reis dos and M. A. G. Jurumenh, "Experimental investigation on the effects of recycled aggregate on fracture behavior of polymer concrete," *Materials Research*, vol. 14, no. 3, pp. 326–330, 2011.

[10] M. Safiuddin, U. J. Alengaram, M. A. Salam, M. Z. Jumaat, F. F. Jaafar, and H. B. Saad, "Properties of high-workability concrete with recycled concrete aggregate," *Materials Research*, vol. 14, no. 2, pp. 248–255, 2011.

[11] American Society for Testing and Materials (ASTM), "Standard test method for 'potential alkali reactivity of aggregates (mortar-bar method)'," ASTM C1260, EUA. ASTM, Philadelphia, Pa, USA, 1994.

[12] K. F. Portella, A. Joukoski, R. Franck, and R. Derksen, "Secondary recycling of electrical insulator porcelain waste in Portland concrete structures: determination of the performance under accelerated aging," *Cerâmica*, vol. 52, no. 323, pp. 155–167, 2006.

[13] C. Freitas, J. C. A. Galvão, K. F. Portella, A. Joukoski, C. V. G. Filho, and E. S. Ferreira, "Physicochemical and mechanical performance of portland cement concrete with recycled styrene-butadiene tyre-rubber waste," *Química Nova*, vol. 32, pp. 913–918, 2009.

[14] American Society for Testing and Materials (ASTM), "Standard test method for 'compressive strength of cylindrical concrete specimens'," ASTM C39/C39M, EUA. ASTM, Philadelphia, Pa, USA, 1999.

[15] American Society for Testing and Materials (ASTM), "Standard test method for 'flexural strength of concrete (using simple beam with third-point loading)'," ASTM C78, EUA. ASTM, Philadelphia, Pa, USA, 1994.

[16] American Society for Testing and Materials (ASTM), "Standard test method for 'abrasion resistance of concrete (underwater method)'," ASTM C1138, EUA. ASTM, Philadelphia, Pa, USA, 1997.

[17] Associação Brasileira de Normas Técnicas (ABNT), "Postes de concreto armado para redes de distribuição de energia elétrica—Especificação," ABNT NBR 8451, NBR, Brazil, 1985.

[18] American Society for Testing and Materials (ASTM), "Standard test method for 'half cells potentials of uncoated reinforcing steel in concrete'," ASTM C876, EUA. ASTM, Philadelphia, Pa, USA, 1991.

[19] A. Joukoski, K. F. Portella, O. Baron et al., "The influence of cement type and admixture on life span of reinforced concrete utility poles subjected to the high salinity environment of Northeastern Brazil, studied by corrosion potential testing," *Cerâmica*, vol. 50, no. 313, pp. 12–20, 2004.

[20] A. M. Neville and S. E. Giammusso, *Propriedades do Concreto*, PINI, São Paulo, Brazil, 2nd edition, 1997.

[21] Recifes artificiais marinhos: uma proposta de conservação da biodiversidade e desenvolvimento da pesca artesanal, http://www.brasilmergulho.com.br/port/artigos/2003/002.shtml.

[22] K. M. Dooley, C. F. Knopf, and R. P. Gambrell, pH-neutral concrete for attached microalgae and enhanced carbon dioxide fixation DE-AC26-98FT40411-01, Department of Energy, Louisiana State University, 1998.

Mineral-Oxide-Doped Aluminum Titanate Ceramics with Improved Thermomechanical Properties

R. Papitha,[1] **M. Buchi Suresh,**[1] **Dibakar Das,**[2] **and Roy Johnson**[1]

[1] *Center for Ceramic Processing, International Advanced Research Centre for Powder Metallurgy and New Materials (ARCI), Balapur, Hyderabad 500005, India*
[2] *School of Engineering Sciences and Technology, University of Hyderabad, Hyderabad 500046, India*

Correspondence should be addressed to Roy Johnson; royjohnson@arci.res.in

Academic Editor: Zhenxing Yue

Investigations were carried out, on the effect of addition of kaolinite ($2Al_2O_3 \cdot 3SiO_2 \cdot 2H_2O$) and talc ($Mg_3Si_4O_{10}(OH)_2$) in terms of bulk density, XRD phases, microstructure, as well as thermal and mechanical properties of the aluminium titanate (AT) ceramics. AT ceramics with additives have shown enhanced sinterability at 1550°C, achieving close to 99% of TD (theoretical density) in comparison to 87% TD, exhibited with pure AT samples sintered at 1600°C, and found to be in agreement with the microstructural observations. XRD phase analysis of samples with maximum densities resulted in pure AT phase with a shift in unit cell parameters suggesting the formation of solid solutions. TG-DSC study indicated a clear shift in AT formation temperature with talc addition. Sintered specimens exhibited significant reduction in linear thermal expansion values by 63% (0.42×10^{-6}/C, (30–1000°C)) with talc addition. Thermal hysteresis of talc-doped AT specimens showed a substantial increase in hysteresis area corresponding to enhanced microcrack densities which in turn was responsible to maintain the low expansion values. Microstructural evaluation revealed a sizable decrease in crack lengths and 200% increase in flexural strength with talc addition. Results are encouraging providing a stable formulation with substantially enhanced thermomechanical properties.

1. Introduction

Aluminum Titanate (Al_2TiO_5, designated as AT) ceramics exhibit excellent thermal properties such as low thermal expansion coefficient in the range of $1.0–1.5 \times 10^{-6}$/C (RT-1000°C) in combination with low thermal conductivity ($\sim 1.5\,Wm^{-1}K^{-1}$) and a high melting temperature of 1860°C [1–4]. This makes Al_2TiO_5, a material of choice for many refractory applications. Some of these applications include wall flow filters for diesel particulate emission control, exhaust port liners in automotive engines and thermal shock resistant refractory parts for nonferrous metallurgical industries [5–8]. One of the disadvantages of this ceramic material for practical applications is the low flexural strength due to the extensive microcracks generated whil processing [3, 4, 9]. Crystal structure of aluminum titanate (β-Al_2TiO_5) is pseudo-brookite and is associated with the strong anisotropy in crystallographic axis while heating, which is responsible for microcracking on cooling [9, 10]. The microcracking phenomenon is closely related to the material microstructure. Below a critical grain size, the elastic energy of the system is insufficient for microcrack formation during cooling and thus the mechanical properties are considerably enhanced. Additionally, eutectoid decomposition to its parent oxides such as α-Al_2O_3 and TiO_2 between 750 to 1280°C is also an issue leading to thermal instability of the Al_2TiO_5 phase. This decomposition occurs when the adjacent Al^{3+} (0.54 Å) and Ti^{4+} (0.67 Å) octahedra collapse, because of the lattice site occupied by the Al^{3+} ion is too large. The thermal energy available from this collapse allows Al^{3+} to migrate from its position and causes structural dissolution to rutile (TiO_2) and corundum (α-Al_2O_3) [9–12].

It is also well known that, thermomechanical properties of microcracked ceramics is a function of dopants and many attempts to dope with various oxide compounds (such as MgO, ZrO_2, Fe_2O_3, SiO_2, $ZrSiO_4$) are made so far [13–19]. These oxides are reported to be effective in densification and improving the thermal stability. However, they need to be doped in high-volume fractions, deteriorating the thermal properties and long-term stability [15]. Doping with spodumene (LiAlSiO$_4$), mullite (3Al$_2$O$_3$·2SiO$_2$), cordierite (Mg$_2$Al$_4$Si$_5$O$_{18}$), and feldspar have also been attempted, but and no significant enhancement in thermomechanical properties were reported [19–25]. Studies have also attempted to dope with SiO$_2$ (9 wt%) and co-doping with MgO (10 wt%) and SiO$_2$ (8 wt%), which have reported significant enhancement in stability and mechanical properties. However, doping with SiO$_2$ resulted in unreacted residual TiO$_2$ phase and the formation of additional phases such as MgAl$_2$O$_4$ and Mg$_2$SiO$_4$ with co-doping [17].

In view of the above, the objective of this paper is to elucidate systematically the effect of additives such as kaolinite and talc into Al$_2$TiO$_5$ formulation with varying concentrations followed by the evaluation of thermomechanical properties. Accordingly, precursor oxide samples were compacted with and without additives and were subjected to sintering at varying temperatures to achieve close to theoretical density. Sintering behavior of AT was elucidated with dilatometric shrinkage curve and DSC studies. Sintered samples were characterized for their density, phase, microstructure, and CTE measurements. Further, the thermal hysteresis was recorded while heating and cooling using dilatometer. Flexural (3-point bend) strength was also determined and correlated with microcrack densities, observed with fractograph.

2. Experimental Procedure

Basic raw materials such as alumina (Al$_2$O$_3$, Baikowski, France) and titania (TiO$_2$, Qualigens) powders along with the additives kaolinite (2Al$_2$O$_3$·3SiO$_2$·2H$_2$O) and talc (Mg$_3$Si$_4$O$_{10}$(OH)$_2$) were used for the investigations. XRD phase analysis of the raw materials was carried out by Bruker's D8 advanced system and morphology and particle size analysis by scanning electron microscope (S-4300SE/N, Hitachi, Tokyo, Japan). Physiochemical properties of the raw materials are depicted in Table 1. The concentration of silica (1.5, 3, and 4.5 wt%) in aluminum titanate and sample ID's (C$_0$, C$_1$, C$_2$, C$_3$, C$_4$, C$_5$, and C$_6$) were depicted in Table 2. The formulations were granulated with 2 wt% of poly vinyl alcohol as a binder and compacted into green compacts of 65 mm × 65 mm × 8 mm using a hydraulic press. The green density of the compacted samples was measured by the dimensional method and was found to be greater than 50% of the theoretical density (estimated by the rule of mixtures). In order to study, the extent of physiochemical changes occurring in green specimens over the processing temperature ranges, the specimens were subjected to the differential scanning calorimetric (DSC) analysis using TG-DSC analyzer (Netzsch, Germany) from room temperature to 1550°C. Basic formulation of Al$_2$TiO$_5$ (C$_0$) has been subjected to dilatometry (Netzsch 402C, Germany) and the shrinkage profile was recorded with respect to the temperature. Three sintering temperatures of 1500°C, 1525°C, and 1550°C were selected for the sintering of all formulations and selection of optimum sintering temperature.

Densities of the samples sintered at different temperatures were evaluated by widely used Archimedes principle (ASTM 792) and phases were analysed by XRD (Bruker's D8 advanced XRD). The polished samples were thermally etched at a temperature, 50°C below the sintering temperature for microstructural observations using a scanning electron microscope (S-4300SE/N, Hitachi, Tokyo, Japan). The specimens were also subjected to dilatometric analysis using a push-rod-type dilatometer (Netzsch 402C, Germany) incorporating the sample holder correction to determine coefficient of linear expansion (CTE). Thermal hysteresis was recorded for the samples of C$_0$, C$_3$, and C$_5$ formulations with additives which have exhibited the highest densities. Sintered samples with the highest densities were also machined to the rectangular specimens of 45 mm × 4 mm × 3 mm for the determination of the flexural strength using 3-point bend test following ASTM C-1161-02C (Instron).

3. Results and Discussion

3.1. Characterisation of Raw Materials. XRD patterns and SEM micrographs of the basic raw materials recorded for Al$_2$O$_3$ and TiO$_2$ used for the present investigations are shown in Figures 1(a), 1(b), 1(c), and 1(d), respectively. XRD pattern of Al$_2$O$_3$ indicated the coexistence of α-phase (~85%) as a major phase along with γ-phase (~15%) and TiO$_2$ powder has shown single anatase phase. SEM micrographs of both the basic powders have shown an irregular morphology with an average particle size of 150 and 300 nm, respectively.

3.2. Characterisation of Sintered Specimens

3.2.1. Densification of Al$_2$TiO$_5$ with and without Additives. Variation of bulk densities with sintering temperature for all the compositions is shown in Figures 2(a) and 2(b). It is evident that the bulk density of Al$_2$TiO$_5$ formulation without additives has shown only a marginal increase from 84 to 85% of TD with increase of sintering temperature from 1500 to 1550°C. No significant increase in density is observed even at a sintering temperature of 1600°C. Composition, concentration, and sintering temperature are to found have a significant effect on the final density values. Concentrations of both the additives are fixed based on their silica content (Table 1) in such a way that the final formulation corresponds to a silica addition of 1.5, 3, and 4.5% (Table 2). Kaolinite (2Al$_2$O$_3$·3SiO$_2$·2H$_2$O) addition from 2.9 to 8.8% has resulted in the consistent increase in density from 90% to a maximum density of 99% of TD. Talc (Mg$_3$Si$_4$O$_{10}$(OH)$_2$) addition from 2.5 to 5% increases the density from 92 to 98.63%; however, unlike kaolinite, a higher concentration of talc (Mg$_3$Si$_4$O$_{10}$(OH)$_2$) has not shown any significant increase in density beyond 92%.

Dilatometric curve of Al$_2$TiO$_5$ depicted in Figure 3(a) clearly shows an initial shrinkage corresponding to the in-situ reaction sintering of the starting mixture of Al$_2$O$_3$ and TiO$_2$ followed by an expansion regime corresponding to formation of Al$_2$TiO$_5$ and is in good agreement with endothermic peak at 1375°C (Figure 3(b)). This expansion regime in

(a)

(b)

(c)

(d)

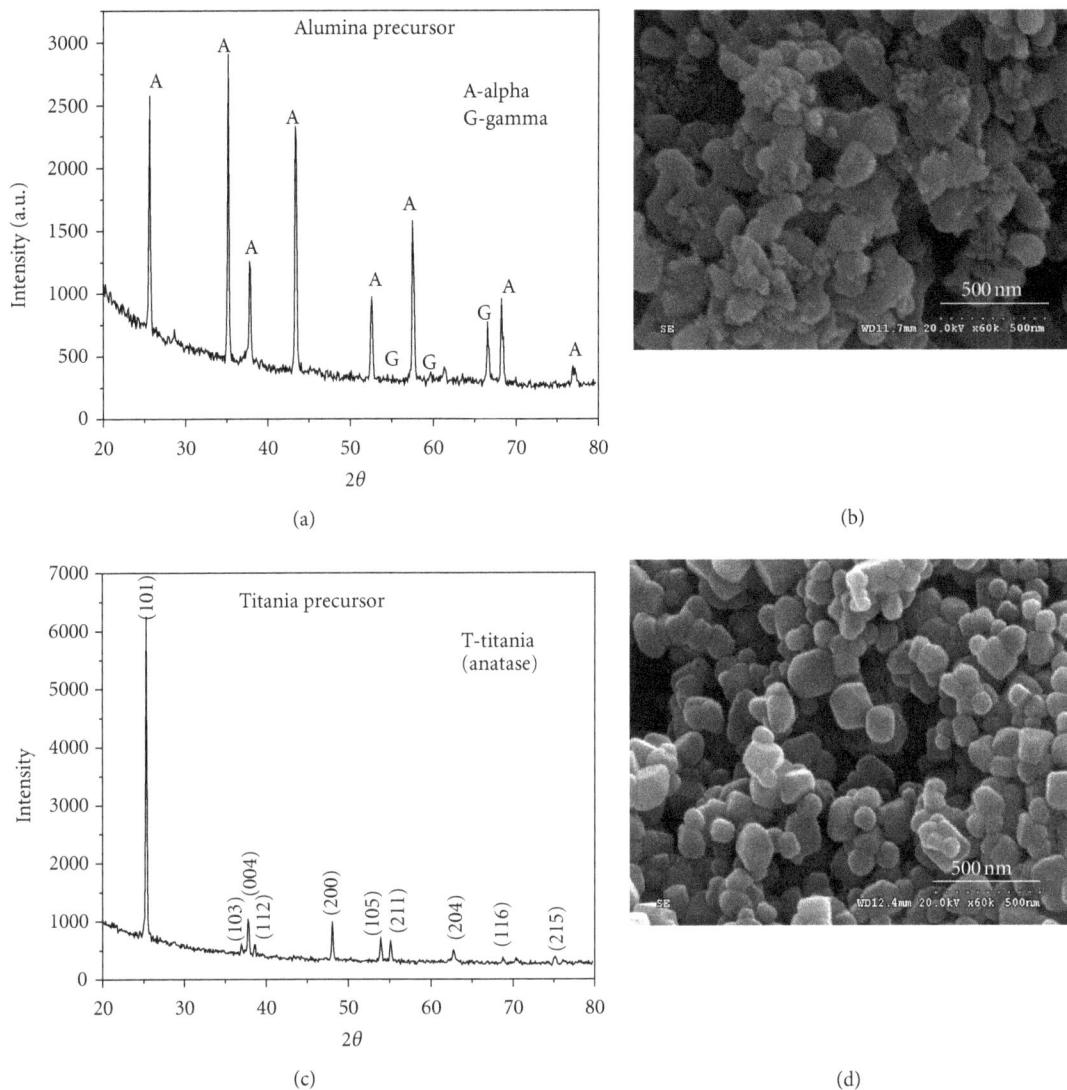

FIGURE 1: (a) XRD pattern of Al_2O_3 powder, (b) morphology of Al_2O_3 powder, (c) XRD pattern TiO_2 powder, and (d) morphology of TiO_2 powder.

TABLE 1: Physiochemical properties of raw materials.

Property	Alumina	Titania	Talc	Kaolinite
Chemical composition wt%				
SiO_2			60.0	44.0
Al_2O_3	99.95		2.1	50.2
TiO_2		99.5	0.08	0.4
MgO	0.005		31.1	<0.1
Na_2O			<0.01	0.11
K_2O			<0.01	<0.01
Physical properties				
Average particle size (D_{50})	200 nm	300 nm	21.62 μm	2.86 μm
Crystalline phase	α and γ	Anatase	Talc	Kaolinite

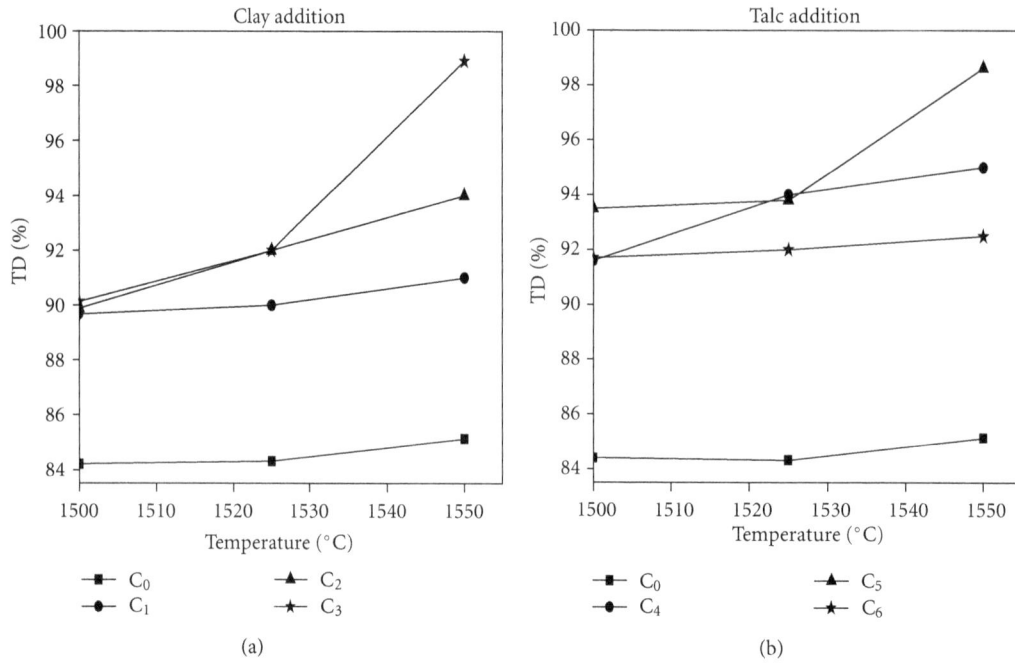

FIGURE 2: (a) Density variation with temperature of C_0, C_1, C_2, C_3, and (b) Density variation with temperature of C_0, C_4, C_5, and C_6.

TABLE 2: Formulation investigated (wt%).

Sr. No.	Sample. ID	Al_2O_3	TiO_2	Clay	Talc
1	C_0	56	44	0	0
2	C_1	55.16	43.34	2.94%	0
3	C_2	54.32	42.68	5.88%	0
4	C_3	53.48	42.02	8.82%	0
5	C_4	55.16	43.34	0	2.5%
6	C_5	54.32	42.68	0	5%
7	C_6	53.48	42.02	0	7.5%

fact offsets the densification of AT matrix by around 20%, resulting in poor sintered density of 87% even at a sintering temperature of 1600°C. Endothermic peak at 1386°C for C_3, indicate that there is no significant shift in Al_2TiO_5 phase formation compared to undoped (C_0) formulation (Figure 3(b)). Al_2TiO_5 phase formation temperature is shifted to a lower temperature range of 1274°C for C_5 compared to 1375°C observed with (C_0) un-doped formulation as is evident from the DSC peak (Figure 3(b)).

XRD pattern of the maximum dense sample (C_3) depicted in Figure 4 has not shown any additional peaks and all the peaks could be indexed with standard Al_2TiO_5 phase. However, a shift in unit cell constants is observed in comparison to the pure Al_2TiO_5 samples, especially in the case of "c" parameter [10, 17]. The cell parameters and volume of C_0 sample are $a = 9.4315$ Å, $b = 9.6385$ Å, $c = 3.590$ Å, and 326.35 (Å)³, where as cell parameters and volume of sample C_3 are $a = 9.4342$ Å, $b = 9.6536$ Å, $c = 3.5940$ Å, and

327.32 (Å)³. The change in unit cell parameter is probably due to silicon (Si^{4+}, ionic radius = 0.41 Å) replacing Al^{3+} in the lattice of Al_2TiO_5 matrix. This may result in the multivalent state for titanium that is, Ti^{3+}/Ti^{4+}, corresponding to a stoichiometry of $((Al^{3+},Ti^{3+})_2(Ti^{4+})_1(O^{2-})_5)$ which in turn enhances the sintering process. A closer look at the micrographs also reveals a change in the porous lamellar type of structure in to relatively dense faceted grains. Further, the microcracks are found visible and crack lengths are reduced to a greater extent in comparison to C_0 with no additives.

XRD pattern of C_5 sample with maximum density depicted in Figure 4 has shown ~96% of Al_2TiO_5 phase with minor quantities of Al_2O_3 after sintering at 1550°C. Further, the TiO_2 peaks are completely absent in the pattern. It is well known that talc ($2MgO\cdot2Al_2O_3\cdot5SiO_2$) transform to clinoenstatite ($MgSiO_3$) close to 1100°C and at high temperatures, it decomposes to MgO and SiO_2. In the case of C_5 formulation, Mg^{2+} (MgO: 1.5%) and Si^{4+} (SiO_2: 3%)

TABLE 3: Comparison of CTE values and hysteresis area.

Sample Id	%TD	%AT phase	CTE value (30–1000°C)	Hysteresis area (cm^2)
C$_0$	85.13	92.04	$1.09 \times 10^{-6}/°C$	91
C$_3$	98.9	98	$0.94 \times 10^{-6}/°C$	68
C$_5$	98.63	95.5	$0.4 \times 10^{-6}/°C$	95

TABLE 4: Comparison of flexural strength and hardness.

Sample	Flexural strength (MPa)	%increase in flexural strength	Hardness (Kg/mm^2)	%increase in hardness
C$_0$	8.53	—	150	—
C$_3$	8.61	0.94	180	20%
C$_5$	25.89	203	200	33.33%

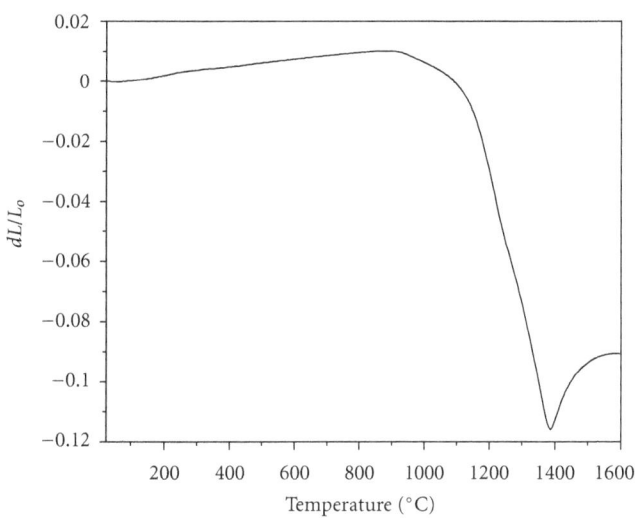

(a)

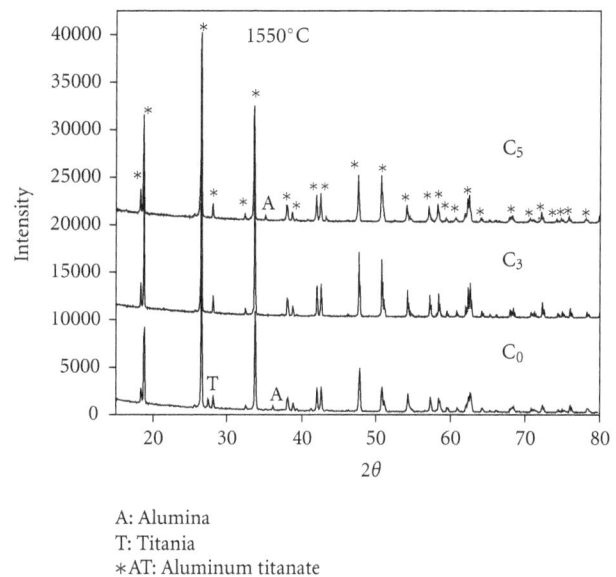

— C$_3$
-·- C$_5$
······ C$_0$

(b)

FIGURE 3: (a) Shrinkage curve of C$_0$ and (b) DSC curves C$_0$, C$_3$, and C$_5$.

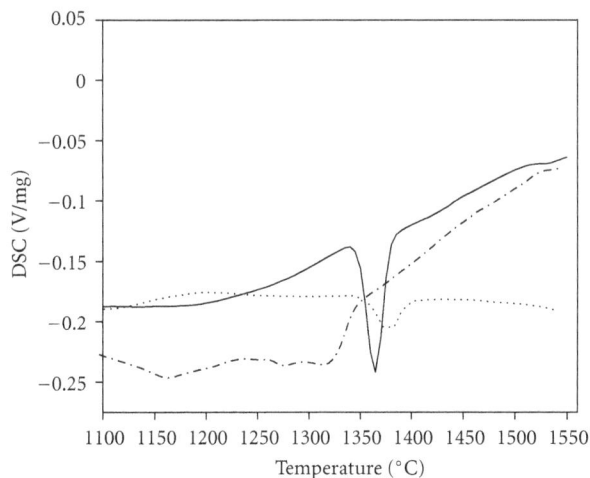

A: Alumina
T: Titania
∗AT: Aluminum titanate

FIGURE 4: XRD patterns of sintered specimens 1550°C.

ions undergo simultaneous lattice substitutions for Al^{3+} to stabilize Al$_2$TiO$_5$ stoichiometry. This leads to change in unit cell parameters due to addition of talc. The unit cell parameters and volume of sample C$_0$ are a = 9.4315 Å, b = 9.6385 Å, c = 3.590 Å, and 326.35 (Å)^3. Whereas in sample C$_5$ unit cell parameters and volume are a = 9.4651 Å, b = 9.6715 Å, c = 3.5981 Å and 329.37 (Å)^3, which displaces the interplanar spaces. The dilation along crystallographic c-axis in talc added samples, further expected to improve phase stability of AT [17]. A larger distortion from the C$_0$ composition can be attributed to the simultaneous substitution of Si^{4+} (ionic radius = 0.41 Å) and Mg^{2+} (ionic radius = 0.65 Å) replacing Al^{3+} in the lattice of Al$_2$TiO$_5$ matrix. Al$_2$TiO$_5$ formation process was led by nucleation and growth of grains and finally diffusion of reactant through the matrix and is controlled by the very slow reacting species. In addition to the multivalent state for titanium, that is, Ti^{3+}/Ti^{4+} as a result of Si^{4+} substitution, the presence of Mg^{2+} ions

(a)

(b)

(c)

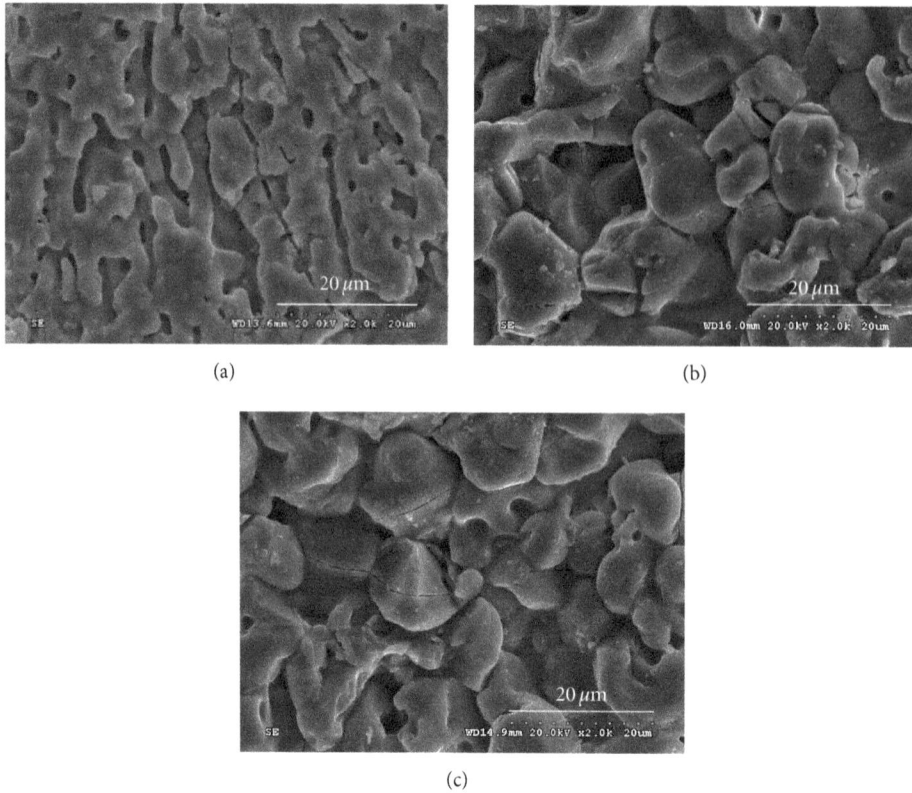

FIGURE 5: (a) SEM microstructure of C_0, (b) SEM microstructure of C_3, and (c) SEM microstructure of C_5.

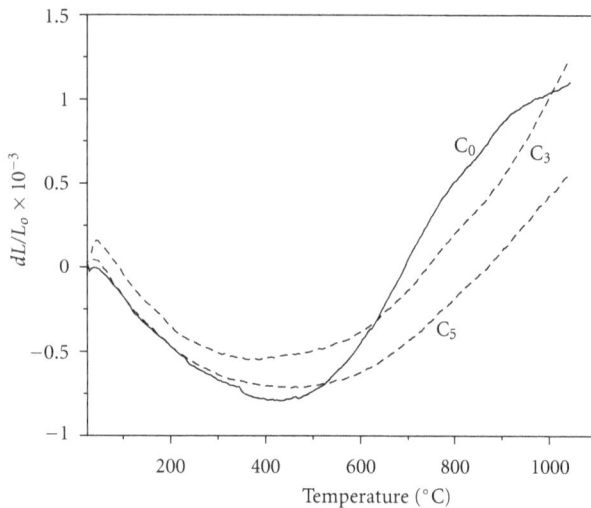

FIGURE 6: Thermal expansion curves of C_0, C_3, and C_5.

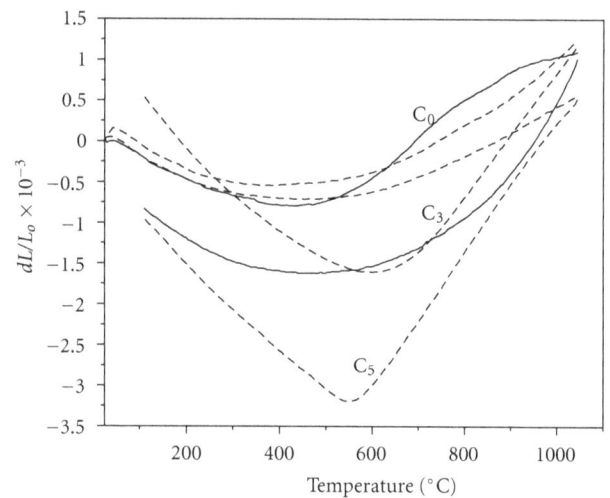

FIGURE 7: Thermal hysteresis of C_0, C_3, and C_5.

that replaces Al^{3+} may result in oxygen vacancy formation $((Al^{3+},Ti^{3+})_2(Ti^{4+})_1(O_{5-\delta}^{2-})$. These vacancies promote the diffusion in the solid state enhancing densification.

Figures 5(a)–5(c) showed the microstructures of C_0, C_3, and C_5 samples. From the microstructure of C_3 and C_5 samples, it is clear that they possess single AT phase, where as in the case of C_0 some parent residual oxides still exist. Microstructural features do not show any significant

variations in the grain size with both the additives (C_3, C_5). However, considerable reduction in intergranular pores is observed with the addition of talc (C_5). The presence of additives enhanced the grain growth in AT ceramics.

3.2.2. Thermal Expansion and Thermal Hysteresis. Dilatometric expansion curve and thermal hysteresis were recorded while heating and cooling of the C_0, C_3, and C_5 samples

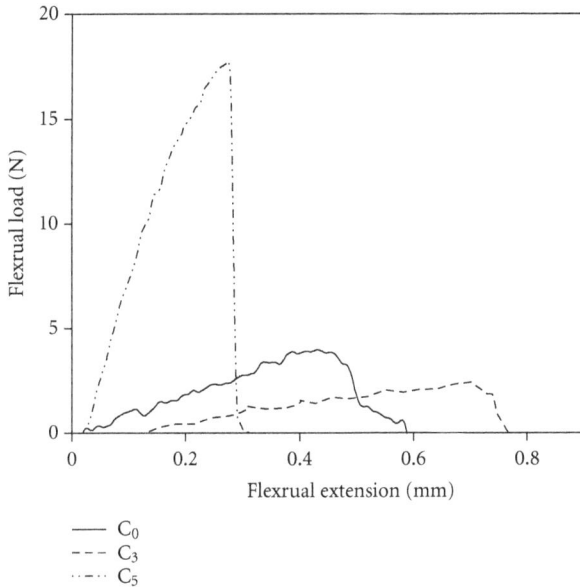

FIGURE 8: Load-displacement curves of C_0, C_3 and C_5.

were shown in Figures 6 and 7, respectively. Table 3 lists the thermal expansion values of C_0, C_3, and C_5 formulations from 30–1000°C. It is evident that, the formulation C_0 has a CTE value of 1.09×10^{-6}/°C followed by C_3 with a marginally lower CTE value of 0.94×10^{-6}/°C(13% less than C_0) and the lowest value of 0.42×10^{-6}/°C (63% less than C_0) for C_5 formulation. Thermal expansion curves of the entire samples initially contract till 450°C. Thermal expansion curves of C_0 exhibit a dissimilar behavior in comparison with C_3 and C_5 formulation. C_0 sample exhibited a low expansion of -1.97×10^{-6}/°C in comparison to C_3 and C_5 for which CTE values are -1.71×10^{-6}/°C and -1.74×10^{-6}/°C, respectively. All samples have shown an expansion behavior beyond 450°C and C_0 exhibited a significant slope change leading to the maximum expansion value. The temperature, beyond which the expansion becomes prominent, was regarded as the temperature at which healing of microcracks occurs that compensates the expansion effectively. Further, the curves corresponding to C_0 are tapered into a plateau beyond 900°C unlike other samples with a positive slope.

Cooling curves also behaved differently with all the three samples. C_0 formulation has shown a negative slope tapering into a plateau followed by the expansion. Initial contraction had α values of 6.08, 7.75, and 3.04×10^{-6}/°C (temperature regime, 1000–450°C) and final expansion region had α values of -4.37, -7.75, and -2.3×10^{-6}/°C (temperature regime, 450–100°C) for C_5, C_3, and C_0 respectively. Expansion below 450°C can be attributed to the reintroduction of micro-cracks healed while heating.

It is obvious that thermal properties of doped and undoped Al_2TiO_5 ceramics are mainly governed by the presence of microcracks and are affected by the crack closure or healing. Areas of the thermal hysteresis were recorded

during heating up to 1000°C and cooling up to 100°C for formulations of C_5, C_3, and C_0 were 95, 68 and 91 cm^2, respectively. Unlike other samples, C_3 showed two closed loops where first loop is in the temperature regime of 1000 to 300°C, which had dL/L_0 value less than dL/L_0 value of sample while heating, as expected. However, the second loop formed at 300°C, exhibited a higher dL/L_0 value which is unusual. Probably this may be due to the transformation of microcracks into macro cracks due to high silica content (4.5%) leading to a relatively low hysteresis area of 68 cm^2 observed with C_3 sample in comparison to 91 cm^2 observed with C_0 sample having close CTE values [17]. C_5 sample exhibited hysteresis area of 95 cm^2, which correlates well with the low CTE value (0.42×10^{-6}/°C) as a result of compensation of expansion values. Microcrack density (number of microcracks/unit area estimated using several SEM images) of C_5 sample was almost 3 times of that of the C_0 sample confirming the above observations.

3.2.3. Flexural Strength and Hardness Measurements. Typical load-displacement curves obtained from the specimens C_0, C_3 and C_5 which were subjected to 3-point bend loading and is shown in Figure 8. Though a minimum of 4-5 specimens were tested in each case, for the sake of clarity, only one load-displacement curve for each condition are shown. Flexural strength (σ_f) is calculated from the load displacement data as $\sigma_f = 3/2PL/bd^2$, where P_{max} is the maximum load, L the span length, b the width, and d the thickness of the specimens. The flexural strength values are given in Table 4. C_5 with talc doping exhibited the peak flexural strength before fracture (average, 28 ± 3 MPa) which is ~200% higher than the corresponding flexural strength values (average 8 ± 4 MPa) of the C_3 and C_0 samples. It is evident from the data in Figure 8 that C_0 and C_3 materials show a similar unstable crack extension with load drop after attaining peak flexural stress. A close look at the load-displacement curves obtained from the flexural loading of the C_3 and C_0 samples (shown in Figure 8) has shown several load excursions while on attaining peak load depicting fracture over a wide range of strain followed by a gradual load drop. Poor strength in fracture behavior even with a high density (99% of TD) can be attributed to increase in crack length and more number of macrocracks (crack length up to 30 μm, Figure 5(b)). It is obvious that poor density (85%) and the high porosity are the factors that contribute in addition to the microcracks to the identical fracture behavior observed with C_0 compositions.

Unlike the other samples, C_5 composition exhibited initial rapid increase in the stress, reaching a peak value followed by the fracture. It is clear that, prior to the attainment of the peak flexural stresses, the material showed almost a linear increase in flexural stress with strain, indicating elastic deformation, before final failure. This can be attributed to the association of the existing large number of micro cracks (Figure 5(c), microcracks are restricted within the grain with maximum crack length <10 μm) leading to the stable crack extension. Finally, the material fails with sudden load drop, which is an indication of rapid propagation of the macrocracks that resulted from coalescence of large number

(a)

(b)

(c)

(d)

(e)

(f)

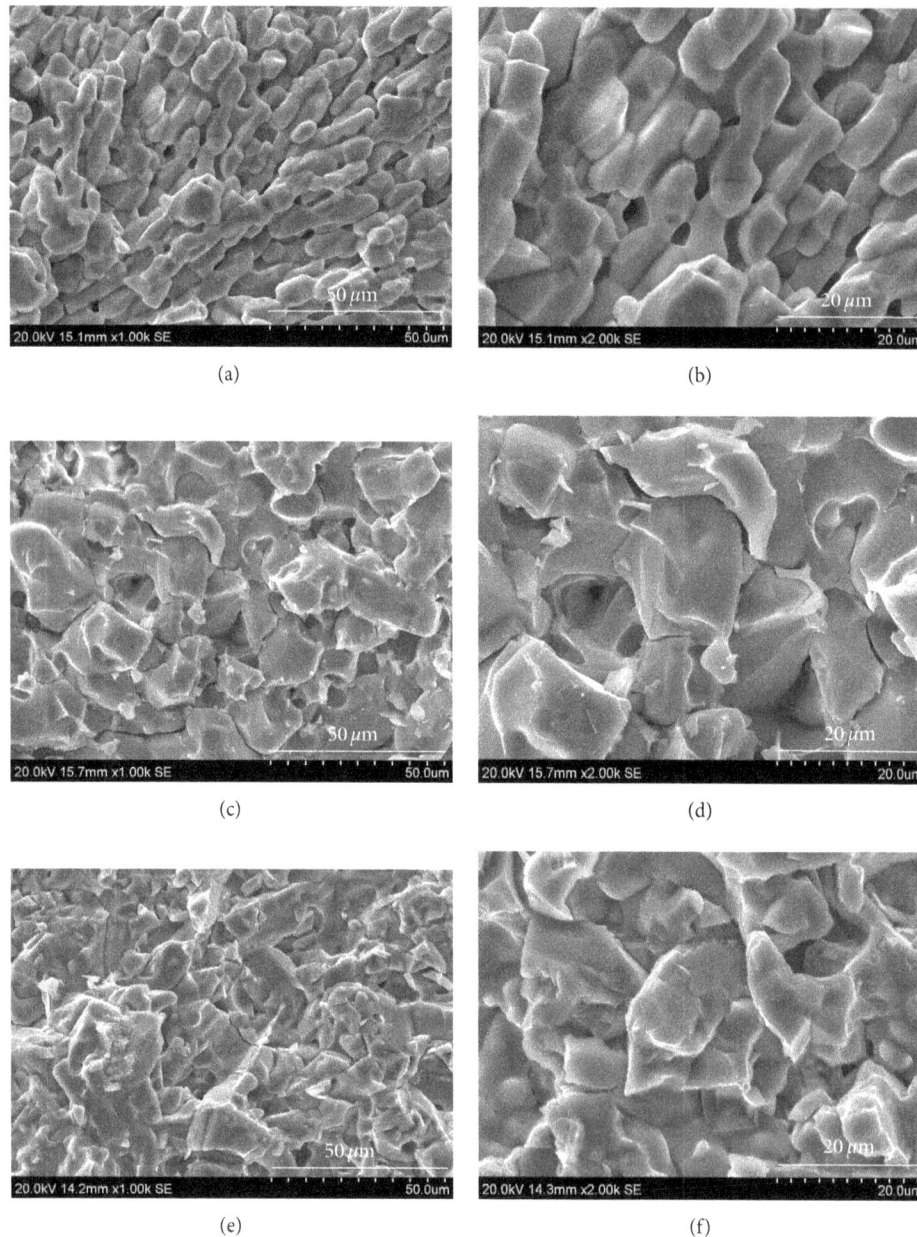

FIGURE 9: (a)-(b) Fractography microstructure of C_0 at different magnifications, (c)-(d) fractography microstructure of C_3 at different magnifications, and (e)-(f) fractography microstructure of C_5 at different magnifications.

of micro cracks under local tensile loading. Hardness of the specimens C_3 and C_5 has shown an enhancement 20% and 33.3% with respect to C_0. A relatively high hardness under identical densities could attribute to the marginal decrease in grain size observed with C_5 samples.

SEM fractographs obtained from the specimens C_0, C_3, and C_5 tested till the failure under flexural (3-point bend loading) testing are shown in Figures 9(a)–9(f). Though large number of fractographs at different magnifications is obtained in each of the specimen, for the sake of clarity, only one representative fractograph is depicted. These fractographs show an identical transgranular fracture. As discussed above, C_5 composition is associated with more

microcracks and the crack path while propagation needs to relocate for further crack extension as it encounters more and more microcracks. Such process needs higher fracture energy as compared to the C_0 and C_3 where microcrack densities were low.

4. Conclusions

Addition of kaolinite ($2Al_2O_3.3SiO_2.2H_2O$) and talc ($Mg_3Si_4O_{10}(OH)_2$) in relatively low concentration of 8.8 wt% and 5 wt%, respectively, resulted in enhanced densification leading to a substantial increase in percentage theoretical density to 99% in comparison with pure Al_2TiO_5 with 87%

processed under identical conditions. Kaolinite and/or talc substitution results in multivalent titanium (Ti^{3+}/Ti^{4+}) and oxygen vacancies in Al_2TiO_5 formulations promoting the enhanced diffusion and densification. Microstructural evaluations revealed transformation of porous lamellar type of structure of pure Al_2TiO_5 into relatively dense faceted grains with kaolinite and talc confirming higher density values. XRD studies have shown an improvement in Al_2TiO_5 phase content from 92% to a maximum of 98% with the addition of kaolinite. The enhancement in phase content was moderate in case of talc with a maximum of 95.5%; however, DSC studies indicated a drop of ~85°C in phase formation temperature. Substantial improvement in thermomechanical properties was observed with talc addition in comparison to kaolinite. A decrease in thermal expansion value of Talc-doped AT by 63% and an enhancement of flexural strength value by 200% is demonstrated in the present study. These properties found to have a good correlation with the presence and the mode of microcracks as revealed by microstructure and thermal hysteresis. Talc-doped Al_2TiO_5 formulation presently developed with superior thermomechanical properties can be explored for various potential applications.

References

[1] M. Nagano, S. Nagashima, H. Maeda, and A. Kato, "Sintering behavior of Al_2TiO_5 base ceramics and their thermal properties," *Ceramics International*, vol. 25, no. 8, pp. 681–687, 1999.

[2] G. Bruno, A. M. Efremov, B. R. Wheaton, and J. E. Webb, "Microcrack orientation in porous aluminum titanate," *Acta Materialia*, vol. 58, no. 20, pp. 6649–6655, 2010.

[3] R. Naghizadeh, H. R. Rezaie, and F. Golestani-fard, "The influence of composition, cooling rate and atmosphere on the synthesis and thermal stability of aluminum titanate," *Materials Science and Engineering B*, vol. 157, no. 1–3, pp. 20–25, 2009.

[4] K. Hamano, Y. Ohya, and Z. E. Nakagawa, "Crack propagation resistance of aluminium titanate ceramics," *International Journal of High Technology Ceramics*, vol. 1, no. 2, pp. 129–137, 1985.

[5] H. C. Kim, K. S. Lee, O. S. Kweon, C. G. Aneziris, and I. J. Kim, "Crack healing, reopening and thermal expansion behavior of Al_2TiO_5 ceramics at high temperature," *Journal of the European Ceramic Society*, vol. 27, no. 2-3, pp. 1431–1434, 2007.

[6] Y. X. Huang and A. M. R. Senos, "Effect of the powder precursor characteristics in the reaction sintering of aluminum titanate," *Materials Research Bulletin*, vol. 37, no. 1, pp. 99–111, 2002.

[7] G. Bruno, A. Efremov, B. Wheaton, I. Bobrikov, V. G. Simkin, and S. Misture, "Micro- and macroscopic thermal expansion of stabilized aluminum titanate," *Journal of the European Ceramic Society*, vol. 30, no. 12, pp. 2555–2562, 2010.

[8] C. H. Chen and H. Awaji, "Temperature dependence of mechanical properties of aluminum titanate ceramics," *Journal of the European Ceramic Society*, vol. 27, no. 1, pp. 13–18, 2007.

[9] B. Morosin and R. W. Lynch, "Structure studies on Al_2TiO_5 at room temperature and at 600°C," *Acta Crystallographica B*, vol. 28, pp. 1040–1046, 1972.

[10] R. D. Skala, D. Li, and I. M. Low, "Diffraction, structure and phase stability studies on aluminium titanate," *Journal of the European Ceramic Society*, vol. 29, no. 1, pp. 67–75, 2009.

[11] A. Durán, H. Wohlfromm, and P. Pena, "Study of the behaviour of Al_2TiO_5 materials in reducing atmosphere by spectroscopic techniques," *Journal of the European Ceramic Society*, vol. 13, no. 1, pp. 73–80, 1994.

[12] I. M. Low, Z. Oo, and B. H. O'Connor, "Effect of atmospheres on the thermal stability of aluminium titanate," *Physica B*, vol. 385-386, pp. 502–504, 2006.

[13] V. Buscaglia, P. Nanni, G. Battilana, G. Aliprandi, and C. Carry, "Reaction sintering of aluminium titanate: i-effect of MgO addition," *Journal of the European Ceramic Society*, vol. 13, no. 5, pp. 411–417, 1994.

[14] A. Yoleva, S. Djambazo, D. Arseno, and V. Hristo, "Effect of SiO_2 addition on thermal hysteresis of aluminum titanate," *Journal of University of Chemical Technology and Metallurgy*, vol. 45, no. 3, pp. 269–274, 2010.

[15] H. A. J. Thomas, R. Stevens, and E. Gilbart, "Effect of zirconia additions on the reaction sintering of aluminium titanate," *Journal of Materials Science*, vol. 26, no. 13, pp. 3613–3616, 1991.

[16] T. Korim, "Effect of Mg^{2+}- and Fe^{3+}-ions on formation mechanism of aluminium titanate," *Ceramics International*, vol. 35, no. 4, pp. 1671–1675, 2009.

[17] J. Lan, C. Xiao-Yan, H. Guo-Ming, and M. Yu, "Effect of additives on properties of aluminium titanate ceramics," *Transactions of Nonferrous Metals Society of China*, vol. 21, no. 7, pp. 1574–1579, 2011.

[18] I. J. Kim, K. S. Lee, and G. Cao, "Low thermal expansion behavior of $ZrTiO_4$-Al_2TiO_5 ceramics having high thermal durability between 750 and 1400°c," *Key Engineering Materials*, vol. 22, pp. 2627–2632, 2002.

[19] L. Giordano, M. Viviani, C. Bottino, M. T. Buscaglia, V. Buscaglia, and P. Nanni, "Microstructure and thermal expansion of Al_2TiO_5-$MgTi_2O_5$ solid solutions obtained by reaction sintering," *Journal of the European Ceramic Society*, vol. 22, no. 11, pp. 1811–1822, 2002.

[20] T. Shimada, M. Mizuno, K. Katou et al., "Aluminum titanate-tetragonal zirconia composite with low thermal expansion and high strength simultaneously," *Solid State Ionics*, vol. 101–103, no. 1, pp. 1127–1133, 1997.

[21] P. Oikonomou, C. Dedeloudis, C. J. Stournaras, and C. Ftikos, "Stabilized tialite-mullite composites with low thermal expansion and high strength for catalytic converters," *Journal of the European Ceramic Society*, vol. 27, no. 12, pp. 3475–3482, 2007.

[22] A. Tsetsekou, "A comparison study of tialite ceramics doped with various materials and tialite-mullite composites: microstructural, thermal and mechanical properties," *Journal of the European Ceramic Society*, vol. 25, no. 4, pp. 335–348, 2005.

[23] F. H. Perera, A. Pajares, and J. J. Meléndez, "Strength of aluminium titanate/mullite composites containing thermal stabilizers," *Journal of the European Ceramic Society*, vol. 31, no. 9, pp. 1695–1701, 2011.

[24] C. G. Shi and I. M. Low, "Use of spodumene for liquid-phase-sintering of aluminium titanate," *Materials Letters*, vol. 36, no. 1–4, pp. 118–122, 1998.

[25] C. G. Shi and I. M. Low, "Effect of spodumene additions on the sintering and densification of aluminum titanate," *Materials Research Bulletin*, vol. 33, no. 6, pp. 817–824, 1998.

3

Effect of Anodic Current Density on Characteristics and Low Temperature IR Emissivity of Ceramic Coating on Aluminium 6061 Alloy Prepared by Microarc Oxidation

Mohannad M. S. Al Bosta,[1] Keng-Jeng Ma,[2] and Hsi-Hsin Chien[2]

[1] *Ph.D. Program in Engineering Science, College of Engineering, Chung Hua University, Hsinchu 30012, Taiwan*
[2] *College of Engineering, Chung Hua University, Hsinchu 30012, Taiwan*

Correspondence should be addressed to Mohannad M. S. Al Bosta; mmbosta2005@yahoo.com

Academic Editor: Baolin Wang

High emitter MAO ceramic coatings were fabricated on the Al 6061 alloy, using different bipolar anodic current densities, in an alkali silicate electrolyte. We found that, as the current density increased from 10.94 A/dm^2 to 43.75 A/dm^2, the layer thickness was increased from 10.9 μm to 18.5 μm, the surface roughness was increased from 0.79 μm to 1.27 μm, the area ratio of volcano-like microstructure was increased from 55.6% to 59.6%, the volcano-like density was decreased from 2620 mm^{-2} to 1420 mm^{-2}, and the γ-alumina phase was decreased from 66.6 wt.% to 26.2 wt.%, while the α-alumina phase was increased from 3.9 wt.% to 27.6 wt.%. The sillimanite and cristobalite phases were around 20 wt.% and 9 wt.%, respectively, for 10.94 A/dm^2 and approximately constant around 40 wt.% and less than 5 wt.%, respectively, for the anodic current densities 14.58, 21.88, and 43.75 A/dm^2. The ceramic surface roughness and thickness slightly enhanced the IR emissivity in the semitransparent region (4.0–7.8 μm), while the existing phases contributed together to raise the emissivity in the opaque region (8.6–16.0 μm) to higher but approximately the same emissivities.

1. Introduction

The growing demand of limited energy sources requires adoption of effective methods to save energy consumption and to prevent unwanted energy loss. The high emitter surfaces enhance the thermal performance of heating and cooling systems and consequently reduce the needed energy [1, 2]. Several methods are applied to the material surfaces to enhance their emissivities, such as coating by thin tape films, paint and lacquer, roughening, and surface anodizing.

One of the promising coating methods is the microarc oxidation, MAO. The MAO ceramic coating has perfect properties such as wear resistance, corrosion resistance, hardness, strong adhesion, and thermal shock resistance and can be fabricated on the surfaces of aluminium, magnesium, and titanium alloys [3–62]. The properties of the MAO ceramic coating are affected by the electrolyte compositions [63–71], treatment time [14, 28, 70–75], electrolyte temperature [70, 76], voltage [72, 77–80], current density [6, 47, 55, 70, 81–83], current mode [84, 85], and electrode geometry [86].

The applied current modes in the MAO process are DC, AC, unipolar, and pulsed bipolar. The pulsed bipolar mode was found to enhance the surface properties of treated metals, improve the thickness and homogeneity of oxide layers, and limit the growth of the porous layer [6–8, 54].

Coating aluminium with a high thermal radiator of MAO ceramic will find its applications in low profile devices, heat sinks, electronic parts, LED, lasers, refrigeration, air conditioning, transport industries, liquefaction plants, power plants, petroleum refineries, and others. Previous studies rarely demonstrated the IR emissivity of the MAO coatings. For example, Wang et al. [87] studied the IR emissivity at 500°C of MAO ceramic coating on 2024 aluminium substrates and illustrated that γ-Al$_2$O$_3$, silicon oxides, and phosphate oxides contributed to high emissivity at wavelengths 8–20 μm, while the surface roughness was responsible for increasing emissivity at wavelength range 3–5 μm. Tang et al. [39] studied the emissivity of MAO ceramic prepared on Ti6Al4V substrate and reported that more Co contents enhanced the emissivity of the ceramic surface at 700°C and at wavelengths 3–20 μm. A relatively high emissivity was found by Wang et al. [88] when they studied the emissivity (8–14 μm, 700°C) of the surface of ceramic coating on Ti6Al2Zr1Mo1V alloy, and they pointed out that the TiO_2 phases were contributed to enhance the emissivity.

The geometry of the MAO-treated samples plays a main role in the industrial applications as a finishing process. The first effect comes to mind (by keeping the current output of main power supply constant and changing the sample dimensions) is the variation of current density. Previous works kept the sample dimensions constant and changed the input current density. Khan et al. [6] fabricated MAO alumina ceramic coatings on Al 6082 alloy using a DC current mode of densities ranged from 5 A/dm^2 to 20 A/dm^2 in various concentrations of KOH electrolytes at temperatures ranged between 20°C and 25°C for 30 minutes. They reported that the denser current formed thicker coatings with minimal stress level, dense morphology, and a relatively high content of α-Al$_2$O$_3$ phase. Raj and Mubarak Ali [70] used different direct current densities (5–20 A/dm^2) for MAO treating of aluminium in the alkaline silicate electrolyte at 10°C for 45 minutes and obtained thicker coatings, higher growth rate, and coating ratio at 15 A/dm^2 of current density, but the 20 A/dm^2 of current density decreased these properties due to the dominant of electrolyte dissolution over the coating building. The effects of bipolar current density on MAO ceramic coating formed on titanium alloy were studied by Sun et al. [83] in an electrolyte of sodium aluminate and hypophosphate at 35°C for 70 minutes. The cathodic (j_c) and anodic (j_a) current densities ranged between 6 and 12 A/dm^2. They pointed out that the lowest ratio of j_a/j_c formed a denser and thinner layer with more uniform microstructures and did not contain α-alumina phase, while the highest j_a/j_c ratio formed a thicker ceramic coating comprised of entirely α-alumina phase with poor adhesion to the substrate.

The main aim of the present work is to change the sample dimensions and find out the effect of the anodic current density on the microstructural and compositional properties of the MAO ceramic coating and the resultant low-temperature IR emissivity.

2. Experimental

Rectangular pieces ($4 \times 4 \times 0.2$ cm^3) of aluminium alloy 6061 (Mg 1%, Si 0.65%, Fe 0.7%, Cu 0.3%, Cr 0.2%, Mn 0.15%, Ti 0.15%, and Al balance) were used as substrates in this study. The exposed surface was ground to a 1200 grit SiC finish using the water as a lubricant, followed by rinsing in the doubly distilled water and drying in the air, while the other side was totally insulated using a Teflon tape.

Four different assemblies were applied by mounting a different number of pieces into a 6061-Al alloy clamp to get 16, 32, 48, and 64 cm^2 of exposed working area to the electrolyte and consequently different applied current densities. The preparation of samples was conducted for 10 minutes in a fresh electrolyte which consisted of 0.046 M sodium

silicate and 0.042 M sodium hydroxide, and the PH was 12.8 in a cooling and stirring bath which kept the electrolyte temperature below 17°C. We used the pulse controller SPIK 2000A (Shen Chang Electric Co., Ltd., Taiwan) to generate an asymmetric bipolar pulsing mode with parameters of 200, 200, 360, and 200 μs for t_{on}^+, t_{off}^+, t_{on}^-, and t_{off}^-, respectively. Two electrical power supplies (PR Series, 650 V, 7.7 A, Matsusada) were connected to the pulse controller with 100 V, 3.5 A for DC1 and 500 V, 7.0 A for DC2, as shown in Figure 1. The samples were connected to the output E2, while another 243 cm^2 aluminium 6061 plate was connected to E1. Only Al 6061 alloy touched the electrolysis solution to avoid any contamination due to diffusion of wires, electrodes, or screws. According to Figure 1, both electrodes (E1 and E2) were alternating between the cathodic and anodic biasing according to the output bipolar pulse; the DC2 supplied the sample in the anodic period, and the DC1 supplied it in the cathodic period. The resultant peak anodic biasing current densities, J_a, were 10.94, 14.58, 21.88, and 43.75 A/dm^2.

Unlike the sample, no microarc oxidation was occurred on the aluminium plate due to the relatively low voltage during its anodic period. To avoid the effect of chemical reaction with the plate surface, we renewed the plate for each MAO experiment, despite being in a good situation. After completing every treatment, the samples were immediately immersed and cleaned with distilled water and then dried and stored in a sealed container. Before any use or test, samples were cleaned using an ultrasonic bath of doubly distilled water, propanol, acetone, and again doubly distilled water, 10 minutes for each, followed by air drying in the fume hood.

Different 27 randomly selected places were captured for each sample by the SEM (S4160, Hitachi). We performed the image analysis using Image Pro Plus 7.0 software (Media Cybernetics, Inc.) to estimate the volcano-like microstructure area ratio. The density of volcano-like was estimated by dividing its total number over the total area of SEM frames for the same sample.

The elemental composition analysis of ceramic coating layers was carried out using Hitachi EDX S4800. Low-angle X-ray diffraction (CuKα; XRD, X'Pert PRO MPD, PANalytical, The Netherlands) was used to obtain the XRD patterns of as-deposited ceramic coatings. We performed the semi-quantitative analysis of the XRD spectra using the so-called reference intensity ratio method (RiR method, de Woolf and Visser [89]) by MATCH! software (V.1.11f, Crystal Impact GbR) and selected the best fit phases and compositions, in addition to estimating their weight percentage.

For each sample, at least 16 different locations were studied to determine the average of surface roughness using a surface roughness tester (Mitutoyo, SJ310) and layer thickness using an eddy-current coating-thickness tester (Extech CG204).

We used a modified Fourier transform infrared spectroscopy (FTLA2000 Series Spectrometer, ABB) to analyze the IR emissivity of the samples at 70°C, by comparing with a reference blackbody radiation in the wavelength ranging between 4 μm and 16 μm.

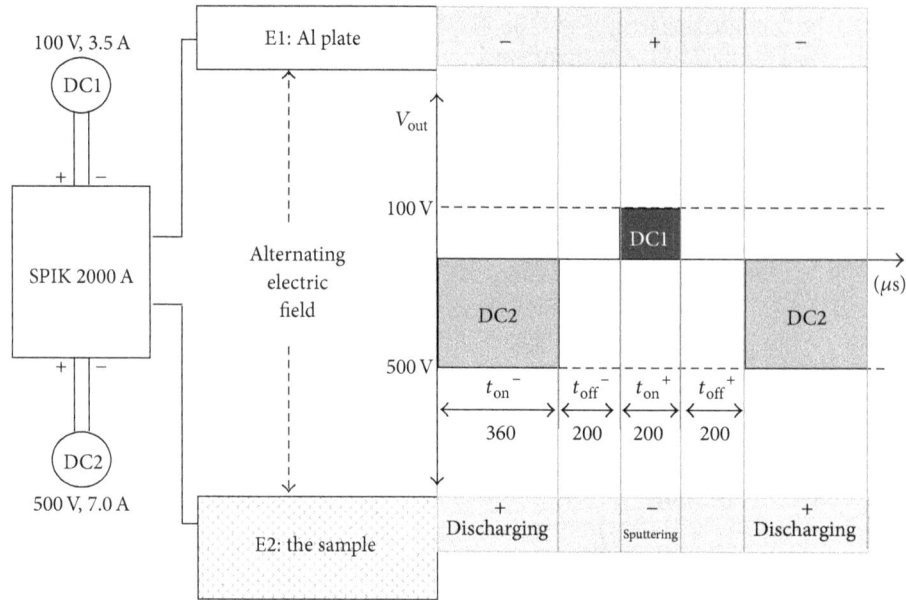

FIGURE 1: A scheme of the arrangement of power supplies and the pulse generator SPIK 2000 A, in addition to the polarity of alternating electrodes according to the bipolar output pulse.

3. Possible Chemical, Electrochemical Reactions and Phase Transformation

Our MAO treatment was conducted in an electrolyte contained sodium hydroxide and sodium silicate. The samples were alternating between the cathodic and anodic polarizations as a result of the asymmetrical bipolar pulse mode, Figure 1. The forming of compositions and phases can be classified into four periods according to the bipolar pulse mode.

(a) t_{off}^{+}, a neutral period: where there is no applied voltage, the chemical reactions will etch the aluminium and release the aluminate ions AlO_2^{-} and $Al(OH)_4^{-}$ into the electrolyte [90, 91]:

$$2Al + 2H_2O + 2OH^{-} = 2AlO_2^{-}\,(aq) + 3H_2 \tag{1}$$

$$Al + 4OH^{-} \longrightarrow Al(OH)_4^{-}\,(gel) \tag{2}$$

The chemical dissolution for the alumina ceramic reduces its thickness as given by the following reactions [91]:

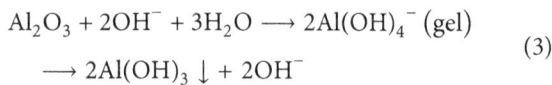

$$Al_2O_3 + 2OH^{-} + 3H_2O \longrightarrow 2Al(OH)_4^{-}\,(gel)$$
$$\longrightarrow 2Al(OH)_3 \downarrow + 2OH^{-} \tag{3}$$

The boehmite AlO_2H also may be produced by the following reaction [92]:

$$Al(OH)_4^{-} + H_2O \longrightarrow AlO_2H \downarrow +2H_2O + OH^{-} \tag{4}$$

The OH^{-} combines the aluminium hydroxide [8]:

$$Al(OH)_3 + OH^{-} \longrightarrow Al(OH)_4^{-} \tag{5}$$

The aluminium oxidation [8] is

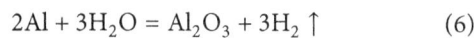

$$2Al + 3H_2O = Al_2O_3 + 3H_2 \uparrow \tag{6}$$

Based on the above, the alumina ceramic surface is chemically dissolved by the attack of OH^{-} [8], and this dissolution is more aggressive with the increment of liberated heat into the electrolyte from the treated surface.

(b) t_{on}^{+}, the cathodic period: in this period, the negatively charged sample attracts the electrolyte cations; the anions including the products of the neutral period will be repelled away into the electrolyte, and the possible reactions are as follows.

The inward deposition of cations into the ceramic surface equations is following

$$Na^{+} + e^{-} \longrightarrow Na \tag{7}$$

$$Al^{3+} + 3e^{-} \longrightarrow Al \tag{8}$$

The sodium is a highly reactive metal and it will immediately react and dissolve into the electrolyte, (9), except that which penetrates into the inner layers of the ceramic:

$$Na + H_2O = Na^{+} + OH^{-} + H_2 \uparrow \tag{9}$$

The water cathodic electrolysis is

$$2H_2O + 2e^{-} \longrightarrow H_2 \uparrow + 2OH^{-} \tag{10}$$

(c) t_{off}^{-}, a neutral period: it has the same chemical reactions for the neutral period, t_{off}^{+}, (1)–(6).

(d) t_{on}^{-}, the anodic period: this period is characterized by the MAO discharging due to the effect of high electric field and the production of alumina ceramic by the reaction between the inward immigrant O^{2-} [93] and the outward immigrant Al^{3+} according to

$$2Al^{3+} + 3O^{2-} \longrightarrow Al_2O_3 \tag{11}$$

(a)   (b)

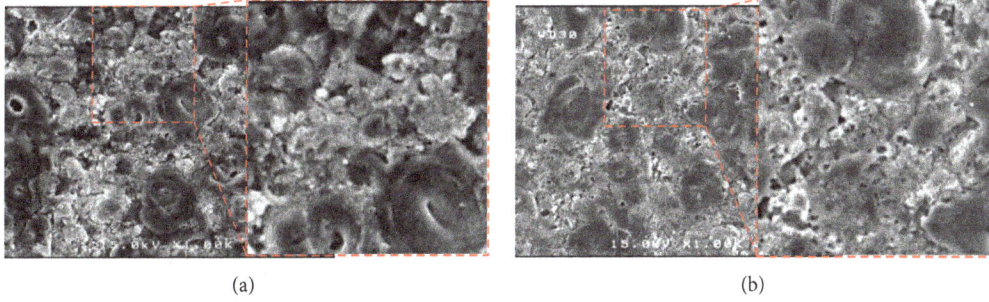

FIGURE 2: Two SEM micrographs for the same sample after MAO treatment by (a) a month and (b) six months.

Some of the immigrant Al^{3+} will be ejected into the electrolyte and combine with the hydroxide or silicate:

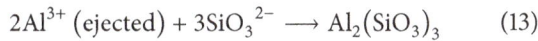

$$Al^{3+} \text{ (ejected)} + 3OH^- \longrightarrow Al(OH)_3 \downarrow \qquad (12)$$

$$2Al^{3+} \text{ (ejected)} + 3SiO_3^{2-} \longrightarrow Al_2(SiO_3)_3 \qquad (13)$$

A part of silica will attach (without reaction) to the molten alumina and transfer into one of the silica phases, while another silica quantity combines with the ejected molten alumina and produces an aluminosilicate phase according to the transfer temperature and pressure:

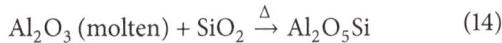

$$Al_2O_3 \text{ (molten)} + SiO_2 \xrightarrow{\Delta} Al_2O_5Si \qquad (14)$$

The ejected molten alumina will contact the surrounding electrolyte, rapidly quenched and solidified, to form the γ-alumina phase [94]:

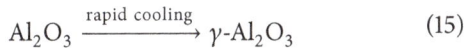

$$Al_2O_3 \xrightarrow{\text{rapid cooling}} \gamma\text{-}Al_2O_3 \qquad (15)$$

While the slower cooling rate in the inner layers of the sparking channels is favored to form the α-alumina phase [45, 94],

$$Al_2O_3 \xrightarrow{\text{slow cooling}} \alpha\text{-}Al_2O_3 \qquad (16)$$

Heating the attached aluminium hydroxide and boehmite to an elevated temperature will transfer it into one of the alumina phases [95]:

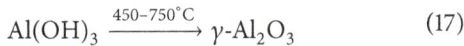

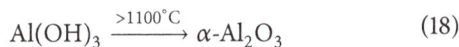

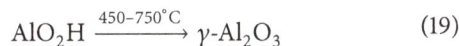

$$Al(OH)_3 \xrightarrow{450-750^\circ C} \gamma\text{-}Al_2O_3 \qquad (17)$$

$$Al(OH)_3 \xrightarrow{>1100^\circ C} \alpha\text{-}Al_2O_3 \qquad (18)$$

$$AlO_2H \xrightarrow{450-750^\circ C} \gamma\text{-}Al_2O_3 \qquad (19)$$

$$AlO_2H \xrightarrow{>1100^\circ C} \alpha\text{-}Al_2O_3 \qquad (20)$$

The anions will be attracted toward the surface during the anodic polarization and produce alumina and other compositions [10, 70]:

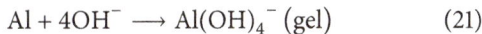

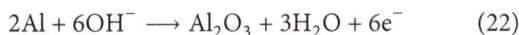

$$Al + 4OH^- \longrightarrow Al(OH)_4^- \text{ (gel)} \qquad (21)$$

$$2Al + 6OH^- \longrightarrow Al_2O_3 + 3H_2O + 6e^- \qquad (22)$$

$$4OH^- \longrightarrow O_2 + 2H_2O + 4e^- \qquad (23)$$

The silica is produced (24) by the combination between attracted silicate anions and H^+ which is produced by the water electrolysis on the anode (25)

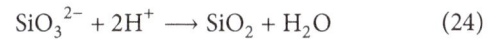

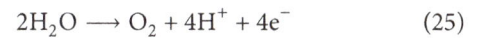

$$SiO_3^{2-} + 2H^+ \longrightarrow SiO_2 + H_2O \qquad (24)$$

$$2H_2O \longrightarrow O_2 + 4H^+ + 4e^- \qquad (25)$$

Generally, the chemical dissolution occurs during the anodic and neutral periods, the MAO only occurs during the anodic period, and the cathodic period does not contribute to the ceramic building or dissolution.

As a result of the previous reactions, the most likely compositions and phases on the surface of MAO-treated aluminium in the alkaline sodium silicate electrolyte are alumina phase/s, silica phase/s, and aluminosilicate phase/s, while other compositions mostly will be released or dissolved in the electrolyte or in water during the cleaning process.

4. Results

After the preparation by a month and six months, the samples were studied by the SEM and XRD. A disappearance of some small particles was notified in the accumulated particles region, as shown in Figure 2. By comparing the XRD patterns, a strong peak for cristobalite (an SiO_2 phase) and other small peaks were almost vanished, Figure 3, which suggested that these released particles mostly consist of the cristobalite phase. The small size cristobalite particles provide more surface area or surface energy, which are prone to be dissolved in water or other solvents during cleaning process, which tends to enhance the detachment of cristobalite particles [96]. Due to this phenomenon, all presented measurements and results in this study belong to the samples after six months of treatment.

A linear correlation occurred for the layer thickness and surface roughness with the current densities $J_a \leq$ 21.88 A/dm^2, followed by a deflection from the linearity for the highest current density, as shown in Figures 4(a) and 4(b). The thinnest coating 10.9 μm was formed at the lowest current density 10.94 A/dm^2, while the thickest coating 18.5 μm was formed at the highest current density 43.75 A/dm^2. The surface roughness ranged between 0.79 μm and 1.27 μm for current densities 10.94 A/dm^2 and 43.75 A/dm^2, respectively. Similar response of thickness-current density was found

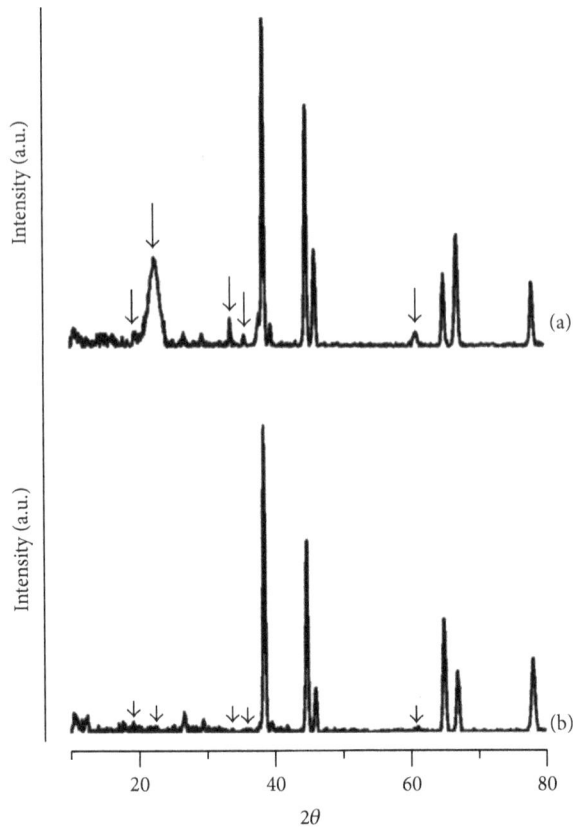

FIGURE 3: XRD patterns of a sample after the MAO treatment by (a) a month and (b) six months. The arrows are pointing to the peaks which were decreased significantly after six months of storage.

using DC current by Raj and Mubarak Ali [70]. A linear correlation between the MAO alumina ceramic thickness and roughness is illustrated in Figure 5, which is consistent with some previously published works [87, 97, 98].

Figure 6 illustrates the apparent microstructures in the SEM micrographs, which can be classified into accumulated particles, microcracks, and volcano-like microstructures. A volcano-like microstructure includes a solidified pool and a centered open/blind crater. The volcano-like microstructures are formed by the discrete localized microdischarge events. The rapid solidification of the molten alumina forms the microcracks and accumulated particles on and around the discharge channels [99]. Wei et al. [45] reported that more α-alumina can be formed in the inner layers due to the low cooling rate, while the high cooling rate on the outer layer surface was favorable to form more γ-alumina during the solidification.

The increment of anodic current density significantly affected the size of volcano-like as presented in Figures 7(a)–7(d). Smallest volcano-like microstructures with more population occurred for the lowest J_a (10.94 A/dm^2), while the largest volcano-like microstructures associated with the highest J_a (43.75 A/dm^2). To get more comprehensive results, an average of 27 different randomly selected positions were captured by the SEM for each sample and digitally analyzed

by the Image-Pro Plus software to estimate the ratio of the occupied area by the volcano-like microstructures to the micrograph frame, and the results are presented in Figure 8. A slight increment in the area ratio from 55.6% to 59.6% occurred for the occupied area ratio of the volcano-like microstructures due to the increment of the J_a from 10.94 A/dm^2 to 43.75 A/dm^2. On the contrary, a significant decrement from 2620 mm^{-2} to 1420 mm^{-2} was accomplished in the volcano-like density due to the increment of anodic current density, as shown in Figure 9.

Figure 10 shows the elemental compositions of several EDX study points over samples prepared at various anodic current densities. For all samples, the aluminium was dominant near to the craters, while its concentration was decreased as the study point moved away from the craters. More silicon was found in the accumulated particles, Figures 10(a)–10(d). Few sodium amounts were found in the accumulated particles of the MAO ceramic surface prepared at 43.75 A/dm^2.

Figure 11 shows the XRD patterns of MAO alumina ceramics prepared at various anodic biasing current densities. The most apparent peaks belong to aluminium due to the X-ray penetration into the substrate through the ceramic layer. To identify the compositions and phases related to other shorter peaks, we applied the RiR method by MATCH! software. The other major phases were α-alumina, γ-alumina, cristobalite (an SiO$_2$ phase), and sillimanite (an Al$_2$SiO$_5$ phase), while there were few amounts of Si and Na which are localized in the inner ceramic layer. The weight percentages of the major phases are presented in Figure 12 after subtracting the weight percentages of Al, Si, and Na. The γ-alumina phase was the dominant in the ceramic coating at low anodic current density. As the anodic current density increased, the γ-alumina decreased from 66.6 wt.% to 26.2 wt.%, while the α-alumina phase increased from 3.9 wt.% to 27.6 wt.%. The sillimanite was around 20 wt.% for the lowest anodic current density of 10.94 A/dm^2 and approximately constant around 40 wt.% for anodic current densities of 14.58, 21.88, and 43.75 A/dm^2. The cristobalite was 9 wt.% for anodic current density of 10.94 A/dm^2 and less than 5 wt.% for $J_a \geq$ 14.58 A/dm^2.

Figure 13 shows the infrared spectra of MAO ceramic surfaces prepared at different anodic current densities, in addition to an untreated saw cut aluminium surface. The measurements were conducted at 70°C; the shaded rectangles belong to the absorption bands of CO$_2$ (centered at 4.3 and 14.9 μm) and H$_2$O (centered at 6.1 μm) in the ambient air [100–104]. The MAO ceramic surfaces have higher emissivity values compared to the untreated saw cut aluminium surface.

The emissivity spectrum has three distinguishable wavelength regions. The emissivity increased with the wavelength increment in the first region, between 4.0 μm and 7.8 μm. The second region is characterized by a rapid ascending that starts from 7.8 μm and ends at 8.6 μm. The third region for wavelengths longer than 8.6 μm is characterized by a relatively high emissivity with three apparent peaks centered at 10.2 μm, 12.8 μm, and 15.5 μm. The high intensity values of the peak at 10.2 μm ranged between 96.6% and 97.4% for 10.94 A/dm^2 and 43.75 A/dm^2, respectively.

(a)   (b)

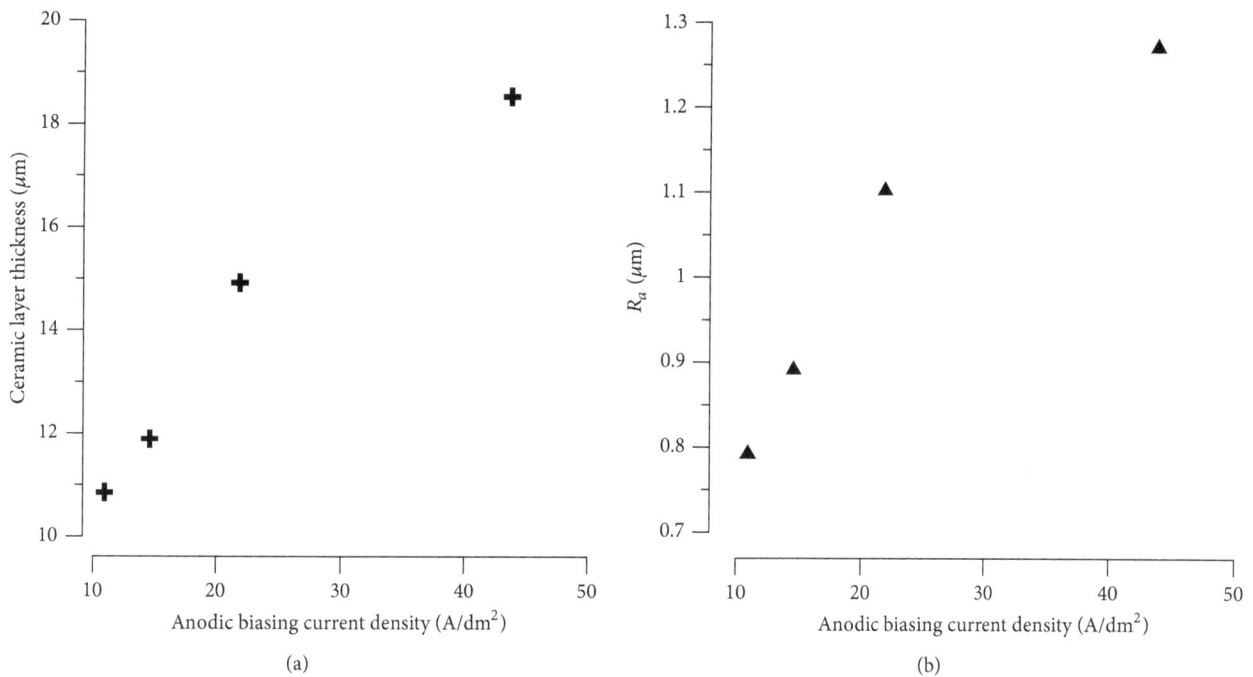

FIGURE 4: The effect of anodic biasing current density on the (a) layer thickness and (b) surface roughness of the MAO ceramic coating.

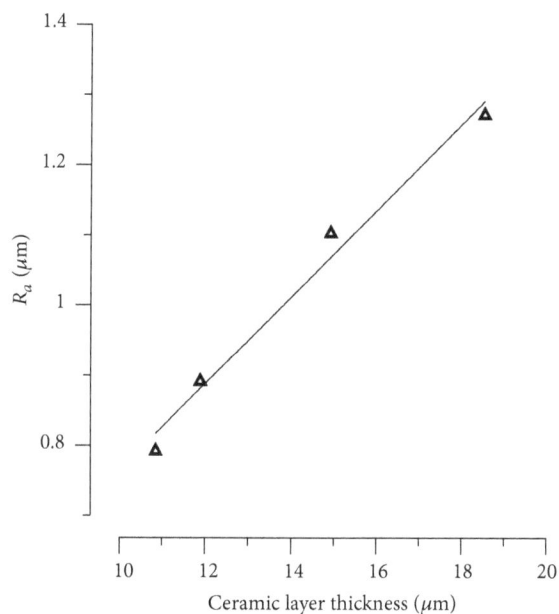

FIGURE 5: The linear correlation between the MAO ceramic thickness and surface roughness.

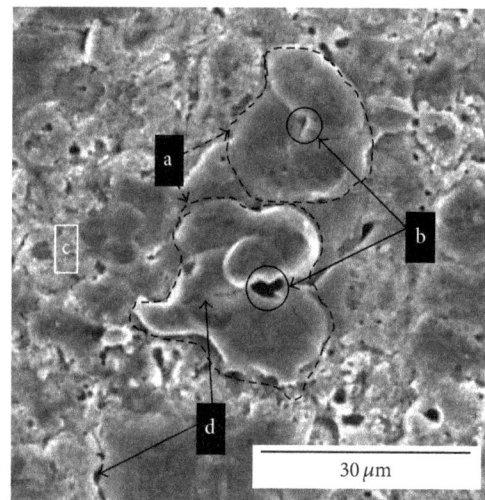

FIGURE 6: The main microstructures in the SEM micrographs are (a) resolidified pools, (b) craters, (c) accumulated particles, and (d) microcracks.

5. Discussions

At the first beginning of the MAO process, the relatively high anodic voltage activates the inward oxidation of the aluminium substrate by the reaction between the inward immigrant O^{2-} and the outward immigrant Al^{3+} according to (11) [93]. The growth of aluminium oxide layer increases the electrical resistance, and at a specific thickness, the current begins to flow through weak points on the oxide layer and strengthens the electrical field, which in turn significantly increases the reionization of the surrounding electrolyte and the aluminium substrate and consequently triggers the plasma microdischarges. The elevated temperature of localized plasma (which ranges between 2 kK [105] and 11 kK [106]) melts the alumina and vaporizes the electrolyte in the discharging circumference and evacuates the region of plasma discharging. Due to the pressure reduction, the molten alumina will be suctioned out of the discharging channel and the electrolyte flows toward the low pressure region and will be vaporized by the molten alumina. Some of

(a)

(b)

(c)

(d)

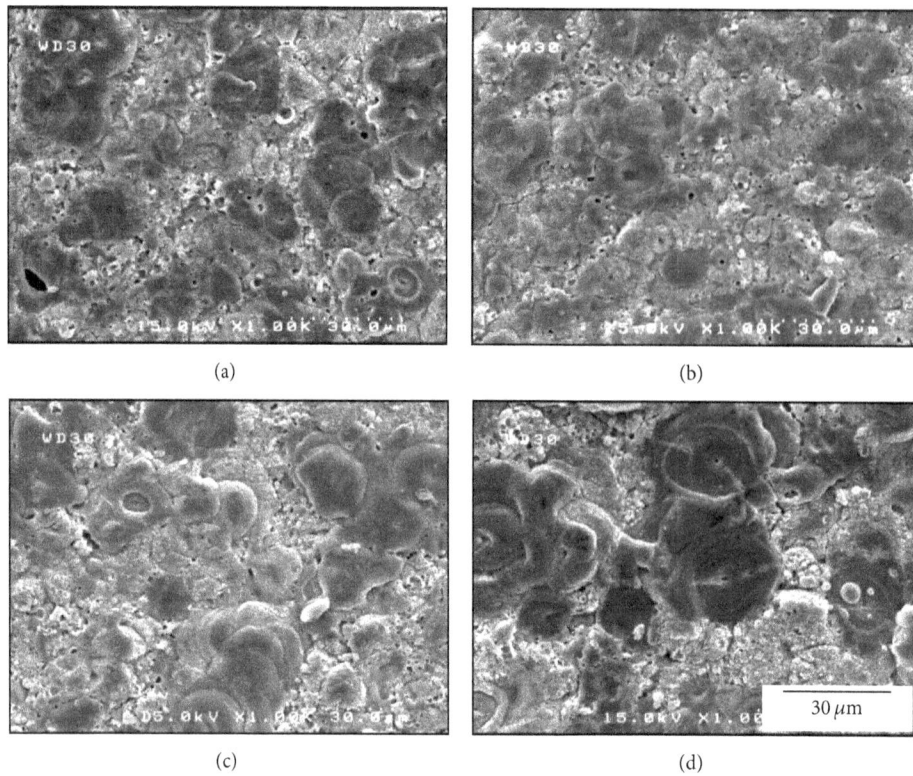

FIGURE 7: Surface morphologies of MAO ceramic coatings prepared at different anodic biasing current densities: (a) 10.94 A/dm^2, (b) 14.58 A/dm^2, (c) 21.88 A/dm^2, and (d) 43.75 A/dm^2.

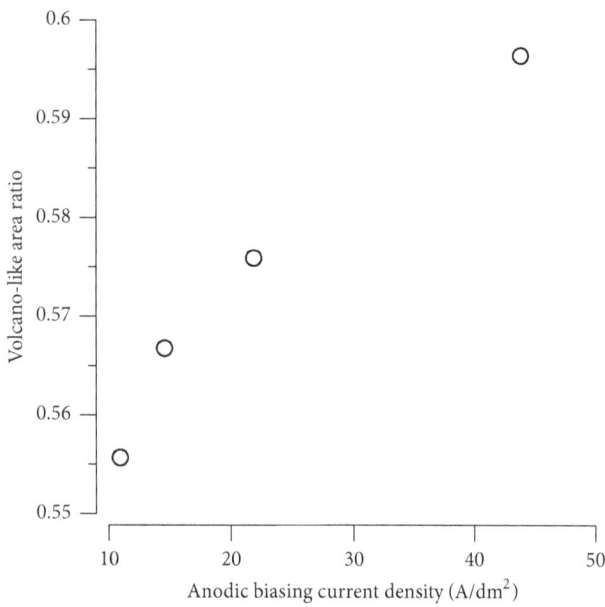

FIGURE 8: The effect of anodic current density on the relatively occupied area by volcano-like microstructures.

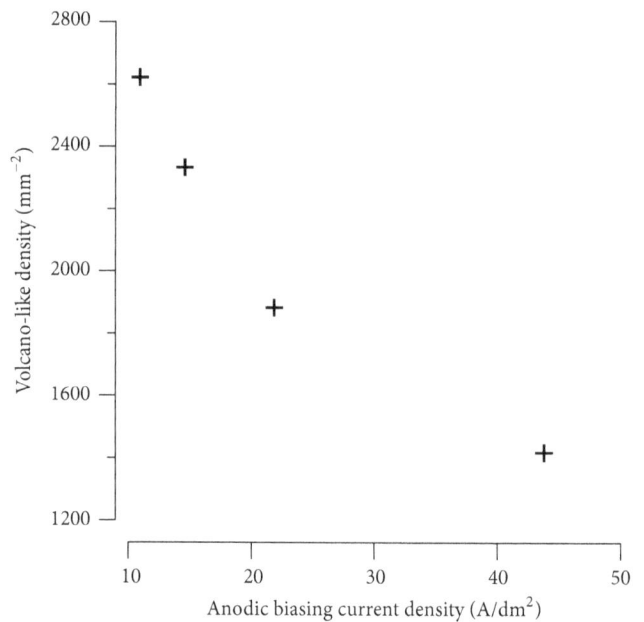

FIGURE 9: The inverse proportional of volcano-like density with the anodic current density.

solid molecules, such as silica, and ions may be attached on the surface of molten alumina, which forms the cristobalite and sillimanite phases which were found on the accumulated particles and on the solidified pools according to the EDX and

XRD results. The volcano-like microstructures are created by the solidification of the molten alumina which spread over the surface around the discharging channels forming the solidified pools and the craters. According to the EDX

Sample	Study point	Al	O	Si	Na	Sample	Study point	Al	O	Si	Na
	a	52.3	47.7	—	—		a	59.4	40.6	—	—
	b	45.9	48.8	5.3	—		b	52.3	41.8	5.9	—
	c	16.2	55.0	28.8	—		c	49.3	46.1	4.6	—
	d	27.2	54.0	18.8	—		d	48.5	44.6	6.9	—
	e	39.2	47.6	13.2	—		e	26.8	50.3	22.9	—
							f	9.9	49.0	41.1	—
							g	19.6	42.9	37.5	—
(a)						(b)					
	a	54.9	36.1	9.0	—		a	56.4	43.6	—	—
	b	54.2	35.2	10.6	—		b	62.0	34.3	3.7	—
	c	48.1	45.8	6.1	—		c	56.9	39.6	3.5	—
	d	13.7	48.9	37.4	—		d	62.3	27.2	10.5	—
	e	29.2	30.2	40.6	—		e	9.2	43.7	47.1	—
	f	23.0	42.0	35.0	—		f	26.0	45.2	28.8	—
	g	41.3	46.2	12.5	—		g	18.2	52.1	27.8	1.9
(c)						(d)					

FIGURE 10: The results of the EDX analysis at several locations on the MAO ceramic surfaces prepared at different anodic current densities: (a) 10.94, (b) 14.58, (c) 21.88, and (d) 43.75 A/dm^2.

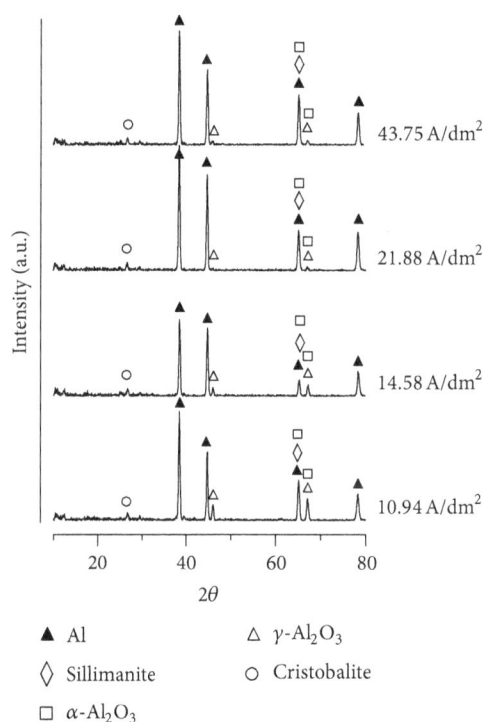

FIGURE 11: XRD spectra for MAO alumina ceramic coatings prepared at different anodic current densities. The most apparent peaks are for aluminium, which minimize the appeared intensities for other phases and compositions.

results, more silicon was found on the accumulated particles and the edges of the solidified pools, which were identified by the XRD results to be the cristobalite and sillimanite phases. The increment of cristobalite and sillimanite phases on the accumulated particles and edges of solidified pools was

because of the longer travelling distance during the ejection of alumina out of the discharging channel, which allowed more silica to be attached. After the solidification, the surface tension forms the microcracks and the accumulated small particles on the solidified pool edges which contain more cristobalite and sillimanite phases [107]. Also, the EDX results stated that the aluminium concentration increased as the study point approached the crater; this reflects that less reaction happened between the center of ejected molten alumina and the silica due to the flow of the vaporized electrolyte which might start from the edges and consequently more silica was attached on the edges.

Figure 4 states the effect of anodic current density on the layer thickness and surface roughness. The small working area increased the current density which resulted in stronger microdischarges and consequently ejected more molten alumina out of the discharging channels, which increased the growth of the MAO ceramic coating, Figure 4(a). As the MAO ceramic grew, the current flowed through less weak points, which strengthened the electrical plasma discharges and produced wider volcano-like microstructures, Figure 7(d), less volcano-like density, Figure 9, and a rougher surface, Figure 4(b). The stronger microdischarges liberated more elevated temperature and vaporized more electrolytes which in turn formed an envelope of gas which surrounded the working area and slowed the cooling rate and was favorable to form more α-alumina phases, Figure 12(a). The growth of the ceramic layer, Figure 4(a), was less at the highest current density, 43.75 A/dm^2, due to the increment of liberated heat which increased the chemical dissolution around the working area [70].

The alumina ceramic is semitransparent in the range 4–7.7 μm and opaque in the 7.7–25.0 μm range as reported by Rozenbaum et al. [108]. Boumaza et al. found that the γ-alumina has a broadband between 11.1 μm and 33.3 μm [109],

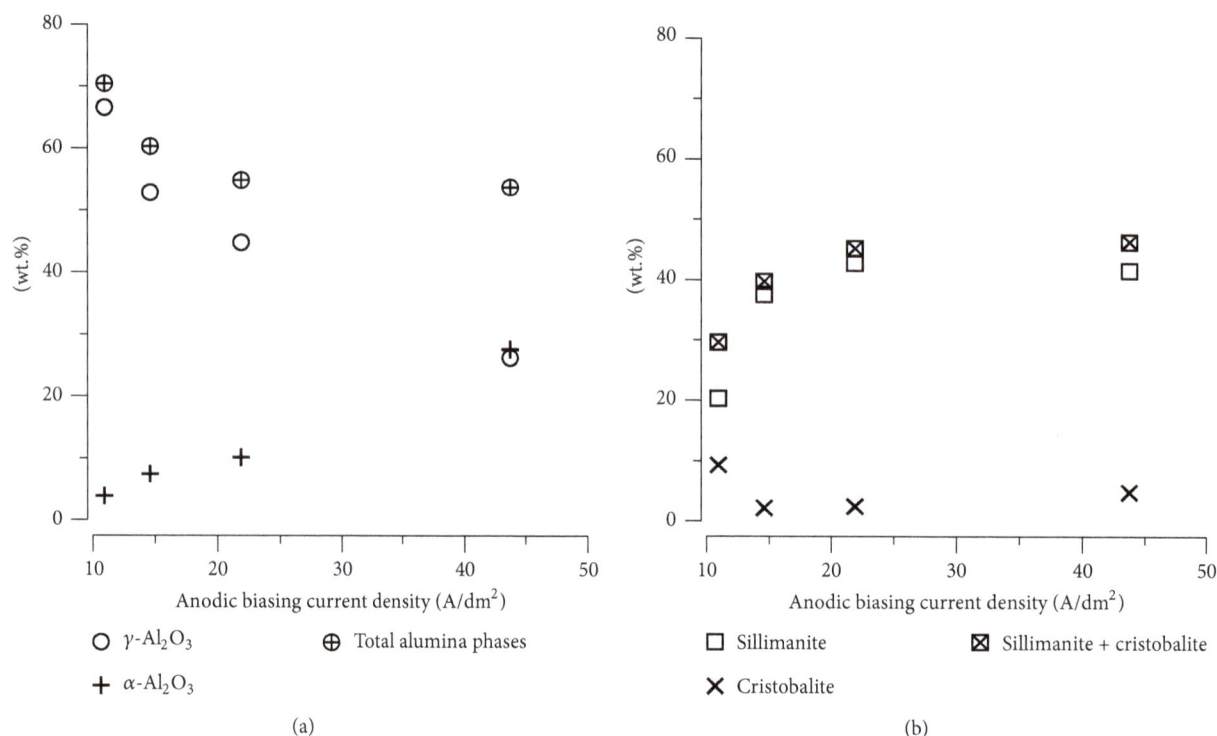

FIGURE 12: Effect of anodic current density on the MAO ceramic: (a) alumina phases and (b) sillimanite and cristobalite phases.

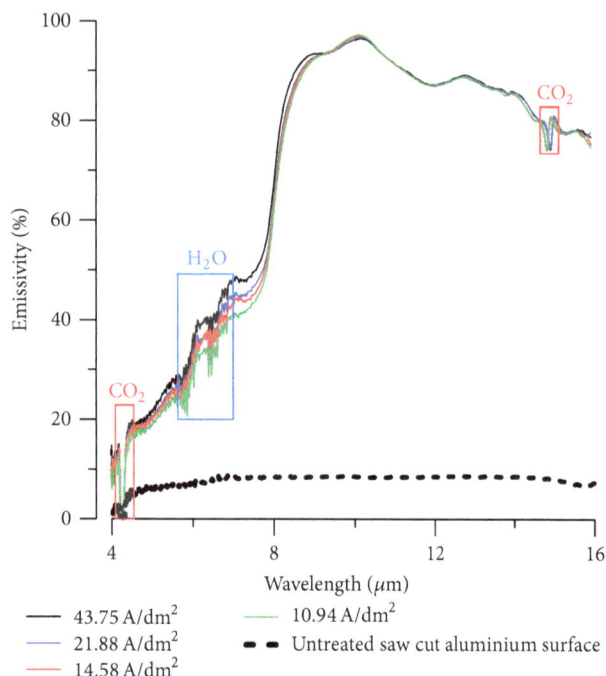

FIGURE 13: IR spectral emissivity of the MAO ceramic surfaces prepared at different anodic current densities.

and the α-alumina has IR band positions at 15.50 and 16.45 μm [109]. The cristobalite phase has IR peaks at 9.6, 11.4, 12.5, and 15.3 μm, measured at 300°C, which shift into longer wavelengths at lower temperatures [110]. The sillimanite phase has IR peaks at 10.75, 11.63, 13.16, and 15.15 μm, and

a broadband ranges from 8.55 μm to 9.80 μm [111]. These facts explain the behavior of IR emissivity spectra in Figure 13. For the range between 4.0 μm and 7.8 μm, the ceramic alumina is semitransparent and its emissivity is lower than that in the region (8.6–16.0 μm) where the ceramic turns to be opaque and stronger emitter due to the existence of γ-alumina and other phases. The existing phases contributed to enhancing the emissivity in the opaque region, and the apparent peaks centered at 10.2 μm and 12.8 μm were formed by the cristobalite and sillimanite, while the 15.5 μm was formed by the α-alumina and cristobalite phases.

Previous studies showed that the infrared emissivity of ceramic surfaces depends on its chemical and physical properties such as compositions, phases, roughness, porosity, grain size, pore size, spatial repetition of the grains, and layer thickness, and each wavelength region has its own effective factors [88, 98, 108, 112]. According to Figure 13, the intensity of emissivity has a slight variation in the semitransparent region, whereas no significant variation occurred in the opaque region. Referring to the phase concentrations in Figure 12, the lowest current density formed 9 and 20 wt.% of cristobalite and sillimanite phases, respectively, while their weight percentages were approximately constants for $J_a \geq 14.58$ A/dm^2. The α- and γ-alumina phases are semitransparent for the wavelengths shorter than 7.7 μm. On the other hand, both of the surface roughness and layer thickness were increased with the J_a. This leads to the conclusion that the slight variation of the emissivity in the semitransparent region was due to the surface roughness and thickness, and the existing phases did not contribute to the variations of emissivity in this region.

The emissivity averages of long wavelength infrared (LWIR) region (8–15 μm) were 88.4, 88.6, 88.6, and 89.4% for the MAO alumina ceramics prepared at anodic current densities of 10.94, 14.58, 21.88, and 43.75 A/dm^2, respectively. The anodic current density did not change the LWIR emissivity significantly, which has a good application in the industrial field to fabricate a relatively high emitter MAO alumina ceramic surface which also provides a strong adhesion to the aluminium substrate, corrosion resistance, wear resistance, thermal shock resistance, and high hardness as found in previously published works. The high emitter MAO alumina ceramic will enhance the heat dissipation of different cooling and heating systems which use aluminium, such as LED, laser, electronic circuits, CPU heat sink, vehicle, indoor radiators, and others. Further works are needed to study the IR emissivity of different MAO alumina ceramics fabricated on the aluminium surfaces using different MAO treatment conditions and electrolyte compositions.

6. Conclusions

The increment of anodic current density widened the volcano-like microstructures and lowered its density, while the relatively occupied area by the volcano-like microstructures was approximately constant.

The high anodic current density thickened and roughened the MAO alumina ceramic. The major phases in the MAO alumina ceramic were α-alumina, γ-alumina, sillimanite, and cristobalite. The increment of current density, from 10.94 A/dm^2 to 43.75 A/dm^2, increased the concentration of α-alumina from 3.9 wt.% to 27.6 wt.% and decreased the concentration of γ-alumina from 66.6 wt.% to 26.2 wt.%. For $J_a \geq$ 14.58 A/dm^2, the sillimanite and cristobalite concentrations were approximately constant around 40 wt.% and less than 5 wt.%, respectively.

The spectral IR emissivity was slightly varied in the semitransparent region, whereas no significant variation occurred in the opaque region. The slight variation of the emissivity in the semitransparent region was due to the surface roughness and thickness, and the existing phases did not contribute to the variations of emissivity in the semitransparent region. The existing phases contributed to enhancing the emissivity in the opaque region. The apparent peaks centered at 10.2 μm and 12.8 μm were formed by the cristobalite and sillimanite phases, while another peak at 15.5 μm was formed by the α-alumina and cristobalite phases.

Using an ecofriendly alkaline silicate electrolyte, a relatively high emissivity MAO alumina ceramic was fabricated on the aluminium substrate. The MAO alumina ceramic has an emissivity peak of 97% at 12.8 μm, measured at 70°C, and an average of LWIR emissivity ranged between 88.4% and 89.4%. More future studies are required to fabricate high emissivity ceramic surfaces using different MAO processing conditions and electrolytes.

Acknowledgments

The authors wish to express their sincere gratitude to Dr. Hong-Jen Li for his cooperation and help in using FTIR spectroscopy. The technical support of Shi-Rui Li, Shu-Wei Huang, Yuan-Yi Sung, Chan-Hsuan Chen, Shi-Chung Chen, Wei-Tie Wu, Chien-Huang Kuo, Chih-Yeh Lin, Jie-Mine Wu, and Wan-Yi Chen is gratefully acknowledged. Also, the authors wish to thank Asian Vital Component Company, Taipei, for providing the aluminium pieces.

References

[1] A. K. A. Shati, S. G. Blakey, and S. B. M. Beck, "The effect of surface roughness and emissivity on radiator output," *Energy and Buildings*, vol. 43, no. 2-3, pp. 400–406, 2011.

[2] S.-H. Yu, D. Jang, and K.-S. Lee, "Effect of radiation in a radial heat sink under natural convection," *International Journal of Heat and Mass Transfer*, vol. 55, no. 1-3, pp. 505–509, 2012.

[3] F. Mécuson, T. Czerwiec, T. Belmonte, L. Dujardin, A. Viola, and G. Henrion, "Diagnostics of an electrolytic microarc process for aluminium alloy oxidation," *Surface and Coatings Technology*, vol. 200, no. 1-4, pp. 804–808, 2005.

[4] S. Stojadinovic, R. Vasilic, I. Belca et al., "Characterization of the plasma electrolytic oxidation of aluminium in sodium tungstate," *Corrosion Science*, vol. 52, no. 10, pp. 3258–3265, 2010.

[5] Z. Wang, L. Wu, Y. Qi, W. Cai, and Z. Jiang, "Self-lubricating Al$_2$O$_3$/PTFE composite coating formation on surface of aluminium alloy," *Surface and Coatings Technology*, vol. 204, no. 20, pp. 3315–3318, 2010.

[6] R. H. U. Khan, A. Yerokhin, X. Li, H. Dong, and A. Matthews, "Surface characterisation of DC plasma electrolytic oxidation treated 6082 aluminium alloy: effect of current density and electrolyte concentration," *Surface and Coatings Technology*, vol. 205, no. 6, pp. 1679–1688, 2010.

[7] L. O. Snizhko, A. L. Yerokhin, A. Pilkington et al., "Anodic processes in plasma electrolytic oxidation of aluminium in alkaline solutions," *Electrochimica Acta*, vol. 49, no. 13, pp. 2085–2095, 2004.

[8] S.-M. Moon and S.-I. Pyun, "The corrosion of pure aluminium during cathodic polarization in aqueous solutions," *Corrosion Science*, vol. 39, no. 2, pp. 399–408, 1997.

[9] L. Wen, Y. Wang, Y. Zhou, J.-H. Ouyang, L. Guo, and D. Jia, "Corrosion evaluation of microarc oxidation coatings formed on 2024 aluminium alloy," *Corrosion Science*, vol. 52, no. 8, pp. 2687–2696, 2010.

[10] J. R. Morlidge, P. Skeldon, G. E. Thompson, H. Habazaki, K. Shimizu, and G. C. Wood, "Gel formation and the efficiency of anodic film growth on aluminium," *Electrochimica Acta*, vol. 44, no. 14, pp. 2423–2435, 1999.

[11] P. I. Butyagin, Y. V. Khokhryakov, and A. I. Mamaev, "Microplasma systems for creating coatings on aluminium alloys," *Materials Letters*, vol. 57, no. 11, pp. 1748–1751, 2003.

[12] J. Jovović, S. Stojadinović, N. M. Šišović, and N. Konjević, "Spectroscopic characterization of plasma during electrolytic oxidation (PEO) of aluminium," *Surface and Coatings Technology*, vol. 206, pp. 24–28, 2011.

[13] A. L. Yerokhin, L. O. Snizhko, N. L. Gurevina, A. Leyland, A. Pilkington, and A. Matthews, "Spatial characteristics of discharge phenomena in plasma electrolytic oxidation of aluminium alloy," *Surface and Coatings Technology*, vol. 177-178, pp. 779–783, 2004.

[14] T. Abdulla, A. Yerokhin, and R. Goodall, "Effect of Plasma Electrolytic Oxidation coating on the specific strength of open-cell

aluminium foams," *Materials and Design*, vol. 32, no. 7, pp. 3742–3749, 2011.

[15] A. L. Yerokhin, A. A. Voevodin, V. V. Lyubimov, J. Zabinski, and M. Donley, "Plasma electrolytic fabrication of oxide ceramic surface layers for tribotechnical purposes on aluminium alloys," *Surface and Coatings Technology*, vol. 110, no. 3, pp. 140–146, 1998.

[16] E. Matykina, R. Arrabal, P. Skeldon, G. E. Thompson, and P. Belenguer, "AC PEO of aluminium with porous alumina precursor films," *Surface and Coatings Technology*, vol. 205, no. 6, pp. 1668–1678, 2010.

[17] S. V. Gnedenkov, O. A. Khrisanfova, A. G. Zavidnaya et al., "Production of hard and heat-resistant coatings on aluminium using a plasma micro-discharge," *Surface and Coatings Technology*, vol. 123, no. 1, pp. 24–28, 2000.

[18] M. Treviño, N. F. Garza-Montes-de-Oca, A. Pérez, M. A. L. Hernández-Rodríguez, A. Juárez, and R. Colás, "Wear of an aluminium alloy coated by plasma electrolytic oxidation," *Surface and Coatings Technology*, vol. 206, no. 8-9, pp. 2213–2219, 2012.

[19] T. S. Lim, H. S. Ryu, and S.-H. Hong, "Electrochemical corrosion properties of CeO$_2$-containing coatings on AZ31 magnesium alloys prepared by plasma electrolytic oxidation," *Corrosion Science*, vol. 62, pp. 104–111, 2012.

[20] A. V. Timoshenko and Y. V. Magurova, "Investigation of plasma electrolytic oxidation processes of magnesium alloy MA2-1 under pulse polarisation modes," *Surface and Coatings Technology*, vol. 199, no. 2-3, pp. 135–140, 2005.

[21] J. Liang, P. B. Srinivasan, C. Blawert, and W. Dietzel, "Comparison of electrochemical corrosion behaviour of MgO and ZrO$_2$ coatings on AM50 magnesium alloy formed by plasma electrolytic oxidation," *Corrosion Science*, vol. 51, no. 10, pp. 2483–2492, 2009.

[22] F. Liu, D. Shan, Y. Song, E.-H. Han, and W. Ke, "Corrosion behavior of the composite ceramic coating containing zirconium oxides on AM30 magnesium alloy by plasma electrolytic oxidation," *Corrosion Science*, vol. 53, no. 11, pp. 3845–3852, 2011.

[23] G.-H. Lv, H. Chen, L. Li et al., "Investigation of plasma electrolytic oxidation process on AZ91D magnesium alloy," *Current Applied Physics*, vol. 9, no. 1, pp. 126–130, 2009.

[24] P. Bala Srinivasan, R. Zettler, C. Blawert, and W. Dietzel, "A study on the effect of plasma electrolytic oxidation on the stress corrosion cracking behaviour of a wrought AZ61 magnesium alloy and its friction stir weldment," *Materials Characterization*, vol. 60, no. 5, pp. 389–396, 2009.

[25] P. Zhang, X. Nie, H. Hu, and Y. Liu, "TEM analysis and tribological properties of Plasma Electrolytic Oxidation (PEO) coatings on a magnesium engine AJ62 alloy," *Surface and Coatings Technology*, vol. 205, no. 5, pp. 1508–1514, 2010.

[26] F. Liu, D. Shan, Y. Song, and E.-H. Han, "Effect of additives on the properties of plasma electrolytic oxidation coatings formed on AM50 magnesium alloy in electrolytes containing K2ZrF6," *Surface and Coatings Technology*, vol. 206, no. 2-3, pp. 455–463, 2011.

[27] P. Wang, J. Li, Y. Guo, and Z. Yang, "Growth process and corrosion resistance of ceramic coatings of micro-arc oxidation on Mg-Gd-Y magnesium alloys," *Journal of Rare Earths*, vol. 28, no. 5, pp. 798–802, 2010.

[28] J. Liang, L. Hu, and J. Hao, "Characterization of microarc oxidation coatings formed on AM60B magnesium alloy in silicate and phosphate electrolytes," *Applied Surface Science*, vol. 253, no. 10, pp. 4490–4496, 2007.

[29] F. Jin, P. K. Chu, G. Xu, J. Zhao, D. Tang, and H. Tong, "Structure and mechanical properties of magnesium alloy treated by micro-arc discharge oxidation using direct current and high-frequency bipolar pulsing modes," *Materials Science and Engineering A*, vol. 435-436, pp. 123–126, 2006.

[30] K. R. Shin, Y. G. Ko, and D. H. Shin, "Effect of electrolyte on surface properties of pure titanium coated by plasma electrolytic oxidation," *Journal of Alloys and Compounds*, vol. 509, supplement 1, pp. S478–S481, 2011.

[31] X. Wang, X. Pan, W. Ye, Y. Wei, and Y. Chen, "Preparation and properties of TiC$_x$N$_{1-x}$ coatings containing calcium on titanium surface by plasma electrolytic carbonitriding," *Surface and Coatings Technology*, vol. 228, supplement 1, pp. S194–S197, 2013.

[32] D. Wei, Y. Zhou, Y. Wang, and D. Jia, "Characteristic of microarc oxidized coatings on titanium alloy formed in electrolytes containing chelate complex and nano-HA," *Applied Surface Science*, vol. 253, no. 11, pp. 5045–5050, 2007.

[33] W. Zhang, K. Du, C. Yan, and F. Wang, "Preparation and characterization of a novel Si-incorporated ceramic film on pure titanium by plasma electrolytic oxidation," *Applied Surface Science*, vol. 254, no. 16, pp. 5216–5223, 2008.

[34] K. R. Shin, Y. G. Ko, and D. H. Shin, "Surface characteristics of ZrO$_2$-containing oxide layer in titanium by plasma electrolytic oxidation in K$_4$P$_2$O$_7$ electrolyte," *Journal of Alloys and Compounds*, vol. 536, supplement 1, pp. S226–S230, 2012.

[35] Y. M. Wang, L. X. Guo, J. H. Ouyang, Y. Zhou, and D. C. Jia, "Interface adhesion properties of functional coatings on titanium alloy formed by microarc oxidation method," *Applied Surface Science*, vol. 255, no. 15, pp. 6875–6880, 2009.

[36] P. Huang, K.-W. Xu, and Y. Han, "Preparation and apatite layer formation of plasma electrolytic oxidation film on titanium for biomedical application," *Materials Letters*, vol. 59, no. 2-3, pp. 185–189, 2005.

[37] Y. Wang, T. Lei, L. Guo, and B. Jiang, "Fretting wear behaviour of microarc oxidation coatings formed on titanium alloy against steel in unlubrication and oil lubrication," *Applied Surface Science*, vol. 252, no. 23, pp. 8113–8120, 2006.

[38] S. Stojadinović, R. Vasilić, M. Petković, and L. Zeković, "Plasma electrolytic oxidation of titanium in heteropolytungstate acids," *Surface and Coatings Technology*, vol. 206, pp. 575–581, 2011.

[39] H. Tang, Q. Sun, T. Xin, C. Yi, Z. Jiang, and F. Wang, "Influence of Co(CH$_3$COO)$_2$ concentration on thermal emissivity of coatings formed on titanium alloy by micro-arc oxidation," *Current Applied Physics*, vol. 12, no. 1, pp. 284–290, 2012.

[40] S. Cui, J. Han, Y. Du, and W. Li, "Corrosion resistance and wear resistance of plasma electrolytic oxidation coatings on metal matrix composites," *Surface and Coatings Technology*, vol. 201, no. 9–11, pp. 5306–5309, 2007.

[41] G. Rapheal, S. Kumar, C. Blawert, and N. B. Dahotre, "Wear behavior of plasma electrolytic oxidation (PEO) and hybrid coatings of PEO and laser on MRI 230D magnesium alloy," *Wear*, vol. 271, no. 9-10, pp. 1987–1997, 2011.

[42] X. Nie, E. I. Meletis, J. C. Jiang, A. Leyland, A. L. Yerokhin, and A. Matthews, "Abrasive wear/corrosion properties and TEM analysis of Al$_2$O$_3$ coatings fabricated using plasma electrolysis," *Surface and Coatings Technology*, vol. 149, no. 2-3, pp. 245–251, 2002.

[43] Y. Jiang, Y. Zhang, Y. Bao, and K. Yang, "Sliding wear behaviour of plasma electrolytic oxidation coating on pure aluminium," *Wear*, vol. 271, no. 9-10, pp. 1667–1670, 2011.

[44] M. Aliofkhazraei, A. Sabour Rouhaghdam, and T. Shahrabi, "Abrasive wear behaviour of Si_3N_4/TiO_2 nanocomposite coatings fabricated by plasma electrolytic oxidation," *Surface and Coatings Technology*, vol. 205, supplement 1, pp. S41–S46, 2010.

[45] T. Wei, F. Yan, and J. Tian, "Characterization and wear- and corrosion-resistance of microarc oxidation ceramic coatings on aluminum alloy," *Journal of Alloys and Compounds*, vol. 389, no. 1-2, pp. 169–176, 2005.

[46] L. Wang, J. Zhou, J. Liang, and J. Chen, "Microstructure and corrosion behavior of plasma electrolytic oxidation coated magnesium alloy pre-treated by laser surface melting," *Surface and Coatings Technology*, vol. 206, no. 13, pp. 3109–3115, 2012.

[47] P. Su, X. Wu, Y. Guo, and Z. Jiang, "Effects of cathode current density on structure and corrosion resistance of plasma electrolytic oxidation coatings formed on ZK60 Mg alloy," *Journal of Alloys and Compounds*, vol. 475, no. 1-2, pp. 773–777, 2009.

[48] R. O. Hussein, D. O. Northwood, and X. Nie, "The influence of pulse timing and current mode on the microstructure and corrosion behaviour of a plasma electrolytic oxidation (PEO) coated AM60B magnesium alloy," *Journal of Alloys and Compounds*, vol. 541, pp. 41–48, 2012.

[49] L. Rama Krishna, G. Poshal, and G. Sundararajan, "Influence of electrolyte chemistry on morphology and corrosion resistance of micro arc oxidation coatings deposited on magnesium," *Metallurgical and Materials Transactions A*, vol. 41, no. 13, pp. 3499–3508, 2010.

[50] J. Liang, L. Hu, and J. Hao, "Improvement of corrosion properties of microarc oxidation coating on magnesium alloy by optimizing current density parameters," *Applied Surface Science*, vol. 253, no. 16, pp. 6939–6945, 2007.

[51] C. Blawert, V. Heitmann, W. Dietzel, H. M. Nykyforchyn, and M. D. Klapkiv, "Influence of electrolyte on corrosion properties of plasma electrolytic conversion coated magnesium alloys," *Surface and Coatings Technology*, vol. 201, no. 21, pp. 8709–8714, 2007.

[52] D. Y. Hwang, Y. M. Kim, D.-Y. Park, B. Yoo, and D. H. Shin, "Corrosion resistance of oxide layers formed on AZ91 Mg alloy in $KMnO_4$ electrolyte by plasma electrolytic oxidation," *Electrochimica Acta*, vol. 54, no. 23, pp. 5479–5485, 2009.

[53] Y. M. Wang, B. L. Jiang, T. Q. Lei, and L. X. Guo, "Microarc oxidation coatings formed on Ti6Al4V in Na_2SiO_3 system solution: microstructure, mechanical and tribological properties," *Surface and Coatings Technology*, vol. 201, no. 1-2, pp. 82–89, 2006.

[54] J. Liang, B. Guo, J. Tian, H. Liu, J. Zhou, and T. Xu, "Effect of potassium fluoride in electrolytic solution on the structure and properties of microarc oxidation coatings on magnesium alloy," *Applied Surface Science*, vol. 252, no. 2, pp. 345–351, 2005.

[55] Y. Guangliang, L. Xianyi, B. Yizhen, C. Haifeng, and J. Zengsun, "The effects of current density on the phase composition and microstructure properties of micro-arc oxidation coating," *Journal of Alloys and Compounds*, vol. 345, no. 1-2, pp. 196–200, 2002.

[56] C.-C. Tseng, J.-L. Lee, T.-H. Kuo, S.-N. Kuo, and K.-H. Tseng, "The influence of sodium tungstate concentration and anodizing conditions on microarc oxidation (MAO) coatings for aluminum alloy," *Surface and Coatings Technology*, vol. 206, no. 16, pp. 3437–3443, 2012.

[57] H. J. Robinson, A. E. Markaki, C. A. Collier, and T. W. Clyne, "Cell adhesion to plasma electrolytic oxidation (PEO) titania coatings, assessed using a centrifuging technique," *Journal of*

the Mechanical Behavior of Biomedical Materials*, vol. 4, no. 8, pp. 2103–2112, 2011.

[58] S. V. Gnedenkov, O. A. Khrisanfova, A. G. Zavidnaya et al., "Composition and adhesion of protective coatings on aluminum," *Surface and Coatings Technology*, vol. 145, no. 1–3, pp. 146–151, 2001.

[59] K. Ramachandran, V. Selvarajan, P. V. Ananthapadmanabhan, and K. P. Sreekumar, "Microstructure, adhesion, microhardness, abrasive wear resistance and electrical resistivity of the plasma sprayed alumina and alumina-titania coatings," *Thin Solid Films*, vol. 315, no. 1-2, pp. 144–152, 1998.

[60] Y. Wang, Z. Jiang, and Z. Yao, "Microstructure, bonding strength and thermal shock resistance of ceramic coatings on steels prepared by plasma electrolytic oxidation," *Applied Surface Science*, vol. 256, no. 3, pp. 650–656, 2009.

[61] Y. Xu, Z. Yao, F. Jia, Y. Wang, Z. Jiang, and H. Bu, "Preparation of PEO ceramic coating on Ti alloy and its high temperature oxidation resistance," *Current Applied Physics*, vol. 10, no. 2, pp. 698–702, 2010.

[62] Y. Wang, Z. Jiang, and Z. Yao, "Formation of titania composite coatings on carbon steel by plasma electrolytic oxidation," *Applied Surface Science*, vol. 256, no. 20, pp. 5818–5823, 2010.

[63] J. Ding, J. Liang, L. Hu, J. Hao, and Q. Xue, "Effects of sodium tungstate on characteristics of microarc oxidation coatings formed on magnesium alloy in silicate-KOH electrolyte," *Transactions of Nonferrous Metals Society of China*, vol. 17, no. 2, pp. 244–249, 2007.

[64] M. Tang, H. Liu, W. Li, and L. Zhu, "Effect of zirconia sol in electrolyte on the characteristics of microarc oxidation coating on AZ91D magnesium," *Materials Letters*, vol. 65, no. 3, pp. 413–415, 2011.

[65] Q. Cai, L. Wang, B. Wei, and Q. Liu, "Electrochemical performance of microarc oxidation films formed on AZ91D magnesium alloy in silicate and phosphate electrolytes," *Surface and Coatings Technology*, vol. 200, no. 12-13, pp. 3727–3733, 2006.

[66] C.-E. Barchiche, E. Rocca, and J. Hazan, "Corrosion behaviour of Sn-containing oxide layer on AZ91D alloy formed by plasma electrolytic oxidation," *Surface and Coatings Technology*, vol. 202, no. 17, pp. 4145–4152, 2008.

[67] M. Tang, W. Li, H. Liu, and L. Zhu, "Preparation Al_2O_3/ZrO_2 composite coating in an alkaline phosphate electrolyte containing K_2ZrF_6 on aluminum alloy by microarc oxidation," *Applied Surface Science*, vol. 258, no. 15, pp. 5869–5875, 2012.

[68] M. Tang, W. Li, H. Liu, and L. Zhu, "Influence of titania sol in the electrolyte on characteristics of the microarc oxidation coating formed on 2A70 aluminum alloy," *Surface and Coatings Technology*, vol. 205, no. 17-18, pp. 4135–4140, 2011.

[69] E. Matykina, R. Arrabal, P. Skeldon, and G. E. Thompson, "Investigation of the growth processes of coatings formed by AC plasma electrolytic oxidation of aluminium," *Electrochimica Acta*, vol. 54, no. 27, pp. 6767–6778, 2009.

[70] V. Raj and M. Mubarak Ali, "Formation of ceramic alumina nanocomposite coatings on aluminium for enhanced corrosion resistance," *Journal of Materials Processing Technology*, vol. 209, no. 12-13, pp. 5341–5352, 2009.

[71] C. S. Dunleavy, J. A. Curran, and T. W. Clyne, "Self-similar scaling of discharge events through PEO coatings on aluminium," *Surface and Coatings Technology*, vol. 206, no. 6, pp. 1051–1061, 2011.

[72] M. Montazeri, C. Dehghanian, M. Shokouhfar, and A. Baradaran, "Investigation of the voltage and time effects on the formation of hydroxyapatite-containing titania prepared by plasma

electrolytic oxidation on Ti-6Al-4V alloy and its corrosion behavior," *Applied Surface Science*, vol. 257, no. 16, pp. 7268–7275, 2011.

[73] W. Xue, Q. Zhu, Q. Jin, and M. Hua, "Characterization of ceramic coatings fabricated on zirconium alloy by plasma electrolytic oxidation in silicate electrolyte," *Materials Chemistry and Physics*, vol. 120, no. 2-3, pp. 656–660, 2010.

[74] J. Li, H. Cai, X. Xue, and B. Jiang, "The outward-inward growth behavior of microarc oxidation coatings in phosphate and silicate solution," *Materials Letters*, vol. 64, no. 19, pp. 2102–2104, 2010.

[75] G. Sundararajan and L. Rama Krishna, "Mechanisms underlying the formation of thick alumina coatings through the MAO coating technology," *Surface and Coatings Technology*, vol. 167, no. 2-3, pp. 269–277, 2003.

[76] H. Habazaki, S. Tsunekawa, E. Tsuji, and T. Nakayama, "Formation and characterization of wear-resistant PEO coatings formed on β-titanium alloy at different electrolyte temperatures," *Applied Surface Science*, vol. 259, pp. 711–718, 2012.

[77] E. V. Parfenov, R. R. Nevyantseva, and S. A. Gorbatkov, "Process control for plasma electrolytic removal of TiN coatings. Part 1: duration control," *Surface and Coatings Technology*, vol. 199, no. 2-3, pp. 189–197, 2005.

[78] P. Huang, F. Wang, K. Xu, and Y. Han, "Mechanical properties of titania prepared by plasma electrolytic oxidation at different voltages," *Surface and Coatings Technology*, vol. 201, no. 9-11, pp. 5168–5171, 2007.

[79] D. Wei, Y. Zhou, D. Jia, and Y. Wang, "Effect of applied voltage on the structure of microarc oxidized TiO_2-based bioceramic films," *Materials Chemistry and Physics*, vol. 104, no. 1, pp. 177–182, 2007.

[80] H. Wu, X. Lu, B. Long, X. Wang, J. Wang, and Z. Jin, "The effects of cathodic and anodic voltages on the characteristics of porous nanocrystalline titania coatings fabricated by microarc oxidation," *Materials Letters*, vol. 59, no. 2-3, pp. 370–375, 2005.

[81] C. B. Wei, X. B. Tian, S. Q. Yang, X. B. Wang, R. K. Y. Fu, and P. K. Chu, "Anode current effects in plasma electrolytic oxidation," *Surface and Coatings Technology*, vol. 201, no. 9-11, pp. 5021–5024, 2007.

[82] X.-M. Zhang, X.-B. Tian, C.-Z. Gong, and S.-Q. Yang, "Effects of current density on coating kinetic and micro-structure of microarc oxidation coatings fabricated on pure aluminum," in *Proceedings of the 3rd IEEE International Nanoelectronics Conference (INEC '10)*, pp. 1482–1483, January 2010.

[83] X. Sun, Z. Jiang, Z. Yao, and X. Zhang, "The effects of anodic and cathodic processes on the characteristics of ceramic coatings formed on titanium alloy through the MAO coating technology," *Applied Surface Science*, vol. 252, no. 2, pp. 441–447, 2005.

[84] P. Bala Srinivasan, J. Liang, R. G. Balajeee, C. Blawert, M. Störmer, and W. Dietzel, "Effect of pulse frequency on the microstructure, phase composition and corrosion performance of a phosphate-based plasma electrolytic oxidation coated AM50 magnesium alloy," *Applied Surface Science*, vol. 256, no. 12, pp. 3928–3935, 2010.

[85] Z. Yao, Y. Liu, Y. Xu, Z. Jiang, and F. Wang, "Effects of cathode pulse at high frequency on structure and composition of Al_2TiO_5 ceramic coatings on Ti alloy by plasma electrolytic oxidation," *Materials Chemistry and Physics*, vol. 126, no. 1-2, pp. 227–231, 2011.

[86] P. Gupta, G. Tenhundfeld, E. O. Daigle, and D. Ryabkov, "Electrolytic plasma technology: science and engineering—an overview," *Surface and Coatings Technology*, vol. 201, no. 21, pp. 8746–8760, 2007.

[87] Y. M. Wang, H. Tian, X. E. Shen et al., "An elevated temperature infrared emissivity ceramic coating formed on 2024 aluminium alloy by microarc oxidation," *Ceramics International*, vol. 39, pp. 2869–2875, 2013.

[88] Z. W. Wang, Y. M. Wang, Y. Liu et al., "Microstructure and infrared emissivity property of coating containing TiO_2 formed on titanium alloy by microarc oxidation," *Current Applied Physics*, vol. 11, no. 6, pp. 1405–1409, 2011.

[89] P. M. de Woolf and J. W. Visser, "Absolute Intensities—outline of a recommended practice," *Powder Diffraction*, vol. 3, pp. 202–204, 1988.

[90] E. P. G. T. van de Ven and H. Koelmans, "The cathodic corrosion of Aluminum," *Journal of The Electrochemical Society*, vol. 123, pp. 143–144, 1976.

[91] E. V. Koroleva, G. E. Thompson, G. Hollrigl, and M. Bloeck, "Surface morphological changes of aluminium alloys in alkaline solution: effect of second phase material," *Corrosion Science*, vol. 41, no. 8, pp. 1475–1495, 1999.

[92] C. Zhu, P. Lu, Z. Zheng, and J. Ganor, "Coupled alkali feldspar dissolution and secondary mineral precipitation in batch systems: 4. Numerical modeling of kinetic reaction paths," *Geochimica et Cosmochimica Acta*, vol. 74, no. 14, pp. 3963–3983, 2010.

[93] Y. K. Pan, C. Z. Chen, D. G. Wang, X. Yu, and Z. Q. Lin, "Influence of additives on microstructure and property of microarc oxidized Mg-Si-O coatings," *Ceramics International*, vol. 38, pp. 5527–5533, 2012.

[94] R. McPherson, "Formation of metastable phases in flame- and plasma-prepared alumina," *Journal of Materials Science*, vol. 8, no. 6, pp. 851–858, 1973.

[95] P. S. Santos, H. S. Santos, and S. P. Toledo, "Standard transition aluminas. Electron microscopy studies," *Materials Research*, vol. 3, pp. 104–114, 2000.

[96] R. K. Iler, *Chemistry of Silica—Solubility, Polymerization, Colloid and Surface Properties and Biochemistry*, chapter 1, John Wiley & Sons, 1979.

[97] L. Rama Krishna, K. R. C. Somaraju, and G. Sundararajan, "The tribological performance of ultra-hard ceramic composite coatings obtained through microarc oxidation," *Surface and Coatings Technology*, vol. 163-164, pp. 484–490, 2003.

[98] F.-Y. Jin, K. Wang, M. Zhu et al., "Infrared reflection by alumina films produced on aluminum alloy by plasma electrolytic oxidation," *Materials Chemistry and Physics*, vol. 114, no. 1, pp. 398–401, 2009.

[99] K. Wang, B.-H. Koo, C.-G. Lee, Y.-J. Kim, S.-H. Lee, and E. Byon, "Effects of electrolytes variation on formation of oxide layers of 6061 Al alloys by plasma electrolytic oxidation," *Transactions of Nonferrous Metals Society of China*, vol. 19, no. 4, pp. 866–870, 2009.

[100] G. E. Walrafen and E. Pugh, "Raman combinations and stretching overtones from water, heavy water, and NaCl in water at shifts to ca. 7000 cm-1," *Journal of Solution Chemistry*, vol. 33, no. 1, pp. 81–97, 2004.

[101] S. P. Langley, "XXII. The selective absorption of solar energy," *Philosophical Magazine Series 5*, vol. 15, no. 93, pp. 153–183, 1883.

[102] R. D. Aines and G. R. Rossman, "The high temperature behavior of water and carbon dioxide in cordierite and beryl," *American Mineralogist*, vol. 69, no. 3-4, pp. 319–327, 1984.

[103] H. Rubens and E. Aschkinass, "Observations on the absorption and emission of aqueous vapor and carbon dioxide in the infrared spectrum," *The Astrophysical Journal*, vol. 8, p. 176, 1898.

[104] J. Ryczkowski, "IR spectroscopy in catalysis," *Catalysis Today*, vol. 68, no. 4, pp. 263–381, 2001.

[105] K. M. Lee, B. U. Lee, S. I. Yoon, E. S. Lee, B. Yoo, and D. H. Shin, "Evaluation of plasma temperature during plasma oxidation processing of AZ91 Mg alloy through analysis of the melting behavior of incorporated particles," *Electrochimica Acta*, vol. 67, pp. 6–11, 2012.

[106] R. O. Hussein, X. Nie, D. O. Northwood, A. Yerokhin, and A. Matthews, "Spectroscopic study of electrolytic plasma and discharging behaviour during the plasma electrolytic oxidation (PEO) process," *Journal of Physics D*, vol. 43, no. 10, Article ID 105203, 2010.

[107] Y. Wang, Z. Jiang, X. Liu, and Z. Yao, "Influence of treating frequency on microstructure and properties of Al_2O_3 coating on 304 stainless steel by cathodic plasma electrolytic deposition," *Applied Surface Science*, vol. 255, no. 21, pp. 8836–8840, 2009.

[108] O. Rozenbaum, D. De Sousa Meneses, and P. Echegut, "Texture and porosity effects on the thermal radiative behavior of alumina ceramics," *International Journal of Thermophysics*, vol. 30, no. 2, pp. 580–590, 2009.

[109] A. Boumaza, L. Favaro, J. Lédion et al., "Transition alumina phases induced by heat treatment of boehmite: an X-ray diffraction and infrared spectroscopy study," *Journal of Solid State Chemistry*, vol. 182, no. 5, pp. 1171–1176, 2009.

[110] T. Morioka, S. Kimura, N. Tsuda, C. Kaito, Y. Saito, and C. Koike, "Study of the structure of silica film by infrared spectroscopy and electron diffraction analyses," *Monthly Notices of the Royal Astronomical Society*, vol. 299, no. 1, pp. 78–82, 1998.

[111] C. H. Rüscher, "Thermic transformation of sillimanite single crystals to 3:2 mullite plus melt: Investigations by polarized IR-reflection micro spectroscopy," *Journal of the European Ceramic Society*, vol. 21, no. 14, pp. 2463–2469, 2001.

[112] K. Nouneh, I. V. Kityk, R. Viennois et al., "Influence of an electron-phonon subsystem on specific heat and two-photon absorption of the semimagnetic semiconductors $Pb_{1-x}Yb_x X$ (X=S, Se,Te) near the semiconductor-isolator phase transformation," *Physical Review B*, vol. 73, no. 3, Article ID 035329, 2006.

4

The Effect of Thickness of Aluminium Films on Optical Reflectance

Robert Lugolole and Sam Kinyera Obwoya

Department of Physics, Kyambogo University, P.O. Box 1, Kyambogo, Kampala, Uganda

Correspondence should be addressed to Sam Kinyera Obwoya; ksobwoya@yahoo.co.uk

Academic Editor: Yuan-hua Lin

In Uganda and Africa at large, up to 90% of the total energy used for food preparation and water pasteurization is from fossil fuels particularly firewood and kerosene which pollute the environment, yet there is abundant solar energy throughout the year, which could also be used. Uganda is abundantly rich in clay minerals such as ball clay, kaolin, feldspar, and quartz from which ceramic substrates were developed. Aluminium films of different thicknesses were deposited on different substrates in the diffusion pump microprocessor vacuum coater (Edwards AUTO 306). The optical reflectance of the aluminium films was obtained using a spectrophotometer (SolidSpec-3700/DUV-UV-VIS-NIR) at various wave lengths. The analysis of the results of the study revealed that the optical reflectance of the aluminium films was above 50% and increased with increasing film thickness and wavelength. Thus, this method can be used to produce reflector systems in the technology of solar cooking and other appliances which use solar energy.

1. Introduction

The need for reflectance of light energy and heat energy is continually increasing due to the global rapid growing population, industrial development, and domestic needs. The reflectance of solar energy can reduce use of nonrenewable energy sources such as fossil fuels, petroleum fuels, and fissionable minerals. According to Intergovernmental Panel on Climate Change Plenary xxxvii [1], such energy sources lead to the emission of greenhouse gases like water vapour, carbon dioxide, methane, nitrous oxide, ozone, particulate matter, nitrogen oxide, sulphur dioxide, and arsenic and fluorinated compounds, which have increased in the atmosphere since the start of the industrial era in 1750 leading to atmospheric air pollution. Due to prolonged emission of greenhouse gases, greenhouse effect has brought about thinning of the ozone layer in the stratosphere. The global average temperature has increased by $0.6°C$ since the mid-20th century due to anthropogenic activity [2–4]. Reflectance of solar energy is used in solar thermal devices such as solar cookers, for cooking, water heating, space heating, space cooling, and heat generation process. Solar cookers are easy to build, are smoke free, nonpollutant [5] and can conserve the environment by using heat-reflecting mirrors [6]. The mean daily illumination intensity of the sun in the equatorial zone is in the range of $5–7\,kWh/m^2$ and has more than 275 sunny days in the year. This can make solar cooker use possible in Uganda because it is in the equatorial zone. The use of solar cookers will save women and children walking long distances looking for wood and significantly reduce the amount of time women spent tending open fires each day for other developmental works [7]. Wood and charcoal supply over 90% of Uganda's cooking energy requirements; thus, the use of solar cookers will also reduce deforestation, which currently is at an alarming rate. This paper therefore investigated the effectiveness of solar cookers made by depositing thin aluminium films on ceramic substrates for domestic purposes.

2. Experimental Procedures

2.1. Material Processing. Aluminium films were deposited on to ceramic substrates that were developed from predetermined minerals from Uganda. The minerals that were used include ball clay, kaolin, quartz, and feldspar. Ball clay was obtained from Ntawo deposit in Mukono district. The site has clay of finite particle sizes that form readily mouldable sticky mass when mixed with water [8]. It becomes hard and brittle, retains its shape when heated [9], and is no longer susceptible to the action of water. Kaolin and feldspar were collected from Mutaka deposit in Bushenyi district because it has high content of alumina while quartz was collected from Dimu deposit in Masaka district because the site has relatively pure fine quartz particle sizes [10]. The mineral particle sizes of 32 μm for ball clay, 45 μm for kaolin, 53 μm for feldspar, and finally 32 μm for quartz were chosen to minimise porosity and increase densification of the samples prepared. This is because small particle sizes increase particle content per unit volume which decreases the average interparticle distance of the clay matrix resulting in close packing of particles of the ceramic [11]. In the study, the thicknesses of aluminium films that were deposited on the ceramic substrates and the wave lengths of the radiation were the independent variables, while the dependent quantity was the optical reflectance of the aluminium films.

The ball clay slip was sieved mechanically and carefully through 80 μm, 53 μm, and 32 μm sieves in order to get the required fine clay mineral size that could form smooth hard ceramic bodies. The slurry finally obtained after sieving was poured on the plaster of Paris mould where excess water was removed, thus, forming a semidry cast. The semidry cast was again left to continue drying in air at room temperature for seven days [12] to get rid of some of the remaining water in the cast. A drying oven was later used to completely dry the clay samples at a temperature of 105°C for five hours in order to be certain of driving away all the water in the pores of the clay mass. The required dry body was removed from the oven after cooling. It was crushed and then ground in a ball mill to obtain fine powders, by Kingery et al. [13], which were pressed into rectangular samples and then fired to yield mechanically strong ceramic bases.

Kaolin, feldspar, and quartz were dry milled for three days in a ball mill to reduce their particle sizes to ease sieving. The fine powder of kaolin was sieved mechanically through standard sieves of particle sizes 150 μm, 80 μm, and 45 μm, while that of feldspar was similarly sieved through 150 μm, 80 μm, and 53 μm sieve meshes, and finally that for quartz was sieved through the 150 μm, 80 μm, 53 μm, and 32 μm sieve meshes.

The 32 μm, 45 μm, 53 μm, and 32 μm of the fine powders were weighed separately in the proportions of 30% ball clay, 25% kaolin, 30% feldspar, and 15% quartz, respectively, and then mixed thoroughly to form a blended mixture. Each ceramic substrate was made from a mass of 2.03×10^{-2} kg drawn from the mixture formed. In addition to the ball clay, bentonite organic binder amounting to 3% of 2.03×10^{-2} kg mass was added to each sample as an auxiliary material to form a colloidal mixture, in order to increase bonding of the particles. Each colloidal mixture was thoroughly mixed using a clean automated mortar for fifteen minutes to obtain uniform particle distribution, which greatly improves the forming process of ceramic materials [13]. The colloidal samples were dried under the sun for seven days and then crushed into finer powder using a roller on a clean flat metallic surface. Each sample, as shown in Figure 2, was slowly compacted in

FIGURE 1: Compaction mould and die.

FIGURE 2: Green body.

FIGURE 3: Ceramic substrate after firing.

a die mould as shown in Figure 1, at a pressure of 152 MPa by use of a hydraulic press (PW-40). A total of twenty four rectangular slides of length 7.70×10^{-2} m, width 2.87×10^{-2} m, and thickness 5.34×10^{-3} m were made. The green slides were carefully removed from the mould and air-dried for four days to drive out some remaining water and to become hard before firing.

The dry samples were fired in an electrical furnace at a rate of 6°C per minute under controlled temperatures and varying pressures. The furnace temperature was held in stages, at a temperature of 110°C for two hours to drive out the ordinary water and at 450°C to get rid of the hydroxyl ions. According to Ryan [14], some crystalline changes take place at the α-β quartz inversion due to quartz expansion, so the rate of temperature rise was slow near the inversion temperature of 573°C to avoid cracking of the body. The temperature was then raised to 1250°C and then held there for two hours for the body to mature. Finally, the samples were air-cooled in the furnace and then removed carefully in order to avoid scratches before aluminium was deposited on them as shown in Figure 3. The slides which got out of the kiln were of length 6.92×10^{-2} m, width 2.54×10^{-2} m, thickness 4.19×10^{-3} m, and mass 1.627×10^{-2} kg.

Aluminium was used to form the thin film reflecting surfaces on ceramic substrates to concentrate light and heat because it is ductile, has a low density of 2.70×10^{3} kgm^{-3} with an acoustic impedance of 8.17×10^{6} kgm^{-2}s^{-1}, and is resistant to oxidation. Similarly, aluminium was used because it is widely available primarily as ore bauxite that makes 8% of the earth's solid surface [15]. It is cheap and nonmagnetic, does not easily ignite, is silvery white, and has a melting point of 660.4°C and evaporation temperature of 1390°C which make it a good reflector of both visible light and heat. Aluminium films used as metallization contacts have low specific resistivity, good thermal stability, high uniformity across the flat substrate, low particle contamination, and good adherence to substrate. These properties have led aluminium to be irreplaceable and its demand is on increase in many areas of today's rapidly developing technology especially optical industries [16]. Highly specular aluminium films made in an ultrahigh vacuum deposition process have a solar reflectance of 92%. The problem is that the optical reflectance of aluminium deteriorates upon exposure to the outdoor environment at a time scale of less than one year. Aluminium foil which is currently used as the reflecting surface on solar cookers is quite easy to damage and is also known to darken when exposed to moisture. This creates wear of aluminium foil leading to the formation of small pinholes in it when it comes in contact with different metals or food that is highly salted or acidic [17].

Vulnerable aluminium surface must, therefore, be protected by some kind of nonabsorbing coating from degradation. Several techniques have been applied; anodic oxidation coating [18], lacquering with PVF_2, PVF, and PMMA [19, 20], vacuum deposited thin dielectric films [21], and sol-gel deposition of thin dielectric films. Aluminium surface can also be protected by vacuum deposition of the metal onto PMMA [22, 23] whereas, to obtain its long-term stability, a second surface mirror is deposited and thereafter sealed. Other surface treatments that can be designed in order to improve the surface properties of aluminium films such as wear resistance, corrosion resistance, and reflectivity are such as electrochemical brightening, electropolishing, annealing, and glazing [24]. Silicon dioxide overcoating protects aluminium and enables careful cleaning to take place [25]. This is fortunate because most materials, including aluminum, have better adherence if they are evaporated onto heated substrates [26].

2.2. Deposition of Aluminium Films. Aluminium adheres well to both silicon and silicon dioxide and can be easily vacuum deposited since it has a low boiling point and has high conductivity [27]. Aluminium films of specific different thicknesses were deposited on different ceramic slide substrates in the diffusion pump microprocessor vacuum coater (Edwards AUTO 306) as shown in Figure 4. The power was regulated at a value of 1.30 kW to heat, melt, and vaporize aluminium at a temperature of 660°C. The aluminium film thicknesses, T, deposited on the slides A, B, C, D, E, and F are shown in Table 1.

2.3. Measurement of Optical Reflectance. The optical reflectance of specific thicknesses, T, of aluminium films deposited on the ceramic slides was measured and recorded over a range of wavelength, λ, 500 nm to 2500 nm using a SolidSpec-3700/DUV-UV-VIS-NIR. Four samples were considered to

FIGURE 4: Aluminium coated ceramic substrate.

TABLE 1: Thicknesses of aluminium films deposited on the slides.

Slides	A	B	C	D	E	F
T/nm	12.3	100	136	396	480	750

obtain the average optical reflectance as shown in Table 2. The data in Table 2 was inserted into Matlab software to generate graphs of optical reflectance against wave length as shown in Figure 5.

3. Experimental Results

3.1. Optical Reflectance of Aluminium Films. The percentage of optical reflectance of aluminium films deposited on ceramic substrates at various wave lengths is shown in Table 2.

4. Discussion and Conclusions of Results

4.1. Discussion of Results

4.1.1. Aluminium Films on Ceramic Bases. The various thicknesses of aluminium films deposited on the ceramic bases were 12.3 nm, 100 nm, 136 nm, 396 nm, 480 nm, and finally 750 nm. This range was chosen because when the films deposited were less than 5 nm, the reflectance was mainly due to the base after absorbing more of the energy. When the films built were above 1 μm, their adhesion to the ceramic substrate could no longer balance the stress when deposited by evaporation. Such films led to the clouding effect which limited their use in the infrared range. The deposition of films on the substrates was at a pressure of 9×10^{-7} MB because aluminium films strongly adhere on the substrates when deposition is carried out at a pressure less than 10^{-7} of atmospheric pressure, to avoid oxidation, and is used to prepare thin films with controlled chemical composition [28]. The vacuum deposition process was selected for the deposition of thin films over other processes such as electrochemical deposition and flame spraying. This is because it gave balanced homogeneous films onto the ceramic bases particularly determined by the smoothness and cleaning level of the substrates, the deposition temperature of 660°C, and the chamber pressure of 6×10^{-7} MB. This prevented aluminium from being removed by acids or alkali solutions because it was chemically inert due to the aluminium modulus of elasticity of 7.1×10^{10} Pa, kaolin, which has alumina that was useful in reinforcing the films, and aluminium would balance the mechanical stress which is the principal

TABLE 2: Optical reflectance of aluminium films deposited on ceramic surfaces.

λ/nm	500	700	900	1100	1300	1500	1700	1900	2100	2300	2500
T/nm					Percentage of optical reflectance of aluminium films						
A	21.435	26.895	26.416	26.139	27.961	28.890	29.866	31.191	32.421	33.560	34.341
B	53.707	61.015	63.591	67.753	70.279	71.767	74.008	73.628	75.914	75.934	77.342
C	69.174	71.416	70.952	73.876	75.257	75.984	76.442	76.896	78.898	78.617	79.799
D	76.330	76.215	72.713	78.480	78.008	80.082	79.009	79.631	81.514	81.699	82.657
E	81.634	79.898	76.222	79.444	80.294	80.664	81.071	81.297	83.137	82.717	83.807
F	93.074	91.827	93.410	93.950	94.221	94.321	94.687	95.505	95.253	95.851	95.953

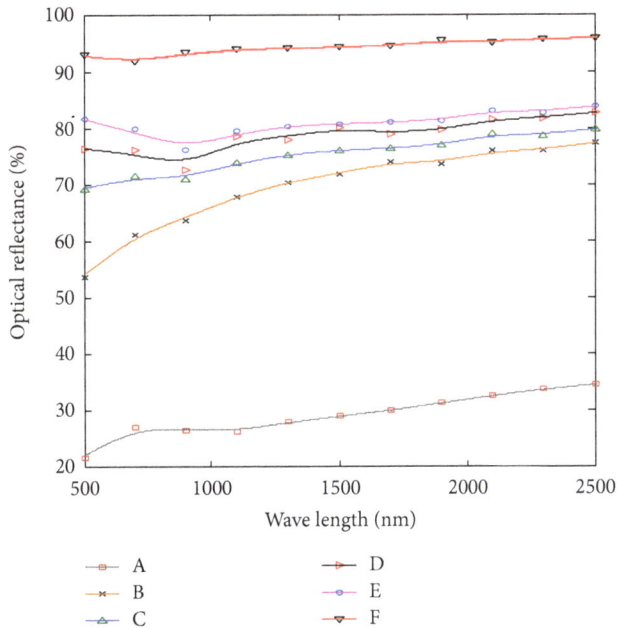

FIGURE 5: Optical reflectance of samples A, B, C, D, E, and F.

factor for limiting the thickness of films when deposited by evaporation on substrates. It is the reason why aluminium films deposited on the fabricated ceramic bases had no effect even after five months of study. It is also because physical vapour deposition produces reflective enhancing oxide layers which avoid further oxidation and corrosion.

4.1.2. Optical Reflectance. In this particular study, over 50% of the incident radiation falling onto the films deposited on the ceramics was reflected as compared to the total light absorbed and transmitted except for sample A. This so far is much better than aluminium foil which on its own reflects about 55% of visible light that even lowers to about 35% when creased and exposed to harsh atmospheric conditions [29]. The graphs of optical reflectance against wave length are shown in Figure 5. The best fit used for the graphs at a 95% level of confidence from a smoothing parameter was a smoothing spline function $f(x)$ in contrast to Gaussian, interpolant, power, linear functions and so forth. The polynomial $f(x)$ for each graph had a high R-square value between 0.9000 and 1.0000 to give a working standard

square error (SSE) and root mean square error (RMSE) which produced smooth curves with goodness of fit due to reduced kinks. Some kinks arose because of change of the spectrophotometer detectors especially when the photomultiplier changed to indium gallium arsenide between 700 nm and 1000 nm and to lead selenide between 1500 nm and 1800 nm, respectively.

The reflectance of aluminium generally increased with increase in thickness of the aluminium films and with increase in the radiation wave length as per Figure 5. This is because aluminium exhibits free electron behaviour at various solar wavelengths over the entire solar wavelength interval [30], due to its small interband transitions. At a 95% level of confidence, the standard deviation obtained for the experimental study on samples A to F was lowest at 1.009 for sample D at a wave length of 2100 nm and highest at 8.885 for sample A at a wave length of 2100 nm which shows the consistency of the results of all samples.

The reflectance of sample A at all wave lengths was the lowest below 35% because much of the light was transmitted by the film and absorbed by the base. The low reflectance of A was also attributed to the nonlinearity and nonsmoothness of the surface. Between 700 nm and 900 nm, the reflectance of all the samples decreased slightly apart from A and B. This is because aluminium maintains high reflectance over the entire solar wavelength interval with the smallest value around 0.82 μm [31, 32]. The inconsistency in the reflectance for all samples below 600 nm wave length was because of the absence of the nitrogen pumping accessory which was not integrated in the spectrophotometer used during the study. Reflectance became stable between 1000 nm and 2500 nm for all samples due to increasing wavelength of the radiation into the infrared band. Maximum steady reflectance of above 90% was achieved by sample F which showed a markedly higher jump in reflectance than other samples for all wave lengths between 800 nm and 2500 nm due to minimal absorption of the radiation by the base because it had 750 nm aluminium film thickness. It is also an attribute of the radiation of high wave length being less penetrative than for the short wave length band. The 750 nm aluminium film thickness offers a smooth surface with reduced transmission and absorption of light by the aluminium film and its base. Any radiation lost in sample F was due to tolerable multiple internal reflections and absorption of light and heat by the film itself and the impurities therein contained such as iron, nickel, chromium, and manganese rather than the support base.

5. Conclusions

The results of the study showed that aluminium films deposited on the ceramic bases in a vacuum coater have high optical reflectance. However, the various thicknesses of aluminium films directly affected the optical reflectance at different wavelengths. Thus, from the overall experimental analysis carried out, salient conclusions arising from this work are summarized as follows.

(1) The optical reflectance of aluminium increased with increase in film thickness.

(2) The optical reflectance of aluminium increased with increase in the radiation wavelength.

(3) Suitable solar cookers and other appliances using solar energy can be developed by vacuum deposition of aluminium films on the ceramic bases fabricated from ball clay, kaolin, quartz, and feldspar. This may alleviate the problem of lack of fuel for food preparation, pollution by fossil fuels, and emission of greenhouse gasses leading to global warming.

Conflict of Interests

The authors declare that there is no conflict of interests regarding the publication of this paper.

Acknowledgments

The authors would like to thank the staff at Kyambogo University for their guidance and support during the course of the study and research. Further thanks and gratitude go to Makerere and Nairobi University Physics Departments, the Physics Department and IPPS, Uppsala University, Sweden, Uganda Industrial Research Institute, Ministry of Energy, Uganda, Uganda National Bureau of Standards, and Uganda Analytical Department for the provision of quality of equipment, resources, and laboratories for this study. They are grateful to their course mates, relatives, and friends particularly Shema Blessing for their support.

References

[1] Intergovernmental Panel on Climate Change Plenary, *Assessment of Intergovernmental Panel on Climate Change*, 2007, http://www.energyquest.ca.gov.

[2] M. Kandlikar, C. C. O. Reynolds, and A. P. Grieshop, "A perspective paper on black carbon mitigation as a response to climate change," Copenhagen Consensus Center Report, 2010.

[3] J. Greene, *Harnessing the Sun to Benefit People and the Environment*, vol. 1, no. 1, Solar Cookers International, 2014.

[4] A. R. Shashikala, A. K. Sharma, and D. R. Bhandari, "Solar selective black nickel-cobalt coatings on aluminum alloys," *Solar Energy Materials and Solar Cells*, vol. 91, no. 7, pp. 629–635, 2007.

[5] J. Li, D. R. Alexander, H. Zhang et al., "Propagation of ultrashort laser pulses through water," *Optics Express*, vol. 15, no. 4, pp. 1939–1945, 2007.

[6] E. Valkonen and B. Karlsson, "Optimization of metal—based multilayers for transparent heat mirrors," *International Journal of Energy Research*, vol. 11, no. 3, pp. 397–403, 1987.

[7] Solar lifeline saves Darfur women, CNN. 17-09-2007, 2015.

[8] O. K. Sam, *Effects of microstructure on mechanical properties of selected clays from Uganda [Ph.D. thesis]*, Makerere University, Kampala, Uganda, 2004.

[9] A. M. V. Agbayani and A. A. Espinosa, *Ceramic Tiles from Crassostrea iredalei*, Philippine Normal University, Boca Raton, Fla, USA, 2006.

[10] P. O. Wilberforce, *Characterisation of ceramic raw minerals in Uganda for production of electrical porcelain insulators [Doctoral Thesis in Material Science]*, KTH, Stockholm, Sweden, 2010.

[11] S. Zhang, X. Y. Cao, Y. M. Ma, Y. C. Ke, J. K. Zhang, and F. S. Wang, "The effects of particle size and content on the thermal conductivity and mechanical properties of Al_2O_3/high density polyethylene (HDPE) composites," *Express Polymer Letters*, vol. 5, no. 7, pp. 581–590, 2011.

[12] A. J. Moulson and J. M. Herbert, *Electro Ceramics Materials Properties Applications*, Chapman & Hall, London, UK, 1990.

[13] W. D. Kingery, H. K. Bowen, and D. R. Uhlmann, *Introduction to Ceramics*, Wiley-Interscience, New York, NY, USA, 2nd edition, 1976.

[14] W. Ryan, *Properties of Ceramic Raw Materials*, Pergamon Press, Oxford, UK, 1978.

[15] U.S. Geological Survey, *Mineral Commodity Summaries*, U.S. Geological Survey, 2014.

[16] Z. Fekkai, N. Mustapha, and A. Hennache, "Optical, morphological and electrical properties of silver and aluminium metallization contacts for solar cells," *American Journal of Modern Physics*, vol. 3, no. 2, pp. 45–50, 2014.

[17] J. Harrison, *Investigation of Reflective Materials for the Solar Cooker*, Florida Solar Energy Center, 2001.

[18] British Standards Institution, *Methods of Test for Anodic Oxidation Coatings on Aluminium and Its Alloys*, British Standards Institution, 2013.

[19] R. B. Pettit and J. M. Freese, "Wavelength dependent scattering caused by dust accumulation on solar mirrors," *Solar Energy Materials*, vol. 3, no. 1-2, pp. 1–20, 1980.

[20] P. Nostell, A. Roos, and B. Karlsson, "Ageing of solar booster reflector materials," *Solar Energy Materials and Solar Cells*, vol. 54, no. 1–4, pp. 235–246, 1998.

[21] Acmite Market Intelligence, *Market Report: Global Optical Coatings Market*, Acmite Market Intelligence, 2014.

[22] P. Schissel, G. Jorgensen, C. Kennedy, and R. Goggin, "Silvered-PMMA reflectors," *Solar Energy Materials and Solar Cells*, vol. 33, no. 2, pp. 183–197, 1994.

[23] P. Schissel, G. Jorgensen, and R. Pitts, "Application experience and field performance of silvered polymer reflectors," in *Proceedings of the Solar World Congress*, vol. 2, pp. 2076–2081, Pergamon Press, Denver, Colo, USA, 1991.

[24] B. Perers, B. Karlsson, and M. Bergkvist, "Intensity distribution in the collector plane from structured booster reflectors with rolling grooves and corrugations," *Solar Energy*, vol. 53, no. 2, pp. 215–226, 1994.

[25] Starna Optical Coatings, http://www.optiglass.com/ukhome/d_ref/xrefsets.html.

[26] "Radiative heat transfer," *Journal of the Optical Society of America*, vol. 51, no. 7, 1961.

[27] Z. H. Levine and B. Ravel, "Identification of materials in integrated circuit interconnects using x-ray absorption near-edge spectroscopy," *Journal of Applied Physics*, vol. 85, no. 1, pp. 558–564, 1999.

[28] P. Wißmann and H.-U. Finzel, *Electrical Resistivity of Thin Metal Films*, vol. 223 of *Springer Tracts in Modern Physics*, Springer, Berlin, Germany, 2007.

[29] *The European Engineering Property Database for Wrought Aluminium and Aluminium Alloys*, 1992, 2 diskettes.EAA, KTH.

[30] Aluminum Federation, *The Properties of Aluminum and Its Alloys*, Aluminum Federation, Birmingham, UK, 1993.

[31] W. B. Frank, *Ullmann's Encyclopedia of Aluminum Industrial Chemistry*, Wiley-VCH, 2009.

[32] American Society for Metals, *Aluminium—Properties and Physical Metallurgy*, American Society for Metals, Metals Park, Ohio, USA, 1984.

5

Mechanical Behavior of Yttria-Stabilized Zirconia Aqueous Cast Tapes and Laminates

V. Moreno,[1,2] **R. M. Bernardino,**[1,2] **and D. Hotza**[1,3]

[1] *Graduate Program on Materials Science and Engineering (PGMAT), Federal University of Santa Catarina, 88040-900 Florianópolis, SC, Brazil*

[2] *Department of Mechanical Engineering, Federal University of Santa Catarina, 88040-900 Florianópolis, SC, Brazil*

[3] *Department of Chemical Engineering, Federal University of Santa Catarina, 88040-900 Florianópolis, SC, Brazil*

Correspondence should be addressed to D. Hotza; dhotza@gmail.com

Academic Editor: Thomas Graule

Aqueous tape casting was used to produce yttria-stabilized zirconia films for electrolyte-supported solid oxide fuel cell (SOFC). Tape casting slurries were prepared varying the binder content between 20 and 25 wt%. A commercial acrylic emulsion served as binder. Rheological measurements of the two slurries were performed. Both slurries showed a shear-thinning behavior. Tapes with 25 wt% binder exhibited adequate flexibility and a smooth and homogeneous surface, free of cracks and other defects. Suitable conditions of lamination were found and a theoretical density of 54% in the laminates was achieved. Laminated tapes showed higher tensile strength compared to single sheets. Tape orientation has a significant influence on the mechanical properties. Tensile strength, elongation to strain, and Young's modulus measured in samples produced in the direction of casting showed higher property values.

1. Introduction

The tape casting technique has been widely used to produce thin and high dense green films for multilayer ceramic packaging technology [1–6]. This is a low cost process that enables producing ceramic sheets and large flat areas commonly used in electronic applications, such as capacitors, fuel cells, piezoelectric devices, and inert and catalytic substrates [7–9]. This method allows controlling the thickness of the sheets varying between 20 μm and 2 mm. Typical tape cast slurries are prepared by dispersion of a powder in an organic or nonorganic solvent with a dispersant, followed by the addition of binders, plasticizers, and other additives. These slurries are cast through a so-called doctor blade over a flat carrier. Tape casting slips are based traditionally on organic solvent; however, in the last years there is a huge concern about the toxicity and volatility of those solvents. Aqueous systems are environmentally correct and offer lower cost [10].

Water-based slurries have some drawbacks compared with organic solvent based slurries, such as slow drying rate, tendency to flocculate, and poor wetting due to the high surface tension of the water [11–14]. A wide large amount of water soluble substances has been used to prepare tape casting slurries such as cellulose and vinyl/acrylic binders [10]. The right choice of the binder is very important, because it directly influences particle packing during the tape casting process, the mechanical and physical properties of the green tapes, and the adhesion between sheets during lamination [15]. Compared to water soluble binders, latex binders such as acrylic emulsions allowed preparing slurries with high solids and binder content without significantly increasing the viscosity and providing a short drying time and a high mechanical strength to the green tapes [13].

In this work, aqueous tape casting was used to produce yttria-stabilized zirconia films for electrolyte-supported solid oxide fuel cell (SOFC). A commercial acrylic emulsion was used as binder, and two different slurry compositions were tested. Rheological behavior was examined and the microstructure and density of the green tapes were analyzed. Tensile strength and elongation strain of the green tapes and laminates were also evaluated.

TABLE 1: Slurry compositions.

Component	1 (wt%)	2 (wt%)
8YSZ	55	55
Mowilith LDM 6138	20	25
Darvan 821A	1	1
Antifoamer	0.5	0.5
Others	1.5	1.5
Deionized water	22	17

TABLE 2: Warm pressing conditions.

	Level 1	Level 2
Temperature (°C)	60	40
Pressure (MPa)	19	16
Time (min)	5	10

TABLE 3: Factorial experimental design.

Experiment	Pressure (MPa)	Temperature (°C)	Time (min)
1	19	60	5
2	19	60	10
3	19	40	5
4	19	40	10
5	16	60	5
6	16	60	10
7	16	40	5
8	16	40	10

2. Materials and Methods

Tape casting slurries were prepared with 55 wt% yttria-stabilized zirconia powder (8YSZ, 8 mol% Y_2O_3 stabilized ZrO_2, Sigma-Aldrich). The powder was deagglomerated in deionized water with addition of 1 wt% dispersant (Darvan 821A, Vanderbilt) using ball milling for 24 h. After deagglomeration, an acrylic emulsion binder (Mowilith LDM 6138, Clariant), antifoamer (Antifoamer A, Sigma-Aldrich), and surfactant (coconut diethanolamide, Stepan) were added and the slurry was mixed by ball milling for the next 30 min. The slurry was cast at 25°C by a tape cast machine (CC-1200, Mistler) with moving polyethylene terephthalate carrier film coated with a fine silicon layer (Mylar G10JRM, Mistler). A casting speed of 6 cm/min was used. The gap between the blade and the carrier was set manually to obtain a final tape thickness of 90 to 200 μm. The green tapes were dried at 25°C for 24 h. Table 1 shows the composition of the two different slurries. Only the binder content was varied. The chosen slurry composition (powder and additives contents) was presented in [16].

Two tapes were laminated and arranged as follows: one in the cast direction and the other perpendicular to the cast direction. Lamination was carried in a warm press between two metal plates. In order to investigate the adequate conditions for a good adhesion between tapes, pressure (P), time (t), and temperature (T) were varied as shown in Table 2. A 2^3 factorial experimental design was carried out in order to choose the best combination of P, T, t, as shown in Table 3.

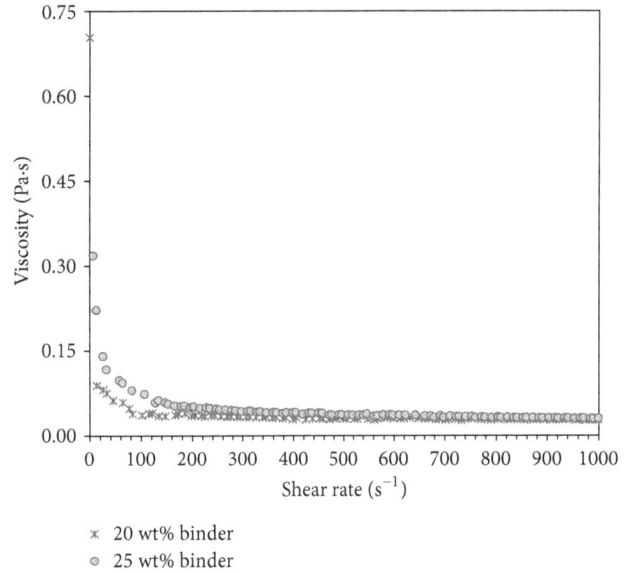

* 20 wt% binder
∘ 25 wt% binder

FIGURE 1: Viscosity versus shear rate of slurries with 20 wt% and 25 wt% binder.

The different P, T, and t levels were chosen according to the previous experience of the research group. Five samples were laminated for each combination. The microstructure of green laminates was analyzed by scanning electron microscopy (SEM XL 30, Philips). Densities of the green tapes were measured geometrically and by Archimedes method. Open porosity and theoretical density were calculated from the bulk and Archimedes density. A texturometer (TA-XT2i, Stable Micro System) was used to measure the mechanical properties of the green tapes with a crosshead speed of 5 mm min^{-1} based on the ISO 527-3 norm [16]. For those mechanical tests, rectangular specimens (100 × 25 mm, 15 samples) were cut using a blade.

3. Results and Discussions

3.1. Slurry Characterization. The viscosity versus shear rate curves of the slurries containing 20 wt% and 25 wt% binder is shown in Figure 1. It can be observed that there is a decrease of the viscosity of the slurries by increasing the shear rate. Both slurries showed similar shear-thinning behavior, but the one with 25 wt% presented a higher viscosity. Tape casting slurries are expected to present a shear-thinning behavior to produce films with a homogeneous and smooth surface. A slip with shear-thinning behavior decreases its viscosity under the shear rate produced by the blade, and after passing through the blade, the shear rate is released and the viscosity increased [17, 18]. Nevertheless, some compositions could not avoid generating defects in the final green tape after drying, such as cracks and bubbles. During tape casting process, the shear rate imposed by the blade was determined as 5 s^{-1}. In Figure 1, the rheograms show viscosity values of ~0.10 Pa·s and 0.30 Pa·s, respectively, for 20 wt% and 25 wt% binder. At higher shear rates both compositions presented similar viscosities. When passing through the blade, the viscosity

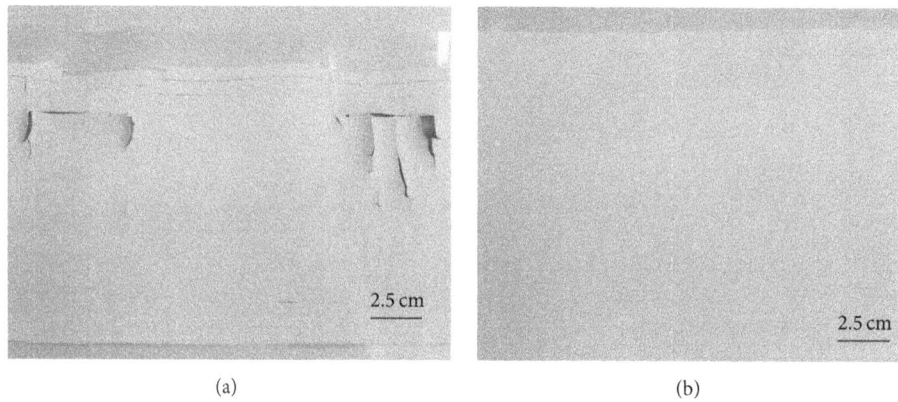

FIGURE 2: Green tapes after 24 h drying: (a) 20 wt% binder; (b) 25 wt% binder.

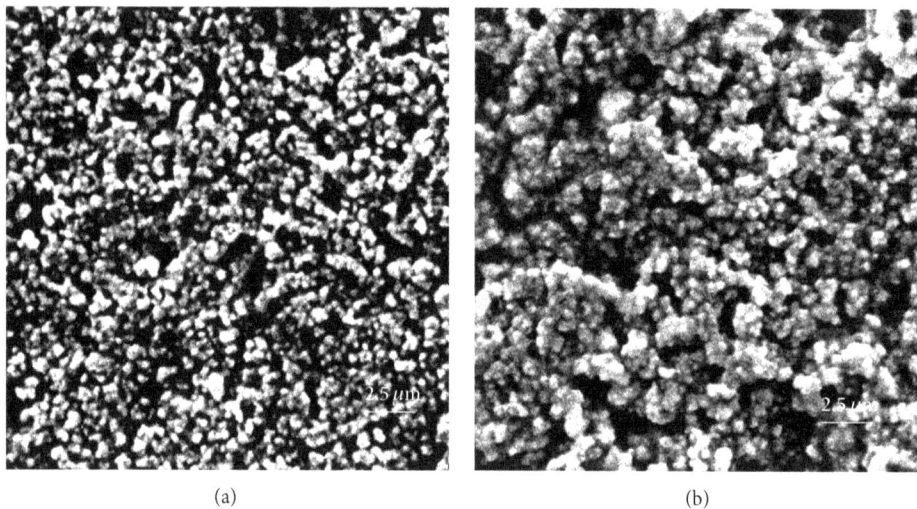

FIGURE 3: Microstructure of green tapes: (a) 20 wt% binder; (b) 25 wt% binder.

of the slurry with 25 wt% binder is slightly higher than that of the 20 wt% binder slurry, though. Slurries with higher viscosity values produce tapes with lower green density [17]. However, the slurries with 20 wt% and 25 wt% binder had an adequate behavior when just the tape casting step is considered.

In order to choose the most suitable composition, both slurries were cast under the same conditions on a polymer carrier with and without silicon coating. Figure 2 presents the tape surfaces of both compositions after 24 h drying. It can be seen from Figure 2(a) that the tape cast with the 20 wt% binder exhibited large cracks and a poor flexibility, whereas 25 wt% binder tape showed a smooth and homogeneous surface, as presented in Figure 2(b).

SEM features of the surface of the green tapes are presented in Figure 3. It can be seen that both microstructures are relatively homogeneous. There are some small pores, but no large defects can be observed. The microstructure of the green tape with 25 wt% binder presented a surface with joined

particles that can be attributed to the higher content of binder compared with the green tape with 20 wt% binder.

3.2. Characterization of Green Tapes. Preliminary evaluation of the green tapes showed that the 20 wt% binder tape was not suitable to be laminated, due to the poor flexibility and the defects exhibited after drying as shown in Figure 2. Lamination was performed according to Table 3, and it was noticed that four different combinations of pressure, temperature, and time showed the highest density of all green laminates. Of the 8 different parameter combinations (Table 3) only 4 laminates reached a green density above 50%. Figure 4 presents the percentage of theoretical density of the best four conditions to laminate the sheets. It can be seen that density values varied between 51% and 54%. Figure 4 shows that temperature and time have no significant effect on the green density, whereas pressure seems to affect more considerably the laminates green density. Although higher pressures can promote better adhesion between layers, it was observed that they also

TABLE 4: Mechanical properties of green tapes (A: parallel and B: perpendicular to casting direction) and green laminate (L).

Sample	Tensile strength (MPa)	Elongation (%)	Young's modulus (MPa)	Thickness (μm)
A	1.53 ± 0.17	5.70 ± 0.87	1.08 ± 0.24	130 ± 8
B	1.52 ± 0.16	4.41 ± 0.52	0.94 ± 0.23	130 ± 8
L	1.56 ± 0.18	5.63 ± 0.64	0.64 ± 0.27	220 ± 12

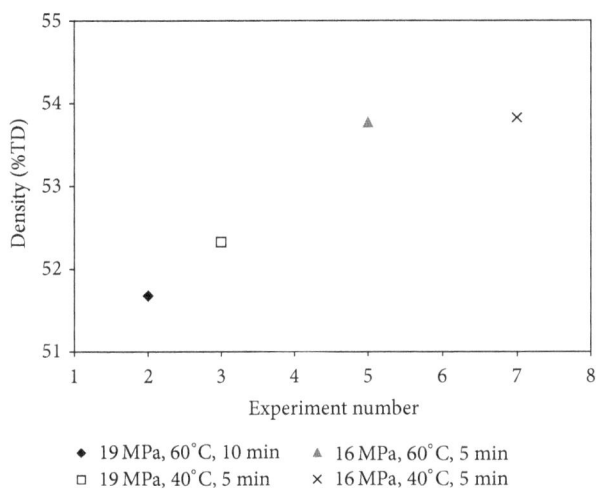

- ♦ 19 MPa, 60°C, 10 min   ▲ 16 MPa, 60°C, 5 min
- □ 19 MPa, 40°C, 5 min   × 16 MPa, 40°C, 5 min

FIGURE 4: Relative density (% theoretical density) of the green laminates at the 4 better conditions.

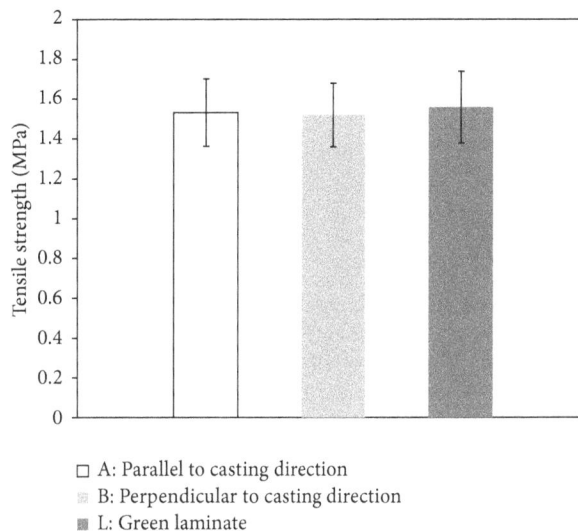

- □ A: Parallel to casting direction
- B: Perpendicular to casting direction
- L: Green laminate

FIGURE 5: Tensile strength of single green tapes (A: parallel and B: perpendicular to casting direction) and green laminate (L).

produce air bubbles between sheets. Suitable lamination conditions must provide homogeneous junction between layers, promoting diffusion between tapes and leading to high green densities. Eventually warm pressing produces inhomogeneous pressure distribution along the piece. In this work, higher pressure (i.e., 16 MPa) resulted in formation of bubbles between layers due to the gradient of pressure generated during lamination. These bubbles prevent the homogeneous junction between layers so that lower green densities were produced. The maximum density achieved was ~54% at a 16 MPa, 40°C, and 5 min. Higher green density results in higher tensile strength in the final product [1, 19]. Gomes et al. [19] produced tapes from a ceramic vitreous material (LSZA) and they found that single tapes with the highest density (1.45 g/cm^3) showed the highest tensile strength values (5.6 ± 0.1 MPa).

The green tensile strength, elongation strain, Young's modulus, and thickness of single green tapes in two different orientations (A: parallel and B: perpendicular to casting direction) and two laminated tapes (L) were measured and the results are presented in Figure 5 and Table 4. From the results it is possible to observe that even though the tensile strength of single green tapes was similar, the samples measured in the direction of casting exhibited higher values, due to particle rearrangement in the polymer matrix. During processing, particles and polymer chains rearranged in the direction of casting providing higher mechanical strength to samples when measured in the same direction. Single layer samples presented a thickness of 130 μm, whereas the laminated samples were 220 μm thick. In Figure 5, it can be

seen that the laminated samples showed the highest tensile strength (~1.56 MPa), as expected. Compared to laminates, single tapes presented lower relative density of about 46 ± 3%. This fact also influences the mechanical strength results, because higher densities provide better mechanical properties to materials. The difference between the density of single tape samples and laminates could be attributed to the diffusion during laminating process and the rearrangement of the particles in the polymeric phase.

4. Conclusions

Aqueous tape casting is an alternative technique to produce flat yttria-stabilized zirconia for electrolyte-supported solid oxide fuel cell. Suitable tapes from slurry with 25 wt% acrylic emulsion binder were successfully prepared. Slurries with 25 wt% binder showed shear-thinning behavior, with a viscosity of ~0.3 Pa·s when passing through the blade during tape casting. A smooth and homogeneous tape surface was produced. Lamination conditions were determined in order to obtain laminates with higher green densities. Adequate laminating conditions (16 MPa, 40°C, and 5 min) were found and laminates with 54% theoretical density were obtained. Laminates showed slightly higher tensile strength of 1.56 ± 0.18 MPa when compared to single tapes but also presented lower elongation to strain. When comparing the mechanical properties of single tapes, samples measured in the direction of casting showed higher mean values of tensile strength, elongation to strain, and Young's modulus, 1.53 ± 0.17 MPa, $5.7 \pm 0.87\%$, and 1.08 MPa, respectively. Lamination provides the increase of the relative density of the samples compared to the single tapes, improving the mechanical strength.

Conflict of Interests

The authors declare that there is no conflict of interests regarding the publication of this paper.

Acknowledgment

The authors gratefully acknowledge the financial support from the CNPq Foundation, Brazil.

References

[1] F. Doreau, G. Tarì, M. Guedes, T. Chartier, C. Pagnoux, and J. M. F. Ferreira, "Mechanical and lamination properties of alumina green tapes obtained by aqueous tape-casting," *Journal of the European Ceramic Society*, vol. 19, no. 16, pp. 2867–2873, 1999.

[2] M. P. Albano and L. B. Garrido, "Influence of the slip composition on the properties of tape-cast alumina substrates," *Ceramics International*, vol. 31, no. 1, pp. 57–66, 2005.

[3] H. Moon, S. D. Kim, S. H. Hyun, and H. S. Kim, "Development of IT-SOFC unit cells with anode-supported thin electrolytes via tape casting and co-firing," *International Journal of Hydrogen Energy*, vol. 33, no. 6, pp. 1758–1768, 2008.

[4] M. P. Albano and L. B. Garrido, "Aqueous tape casting of yttria stabilized zirconia," *Materials Science and Engineering A*, vol. 420, no. 1-2, pp. 171–178, 2006.

[5] M. P. Albano and L. B. Garrido, "Influence of the slip composition on the aqueous processing and properties of yttria stabilized zirconia green tapes," *Ceramics International*, vol. 32, no. 5, pp. 567–574, 2006.

[6] Y. Zhang and J. Binner, "Tape casting aqueous alumina suspensions containing a latex binder," *Journal of Materials Science*, vol. 37, no. 9, pp. 1831–1837, 2002.

[7] J. Feng and F. Dogan, "Aqueous processing and mechanical properties of PLZT green tapes," *Materials Science and Engineering A*, vol. 283, no. 1-2, pp. 56–64, 2000.

[8] H. Jantunen, T. Hu, A. Uusimäki, and S. Leppävuori, "Tape casting of ferroelectric, dielectric, piezoelectric and ferromagnetic materials," *Journal of the European Ceramic Society*, vol. 24, no. 6, pp. 1077–1081, 2004.

[9] P. Vozdecky and A. Roosen, "Direct tape casting of nanosized Al_2O_3 slurries derived from autogenous nanomilling," *Journal of the American Ceramic Society*, vol. 93, no. 5, pp. 1313–1319, 2010.

[10] D. Hotza and P. Greil, "Review: aqueous tape casting of ceramic powders," *Materials Science and Engineering A*, vol. 202, no. 1-2, pp. 206–217, 1995.

[11] T. Chartier and A. Bruneau, "Aqueous tape casting of alumina substrates," *Journal of the European Ceramic Society*, vol. 12, no. 4, pp. 243–247, 1993.

[12] A. Kristoffersson and E. Carlström, "Tape casting of alumina in water with an acrylic latex binder," *Journal of the European Ceramic Society*, vol. 17, no. 2-3, pp. 289–297, 1997.

[13] Y. Zhang, C. Qin, and J. Binner, "Processing multi-channel alumina membranes by tape casting latex-based suspensions," *Ceramics International*, vol. 32, no. 7, pp. 811–818, 2006.

[14] L. Xibao, "Aqueous tape casting of SDC with a multifunctional dispersant," *Journal of Ceramic Processing Research*, vol. 13, pp. 324–329, 2012.

[15] Y. Cho, J. Yeo, Y. Jung, S. Choi, J. Kim, and U. Paik, "Effect of molecular mass of poly(vinyl butyral) and lamination pressure on the pore evolution and microstructure of $BaTiO_3$ laminates," *Materials Science and Engineering A*, vol. 362, no. 1-2, pp. 174–180, 2003.

[16] V. Moreno, J. L. Aguilar, and D. Hotza, "8YSZ tapes produced by aqueous tape casting," *Materials Science Forum*, vol. 727-728, pp. 752–757, 2012.

[17] R. Moreno, "The role of slip additives in tape-casting technology, part II: binders and plasticizers," *American Ceramic Society Bulletin*, vol. 71, pp. 1647–1657, 1992.

[18] Y. Qiao, Y. Liu, A. Liu, and Y. Wang, "Boron carbide green sheet processed by environmental friendly non-aqueous tape casting," *Ceramics International*, vol. 38, no. 3, pp. 2319–2324, 2012.

[19] C. M. Gomes, A. P. N. Oliveira, D. Hotza, N. Travitzky, and P. Greil, "LZSA glass-ceramic laminates: fabrication and mechanical properties," *Journal of Materials Processing Technology*, vol. 206, pp. 194–201, 2008.

6

Alumina-Based Ceramics for Armor Application: Mechanical Characterization and Ballistic Testing

M. V. Silva,[1] **D. Stainer,**[2] **H. A. Al-Qureshi,**[1] **O. R. K. Montedo,**[3] **and D. Hotza**[1]

[1] *Núcleo de Pesquisa em Materiais Cerâmicos e Vítreos (CERMAT), Programa de Pós-Graduação em Ciência e Engenharia de Materiais (PGMAT), Universidade Federal de Santa Catarina (UFSC), 88040-900 Florianópolis, SC, Brazil*

[2] *CMC Tecnologia, Avenida Roberto Galli 1220, 88845-000 Cocal do Sul, SC, Brazil*

[3] *Programa de Pós-Graduação em Ciência e Engenharia de Materiais (PPGCEM), Laboratório de Cerâmica Técnica (CerTec), Universidade do Extremo Sul Catarinense (UNESC), Avenida Universitária, 1105, 88806-000 Criciúma, SC, Brazil*

Correspondence should be addressed to O. R. K. Montedo; oscar.rkm@gmail.com

Academic Editor: Shaomin Liu

The aim of this work is to present results of mechanical characterization and ballistic test of alumina-based armor plates. Three compositions (92, 96, and 99 wt% Al_2O_3) were tested for 10 mm thick plates processed in an industrial plant. Samples were pressed at 110 MPa and sintered at 1600°C for 6 h. Relative density, Vickers hardness, and four-point flexural strength measurements of samples after sintering were performed. Results showed that the strength values ranged from 210 to 300 MPa depending on the porosity, with lower standard deviation for the 92 wt% Al_2O_3 sample. Plates (120 mm × 120 mm × 12 mm) of this composition were selected for ballistic testing according to AISI 1045, using a metallic plate as backing and witness plates in the case of penetration or deformation. Standard NIJ-0108.01 was followed in regard to the type of projectile to be used (7.62 × 51 AP, Level IV, 4068 J). Five alumina plates were used in the ballistic tests (one shot per plate). None of the five shots penetrated or even deformed the metal sheet, showing that the composition containing 92 wt% Al_2O_3 could be considered to be a potential ballistic ceramic, being able to withstand impacts with more than 4000 J of kinetic energy.

1. Introduction

Ceramics have been considered one of the most important materials for lightweight armor applications due to their low density, high compressive strength, and high hardness [1]. Ceramic materials for using as ballistic armor must be sufficiently rigid to fragment the bullet and reduce its speed, transforming it into small fragments that should be stopped by the layer of flexible material that supports the ceramic. Thus, it is necessary that the ceramic material presents high elastic modulus and high hardness [2]. Fracture toughness is also a very important requirement for this application.

The main ceramic materials used commercially in the development of ballistic armors are Al_2O_3, B_4C, SiC, and ceramic matrix composites (CMCs) such as Al_2O_3/ZrO_2 system. High cost, processing hindrances, and restrictions to predict ballistic performance from the properties of the material are some drawbacks of ceramic armors [3].

Alumina provides the best cost-benefit ratio among advanced ceramics, featuring high modulus of elasticity, high refractoriness, high hardness, and relatively lower cost. However, because of its low fracture toughness and low flexural strength, ballistic performance of alumina is lower when compared to SiC and B_4C [4]. The properties of the alumina may be improved, either by introducing zirconia or by the manufacturing CMCs, which increase fracture toughness and flexural strength by introducing tetragonal zirconia particles or ceramic fibers, respectively [5, 6].

Thus, ceramic materials are usually a part of a ballistic personal or vehicle protection system. Even in this case, only with a rigorous control of the microstructure assured a reliable ballistic performance can be. For example, a system consisting of a composite of B_4C and glass fibers or aramid fibers coated with a protective fabric is often used. The replacement of metallic materials by ceramic materials in armored vehicles may lead to weight reduction, autonomy increase, and higher level of protection [7].

A ballistic armor system consists of several layers. The first layer is usually formed of ceramic materials whose function is to cushion the initial impact of the projectile. This layer must fracture the tip of the projectile dissipating much of the kinetic energy of the projectile fragment mass and improve the distribution of impact pressure on the second layer. The second layer is also called backing and is formed of ductile materials. Its function is to absorb the kinetic energy of the fragments derived from the residual projectile and ceramics by plastic deformation [8]. The most important requirement of the backing is no fail during the initial stages of the penetration process of the projectile, that is, the backing must withstand compressive stresses transferred to the ceramic after impact. Thus, it would prevent the penetration of the shrapnel containing high kinetic energy and would not deform excessively, as this would jeopardize the lives of persons protected by the armor or the integrity of the equipment. It is important to mention that impacts at high speeds (high kinetic energies) are high complexity phenomena that present limited reproduction, because parameters such as the incidence of the projectile on the armor, for example, are extremely difficult to be adjusted or prevented [9].

The design of an armor using ceramic materials should consider that the fracture is associated with instantaneous loads in ballistic impacts, which are quite different from those associated with static loads. In static load condition, stresses and strains are distributed throughout the body subjected to the impact and all points are involved in the start of fracture. In the instantaneous loads, stresses and strains are very well localized, so that fractures may occur in an isolated part of the body. This kind of change can dramatically affect the mechanical properties of the material due to the high pressures and loading rates. The impact of high kinetic energy projectiles on the ceramic composite armor usually gives rise to a cone of fractures with radial and circumferential cracks [7].

Projectiles fired over alumina targets, whose tips were flattened, presented higher residual speeds after drilling the targets and therefore greater power of penetration in relation to projectiles with sharp geometry. Moreover, a single layer of ceramic material absorbs more energy than several layers with the same overall thickness [10]. In this case, immediately after the impact on the first plate occurs the formation of an axial crack in the interface between the plates, which causes premature fracture of the assembly. During the first stage of the mechanism of penetration, the most important factor is to keep the integrity of the ceramic so that it can erode the greatest possible amount of mass of the projectile, that is, delaying the startup of the fracture of the ceramic material. This factor is decisive for the choice of the ceramic material to be used.

It is not possible to ensure an effective correlation between ballistic performance and a single characteristic or property of the material, due to the dynamic nature of the event occurring at intervals of time ranging from nano- to microseconds. Thus, ballistic tests under certain conditions are always required to determine the effectiveness of the protection systems. The development of ceramic ballistic requires careful

TABLE 1: Chemical composition of the alumina used in this study (wt%).

Composition	Al_2O_3	SiO_2	CaO	MgO	Na_2O	Fe_2O_3
A92	92.0	2.5	2.3	2.8	0.1	0.03
A96	96.0	3.1	0.1	0.6	0.1	0.04
A99	99.7	0.0	0.0	0.1	0.1	0.02

assessment of physical and mechanical properties of the material in order to obtain adequate performance ballistic plates to the level of protection required.

The aim of this work is to characterize alumina-based ballistic plates and to evaluate their performance in situations where high levels of kinetic energy are required.

2. Materials and Methods

Three alumina compositions were used in this study according to Table 1 (X-ray fluorescence spectrometry, (XRF), Philips PW 2400, The Netherlands). The mean particle sizes were determined using a laser scattering particle size analyzer (CILAS 1064L, France). Each composition was uniaxially compacted at 110 MPa into 120 mm × 120 mm × 10 mm plates in an industrial press (Sacmi PH-300, Italy), resulting in samples presenting an apparent density of 2.44 ± 0.04 g·cm^{-3}. The specimens were then sintered at $1600 \pm 5°C$ for 6 h (holding time) in an industrial kiln (2°C·min^{-1} heating rate). After cooling down to room temperature, crystalline phases developed during heat treatment of the bodies were determined in a X-ray diffractometry (Phillips X'Pert, The Netherlands) using CuKα radiation, within a 2θ range of 5 to 100°. The theoretical densities (ρ_t, g·cm^{-3}) of the sintered samples (using powdered pieces of the samples) were measured by He-pycnometry (AccuPyc 1330, Micromeritics, USA). The apparent density (ρ_{ap}, g·cm^{-3}) was determined by the Archimedes' principle by water immersion at 20°C. The relative density (ρ_r) was calculated from the relationship between apparent density and theoretical density according to (1), while porosity was calculated from (2):

$$\rho_r = \frac{\rho_{ap}}{\rho_t}, \qquad (1)$$

$$\varepsilon = 1 - \rho_r. \qquad (2)$$

Microhardness measurements were performed with a Vickers automatic hardness tester (Shimadzu HMV, Japan) equipped with a diamond Vickers indenter. Tests were carried out with load of 1 kg, according to Karandikar [3], for tests in ballistic ceramics. Each sample was polished for leveling surface, and submitted at least to 5 measurements. The bending strength (σ_f) of the sintered samples was determined in a test machine (Model DL 2000, EMIC, Brazil), which consisted of a four-point test on thirty samples with dimensions of 100 mm × 15 mm × 10 mm at a load rate of 0.5 mm·min^{-1}. Weibull distribution [11] was used to evaluate the reliability of the obtained

TABLE 2: Standard NIJ-0108.01 test parameters for ballistic materials resistant to projectiles at different levels of ballistic protection.

Level of ballistic protection	Ammunition type	Nominal mass (g)	Barrel length (cm)	Projectile velocity (m/s)	Kinetic energy (J)	Shots per panel
I	.22 LRHV Lead	2.6	15–16.5	320 ± 12	133.12	5
	.38 Special RN Lead	10.2	15–16.5	259 ± 15	342.12	
II-A	9 mm FMJ	10.2	10–12	381 ± 12	440.09	5
	.357 Mag JSP	8.0	10–12	332 ± 12	740	
II	9 mm FJM	10.2	15–16.5	425 ± 15	512.66	5
	.357 Mag JSP	8.0	10–12	358 ± 12	921	
III-A	44 Magnum Lead SWC	15.55	14–16	426 ± 15	725.9	5
	9 mm FMJ	8.0	24–26	426 ± 15	1406	
III	7.62 × 51 FJNB 308 Winchester FMJ	9.7	56	838 ± 15	3405	5
IV	30-06 AP	10.8	56	868 ± 15	4068.5	1

results and also to select the composition for the ballistic testing, according to the following for the cumulative probability of fracture:

$$P = 1 - e^{[(-v/v_0)\cdot\{\sigma-\sigma_u/\sigma_0\}]^m}, \qquad (3)$$

where P is probability of fracture; v is tested amount of material; v_0 is standard amount; m is Weibull modulus; σ is tensile strength of the material; σ_o is parameter adjustment; σ_u is strength below which the fracture probability is zero.

Considering a constant amount of specimens, (4) can be applied for a given sampling:

$$P = 1 - e^{[-(\sigma-\sigma_u)/\sigma_0]^m}, \qquad (4)$$

where m, σ_o, and σ_u are known as Weibull parameters. The determination of those parameters is carried out considering firstly $\sigma_u = 0$ and then modifying the equation into the following expression:

$$\ln\left[\ln\left(\frac{1}{1-P}\right)\right] = m\ln\sigma - m\ln\sigma_o. \qquad (5)$$

By using the linear regression method it is possible to determine the values of the parameters m and σ_o.

Samples of sintered alumina (92 wt% Al_2O_3, 120 mm × 120 mm × 12 mm) were fixed in metal plates (backing, 120 mm × 120 mm × 4 mm) of SAE 1045 steel for the ballistic tests. A metal substrate was also designed to engage and support the specimen during the impact of the shot as shown in Figure 1.

There are two main international standards used to evaluate the performance of ballistic protection: NIJ-0101.04 and NIJ-0108.0 (National Institute of Justice) [12] and NATO (North Atlantic Treaty Organization) STANAG 4569 [13]. Tables 2 and 3 show test parameters for ballistic materials resistant to projectiles at different levels of ballistic protection, respectively for standards, NIJ-0108.01 and STANAG 4569. The ballistic testing was carried out based on the standards with the following modifications: a Mauser M1908 riffle with 7.62 AP bullets (STANAG III/NIJ IV) was used in the tests

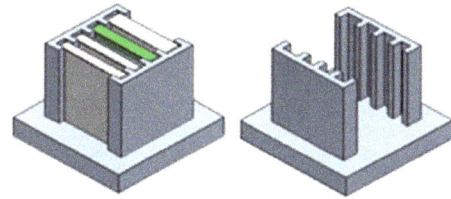

(a)

(b)

FIGURE 1: Schematic pictures of the test stand: (a) metallic substrate of the specimens; (b) specimens inserted into the metallic support.

made at 15 m of distance with one shot per panel. Polyurethane webs of approximately 3 mm thickness were placed in front of ceramic plates to retain fragments of the projectile and the ceramic released during the test. A backing plate was used as control.

3. Results and Discussion

Figure 2 shows the XRD patterns of the used compositions. As expected, the sintered samples were mostly constituted by α-Al_2O_3 (JCPDS 10-0173). However, diffraction pattern of composition A92 showed the following minority crystalline phases: spinel ($MgAl_2O_4$), wollastonite ($CaSiO_3$) and hibonite ($CaAl_{12}O_{19}$). These phases were formed due to the sintering additives employed.

TABLE 3: Standard STANAG 4569 test parameters for armored vehicles at different levels of ballistic protection.

Level of ballistic protection	Threat	Ammunition type	Test distance (m)	Projectile velocity (±15 m/s)
I	Rifle	7.62 × 51 NATO Ball (Ball M80)	30	833
		5.56 × 45 NATO SS109		900
		5.56 × 45 M193		937
II	Infantry rifle	7.62 × 39 APIBZ	30	695
III	Sniper rifle	7.62 × 51 AP (WC core)	30	930
		7.62 × 54 R B32 API (Dragunov)	30	854
IV	Heavy machine gun	14.5 × 114 AP/832	200	911
V	Automatic cannon	25 mm APDS-TM-791 or TLB 073	500	1258

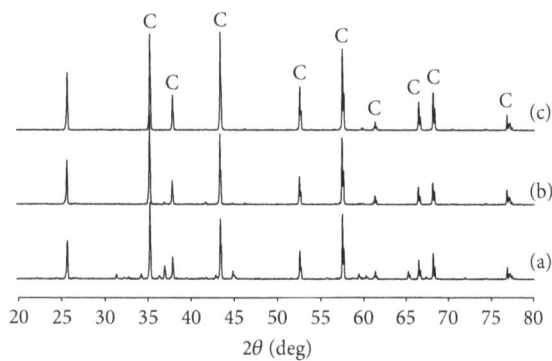

FIGURE 2: XRD patterns of the used compositions: (a) A92, (b) A96, and (c) A99. C: α-Al$_2$O$_3$ (JCPDS card n. 10-0173).

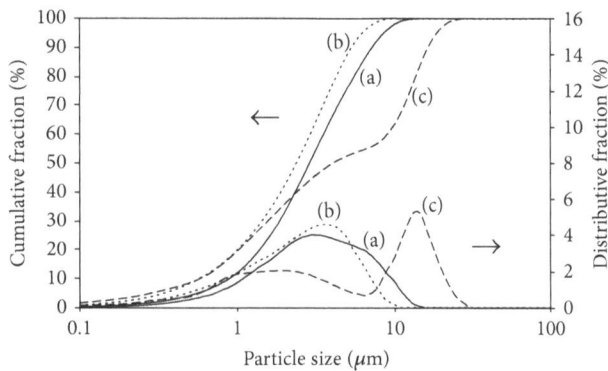

FIGURE 3: Particle size distribution of the used compositions: (a) A92, (b) A96, and (c) A99.

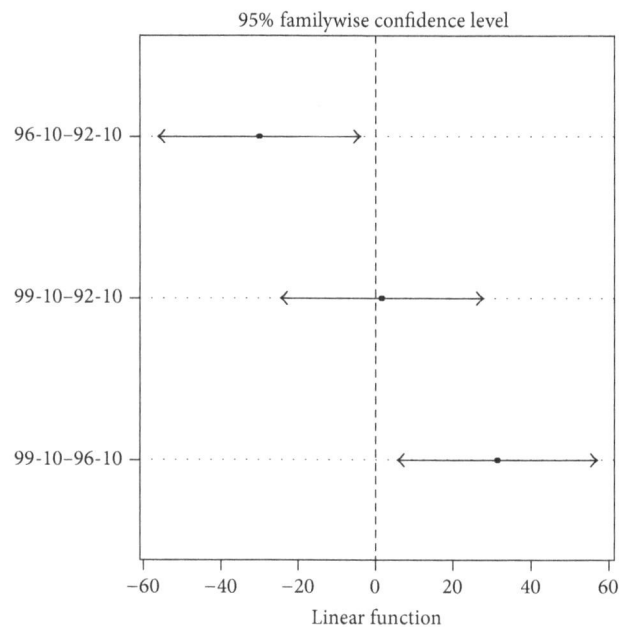

FIGURE 4: Analysis of variance (ANOVA) for a factor.

TABLE 4: Apparent density, theoretical density, relative density, and porosity of the sintered compositions.

Composition	A92	A96	A99
Apparent density (ρ_{ap}, g·cm^{-3})	3.65 ± 0.01	3.70 ± 0.01	3.85 ± 0.02
Theoretical density (ρ_t, g·cm^{-3})	3.70 ± 0.01	3.80 ± 0.01	3.96 ± 0.01
Relative density (ρ_r, %)	98.7	97.6	97.1
Porosity (ε, %)	1.3	2.4	2.9

Figure 3 shows the particle size distribution of the used compositions. As it can be seen, those materials present a practically monomodal distribution, with mean particle size (d_{50}) of 2.92, 2.41, and 4.10 μm, respectively, for A92, A96, and A99.

Table 4 shows values of relative density and porosity of the sintered compositions. Relative densities were found to be around 97-98%. As expected, relative density decreased with increasing α-Al$_2$O$_3$ contents, for the same forming pressure, sintering temperature, and holding time. Moreover, mean particle size of composition A99 was higher than the other

ones that should have caused less accommodation of the particles and consequently more porosity compacts.

Table 5 shows values of Vickers hardness of sintered compositions. Despite of differences in porosity, the investigated compositions showed values of Vickers hardness above 14 GPa in good agreement with values obtained by Karandikar [3] for commercial ballistic ceramics. Hardness is a very important mechanical property for ballistic application

FIGURE 5: Results of the shots carried out on the ballistic plates prepared from composition A92: (a) front view before impact, (b) front view after the impact of the projectile 7.62 × 51 AP, and (c) a detail of the local impact of the projectile on the surface of alumina after removing of the fragments.

TABLE 5: Vickers hardness of sintered compositions.

Compositions	HV (GPa)
A92	14.8 ± 0.6
A96	13.3 ± 0.9
A99	14.8 ± 0.7

TABLE 6: Bending strength (σ_f) and Weibull modulus (m) for the tested compositions.

Composition	A92	A96	A99
σ_f (MPa)	221 ± 40	195 ± 40	227 ± 20
σ_{f50} (MPa)	225	201	231
m	6.4	4.8	8.0

TABLE 7: Results of statistical hypothesis test.

Linear hypotheses	Estimate	SD Error	T value	Pr ($>|t|$)
96-10–92-10	29.676	10.973	−2.704	0.0223
99-10–92-10	1.647	10.973	0.150	0.9877
99-10–96-10	31.323	10.681	2.933	0.0120

TABLE 8: Results of the analysis of variance.

Linear hypotheses	Estimate	Lower	Upper
96-10–92-10	−29.6763	−55.8574	−3.4952
99-10–92-10	1.6465	−24.5346	27.8276
99-10–96-10	31.3229	5.8401	56.8056

since it is expected the rupture of the projectile by the ballistic plate.

Table 6 shows values of bending strength and Weibull modulus for the studied compositions. All compositions presented similar values of bending strength considering the standard deviations presented. Nevertheless, compositions A92 and A99 showed practically the same average values of bending strength, which are higher than the ones for composition A96. This can be statistically confirmed with a 95% of confidence level by the analysis of variance and hypothesis test, as shown in Tables 7 and 8, and Figure 4.

Moreover, composition A96 showed a very lower value of m in comparison with compositions A92 and A99 for a technical application. In fact, technical ceramics must show values of m ranging from 5 to 10 according to Kanno [14]. In this sense and considering the value of σ, composition A96 could not be used for ballistic tests. On the other hand, compositions A92 and A99 showed suitable values of m for technical application and there is no statistical evidence of significant difference between them, as shown by their linear function. However, because of its lower porosity, composition A92 was selected for ballistic tests.

Figures 5 and 6 show the results of the shots on the ballistic plates prepared from composition A92. In fact, a high amount of energy was absorbed by the alumina plate causing its fracture. Nevertheless, its hardness caused the breaking of the projectile into small fragments. Fragments did not penetrate or cause deformation in none of the backing (metallic plates) of the five plates tested. The velocity of the projectile in the impact was estimated to be ~865 m·s^{-1}, while the kinetic energy should be above 4000 J. In this way, all the tested plates were approved in the ballistic tests (Level IV of Standard NIJ-0108.01). Those results are in very good agreement with those obtained by Karandikar [3] (alumina, 10 mm thickness) and above to those obtained by Couto [7] (level III, alumina/zirconia, 11.3 mm thickness).

4. Conclusion

Three compositions of commercial alumina containing 92, 96, and 99.7 wt% Al2O3, named compositions A92, A96, and A99, respectively, were investigated in this work to be used as ballistic ceramics. Compositions were characterized by Vickers hardness measurements and four point flexural bending strength. The reliability of the obtained results was evaluated

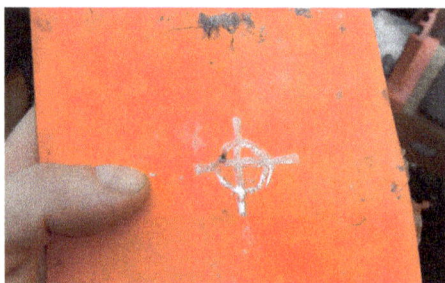

FIGURE 6: View of the hole caused by the shot over the plate of polyurethane.

by Weibull statistics. According to the obtained results, one composition was selected for the ballistic testing. Compositions A92 and A99 showed comparable results with each other and higher Vickers hardness and bending strength than composition A96; however, composition A92 was selected for ballistic tests because of its lower porosity. Alumina plates (12 mm thickness) were submitted to a procedure based on the Standard NIJ-0108.01 (Level IV, $865\,\text{m·s}^{-1}$, more than 4000 J). Fragments did not penetrate or cause deformation in none of the backing (metallic plates) of the five plates tested. All the tested plates were approved in the ballistic tests.

Conflict of Interests

The authors declare that there is no conflict of interests regarding the publication of this paper.

Acknowledgment

The authors are grateful to Coordenação de Aperfeiçoamento de Pessoal de Nível Superior (CAPES/Brazil) for funding this work.

References

[1] S. Yadav and G. Ravichandran, "Penetration resistance of laminated ceramic/polymer structures," *International Journal of Impact Engineering*, vol. 28, no. 5, pp. 557–574, 2003.

[2] C. Xavier and C. R. C. Costa, "Study on mechanical behavior of alumina plates under ballistic impact," *Cerâmica*, vol. 30, no. 175, pp. 161–168, 1984.

[3] P. G. Karandikar, "A review of ceramics for armor applications," in *Advances in Ceramic Armor IV*, vol. 29, pp. 163–175, The American Ceramic Society, 2009.

[4] R. E. Tressler, "An assessment of low cost manufacturing technology for advanced structural ceramics and its impact on ceramic armor," *Ceramic Transactions*, vol. 134, pp. 451–462, 2002.

[5] R. Stevens and P. A. Evans, "Transformation toughening by polycrystalline zirconia," *Transactions and Journal of the British Ceramic Society*, vol. 83, no. 1, pp. 18–31, 1984.

[6] R. B. Heimann, *Classic and Advanced Ceramics: From Fundamentals to Applications*, Wiley-VCH, New York, NY, USA, 2010.

[7] C. A. O. Couto, *Estudo de blindagem para proteção contra impactos de micrometeoróides em satélites artificiais [Mestrado em Engenharia e Tecnologia Espaciais/Materiais e Sensores],* Instituto Nacional de Pesquisas Espaciais, São José dos Campos, Brazil, 20112011.

[8] "Matweb Material Property Data, Alumina, alpha Al_2O_3, 99.5%," http://www.matweb.com/search/PropertySearch.aspx.

[9] L. Neckel, *Modelamento e simulação de impacto balístico em sistema cerâmica-metal [Mestrado em Ciência e Engenharia de Materiais]*, Programa de Pós-Graduação em Ciência e Engenharia de Materiais, Universidade Federal de Santa Catarina, Florianópolis, Brazil, 20122012.

[10] M. L. Wilkins, "Mechanics of penetration and perforation," *International Journal of Engineering Science*, vol. 16, no. 11, pp. 793–807, 1978.

[11] W. Weibull, "A statistical distribution function of wide applicability," *Journal of Applied Mechanics*, vol. 18, pp. 293–297, 1951.

[12] National Institute of Justice Standard, "Ballistic resistance of body armor," Tech. Rep. NIJ-0101.06, 2008.

[13] North Atlantic Treaty Organization, "Standards for Ballistic protection for light armoured vehicles," Stanag 4569, 2008.

[14] W. M. Kanno, *Propriedades Mecânicas do Gesso de Alto Desempenho [Doutor em Ciência e Engenharia de Materiais]*, Área de Concentração, Desenvolvimento, Caracterização e Aplicação de Materiais, Universidade de São Paulo, São Carlos, Brazil, 2009 2009.

7

Cubic Phases in the Gd_2O_3-ZrO_2 and Dy_2O_3-TiO_2 Systems for Nuclear Industry Applications

Maria Teresa Malachevsky,[1,2] Diego Rodríguez Salvador,[1] Sergio Leiva,[1] and Claudio Alberto D'Ovidio[1]

[1]Centro Atómico Bariloche, CNEA, Avenida Bustillo 9500, 8400 San Carlos de Bariloche, Argentina
[2]CONICET, Argentina

Correspondence should be addressed to Maria Teresa Malachevsky; malache@cab.cnea.gov.ar

Academic Editor: Jim Low

Neutron absorbers are elements with a high neutron capture cross section that are employed at nuclear reactors to control excess fuel reactivity. If these absorbers are converted into materials of relatively low absorption cross section as the result of neutron absorption, they consume during the reactor core life and so are called burnable. These elements can be distributed inside an oxide ceramic that is stable under irradiation and thus called inert. Cubic zirconium oxide is one of the preferred materials to be used as inert matrix. It is stable under irradiation, experiments very low swelling, and is isomorphic to uranium oxide. The cubic phase is stabilized by adding small amounts of dopants like Dy_2O_3 and Gd_2O_3. As both dysprosium and gadolinium have a high neutron cross section, they are good candidates to prepare burnable neutron absorbers. Pyrochlores, like $Gd_2Zr_2O_7$ and $Dy_2Ti_2O_7$, allow the solid solution of a large quantity of elements besides being stable under irradiation. These characteristics make them also useful for safe storage of nuclear wastes. We present a preliminary study of the thermal analysis of different compositions in the systems Gd_2O_3-ZrO_2 and Dy_2O_3-TiO_2, investigating the feasibility to obtain oxide ceramics useful for the nuclear industry.

1. Introduction

At the beginning of a reactor cycle, the fuel presents an excess reactivity to compensate future fuel depletion and fission products buildup. This is controlled by using burnable neutron absorbers or neutron poisons, that is, by inclusion into the fuel assembly materials with a large neutron cross section. If the absorbers are incorporated directly into the fuel, they are called continuous or homogeneous [1, 2]. When the burnable neutron absorbers are dispersed into an inert matrix, they are called discrete or inhomogeneous and they are distributed among the fuel pellets. As these can be prepared independently of the fuel and with no limitations in their handling and storage, we focused on their preparation.

Zirconium oxide is a good candidate to be used as inert matrix mainly because of its stability under irradiation and its compatibility with reactor materials [3]. It accommodates both fission materials and neutron absorbers. Actually, CANDU reactor has among the 43 bars that compose the fuel element a central bar that includes a neutron absorber in zirconium oxide inert matrix [4].

Zirconium oxide has three polymorphic phases: it is monoclinic up to 1443 K, tetragonal up to 2643 K, and cubic up to its fusion at 2953 K. It is impossible to prepare a single-phase material with a particular crystalline structure due to these phase transformations that proceed during cooling. Thus the cubic phase is stabilized by adding small amounts of dopants (about 8 mol%) like Y_2O_3, CaO, or MgO [5]. There is an added interest in the cubic phase as it is isomorphic to uranium oxide, even if any of the polymorphs can be employed as inert matrix. The stabilized zirconia is not amorphized under irradiation and its swelling under fast neutron is neglectable, as happens with the fluorite structure [6, 7].

Due to their chemical resemblance with Y_2O_3, Gd_2O_3 stabilizes the cubic phase of the zirconium oxide. As this oxide has a large cross section for neutron capture, it is a good candidate to prepare burnable poisons in zirconia inert matrix [8].

TABLE 1: Selected compositions for the mixtures.

ID	wt%				Characteristic
	Gd_2O_3	Dy_2O_3	ZrO_2	TiO_2	
GdZr1	33.31	—	66.69	—	Eutectoid
GdZr2	59.62	—	40.38	—	Pyrochlore
GdZr3	76.30	—	23.70	—	Eutectoid
GdZr4	87.86	—	11.14	—	Eutectic
GdZr5	94.00	—	6.00	—	Eutectoid
DyTi1	—	82.34	—	17.66	Polymorph
DyTi2	—	76.07	—	23.93	Eutectic
DyTi3	—	70.01	—	29.99	Pyrochlore
DyTi4	—	49.65	—	50.35	Eutectic

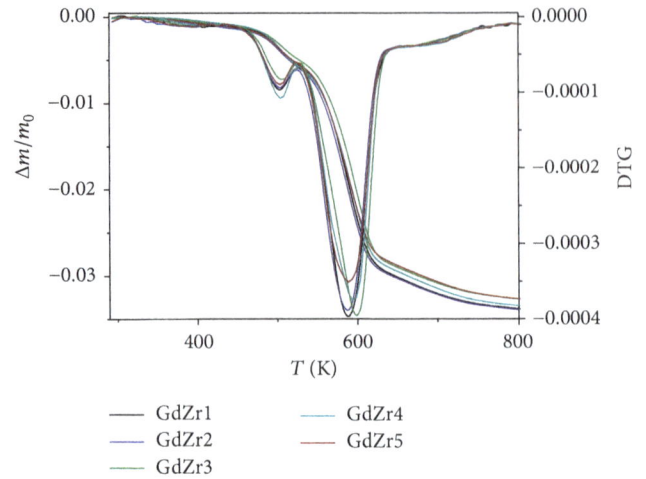

FIGURE 1: TG/DTG from room temperature to 800 K.

Pyrochlores are oxides of the form $A_2B_2O_7$, with a structure that is closely related to the fluorite structure [9]. Its phase stability is basically determined by the A and B cationic radius ratio, as similar cationic radii are more likely to form as disordered fluorites than ordered pyrochlores. Pyrochlores are excellent host matrices for nuclear waste immobilization because they can dissolve various lanthanides, actinides, and other elements which are generated inside nuclear reactors [10–14]. It has been found that stability of the pyrochlores under irradiation increases with the decrease in radius ratio. There is another related phase of interest, dysprosium titanate Dy_2TiO_5, that is an effective burnable poison that has been successfully employed for several years as control rod in Russian reactors [15].

In the present work, we focus on the feasibility to obtain cubic phases that could be employed as burnable neutron absorbers, studying composition and suitable sintering temperatures. We also investigate the feasibility of forming a cubic pyrochlore phase and titanates. Two different systems are investigated: Gd_2O_3-ZrO_2 and Dy_2O_3-TiO_2.

2. Materials and Methods

We started with Gd_2O_3 and Dy_2O_3 99.9% from Alfa Aesar and ZrO_2 and TiO_2 99.9% from Sigma Aldrich. Appropriate amounts of each oxide were mixed in an attrition mill at 200 rpm for 30 minutes with 2 mm yttria stabilized zirconia balls. We prepared the mixtures in stoichiometries corresponding to particular compositions of interest identified in the corresponding binary phase diagram. The selected compositions are presented in Table 1.

We prepared 20 g of each mixture, adding an organic binder so as to reproduce the conditions needed for preparing a pellet. For the compositions in the system Gd_2O_3-ZrO_2, a mixture of 1.5 wt% sodium alginate and 3 wt% methylcellulose was employed. For the Dy_2O_3-TiO_2 systems, 5 wt% of polyvinyl butyral (PVB) was employed as binder.

We carried out simultaneous thermal analysis (STA) in a Netzsch 449 F1 Jupiter to determine sintering conditions and phase formation. Measurements were performed inside a SiC furnace with maximum working temperature of 1828 K, at a rate of 5 K/min in helium atmosphere. The balance resolution

was of 0.025 μg and the temperature was measured with a resolution of 1 μW. After the analysis, the existent phases were identified by X-ray diffraction using the program HighScore Plus version 3.0.4.

Based on these results, we selected three compositions to verify the formation of the desired phases after sintering: GdZr1, GdZr2, and DyTi4. Pellets were prepared by uniaxial pressing at 1000 atm. These were sintered in air with a slow heating rate of 1 K/min up to 773 K to eliminate the binder, followed by a faster heat-up at 2 K/min to the sintering temperature (1973 K for GdZr1, 1873 K for GdZr2, and 1673 K for DyTi4). After a dwell time of 2 hours, the pellets were cooled in the furnace.

3. Results and Discussion

3.1. Gd_2O_3-ZrO_2 System. In Figure 1, we present the mass evolution with temperature TG and its derivative DTG, from room temperature to 800 K. There is a first weight loss starting at about 450 K coincident with the binder decomposition, followed by a marked weight loss that starts at about 530 K and ends at 800 K with the complete elimination of this additive. This behavior is similar for all the samples measured as the binder and its proportion are the same in all of them.

From 800 K up to 1270 K there is a slight mass loss, as can be observed in Figure 2, that can be related to carbon elimination from the binder and/or impurities present in the starting oxides. From 1270 K to the final temperature, about 1 mg of mass is lost. Sample GdZr4 shows an additional mass loss at about 1200 K.

In Figure 3 we present the DTA results. From the phase diagram, several solid-solid phase transformations can be associated with the observed peaks. Pyrochlore transforms to fluorite at 1800 K, there is a eutectoid transformation to fluorite at 1400 K, zirconium oxide transforms from monoclinic to tetragonal at 1500 K, and gadolinium oxide transforms from cubic to monoclinic at 700 K.

For sample GdZr1 a small endothermic peak is detected at 321 K that can be associated with the eutectic transformation

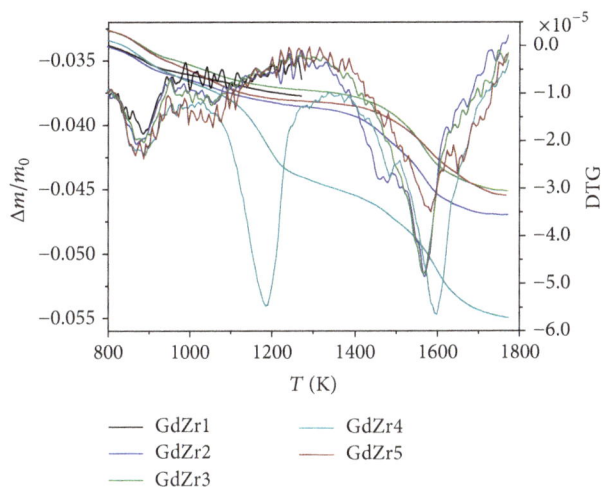

FIGURE 2: TG/DTG from 800 K to the end of the test.

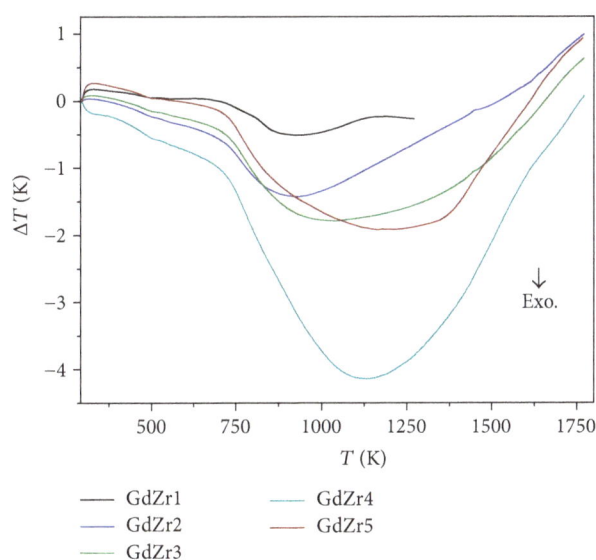

FIGURE 3: DTA as a function of temperature.

to the cubic phase. This temperature is lower than the 400 K reported in the phase diagrams. An X-ray analysis of the powders at the end of the test showed the presence of monoclinic ZrO_2 and cubic Gd_2O_3.

The pyrochlore phase present in sample GdZr2 transforms to fluorite at 1800 K. Both phases were detected by X-ray diffraction at the end of the test. The eutectoid transformation to fluorite of sample GdZr3 takes place at 1400 K and ZrO_2 (present in both GdZr2 and GdZr3) transforms to a tetragonal phase around 1500 K. All these solid-solid transformations are represented at the beginning of the endothermic peaks observed in the studied temperature range. These peaks are consequent to the mass loss observed in the TG measurement from 1270 K. This loss is not detected in sample GdZr1 as the intensity of the endothermic peak is lower. X-ray diffraction confirms the existence of tetragonal ZrO_2 in samples GdZr2 and GdZr3.

A wide exothermic peak is observed in all samples, starting at 700 K that can be originated by a phase transformation from cubic to monoclinic in the gadolinium oxide, coincident with X-ray diffraction phase detection. The behavior of samples GdZr4 and GdZr3 is similar, as expected from the system's phase diagram. In both samples this peak is wider, being better defined in sample GdZr4.

In Figure 4 we present the X-ray diffraction patterns obtained on the powders after the STA measurement. The major phases are indicated, but other minor phases were identified. These are fluorite in GdZr1, tetragonal ZrO_2 and fluorite in GdZr2, GdZr4, and GdZr5, and tetragonal ZrO_2 and pyrochlore in GdZr3.

3.2. Dy_2O_3-TiO_2 System. In Figure 5, the mass loss with temperature can be followed. The loss up to 800 K corresponds to the binder decomposition and elimination.

In Figure 6 we show the DTA measurements. The phases detected by X-ray diffraction after the STA test is completed are those expected from the phase diagram. We present the identified phases for samples DyTi2 and DyTi4 in Figure 7. Initially, a small endothermic peak is observed in all samples associated with the binder decomposition and loss. A wide peak is observed around 600 K. At this fast heating rate, the formation of new phases as dysprosium titanate and pyrochlore is not complete and this peak includes the transformation of TiO_2 from anatase to rutile. The endothermic peak starting at 1600 K observed in samples DyTi1 and DyTi2 corresponds to polymorphic changes of the Dy_2TiO_5 towards cubic structures. The same peak observed in samples DyTi3 and DyTi4 can be associated with the $Dy_2Ti_2O_7$ pyrochlore transforming to the hexagonal structure Dy_2TiO_5.

3.3. Sintered Pellets. Pellets were polished and observed with the scanning electron microscope (SEM). Both samples GdZr1 and GdZr2 are similar and presented a large quantity of pores, as can be observed in Figures 8(a) and 8(b). DyTi4 is quite dense and two different phases homogeneously distributed along the sample can be easily identified (Figure 8(c)). As confirmed by EDS, the darker phase is TiO_2 and the lighter phase corresponds to the pyrochlore $Dy_2Ti_2O_7$, being an excellent candidate for use as burnable poison.

In order to determine the phase formation in the samples of the Gd_2O_3-ZrO_2 system, X-ray diffraction of the pellets was performed. In GdZr1 the major phase identified is cubic ZrO_2, showing that gadolinium oxide effectively stabilized the zirconia even if the transition could not be identified by the STA measurement in this sample. $Gd_2Zr_2O_7$ pyrochlore and cubic Gd_2O_3 are also present in lower quantities. In GdZr2 the major phases are $Gd_2Zr_2O_7$ pyrochlore and tetragonal ZrO_2. Other ZrO_2 polymorphs are also present in low quantities. Even if zirconia is not in the fluorite structure, this material could be employed as burnable poison.

4. Conclusions

Pyrochlores phase stability is basically determined by the A and B cationic radius ratio, as similar cationic radii are

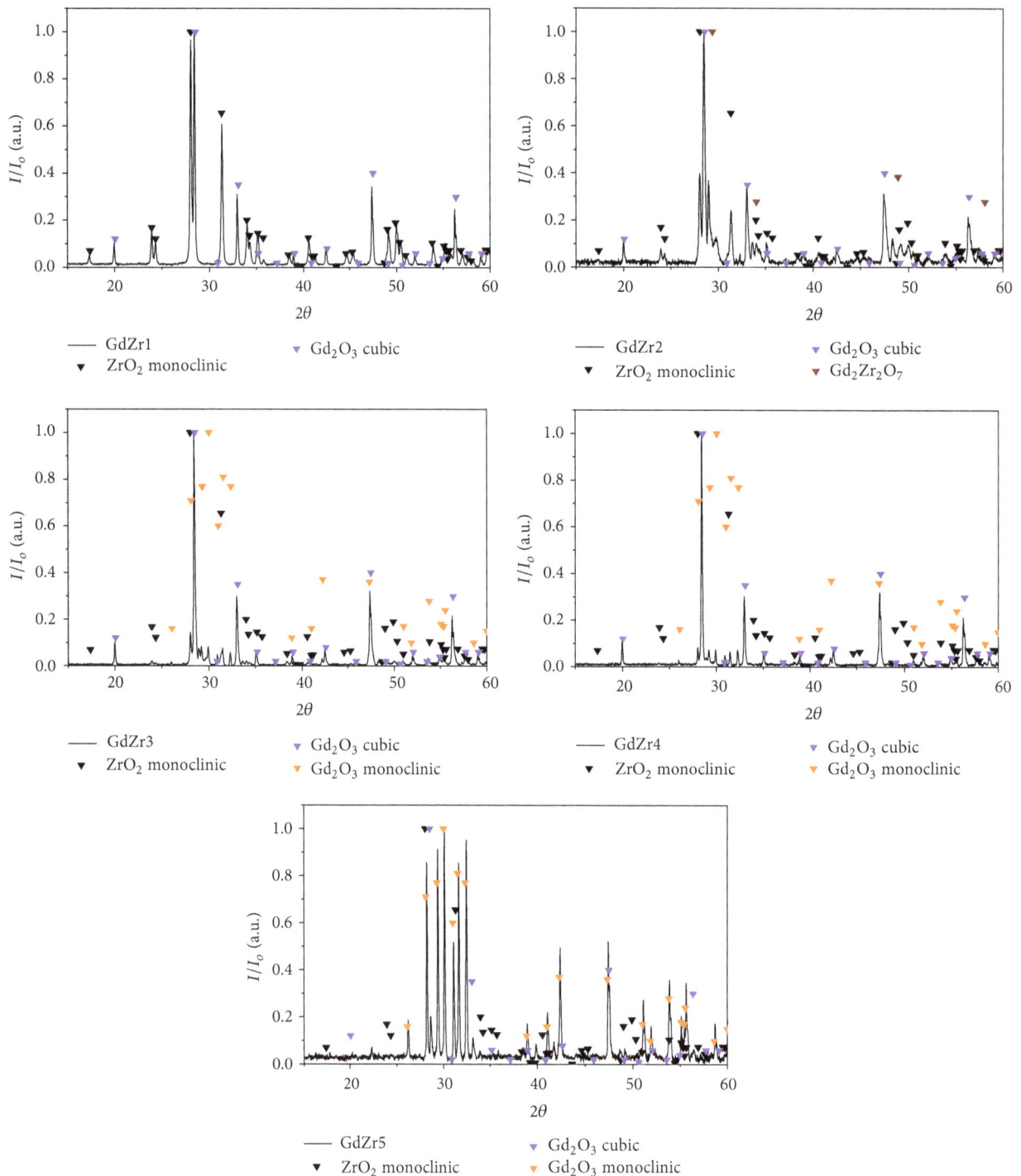

FIGURE 4: X-ray diffraction of the samples after the STA measurement.

more likely to form as disordered fluorites than ordered pyrochlores. Taking into account that $r_{Dy}/r_{Ti} \approx 1.412$ and $r_{Gd}/r_{Zr} \approx 1.253$, a pyrochlore phase should be more easily achieved in the Dy_2O_3-TiO_2 system. We were able to acknowledge that the Gd_2O_3-ZrO_2 system needs higher temperatures than the Dy_2O_3-TiO_2 system to form cubic phases. Excluding sample GdZr1, for all the studied compositions the eutectoid transformation to fluorite was detected.

This indicates a relative ease for cubic phase formation independent of the amount of gadolinium oxide employed, making this system suitable for developing burnable poisons in inert matrix.

For the Dy_2O_3-TiO_2 system, X-ray diffraction reveals that dysprosium titanate Dy_2TiO_5 is in its hexagonal phase but the cubic $Dy_2Ti_2O_7$ pyrochlore can be easily synthesized at low temperatures.

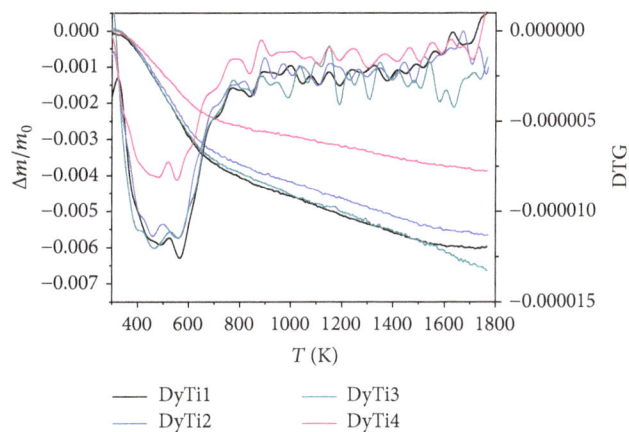

FIGURE 5: TG-DTG as a function of temperature.

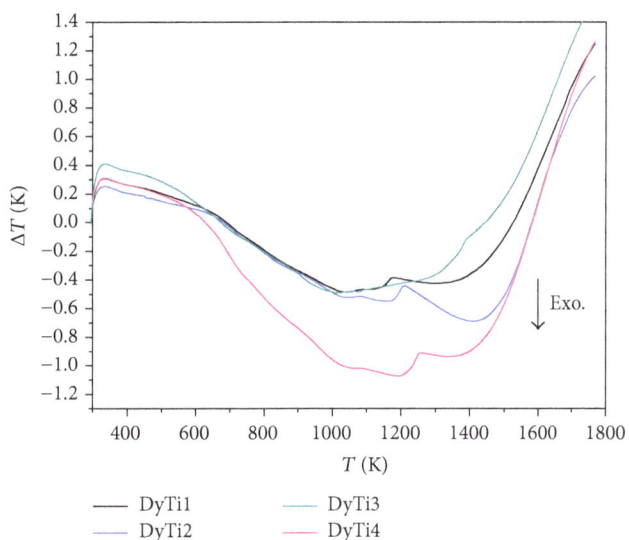

FIGURE 6: DTA as a function of temperature.

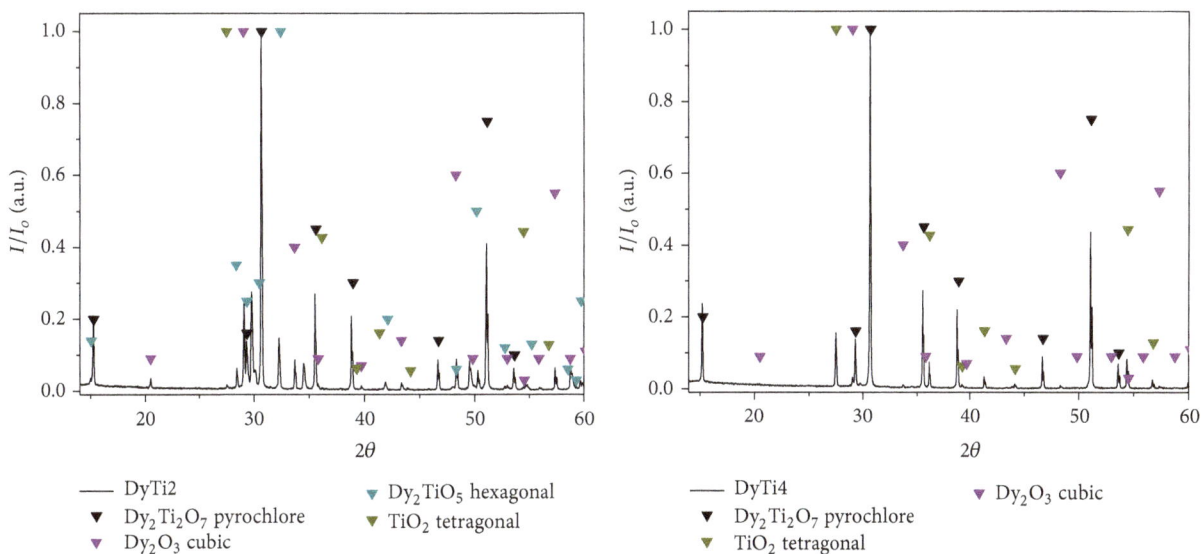

FIGURE 7: X-ray diffraction of the samples after the STA measurement.

(a)

(b)

(c)

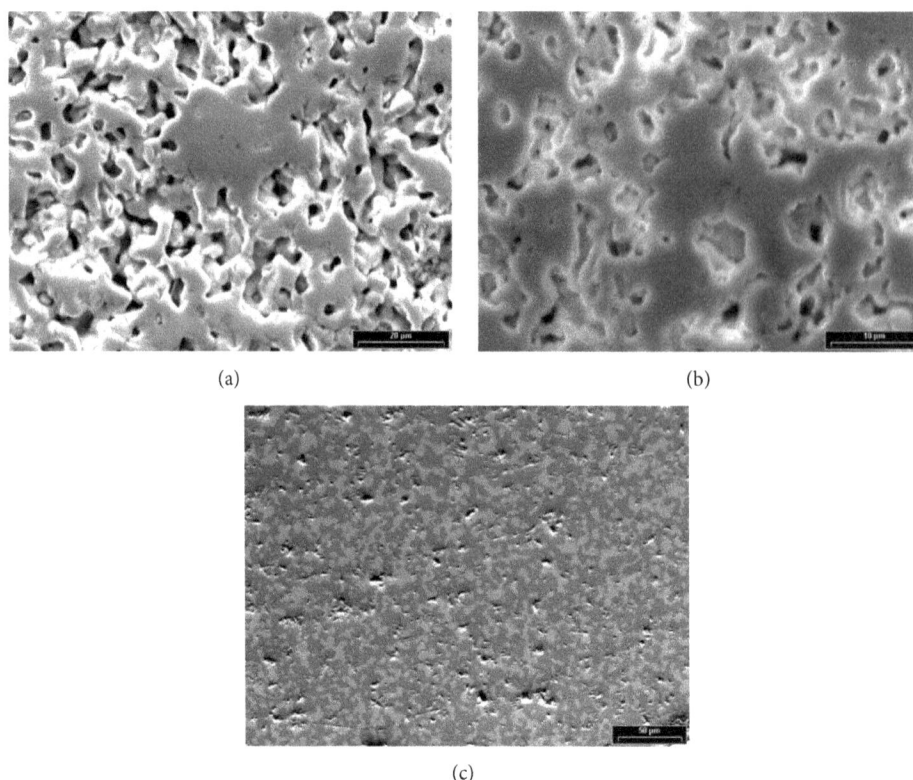

FIGURE 8: SEM photographs of the polished surface of the sintered pellets: (a) GdZr1, (b) GdZr2, and (c) DyTi4.

Some pellets were sintered to verify the STA measurement observations. The cubic zirconia phase was effectively stabilized in sample GdZr1, even if the fluorite transformation could not be identified in the STA measurement. Pyrochlores were successfully sintered in both GdZr2 and DyTi4 samples.

Both systems present adequate conditions to fabricate an oxide ceramic material for nuclear applications.

Conflict of Interests

The authors declare that there is no conflict of interests regarding the publication of this paper.

References

[1] T. Cardinaels, J. Hertog, B. Vos, L. de Tollenaere, C. Delafoy, and M. Verwerft, "Dopant solubility and lattice contraction in gadolinia and gadolinia–chromia doped UO_2 fuels," *Journal of Nuclear Materials*, vol. 424, no. 1–3, pp. 289–300, 2012.

[2] T. Jevremovic, *Nuclear Principles in Engineering*, Springer, 2nd edition, 2009.

[3] G. Velişa, A. Debelle, L. Thomé et al., "Implantation of high concentration noble gases in cubic zirconia and silicon carbide: a contrasted radiation tolerance," *Journal of Nuclear Materials*, vol. 451, no. 1–3, pp. 14–23, 2014.

[4] AECL, "ACR-1000 technical description," AECL Report 10820-01372-230-002, Atomic Energy of Canada, 2010, Revision.

[5] M. Yoshimura, *Ceramic Bulletin*, vol. 34, pp. 1950–1955, 1988.

[6] G. Ledergerber, C. Degueldre, P. Heimgartner, M. A. Pouchon, and U. Kasemeyer, "Inert matrix fuel for the utilisation of plutonium," *Progress in Nuclear Energy*, vol. 38, no. 3-4, pp. 301–308, 2001.

[7] W. J. Carmack, M. Todosow, M. K. Meyer, and K. O. Pasamehmetoglu, "Inert matrix fuel neutronic, thermal-hydraulic, and transient behavior in a light water reactor," *Journal of Nuclear Materials*, vol. 352, no. 1–3, pp. 276–284, 2006.

[8] H. S. Kim, C. Y. Joung, B. H. Lee, S. H. Kim, and D. S. Sohn, "Characteristics of $Gd_xM_yO_z$ (M = Ti, Zr or Al) as a burnable absorber," *Journal of Nuclear Materials*, vol. 372, no. 2-3, pp. 340–349, 2008.

[9] B. P. Mandal, M. Pandey, and A. K. Tyagi, "$Gd_2Zr_2O_7$ pyrochlore: potential host matrix for some constituents of thoria based reactor's waste," *Journal of Nuclear Materials*, vol. 406, no. 2, pp. 238–243, 2010.

[10] M. Lang, F. Zhang, J. Zhang et al., "Review of $A_2B_2O_7$ pyrochlore response to irradiation and pressure," *Nuclear Instruments and Methods in Physics Research B: Beam Interactions with Materials and Atoms*, vol. 268, no. 19, pp. 2951–2959, 2010.

[11] B. P. Mandal, N. Garg, S. M. Sharma, and A. K. Tyagi, "Solubility of ThO_2 in $Gd_2Zr_2O_7$ pyrochlore: XRD, SEM and Raman spectroscopic studies," *Journal of Nuclear Materials*, vol. 392, no. 1, pp. 95–99, 2009.

[12] M. Jafar, P. Sengupta, S. N. Achary, and A. K. Tyagi, "Phase evolution and microstructural studies in $CaZrTi_2O_7$ (zirconolite)–$Sm_2Ti_2O_7$ (pyrochlore) system," *Journal of the European Ceramic Society*, vol. 34, no. 16, pp. 4373–4381, 2014.

[13] E. J. Harvey, K. R. Whittle, G. R. Lumpkin, R. I. Smith, and S. A. T. Redfern, "Solid solubilities of (La Nd,)$_2$(Zr,Ti)$_2$O$_7$

phases deduced by neutron diffraction," *Journal of Solid State Chemistry*, vol. 178, no. 3, pp. 800–810, 2005.

[14] B. P. Mandal, R. Shukla, S. N. Achary, and A. K. Tyagi, "Crucial role of the reaction conditions in isolating several metastable phases in a Gd-Ce-Zr-O system," *Inorganic Chemistry*, vol. 49, no. 22, pp. 10415–10421, 2010.

[15] V. D. Risovany, E. E. Varlashova, and D. N. Suslov, "Dysprosium titanate as an absorber material for control rods," *Journal of Nuclear Materials*, vol. 281, no. 1, pp. 84–89, 2000.

8

Grain Boundary Resistivity of Yttria-Stabilized Zirconia at 1400°C

J. Wang,[1] A. Du,[2] Di Yang,[1] R. Raj,[3] and H. Conrad[1]

[1] *Materials Science and Engineering Department, North Carolina State University, Raleigh, NC 27695-7907, USA*
[2] *Materials and Metallurgy Department, Kunming University of Science and Technology, Kunming, Yunnan 650093, China*
[3] *Department of Mechanical Engineering, Engineering Center, ECME 114, University of Colorado, Boulder, CO 80309-0427, USA*

Correspondence should be addressed to H. Conrad; hans_conrad@ncsu.edu

Academic Editor: Young-Wook Kim

The grain size dependence of the bulk resistivity of 3 mol% yttria-stabilized zirconia at 1400°C was determined from the effect of a dc electric field E_a = 18.1 V/cm on grain growth and the corresponding electric current during isothermal annealing tests. Employing the brick layer model, the present annealing test results were in accordance with extrapolations of the values obtained at lower temperature employing impedance spectroscopy and 4-point-probe dc. The combined values give that the magnitude of the grain boundary resistivity ρ_b = 133 ohm-cm. The electric field across the grain boundary width was 28–43 times the applied field for the grain size and current ranges in the present annealing test.

1. Introduction

In view of its high ionic conductivity, polycrystalline yttria-stabilized zirconia (YSZ) is attractive for application in solid oxide fuel cells (SOFCs) and sensors [1, 2]. The conductivity however, decreases with decrease in grain size d [1–3], which is attributed to the greater resistivity of the grain boundaries compared to that of the grain interior.

The grain boundary resistivity in ceramics is usually determined employing impedance spectroscopy, which along with the cubic brick layer model of the grain microstructure gives for the bulk resistivity ρ at constant temperature [4–7]

$$\rho = \rho_g + \rho_b^* = \rho_g + \frac{\rho_b \delta_b}{d}, \tag{1}$$

where ρ_g is the resistivity of the grain interior, ρ_b^* *the contribution of the grain boundaries* to the *bulk resistivity*, d the grain size, ρ_b the actual grain boundary resistivity (the so-called "specific grain boundary resistivity"), and δ_b (= 1–10 nm) the grain boundary width including the space charge zone. An important feature of impedance spectroscopy is that it provides a measure of both the bulk resistivity and the grain boundary contribution. Employing this technique along with

(1) has given values of ρ_b that are one-to-three orders of magnitude greater than those of ρ_g [1–3].

Impedance spectroscopy measurements on YSZ are difficult to perform at high temperatures and have been limited for the most part to temperatures below ~1000°C. In this paper, we will present a method for determining ρ_b at higher temperatures. The method is based on measuring the grain size and concurrent electric current in isothermal annealing tests with a dc applied electric field. The method also provides the magnitude of the electric field acting across the grain boundary width.

2. Experimental

Test specimens were prepared from 3 mol% yttria-stabilized zirconia (3Y-TZP) powder purchased from Tosoh, having initial grain size d_0 = 26 nm and the following chemical composition in wt% (see Table 1).

The as-received powder was cold-compacted (98 MPa) in a stainless steel die having a dog-bone-shaped cavity. The binder was removed by heating the compacted powder in air to 800°C in 72 hr and then holding at this temperature for 1 hr. Following binder burnout, the lower tab of the specimen was cutoff giving the geometry shown in Figure 1, which was then

TABLE 1

Y_2O_3	Al_2O_3	SiO_2	Fe_2O_3	Na_2O	I_g loss	ZrO_2
5.21	<0.005	0.005	0.002	0.022	0.73	Remainder

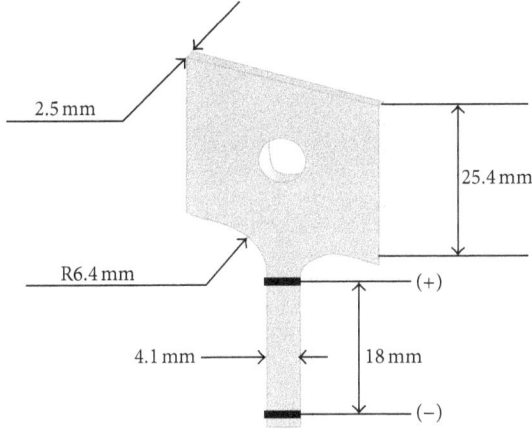

FIGURE 1: Schematic of test specimens geometry and the electric connections.

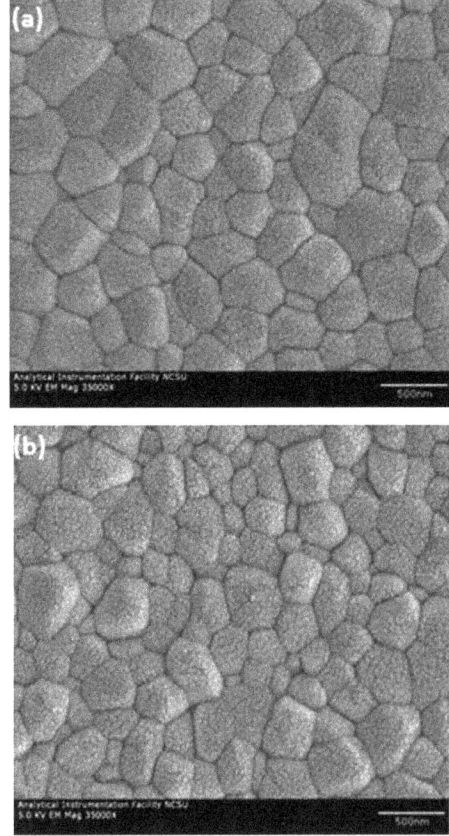

FIGURE 2: SEM micrographs taken at the middle location of specimens annealed 12 hr at 1400°C: (a) $E = 0$ and (b) $E_a = 18.1$ V/cm.

sintered by heating in air at a rate of 11°C/min to 1400 ± 1°C both without and with a constant dc voltage $V = 25 \pm 0.1$ V. The sintered specimens were then directly annealed at this temperature for various times without and with the applied dc electric field, the magnitude of which due to the sintering shrinkage had increased from 13.9 to 18.1 V/cm. The relative density ρ_r at the beginning and throughout the isothermal annealing was determined from measurements of the length change due to the shrinkage of the specimen during sintering and subsequent annealing.

The electric field was applied to the specimen by wrapping fine Pt wire (0.13 mm dia.) around each end of the gage section as shown in Figure 1. Along with the voltage, the electric current I during the anneal was measured to an accuracy of 1 mA. The corresponding current density j was determined by taking

$$j = \frac{I}{A_j} = \frac{I}{\left[A_0(1+\epsilon)^2 \rho_r\right]}, \tag{2}$$

where A_j is the effective cross-section area for the current passage, A_0 is the specimen cross-section area following bake-out, the quantity $(1+\epsilon)$ is the reduction due to shrinkage with ϵ the relative longitudinal contraction, and ρ_r accounts for the reduction in cross-section area due to the porosity.

The magnitude of Joule heating ΔT_J during the annealing with electric field was estimated from the relationship which considers the heat loss by black body radiation, namely [8],

$$\frac{\Delta T_J}{T} = \frac{W_j}{\left(4A_S \sigma T^4\right)}, \tag{3}$$

where W_j is the electric energy in watts, $A_S = 8.7 \times 10^{-5}$ m^2 is the surface area of the specimen between the electrodes, and $\sigma = 5.67 \times 10^{-8}$ W m^{-2} K^{-4} is the Stefan-Boltzmann

constant. Taking $W_j = 5$ watts at $t_a \approx 0$ hr and 8.3 watts at $t_a \approx 12$ hr ($V = 25$ V, with $I = 0.22$ A and 0.33 A, resp.), (2) gives $\Delta T_J = 54$ K and 89 K, respectively, which are ~3% and ~5% of the 1673 K annealing temperature. In view of the heat loss by conduction from the region between the electrodes to the specimen tab (see Figure 1) and by conduction and convection to the surrounding air, it is expected that ΔT_J will be appreciably less than that given by (3). A temperature rise of only ~5°C was measured by a Pt-PtRh thermocouple contacting the midsection between the two electrodes.

Following each scheduled annealing time the specimen was furnace-cooled to room temperature. To determine the grain size, cross-sections were taken at three locations: (a) ~5 mm below the upper positive (+) electrode, (b) midway between the two electrodes, and (c) ~5 mm above the lower negative (−) electrode. The cross-sections were mechanically polished, thermally etched (1 hr at 1200°C), and observed by scanning electron microscopy (SEM). An example of the microstructure without and with field is shown in Figure 2. Approximately 200 linear intercept measurements with a resolution of 5 nm were made on each of the three SEM micrographs for a specific annealing time, giving a total of ~600 measurements for each annealing time. The reported grain size is the mean linear intercept value $\bar{d}$ for the combined three locations.

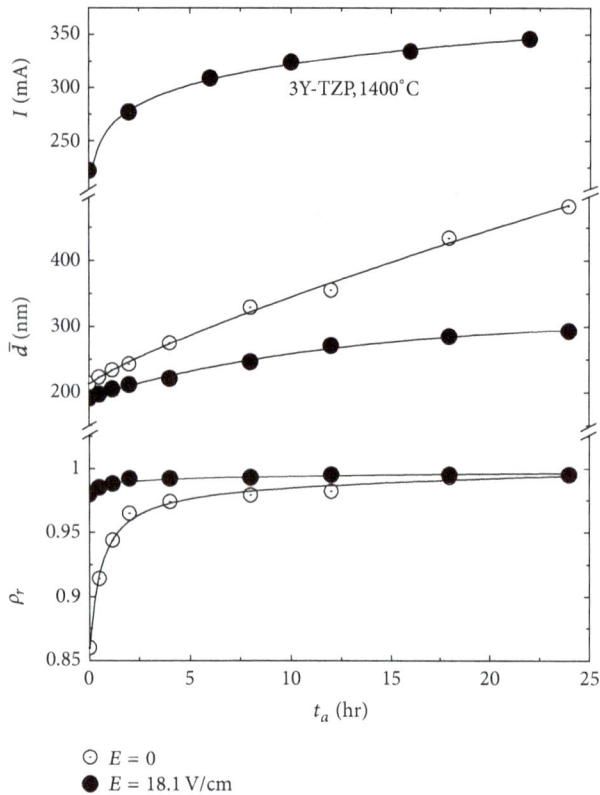

FIGURE 3: Relative density ρ_r, mean linear intercept grain size $\bar{d}$, and corresponding electric current I versus time t_a for annealing 3Y-TZP without and with an externally applied electric field E = 18.1 V/cm.

3. Results and Analysis

Figure 3 presents the mean linear intercept grain size $\bar{d}$ for the combined (three) locations along the gage section versus the annealing time t_a for the tests both without and with the applied electric field E_a. The variation in $\bar{d}$ between the three locations was within 10%. There was however no consistent variation from top to bottom (positive to negative electrode). To be noted in Figure 3 is that the grain growth curves are parabolic in shape and that the curve for annealing with field lies below that without, that is, the field retarded grain growth throughout the annealing. Included in Figure 3 are the corresponding relative density ρ_r (= measured density divided by 6.03 g/cm^3) and the electric current I as a function of the annealing time. The form of the I versus t_a curve is also parabolic. Without field ρ_r increases from 86% to 97% within 2 hrs and then increases more gradually to 99.5% at 24 hr. With field ρ_r increases gradually from 98.0% to 99.5% over the 24 hr period.

In keeping with (1) a plot of the bulk resistivity $\rho = E_a/j$ (j calculated employing (2)) versus the reciprocal of the three-dimensional average grain size d_{3D} (= $1.78\,\bar{d}$ for tetrakaidecahedron [9]) for the present tests is given in Figure 4. Included for comparison are extrapolated values from Arrhenius plots of the resistivity ρ for 3Y-TZP polycrystals obtained by impedance spectroscopy (IS) [10–12] at lower temperatures (from maximum temperatures of

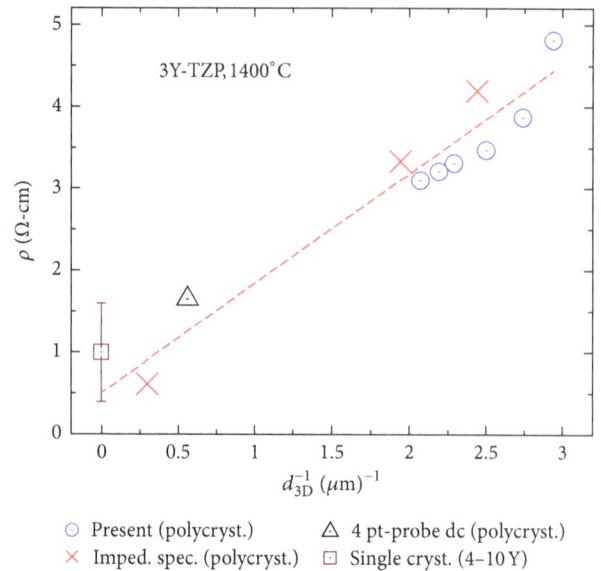

FIGURE 4: Bulk resistivity ρ versus the reciprocal of the 3D grain size d_{3D} for present results, impedance spectroscopy measurements [10–12], and 4-point-probe dc measurements [13] on polycrystals and impedance spectroscopy [14, 16, 17] and 4-point-probe measurements [14, 15] on single crystals.

~1323 K, ~833 K, and ~773 K, resp.) and 4-point-probe dc measurements [13] (from maximum temperature ~1253 K). Also included in Figure 4 are the range and the average values of ρ ($\equiv \rho_g$) for YSZ single crystals with 4–10 mol% yttria employing both 4-point-probe [14, 15] and IS [14, 16, 17] methods. To be noted in Figure 4 is that there exists reasonable agreement between the present results for the bulk resistivity of polycrystalline 3 Y-TZP and those obtained by IS and conventional 4-point-probe methods. Moreover, the combined results give a reasonable fit to a straight line whose intercept is in accordance with the value of ρ ($\equiv \rho_g$) for single crystals. The least-squares values of the intercept and slope of the straight line for the combined measurements on polycrystals are 0.51 (ohm-cm) and 1.33 (ohm-cm) (μm), respectively, with a correlation coefficient of 0.96.

According to (1) the slope of the line in Figure 4 equals the product $\rho_b \delta_b$. Taking δ_b = 10 nm (based on the segregation of yttria to the grain boundaries [18]) gives ρ_b = 133 ohm-cm. Further, taking ρ_g = 0.51 ohm-cm gives for the ratio of the grain boundary resistivity to that in the grain interior ρ_b/ρ_g = 261, which again is in accordance with that reported for impedance spectroscopy measurements [1–3].

Knowing the magnitudes of j and ρ_b one can calculate the value of the electric field E_b across the grain boundary width by employing the conventional relation

$$E_b = \rho_b j. \qquad (4)$$

Taking ρ_b = 133 ohm-cm and the values of j determined from the measured values of the current I, one obtains the magnitude of E_b and the ratio E_b/E_a as a function of grain size presented in Figure 5. The magnitudes of E_b and in turn the ratio E_b/E_a increase from 497 V/cm to 771 V/cm and from 27.5 to 42.6, respectively, with increase in d_{3D} from 0.340 to

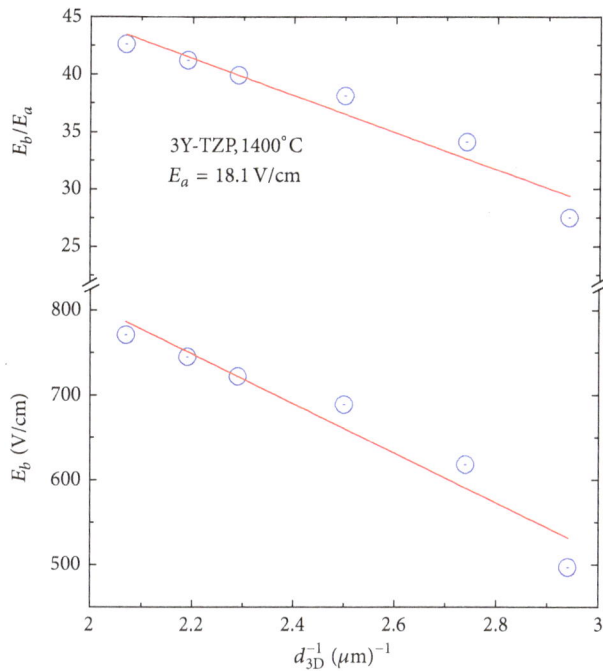

FIGURE 5: Electric field across the grain boundary width E_b and the ratio E_b/E_a versus d_{3D}^{-1} for the present tests.

0.483 μm with $E_a = 18.1$ V/cm and the corresponding values of the electric current.

4. Summary and Conclusions

The bulk resistivity ρ of 3 mol% yttria-stabilized zirconia polycrystals (3Y-TZP) was determined from the effect of an applied dc field on grain growth and the corresponding electric current in isothermal annealing tests at 1400°C. Employing the brick layer model, the results give for the magnitude of the grain boundary resistivity $\rho_b = 133$ Ω-cm, which in turn gives that the electric field across the grain boundary width $E_b \approx 28$–43 times that of the applied electric field for the present test conditions. These values of ρ and ρ_b are in reasonable accordance with values obtained by impedance spectroscopy at lower temperatures.

Acknowledgments

J. Wang and H. Conrad were funded by NSF Grant no. DMR-1002751, Dr. Lynnette Madsen, Manager Ceramics Program, Materials Science Division. D. Yang and R. Raj were funded by the Division of Basic Sciences, Department of Energy, Grant no. DE-F602.ER46403. H. Conrad acknowledges stimulating discussions with Professor H. Näfe, University of Stuttgart.

References

[1] M. C. Martin and M. L. Mecartney, "Grain boundary ionic conductivity of yttrium stabilized zirconia as a function of silica content and grain size," *Solid State Ionics*, vol. 161, no. 1-2, pp. 67–79, 2003.

[2] S. Hui, J. Roller, S. Yick et al., "A brief review of the ionic conductivity enhancement for selected oxide electrolytes," *Journal of Power Sources*, vol. 172, no. 2, pp. 493–502, 2007.

[3] S. H. Chu and M. A. Seitz, "The ac electrical behavior of polycrystalline ZrO$_2$-CaO," *Journal of Solid State Chemistry*, vol. 23, no. 3-4, pp. 297–314, 1978.

[4] J. E. Bauerle, "Study of solid electrolyte polarization by a complex admittance method," *Journal of Physics and Chemistry of Solids*, vol. 30, no. 12, pp. 2657–2670, 1969.

[5] H. Näfe, "Ionic conductivity of ThO$_2$ and ZrO$_2$-based electrolytes between 300 and 2000 K," *Solid State Ionics*, vol. 13, no. 3, pp. 255–263, 1984.

[6] R. Gerhardt and A. Nowick, "Grain-boundary effect in ceria doped with trivalent cations. I, electrical measurements," *Journal of the American Ceramic Society*, vol. 69, no. 9, pp. 641–646, 1986.

[7] J. MacDonald, *Impedance Spectroscopy: Emphasizing Materials and Systems*, John Wiley & Sons, New York, NY, USA, 1987.

[8] D. Yang, R. Raj, and H. Conrad, "Enhanced sintering rate of zirconia (3Y-TZP) through the effect of a weak dc electric field on grain growth," *Journal of the American Ceramic Society*, vol. 93, no. 10, pp. 2935–2937, 2010.

[9] J. H. Han and D. Y. Kim, "Analysis of the proportionality constant correlating the mean intercept length to the average grain size," *Acta Metallurgica Et Materialia*, vol. 43, no. 8, pp. 3185–3188, 1995.

[10] S. P. S. Badwal, F. T. Ciacchi, and V. Zelizko, "The effect of alumina addition on the conductivity, microstructure and mechanical strength of zirconia—yttria electrolytes," *Ionics*, vol. 4, no. 1-2, pp. 25–32, 1998.

[11] T. Uchikoshi, Y. Sakka, and K. Hiraga, "Effect of silica doping on the electrical conductivity of 3 mol% yttria-stabilized tetragonal zirconia prepared by colloidal processing," *Journal of Electroceramics*, vol. 4, no. 1, pp. 113–120, 1999.

[12] X. Guo and J. Maier, "Grain boundary blocking effect in zirconia: a schottky barrier analysis," *Journal of the Electrochemical Society*, vol. 148, no. 3, pp. E121–E126, 2001.

[13] Y. Shiratori, F. Tietz, H. J. Penkalla, J. Q. He, Y. Shiratori, and D. Stöver, "Influence of impurities on the conductivity of composites in the system (3YSZ)$_{1-x}$(MgO)$_x$," *Journal of Power Sources*, vol. 148, no. 1-2, pp. 32–42, 2005.

[14] S. Ikeda, O. Sakurai, K. Uematsu, N. Mizutani, and M. Kato, "Electrical conductivity of Yttria-Stabilized zirconic single crystals," *Journal of Materials Science*, vol. 18, pp. 32–42, 2005.

[15] J. D. Solier, I. Cachadina, and A. Dominguez-Rodriguez, "Ionic Conductivity of ZrO$_2$-12 mol % Y$_2$O$_3$ single crystals," *Physical Review B*, vol. 48, pp. 3704–3712, 1993.

[16] A. Pimenov, J. Ullrich, P. Lunkenheimer, A. Loidl, and C. H. Rüscher, "Ionic conductivity and relaxations in ZrO$_2$-Y$_2$O$_3$ solid solutions," *Solid State Ionics*, vol. 109, no. 1-2, pp. 111–118, 1998.

[17] I. Kosacki, C. M. Rouleau, P. F. Becher, J. Bentley, and D. H. Lowndes, "Nanoscale effects on the ionic conductivity in highly textured YSZ thin films," *Solid State Ionics*, vol. 176, no. 13-14, pp. 1319–1326, 2005.

[18] K. Matsui, H. Yoshida, and Y. Ikuhara, "Grain-boundary structure and microstructure development mechanism in 2~8 mol% yttria-stabilized zirconia polycrystals," *Acta Materialia*, vol. 56, no. 6, pp. 1315–1325, 2008.

9

Study of Gamma Ray Exposure Buildup Factor for Some Ceramics with Photon Energy, Penetration Depth and Chemical Composition

Tejbir Singh,[1] Gurpreet Kaur,[2] and Parjit S. Singh[3]

[1] *Department of Physics, Sri Guru Granth Sahib World University, Fatehgarh Sahib, Punjab 140407, India*
[2] *Department of Physics, Maharishi Markandeshwar University, Mullana, Haryana 133207, India*
[3] *Department of Physics, Punjabi University, Patiala, Punjab 147002, India*

Correspondence should be addressed to Tejbir Singh; dr.tejbir@gmail.com

Academic Editor: Baolin Wang

Gamma ray exposure buildup factor for some ceramics such as boron nitride (BN), magnesium diboride (MgB_2), silicon carbide (SiC), titanium carbide (TiC) and ferrite (Fe_3O_4) has been computed using five parametric geometric progression (G.P.) fitting method in the energy range of 0.015 to 15.0 MeV, up to the penetration of 40 mean free path (mfp). The variation of exposure buildup factors for all the selected ceramics with incident photon energy, penetration depth, and chemical composition has been studied.

1. Introduction

The recent nuclear reactor explosion in Japan emphasized the dire need of systematic and precise studies of dosimetric parameters of different type of materials. In the nuclear reactor, multienergetic photons were released, and for protection from these highly penetrating radiations, thick walls of concrete were built around the nuclear reactor. However, in case of nuclear accident, these highly penetrating radiations can travel longer distances and can cause harm to living organisms. In such a situation, the extent to which building materials can provide shielding from these harmful radiations is of utmost concern. Keeping this in mind, an attempt has been made to visualize the interaction of photons with one of the building material, namely, ceramics.

Ceramics are the composite materials in which the mechanical properties such as strength, modulus, toughness, wear resistance, and hardness are of primary interest. Despite possessing the strength and modulus values which are equal to or better than metals, these materials have chemical inertness and brittle fracture behavior. Considering such properties, ceramics have been selected to visualize the feasibility of using these materials as gamma ray shielding material.

The intensity of a gamma rays beam follows Lambert-Beer law ($I = I_o e^{-\mu x}$) under three conditions which are (i) monochromatic radioactive source, (ii) thin absorbing material, and (iii) narrow beam geometry that should be used. In case, any of the three conditions has been violated, this law no longer holds. However, violation of the law can be maintained using the correction factor B, which is known as buildup factor. Different researchers have conducted experimental and theoretical studies in different type of materials. Several methods (geometric progression (G.P.) fitting method [1–3] and invariant embedding method [4]) have been used for the computation of buildup factors for different materials in different geometrical situations.

American Nuclear Society [2] provided a comprehensive set of standard data for exposure buildup factor which includes twenty three elements, two compounds, and one mixture in the energy range of 0.015 to 15.0 MeV and up to the penetration depth of 40 mfp.

In our previous works [5, 6], the various types of buildup factors and different methods/codes available to compute the buildup factor have been already discussed. Recently, different researchers had contributed in providing gamma ray buildup factor data for different materials such as for thermoluminescent dosimetric materials [7], flyash concretes [8], human tissue [9], teeth [10, 11], some essential amino acids, fatty acids, and carbohydrates [12], and samples from the earth, moon, and mars [13].

In the present work, G.P. fitting method has been adopted to compute exposure buildup factors at some incident photon energies in the range of 0.015 to 15 MeV with penetration depth up to 40 mfp for some ceramics.

2. Computational Work

The computational work of exposure buildup factor for the selected ceramics has been divided into three parts. The first part deals with the computation of equivalent atomic number (Z_{eq}) for the selected ceramics in the energy region of 15.0 keV to 15.0 MeV. The second part concerns with the computation of G.P. fitting parameters, and finally in the third part, exposure buildup factor values have been computed in the same energy region.

2.1. Computations of Equivalent Atomic Numbers (Z_{eq}). For the computation of Z_{eq}, the values of Compton partial attenuation coefficient (μ_{Comp}) and the total attenuation coefficients (μ_{total}) were obtained in cm^2/g for the selected ceramics in the energy range of 0.015 to 15.0 MeV using WinXCom program [14]. The values of Z_{eq} for the selected ceramics were computed by matching the ratio R (μ_{Comp}/μ_{total}) of a particular ceramics at a selected energy with the corresponding ratio of an element at the same energy. In case the value of ratio lies between two ratios for known successive elements, the value Z_{eq} was then interpolated using the following logarithmic interpolation formula [6]:

$$Z_{eq} = \frac{Z_1 \left(\log R_2 - \log R\right) + Z_2 \left(\log R - \log R_1\right)}{\left(\log R_2 - \log R_1\right)}, \quad (1)$$

where Z_1 and Z_2 are the atomic numbers of elements corresponding to the (μ_{Comp}/μ_{total}) ratios, R_1 and R_2, respectively, and R (μ_{Comp}/μ_{total}) is the ratio for the selected ceramic at a particular energy, which lies between ratios R_1 and R_2.

2.2. Computations of G.P. Fitting Parameters. American National Standards [2] provided the exposure G.P. fitting parameters of 23 elements ($_4$Be-$_8$O, $_{11}$Na-$_{16}$S, $_{18}$Ar-$_{20}$Ca, $_{26}$Fe, $_{29}$Cu, Mo, Sn, La, Gd, W, $_{82}$Pb, and $_{92}$U), one compound (water), and two mixtures (air and concrete) in the energy range of 0.015 to 15.0 MeV and up to a penetration depth of 40 mfp. The computed values of Z_{eq} for the selected ceramics were used to interpolate G.P. fitting parameters (b, c, a, X_k, and d) for the exposure buildup factor using the following logarithmic interpolation formula [6]:

$$P = \frac{P_1 \left(\log Z_2 - \log Z_{eq}\right) + P_2 \left(\log Z_{eq} - \log Z_1\right)}{\log Z_2 - \log Z_1}, \quad (2)$$

where Z_1 and Z_2 are the elemental atomic numbers between which the equivalent atomic number Z_{eq} of the chosen ceramic lies. P_1 and P_2 are the values of G.P. fitting parameters corresponding to the atomic numbers Z_1 and Z_2, respectively, at a given energy. Using the interpolation formula, G.P. fitting parameters for exposure buildup factors were computed at the selected incident photon energies for the chosen ceramics.

2.3. Computations of Buildup Factors. The computed G.P. fitting parameters (b, c, a, X_k, and d) were used to compute the exposure buildup factors for the selected ceramics in incident photon energy range of 0.015 to 15.0 MeV and up to the penetration depth of 40 mfp using following equations [2–4]:

$$B(E, x) = 1 + \frac{b-1}{K-1}\left(K^x - 1\right), \quad \text{for } K \neq 1,$$
$$B(E, x) = 1 + (b-1)x, \quad \text{for } K = 1, \quad (3)$$

where

$$K(E, x) = cx^a + d\frac{\tanh\left(x/X_k - 2\right) - \tanh\left(-2\right)}{1 - \tanh\left(-2\right)}, \quad (4)$$
$$\text{for } x \leqslant 40 \text{ mfp}.$$

3. Results and Discussion

The variation of exposure buildup factor with the incident photon energy in the range of 0.015 to 15.0 MeV has been shown in Figure 1 for BN at some of the penetration depths (1, 5, 10, 20, 30, and 40 mfp). For the fixed penetration depth of 1 mfp, at lower incident photon energy (0.015 MeV), the value of energy absorption buildup factor is small, and it increases with the increase in incident photon energy, reaches a maximum value in the intermediate energy region, and after that starts decreasing with the further increase in the incident photon energy.

In lower and higher energy regions photo-electric and pair productions are most dominant processes (in which complete absorption of photon takes place), which result in minimum value of buildup factor. While in the intermediate energy region, Compton scattering is the dominant photon interaction process, which results only in the energy degradation of the photon and not the complete absorption. Hence the photons will pile up and give rise to peak. Similar trend has been observed at higher penetration depths of the ceramic.

The dominant range of Compton scattering process can be expressed in the range of $E_{photo-Comp}$ and $E_{Comp-pair\ prod}$ where $E_{photo-Comp}$ represents the energy for which both photoelectric absorption and Compton scattering show almost equal value for mass attenuation coefficient. Whereas $E_{Comp-pair\ prod}$ represents the energy for which both Compton scattering and pair production processes show equal dominance. For boron nitride (least Z_{eq} ceramic), the value of $E_{photo-Comp}$ is about 23 keV (the corresponding value of mass attenuation coefficient (μ_m) at which energy is 0.160 cm^2/g)

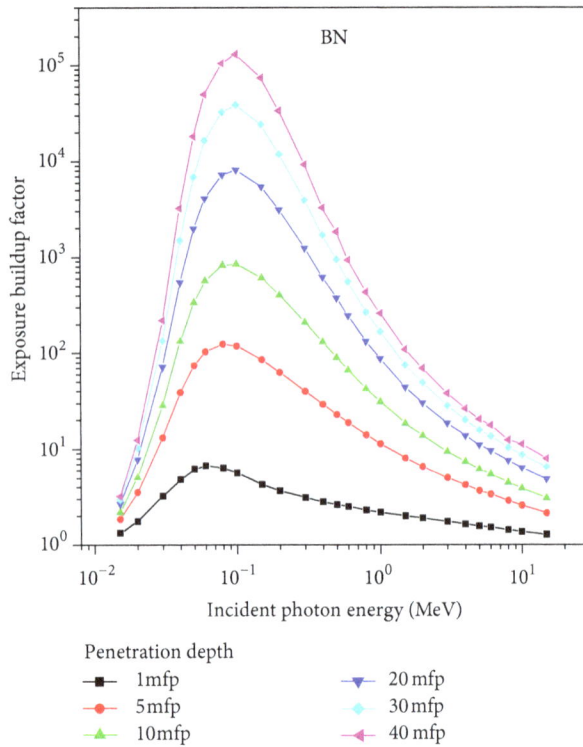

FIGURE 1: Variation of exposure buildup factor with incident photon energy for BN.

and the value of $E_{\text{Comp-pair prod}}$ is 28 MeV (the corresponding value of mass attenuation coefficient (μ_m) at which energy is 0.68×10^{-2} cm^2/g). Whereas for ferrite (highest Z_{eq} ceramic), the value of $E_{\text{photo-Comp}}$ is about 100 keV (the corresponding value of mass attenuation coefficient (μ_m) at which energy is 0.14 cm^2/g), and the value of $E_{\text{Comp-pair prod}}$ is 12 MeV (the corresponding value of mass attenuation coefficient (μ_m) at which energy is 0.13×10^{-1} cm^2/g). Since, the exposure buildup factor is the result of multiple Compton scattering processes, hence the study of buildup factor is of utmost importance between the values of $E_{\text{photo-Comp}}$ and $E_{\text{Comp-pair prod}}$. Similarly, for other ceramics like magnesium diboride, silicon carbide, titanium carbide, and ferrite, the dominant range for Compton scattering process lies in between 15.0 keV and 15.0 MeV.

Further, it has been also observed that the ceramics with low Z_{eq} show large Compton scattering dominant range (as in case of BN), whereas for ceramics with comparatively high Z_{eq}, Compton scattering dominance region is less (as in cases of ferrite and titanium carbide). Exposure buildup factor as a function of penetration depth up to 40 mean free path for the selected ceramics has been shown in Figure 2 for BN at some of the selected incident photon energies (0.015, 0.10, 0.50, 5.00, and 15.0 MeV). It has been observed that at all the selected energies, exposure buildup factor increases with the increase in penetration depth of BN. It may be due to the reason that as thickness of ceramic increases, the probability of multiple Compton scatterings also increases, and hence

the exposure buildup factor increases. Similar trend has been observed for other ceramics.

However, the increasing rate was found to be slow for lower and higher incident photon energies, and rapid increase was observed in case of intermediate energy region. The slower increasing rate in the lower and higher energy regions was due to the dominance of different photon absorption processes in these energy regions (photoelectric effect in the lower energy region and pair production in the higher energy region) which results in the complete absorption of gamma photons in the interacting medium, whereas in the intermediate energy region the dominant process is the Compton scattering, which results only in the energy degradation of photons. Hence, there is a finite possibility of the photon to reach the detector even for the large penetration depths of the ceramics, and hence maximum violation of Lambert-Beer equation has been observed.

Further, the increasing rate of exposure buildup factor with the penetration depth is more rapid up to the certain incident photon energy (0.1 MeV), where the Compton scattering process is most dominant process, and after this the increasing rate of exposure buildup factor becomes slower for higher energies.

All the selected ceramics have different chemical composition and hence different equivalent atomic number (Z_{eq}). So, to study the chemical composition dependence of different ceramics on exposure buildup factor, exposure buildup factor for all the selected ceramics has been plotted against the incident photon energy at fixed penetration depths of 1, 5, 10, and 40 mfp and has been shown in Figures 3, 4, 5, and 6. From these figures, it has been observed that for all the selected ceramics, exposure buildup factor values are small at lower incident photon energies as well as higher incident photon energies and show maximum values in the intermediate energy region. It may be due to the same reason of dominance of different partial photon interaction processes in different energy regions. Among the selected ceramics, ferrite (highest Z_{eq}) shows the minimum value for the exposure buildup factor, whereas maximum values are observed for boron nitride (lowest Z_{eq}). It may be due to the reason that ferrite, which is a ceramic of oxygen ($Z = 8$, weight fraction = 0.30) and iron ($Z = 26$, weight fraction = 0.70), has the maximum equivalent atomic number due to the major contribution of iron. Whereas boron nitride consists of boron ($Z = 5$, weight fraction = 0.44) and nitrogen ($Z = 7$, weight fraction = 0.56) and has the minimum equivalent atomic number. From this observation, it can be concluded that exposure buildup factor is inversely proportional to the equivalent atomic number of the ceramics at lower penetration depths (below 10 mfp).

In Figure 5, for the fixed penetration depth of 10 mfp of all the selected ceramics and for incident photon energy above 3 MeV, different ceramics show almost same values for exposure buildup factor. It signifies that, above certain incident photon energy (about 3 MeV), exposure buildup factor becomes almost independent of the chemical composition of the interacting material. Further, the selected ceramics mostly follow different crystal structures such as BN, SiC, and MgB$_2$ follow hexagonal, TiC follows cubic, and Fe$_2$O$_3$ follows rhombohedral structure. Since different

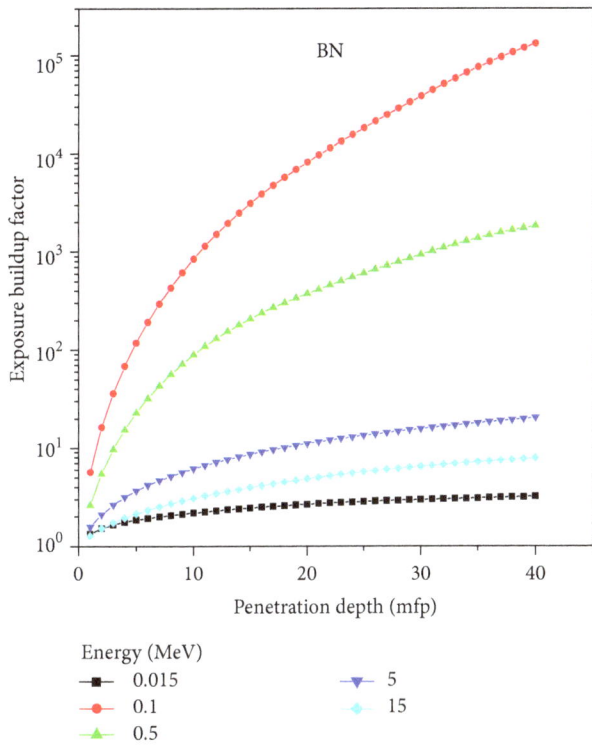

FIGURE 2: Variation of exposure buildup factor with penetration depth of BN.

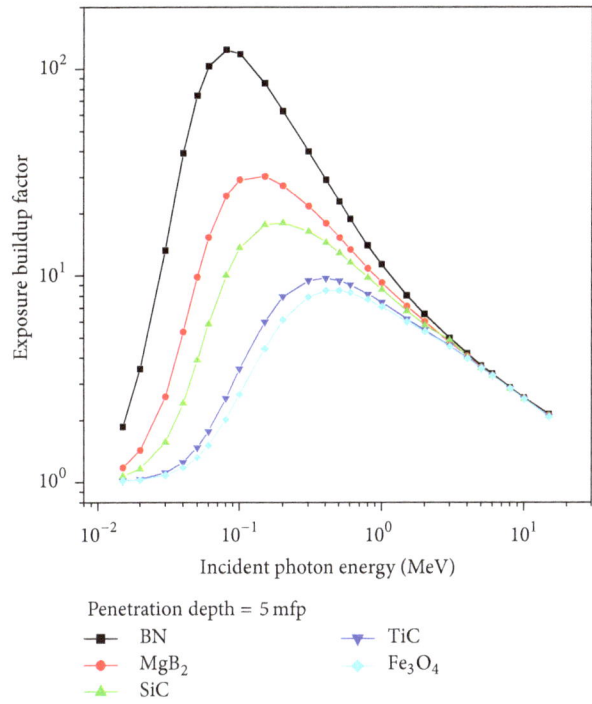

FIGURE 4: Variation of exposure buildup factor with incident photon energy for all ceramics at 5 mfp.

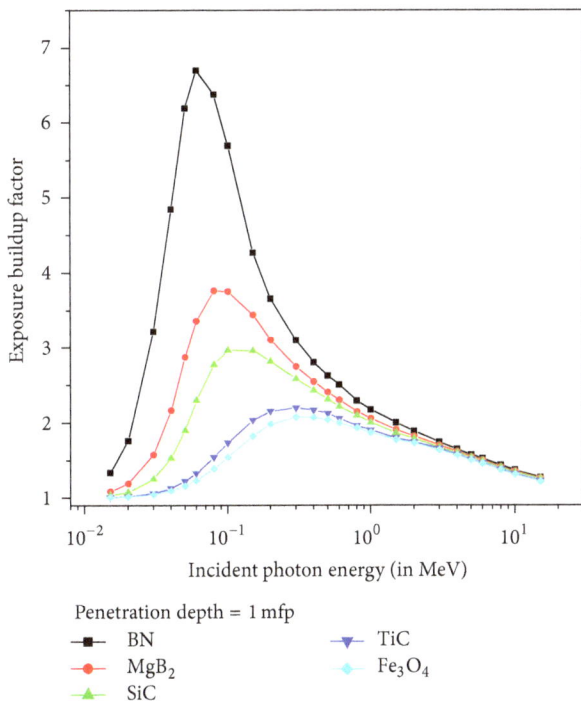

FIGURE 3: Variation of exposure buildup factor with incident photon energy for all ceramics at 1 mfp.

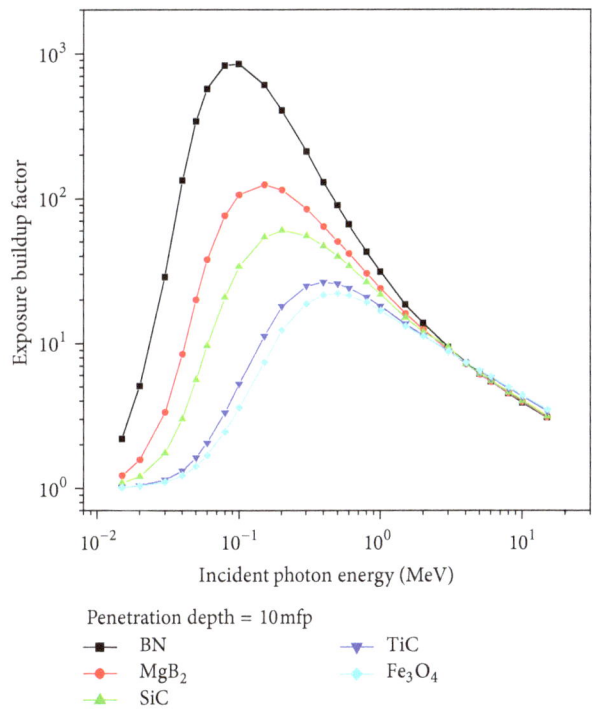

FIGURE 5: Variation of exposure buildup factor with incident photon energy for all ceramics at 10 mfp.

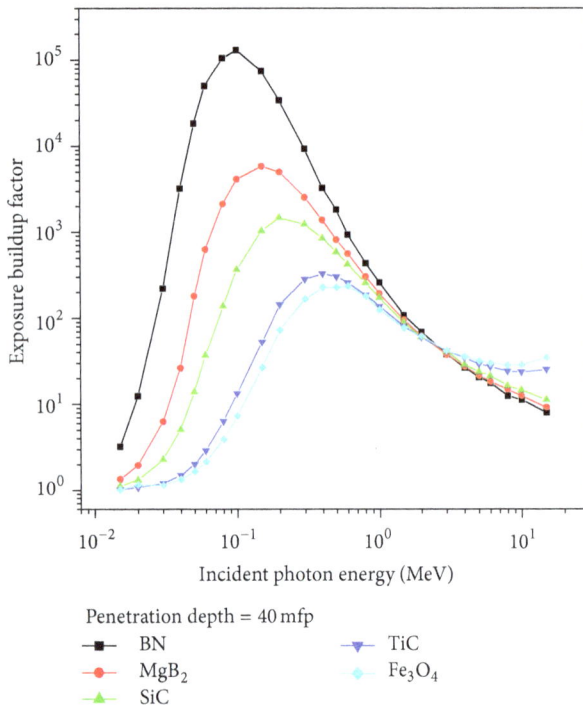

FIGURE 6: Variation of exposure buildup factor with incident photon energy for all ceramics at 40 mfp.

ceramics follow different crystal structure, the same values for exposure buildup factor above 3 MeV photon energy and at the penetration depth of 10 mfp suggest that exposure buildup factor becomes independent of crystal structure.

However, in Figure 6, which shows the variation of exposure buildup factor for all the selected ceramics at the fixed penetration depths of 40 mfp, reversal in the trend of exposure buildup factor values has been observed above 3 MeV. That is above the incident photon energy of 3 MeV, exposure buildup factor shows maximum values for ferrite (highest Z_{eq} ceramic) and minimum values for boron nitride (lowest Z_{eq} ceramic); that is, the exposure buildup factor becomes directly proportional to the equivalent atomic number of the ceramic. It may be due to the reason that pair production initiates from 1.022 MeV, and its dominance increases with the increase in photon energy, and it results in the formation of an electron and a positron. For smaller penetration depths (below 10 mean free path) of the ceramics, these particles escape either from the material or after multiple collision within the ceramic comes to rest and further annihilates, that is, creates two secondary gamma rays of 0.511 MeV, which escapes from the ceramic material. With the increase in penetration depth (above 10 mfp), these secondary gamma rays (due to annihilation) contribute in increasing the intensity of primary gamma rays and try to compensate for the decrease in primary gamma rays due to pair production. With the further increase in the penetration depth, that is, for larger penetration depths, the probability of creation of secondary gamma rays increases, and hence the contribution of these

secondary gamma rays towards exposure buildup factor also increases.

4. Conclusions

From the present studies, the following conclusions can be drawn.

(i) Exposure buildup factor increases with the increase in penetration depth (40 mfp).

(ii) Exposure buildup factor shows the following different trends with incident photon energy.

(a) For the entire energy region (0.015–15.0 MeV), in case of small penetration depths (below 10 mfp), exposure buildup factor is inversely proportional to the Z_{eq}.

(b) In the higher energy region (above 3 MeV for the selected ceramics), there exists a penetration depth (about 10 mfp in the present case), for which exposure buildup factor becomes almost independent of the Z_{eq} or the chemical composition of the ceramics.

(c) In the higher energy region (above 3 MeV), for large penetration depths (above 15 mfp), exposure buildup factor becomes directly proportional to Z_{eq}.

Among the selected ceramics, *ferrite (Fe$_3$O$_4$)* offers better gamma ray shielding.

References

[1] Y. Sakamoto, S. Tanaka, and Y. Harima, "Interpolation of gamma ray build-up factors for point isotropic source with respect to atomic number," *Nuclear Science and Engineering*, vol. 100, pp. 33–42, 1988.

[2] American National Standard Institute, "Gamma-ray attenuation coefficients and buildup factors for engineering materials," Report ANSI/ANS 6.4.3, 1991.

[3] Y. Harima, Y. Sakamoto, S. Tanaka, and M. Kawai, "Validity of the geometrical progression formula in approximating gamma-ray buildup factors," *Nuclear Science and Engineering*, vol. 94, pp. 24–35, 1986.

[4] A. Shimizu and H. Hirayama, "Calculation of gamma-ray buildup factors up to depths of 100 mfp by the method of invariant embedding," *Nuclear Science and Technology*, vol. 40, pp. 192–200, 2003.

[5] P. S. Singh, T. Singh, and P. Kaur, "Variation of energy absorption buildup factors with incident photon energy and penetration depth for some commonly used solvents," *Annals of Nuclear Energy*, vol. 35, pp. 1093–1097, 2008.

[6] T. Singh, N. Kumar, and P. S. Singh, "Chemical composition dependence of exposure buildup factors for some polymers," *Annals of Nuclear Energy*, vol. 36, no. 1, pp. 114–120, 2009.

[7] S. R. Manohara, S. M. Hanagodimatha, and L. Gerwardb, "Energy absorption buildup factors for thermoluminescent dosimetric materials and their tissue equivalence," *Radiation Physics and Chemistry*, vol. 79, no. 5, pp. 575–582, 2010.

[8] S. Singh, S.S. Ghumman, C. Singh, K. S. Thind, and G. S. Mudahar, "Buildup of gamma ray photons in flyash concretes: a study," *Annals of Nuclear Energy*, vol. 37, pp. 681–684, 2010.

[9] M. Kurudirek, B. Doğan, M. İngeç, N. Ekinci, and Y. Özdemir, "Gamma-ray energy absorption and exposure buildup factor studies in some human tissues with endometriosis," *Applied Radiation and Isotopes*, vol. 69, pp. 381–388, 2011.

[10] H. C. Manjunatha and B. Rudraswamy, "Computation of exposure build-up factors in teeth," *Radiation Physics and Chemistry*, vol. 80, no. 1, pp. 14–21, 2011.

[11] M. Kurudirek and S. Topcuoglu, "Investigation of human teeth with respect to the photon interaction, energy absorption and buildup factor," *Nuclear Instruments and Methods in Physics Research B*, vol. 269, no. 10, pp. 1071–1081, 2011.

[12] M. Kurudirek and Y. Ozdemir, "A comprehensive study on energy absorption and exposure buildup factors for some essential amino acids, fatty acids and carbohydrates in the energy range 0.015–15 MeV up to 40 mean free path," *Nuclear Instruments and Methods in Physics Research B*, vol. 269, no. 1, pp. 7–19, 2011.

[13] M. Kurudirek, B. Dogan, Y. Özdemir, A. Camargo Moreira, and C. R. Appoloni, "Analysis of some Earth, Moon and Mars samples in terms of gamma ray energy absorption buildup factors: Penetration depth, weight fraction of constituent elements and photon energy dependence ," *Radiation Physics and Chemistry*, vol. 80, pp. 354–364, 2011.

[14] L. Gerward, N. Guilbert, K. Bjørn Jensen, and H. Levring, "X-ray absorption in matter. Reengineering XCOM," *Radiation Physics and Chemistry*, vol. 60, no. 1-2, pp. 23–24, 2001.

10

Provenance Study of Archaeological Ceramics from Syria Using XRF Multivariate Statistical Analysis and Thermoluminescence Dating

Elias Hanna Bakraji, Rana Abboud, and Haissm Issa

Atomic Energy Commission, Chemistry Department, P.O. Box 609, Damascus, Syria

Correspondence should be addressed to Elias Hanna Bakraji; cscientific@aec.org.sy

Academic Editor: Guillaume Bernard-Granger

Thermoluminescence (TL) dating and multivariate statistical methods based on radioisotope X-ray fluorescence analysis have been utilized to date and classify Syrian archaeological ceramics fragment from Tel Jamous site. 54 samples were analyzed by radioisotope X-ray fluorescence; 51 of them come from Tel Jamous archaeological site in Sahel Akkar region, Syria, which fairly represent ceramics belonging to the Middle Bronze Age (2150 to 1600 B.C.) and the remaining three samples come from Mar-Takla archaeological site fairly representative of the Byzantine ceramics. We have selected four fragments from Tel Jamous site to determinate their age using thermoluminescence (TL) method; the results revealed that the date assigned by archaeologists was good. An annular ^{109}Cd radioactive source was used to irradiate the samples in order to determine their chemical composition and the results were treated statistically using two methods, cluster and factor analysis. This treatment revealed two main groups; the first one contains only the three samples M52, M53, and M54 from Mar-Takla site, and the second one contains samples that belong to Tel Jamous site (local).

1. Introduction

Analysis of archaeological ceramics can confirm the information recorded in historical documents, such as trade routes linking populations of different areas, and help to find out the chronology of events. Establishing databases of Syrian ceramics, by using many techniques, was started a few years ago.

The classification of ceramics based on typology is one of useful methods, but only when applied to whole or reconstructed objects [1, 2]. The chemical composition of the made ceramics is unique and related to sources identification of provenance [3–5], from which they were fashioned. In order to classify ceramics, we need to determine the chemical composition of a large number of samples and they should be from a single site and from a single period.

To reach this goal many techniques were applied, since the initial ceramics study by Sayre and Dodson [6], such as X-ray fluorescence (XRF) [7, 8], proton induced X-ray emission (PIXE) [9, 10], and neutron activation analysis (NAA) [3, 5, 11,

12]. We applied in our laboratory most of these techniques to study archaeological objects. The main aim of our study was to prove to archaeologists the advantage of applying physical techniques and present the effectiveness of the combination of some methods in their studies such as dating of the sites and the provenance studies of ancient ceramics. The other aim was providing new additional data to the Syrian database.

In the present study we applied

(i) thermoluminescence (TL) dating for the age determination of ceramics sherds. This technique is the only available today to determine the age of ceramics. See [13, 14] for principles and mechanism of TL;

(ii) radioisotope X-ray fluorescence spectroscopy for determining the elemental composition of the ceramics, where fourteen chemical elements were determined.

XRF is nondestructive methods and allows fast multielemental analysis.

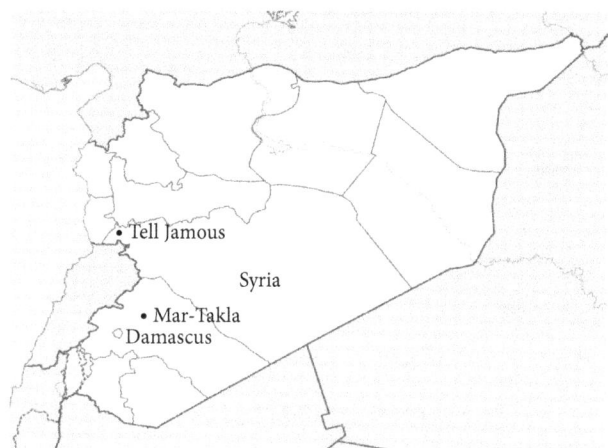

FIGURE 1: Map of Syria with the area considered.

Fifty-four samples were analyzed in this study. Fifty-one of them labeled 1–51 come from the excavation at the site of Tel Jamous in Sahel Akkar region located at the west of Syria (see Figure 1), which fairly represent ceramics belonging to the Middle Bronze Age (2150 to 1600 B.C.) according to archaeologists. The remaining three samples labeled M52–M54, set as "control" samples, come from excavation at the site of Mar-Takla, already studied in [7], located at 20 km north east of Damascus city (see Figure 1) and they are fairly representative of the Byzantine ceramics.

Four samples (samples J-A, J-B, J-C, and J-D) were chosen randomly among the 51 samples from Tel Jamous site, to be analyzed by thermoluminescence for age determination.

The data which consisted of the concentration of fourteen chemical elements have been treated statistically using two methods, cluster and factor analysis, to establish a categorization of the ceramics raw material source from Tel Jamous archaeological site. The chemical groups are assumed to present sources which could present the use of local material or production workshops.

2. Experimental

2.1. TL Experiment

2.1.1. Sample Preparation. The fine grain technique [14] was used, and the preparation procedure was carried out in subdued red light to avoid bleaching effects. We removed about 3 mm of the pottery's outer surface to eliminate the beta dose contamination from the soil.

The next step is to obtain a quantity of powder, about 250 mg, through the drill within the sample; after that we added acetone, about 60 cm^3, to the sample and waited for 2 min. We left the precipitant and took the solution. We added acetone again and waited for 20 min. In the next step, we took this time the precipitant with a small amount of acetone. We put, using micropipette, 3 mL of the solution in each of the glass tubes (1 cm of diameter and 6 cm of height). Finally we leave the tubes for the next day in an oven at 50°C. The separated grains then are allowed to deposit on aluminum discs in a thin layer of a few microns thickness which are

placed at the bottom of individual flat-bottomed glass tubes. We prepared twenty discs for each sample, the whole discs were placed on the tray, using tweezers [15], which contain 48 holes, before measurement.

2.1.2. Instrumentation and Measurements. The age of the ceramics can be calculated by the absorbed dose or Paleodose (ED) in Gy unit divided by the dose rate (DR) in mGy/a or in Gy/ka. The absorbed dose, which is called equivalent dose (ED), is related to the time in which the samples are exposed to natural radiation, and the dose rate is the dose received from natural radiation by the sample for one year (annual dose). The measurements were performed using RISØ TL/OSL reader model DA-20 at atomic energy commission of Syria. The additive dose procedure was used to determine the absorbed dose [15]. Strontium-90 (emitting beta particles) was used as the radioactive source, with a dose rate of 0.135 Gy/s. The annual dose was estimated from measurements of the radioactive elements U, Th, and K within the sample and soil and from cosmic rate. Alpha spectrometry system was used to determine uranium and thorium while atomic absorption spectrometry was used to determine the radioactive element potassium.

2.2. XRF Experiment

2.2.1. Samples Preparation. The sherds were ground, after removal of the surface deposit, using an agate mortar; this step is important to have good homogeneity of the analyzed sherds. The obtained powders were dried at 105°C for 24 hours; after that the powders were converted into pellets with a hydraulic press [16].

2.2.2. Instrumentation and Measurements. The pellets (25 mm diameter) were irradiated for 1000 sec. live time by an annular ^{109}Cd radioactive source, which has the following specifications: activity (~9 10^8 Bq), outer diameter of 2.54 cm, active diameter of 1.9 cm, and thickness of 5 mm. An X-ray spectrometer mounting a Si(Li) detector was used for the measurements of the X-ray fluorescence and the distance between sample and detector was 3 cm. The energy resolution of the detector (full-width-half-maximum) is about 180 eV at 5.9 keV for the Mn-Kα X-ray. The X-ray fluorescence data were collected and analyzed using a personal computer based on the multichannel analyzer (MCA) and a quantitative X-ray analysis system (QXAS) program from the international atomic energy agency (IAEA). The subprogram AXIL [17] was used to fit the spectra in order to calculate the net peak intensities of the Kα and Lα lines. For testing the accuracy of sensitivity curves, Soil-7 (IAEA), SL-1 (IAEA), and rock GSR-3 (China) were used as standards. The repeated analyses of several samples reveal that the relative standard deviation (RSD) was less than 5% for each determined element.

2.3. Statistical Treatment. The final data which consist of observations (samples) and variables (elements) have been treated using two statistical methods, cluster analysis (CA) and factor analysis (FA), by using Statistica 8.0 package.

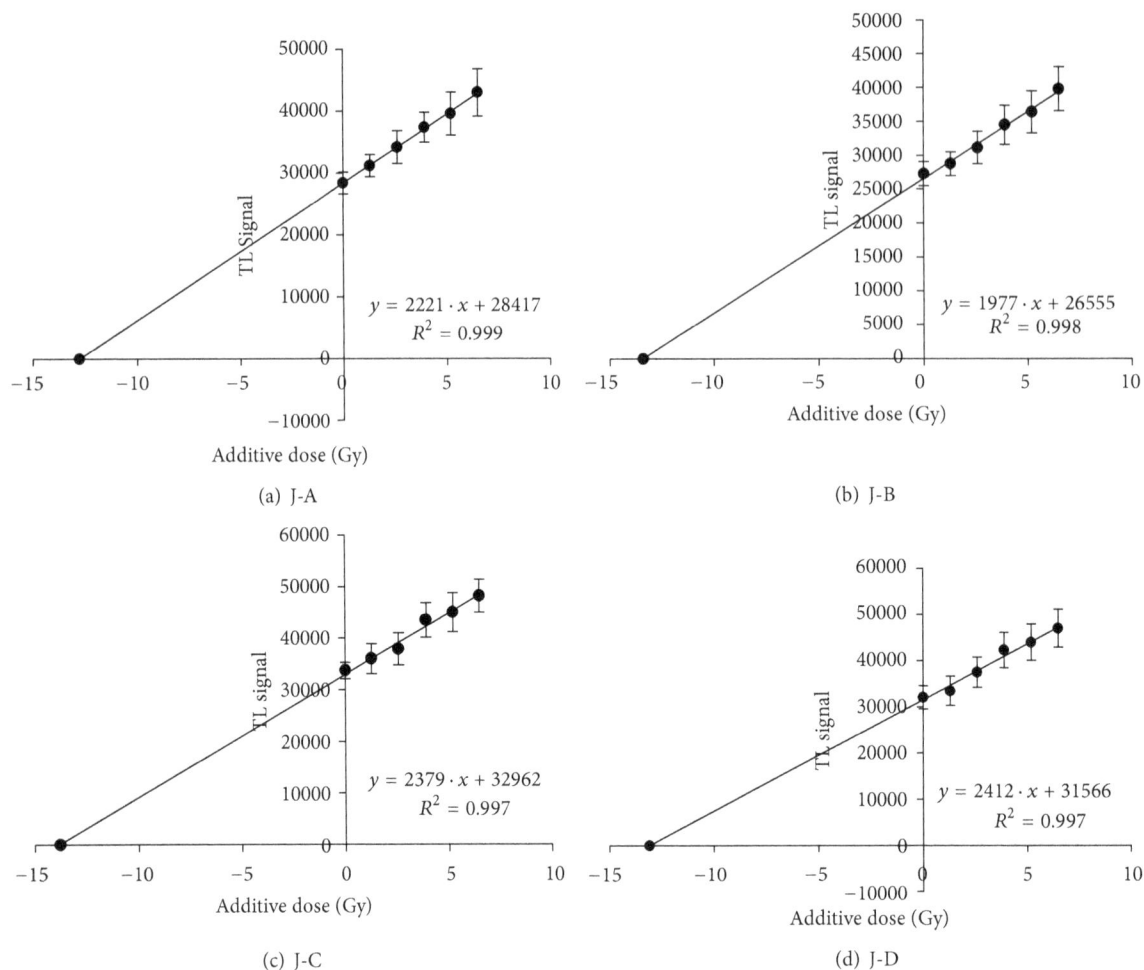

FIGURE 2: Luminescence counts versus additive dose for the four samples J-A, J-B, J-C, and J-D.

Cluster analysis (CA) classifies samples into distinct groups by calculating distance measures between the samples [18]. The results are commonly presented as dendrogram. We used single linkage as a grouping rule, according to Euclidean distances.

Using factor analysis (FA) we can extract a minimum number of factors which explains an acceptable amount of total variance of the data set. In general the two or three first factors are sufficient to reach this goal.

3. Results and Discussion

3.1. TL Results

3.1.1. Accumulated Dose (Paleodose). A growth curve was built on the basis of four additive beta doses using three discs for each dose. The graphs of additive dose versus luminescence counts were plotted. They are shown in Figure 2. The fitting-line of the graphs was as follows: $Y = 2221X + 28417$ (sample J-A), $Y = 1977X + 26555$ (sample J-B), $Y = 2379X + 32962$ (sample J-C), and $Y = 2412X + 31566$ (sample J-D). As it is known the intercept with the dose = 0 axis (X-axis) gives the absorbed dose. The values of the absorbed dose are presented in Table 1 (fifth column).

3.1.2. Annual Dose (Dose Rate). Alpha rays have the shortest range in geological materials (approximately 0.03 mm); beta rays traverse up to 3 mm in solid matter, while gamma rays penetrate about 30 cm, much greater than the dimensions of our pottery samples.

Before TL measurements are carried out on our sherds, the outer 3 mm of the sample is cut away. The remaining portion has therefore received its alpha and beta dose entirely from within the volume of the sherd. This dose, termed the internal dose, can be determined from an examination of the sample alone. Like the alpha and beta rays, gamma radiation derives from the decay of naturally occurring radionuclides present in the ground, such as potassium-40 and members of the uranium and thorium decay series.

The doses rates (annual doses) are presented in Table 1 (fourth column), where the internal contribution, Table 1 (second column), was calculated from alpha and beta activity measures coming from the samples taking into account the alpha correction factor (α-effectiveness, α-0.15) (Aitken, 1985) [14], while the external contribution, Table 1 (third column), was calculated from gamma activity coming from the soil. The conversion of alpha, beta, and gamma has been achieved according to Adamiec and Aitken [19] and the

TABLE 1: U, Th, and K concentrations with internal annual dose, external annual dose, annual dose, equivalent doses, and the ages obtained and ages assigned by archaeologists for the pottery sherds investigated.

Sample	Internal dose (mGy/a)	External dose (mGy/a)	Annual dose (mGy/a)	Archaeodose (Gy)	Age (a) B.P	Age assigned by archaeologists
J-A	2.67 ± 0.27	0.75 ± 0.06	3.42 ± 0.33	12.8 ± 0.3	3750 ± 240	1600–2150 B.C.
J-B	2.44 ± 0.26	0.75 ± 0.06	3.19 ± 0.32	13.4 ± 0.3	4200 ± 220	1600–2150 B.C.
J-C	2.76 ± 0.30	0.75 ± 0.06	3.51 ± 0.36	13.8 ± 0.5	3930 ± 270	1600–2150 B.C.
j-D	2.86 ± 0.29	0.75 ± 0.06	3.61 ± 0.35	13.1 ± 0.3	3630 ± 260	1600–2150 B.C.

The dose rate conversion factors used are given in Adamiec and Aitken [19].

cosmic rate was 0.18 Gy/ka. The calculated age before present (BP) of the sherds is presented in Table 1 (sixth column). The errors for the calculated age vary in the range of 5–10%.

We can notice from Table 1 (sixth column and last column) that the calculated ages are in accordance with the ages supposed by archaeologists.

3.2. XRF and Statistics Results. The statistical analysis was performed after elimination of the elements in the data set which have more than 25% missing values and then transform the data to base log.10. Only 12 elements were considered for the statistical analysis. The elements Cu and Ni were not included because their concentrations values are missing for more than 50% of samples analyzed.

The two elements niobium and lead have three and four missing values, respectively. The procedure used to estimate the missing values for them was to replace any missing value by the minimum detection limits (MDL) determined by XRF. The MDL are 15 ppm (niobium) and 10 ppm (lead). The final data set consisted of 54 observations (samples) and 12 variables (elements) for a total of 648 data entries.

The resulting dendrogram based on the analysis of 12 elements is shown in Figure 3. Two main clusters were found. Cluster 1 contains the three samples M52, M53, and M54 which are from Mar-Takla site and all the remaining samples, which are from Tel Jamous site, form only one group (cluster 2) except three isolated samples (2, 9, and 20).

The factor analysis (FA) was carried out on the same twelve elements used for cluster analysis. The three factors extracted in this study explain 63.8% of the total variance of the data set. Varimax method was used for rotation and maximum likelihood method was used for factor extraction.

Table 2 shows the factor loading for the three extracted factors. Factor scores quantify the relative intensities of factor strength on each sample. Samples with the same factor score patterns can be grouped together into particular categories. Figures 4 and 5 present plots of factor score 1 against factor scores 2 and 3, respectively, for each of the 54 samples. It is clear from Figures 4 and 5 that the classification is similar to the classification in the cluster analysis. Samples from Mar-Takla site identified as group 1 from the cluster analysis in Figure 3 follow a consistent pattern. The 48 samples from Tel Jamous site identified as group 2 from the CA in Figure 3 follow also a consistent pattern.

The results confirm, after statistical analysis using two methods, that there are one principal group which contains 48 samples derived from Tel Jamous site and another group which contains "control" samples, that is, samples M51, M52,

TABLE 2: Factor loading for the samples data set, twelve elements.

Elements	Factor 1	Factor 2	Factor 3
Ca	−0.31	0.84	−0.25
Fe	0.32	0.29	0.52
K	−0.10	0.91	0.02
Mn	0.31	0.04	0.72
Nb	0.25	−0.77	−0.22
Pb	0.04	−0.64	−0.31
Rb	0.35	−0.20	0.51
Sr	−0.51	0.34	−0.09
Ti	0.81	−0.19	0.34
Y	0.94	−0.25	0.03
Zn	0.47	0.27	0.39
Zr	0.70	−0.15	0.38
Total variance %	25.0	24.8	14.0

TABLE 3: Mean values and standard deviation for the two chemical groups in the pottery. All values are in μg/g; n equals the number of samples in each category.

Elements	Group 1 ($n = 3$) Mar-Takla Mean ± SD	Group 2 ($n = 48$) Tel Jamous Mean ± SD
Ca	155000 ± 10000	75000 ± 11000
Cu	39 ± 10	45 ± 13
Fe	32000 ± 3000	49000 ± 10000
K	17000 ± 1000	11000 ± 1000
Mn	260 ± 16	600 ± 67
Nb	18 ± 2	20 ± 4
Ni	58 ± 10	77 ± 26
Pb	19 ± 3	19 ± 4
Rb	21 ± 3	47 ± 7
Sr	240 ± 28	170 ± 39
Ti	4000 ± 200	7600 ± 1200
Y	22 ± 2	32 ± 4
Zn	110 ± 13	133 ± 10
Zr	195 ± 9	310 ± 41

and M53, derived from Mar-Takla site. The 48 samples of category 2 are thus considered to correspond to wares manufactured in Tel Jamous site.

Finally Table 3 presents the average elemental concentration and standard deviation for the samples in categories 1 and

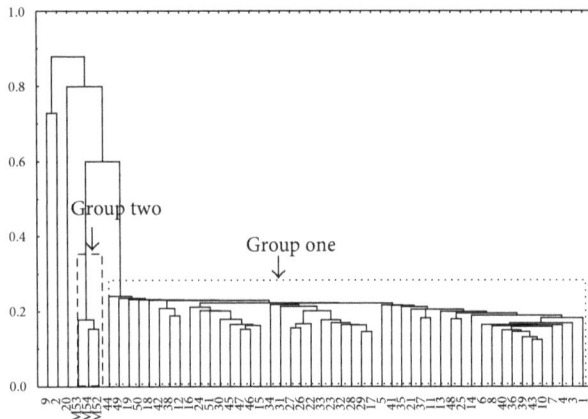

FIGURE 3: Grouping of pottery samples by cluster analysis.

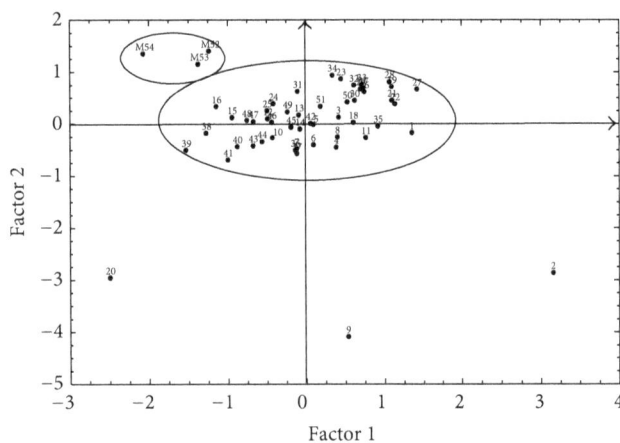

FIGURE 5: Factor score 1 against factor score 3 of ceramics samples.

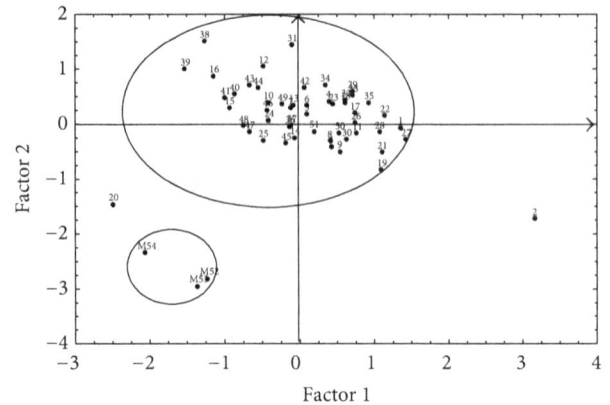

FIGURE 4: Factor score 1 against factor score 2 of ceramics samples.

2. It is clear from the standard deviation that there are large variations (scattering) in the elemental concentrations among the samples in the same category. It is clear from Table 3 that the concentrations of Ca, K, and Sr in Mar-Takla site are roughly 1.5 to 2 times higher than those for Tel Jamous site. It is also evident, from this table, that the concentrations of Fe, Mn, Rb, Ti, Y, and Zr in Tel Jamous site are higher than those in Mar-Takla site.

4. Conclusion

Thermoluminescence (TL) dating and X-ray fluorescence (XRF) analysis combined with multivariate statistical method have been utilized to analyze 51 ceramics samples from Tel Jamous site, Syria. Four samples among them were chosen for dating. Three additional ceramics samples from Mar-Takla archaeological site were used as "control" samples. The ceramics sherds of Tel Jamous date back to the Middle Bronze Age (2150 to 1600 B.C.) according to archaeologists. The date obtained by TL technique was in good agreement with the date assigned by archaeologists. Up to 14 elements were determined by XRF and the concentrations of 12 of them have been taken into consideration for statistical analysis

where two methods were applied, cluster analysis and factor analysis, in order to classify the ceramic sherds. Statistical results separate samples 2, 9, and 20, put samples M52, M53, and M54 from Mar-Takla archaeological site in an isolated group, and confirm that there is one principal group of ceramics from Tel Jamous site. Application of XRF, TL techniques, and statistical analysis has proved to be very helpful for Syrian archaeologists to study ancient ceramics. This study could be beneficial in geological studies such as sediment studies where optically stimulated luminescence (OSL) technique is used in general instead of TL technique for chronological dating t of the sediments.

Conflict of Interests

The authors declare that there is no conflict of interests regarding the publication of this paper.

Acknowledgments

The authors wish to thank the International Atomic Energy Agency (RAS/1/011) and the General Director of AEC of Syria for their support of this work and the General Director of antiquity and museum in Damascus for supplying the samples discussed in this study.

References

[1] A. P. Aruga, M. Piero, and C. Antonella, "Application of multivariate chemometric techniques to the study of Roman pottery (terra sigillata)," *Analytica Chimica Acta*, vol. 276, no. 1, pp. 197–204, 1993.

[2] C. Punyadeera, A. E. Pillay, L. Jacobson, and G. Whitelaw, "The use of correspondence analysis to compare major and trace elements for provenance studies of iron-age pottery from the Mngeni river area, South Africa," *Journal of Trace and Microprobe Techniques*, vol. 17, no. 1, pp. 63–79, 1999.

[3] E. H. Bakraji, I. Othman, A. Sarhil, and N. Al-somel, "Application of instrumental neutron activation analysis and multivariate statistical methods to archaeological Syrian ceramics," *Journal of Trace and Microprobe Techniques*, vol. 20, no. 1, pp. 57–68, 2002.

[4] M. D. Glascock, "Characterization of archaeological ceramics at MURR by neutron activation analysis and multivariate statistics," in *Chemical Characterization of Ceramic Pastes in Archaeology*, H. Neff, Ed., pp. 11–26, Prehistory, Madison, Wis, USA, 1992.

[5] F. Widemann, "Neutron activation analysis for provenance studies of archaeological artifacts," *Journal of Radioanalytical Chemistry*, vol. 55, no. 2, pp. 271–281, 1980.

[6] E. V. Sayre and R. W. Dodson, "Neutron activation study of mediterranean potsherds," *American Journal of Archaeology*, vol. 61, no. 1, pp. 35–41, 1957.

[7] E. H. Bakraji, I. Othman, and J. Karajou, "Provenance studies of archaeological ceramics from Mar-Takla (Ain-Minin, Syria) using radioisotope X-ray fluorescence method," *Nuclear Science and Techniques*, vol. 12, no. 2, pp. 149–153, 2001.

[8] C. Punyadeera, A. E. Pillay, L. Jacobson, and G. Whitelaw, "Application of XRF and correspondence analysis to provenance studies of coastal and inland archaeological pottery from the Mngeni River Area, South Africa," *X-Ray Spectrometry*, vol. 26, no. 5, pp. 249–256, 1997.

[9] N. Hagihara, S. Miono, Z. Chengzhi, Y. Nakayama, K. Hanamoto, and S. Manabe, "The combined application of PIXE analysis and thermoluminescence (TL) dating for elucidating the origin of iron manufacturing in Japan," *Nuclear Instruments and Methods in Physics Research, Section B: Beam Interactions with Materials and Atoms*, vol. 150, no. 1–4, pp. 635–639, 1999.

[10] I. E. Kieft, D. N. Jamieson, B. Rout, R. Szymanski, and A. S. Jamieson, "PIXE cluster analysis of ancient ceramics from North Syria," *Nuclear Instruments and Methods in Physics Research, Section B: Beam Interactions with Materials and Atoms*, vol. 190, no. 1–4, pp. 492–496, 2002.

[11] J. W. Beal and I. Olmez, "Provenance studies of pottery fragments from medieval Cairo, Egypt," *Journal of Radioanalytical and Nuclear Chemistry*, vol. 221, no. 1-2, pp. 9–17, 1997.

[12] E. de Sena, S. Landsberger, J. T. Pena, and S. Wisseman, "Analysis of ancient pottery from the Palatine hill in Rome," *Journal of Radioanalytical and Nuclear Chemistry*, vol. 196, no. 2, pp. 223–234, 1995.

[13] D. Curie, *Luminescence in Crystal*, John Wiley & Sons, New York, NY, USA, 1960.

[14] M. A. Aitken, *Thermoluminescence Dating*, Academic Press, 1985.

[15] A. G. Wintle, "Luminescence dating: laboratory procedures and protocols," *Radiation Measurements*, vol. 27, no. 5-6, pp. 769–817, 1997.

[16] E. H. Bakraji, "Study of Syrian archaeological pottery by the combined application of thermoluminescence (TL) dating, X-ray fluorescence analysis and statistical multivariate analysis," *Nuclear Instruments and Methods in Physics Research, Section B: Beam Interactions with Materials and Atoms*, vol. 269, no. 19, pp. 2052–2056, 2011.

[17] Quantitative X-ray analysis system, QXAS, Doc, Version 2. 0, IAEA, 2005.

[18] A. M. Bieber, D. W. Brooks, G. Harbottle, and E. V. Sayre, "Application of multivariate techniques to analytical data on Aegean ceramics," *Archaeometry*, vol. 18, no. 1, pp. 59–74, 1976.

[19] G. Adamiec and M. J. Aitken, "Dose-rate conversion factors: update," *Ancient TL*, vol. 16, no. 2, pp. 37–50, 1998.

Basic Elastic Properties Predictions of Cubic Cerium Oxide Using First-Principles Methods

Jon C. Goldsby

Glenn Research Center, National Aeronautics and Space Administration, 21000 Brookpark Road, Cleveland, OH 44135, USA

Correspondence should be addressed to Jon C. Goldsby; jon.c.goldsby@nasa.gov

Academic Editor: Young-Wook Kim

Computational material methods were used to predict and investigate electrical and structural properties of cerium oxide (CeO_2). Density functional theory was used to obtain the optimized crystal structure and simulate the material's electronic and elastic responses. Oxygen to oxygen nearest neighbor distance is 2.628 Å, while oxygen to cerium distance is calculated to be 2.276 Å. The conduction band has a prominent set of bands, which exists between 6 and 17 eV. An indirect energy gap (6.04 eV) exists between the valence and conduction bands. The independent elastic constants allow a mechanical assessment on the suitability of cubic cerium oxide as a substrate for advanced electronic devices. The calculated results of phonon dispersion curves are also given.

1. Introduction

A significant need exists to develop materials not only capable of providing desired electronic properties but also capable of surviving at extreme temperatures during device fabrication, such as in cofiring. An excellent candidate material is the oxide-based ceramic ceria. Ceria with its cubic structure can provide a platform for the introduction of dopants that can alter its electrical properties from insulating to semiconducting though oxygen vacancies [1–4]. This oxide-based ceramic material also possesses the refractory nature needed for high-temperature operation. However, the proper selection of dopants has to this point been somewhat speculative in nature based upon empirical experimentation. With the use of computational methods, material properties can be evaluated and optimized before the material is processed allowing an optimized structure to be realized quickly. This saves enormous time and efforts of experimentally assessing properties of multiple changes in composition and structure. The purpose of this investigation is to predict electronic and static and dynamic lattice response of pure ceria.

2. Computational Methods

The commercial software package Material Studio, including the Cambridge Serial Total Energy Package CASTEP module was used to conduct these simulations [5]. The calculations were performed on a parallel computer cluster consisting of 14 AMD Opteron 64-bit processors utilizing a LINUX-based operating system. Starting with the base structure, a geometric optimization was performed to determine the initial lattice parameters and density. In this calculation, the generalized gradient approximation (GGA) [6] was used along with ultrasoft pseudopotentials to represent the atoms. Specifically, for this geometric optimization, the Perdew-Burke-Ernzerhof potential for solids (PBEsol) was used [7]. To obtain accurate band structures, calculations using the GGA PBEsol, as well as the hybrid functional Becke, three-parameter, Lee-Yang-Parr (B3LYP) potentials, were used [8, 9]. The finite element displacement method applied to a supercell of the geometrically optimized primitive lattice using norm-conserving potentials [10] to calculate the phonon dispersion curves was used. The real space cutoff radius was set to 5 Å which resulted in a supercell volume 27

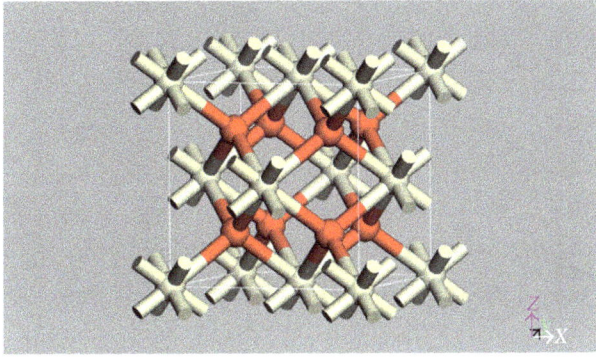

FIGURE 1: Cerium oxide Fm-3m oxygen atoms denoted in red.

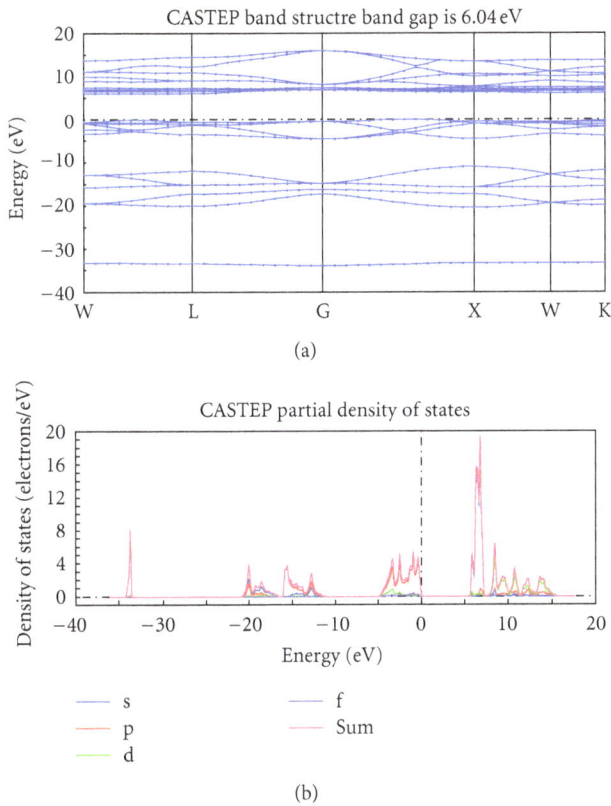

(a)

(b)

FIGURE 2: Electronic distribution in cubic ceria: band structure (a) and density of electron states (b).

times larger than the original unit cell. The quality of the q-factor spacing was set to $0.015\,\text{Å}^{-1}$.

3. Results and Discussion

3.1. Crystal Structure. The ceria structure used for these calculations possesses the Fm-3m space group as seen in Figure 1. The geometric and electronic band structure calculations do not include thermal entropy effects. Therefore, these results are interpreted to occur at absolute zero temperature. The results of the calculation yielded a lattice parameter of $5.256\,\text{Å}$ compared to their room temperature expected value

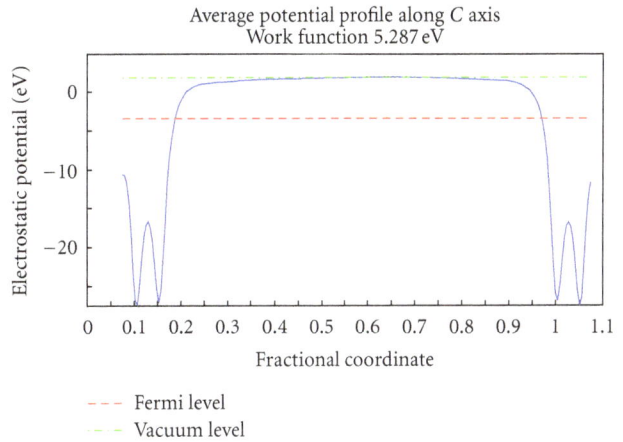

FIGURE 3: Distribution of the electrostatic potential in a vacuum between two (110) planes.

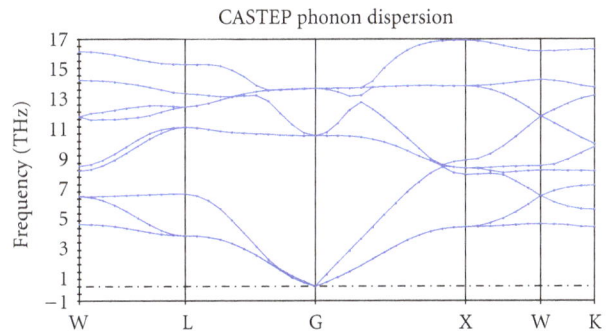

FIGURE 4: Phonon dispersion curves of cubic ceria.

of $5.410\,\text{Å}$ [11]. The cerium to cerium distance is $3.717\,\text{Å}$ while the next nearest neighbor cerium to cerium distance corresponds to the lattice parameter. The oxygen to oxygen nearest neighbor distance is $2.628\,\text{Å}$ while oxygen to cerium distance was calculated to be $2.276\,\text{Å}$. All lattice parameters remained orthogonal with the angles all being $90°$. The total volume is $145.260\,\text{Å}^3$ with an overall density of $7.870\,\text{g/cm}^3$.

3.2. Electronic Properties. The starting electronic configurations are Ce $[\text{Xe}]\,4f^1 5d^1 6s^2$ and O $[\text{He}]\,2s^2 2p^4$. Figure 2 shows the band structure for a standard ceria unit cell. Three major band groups below the Fermi level are found centered at the following positions at $-34, -15,$ and $-2\,\text{eV}$, respectively, while other major band groups are located at 4 and $10\,\text{eV}$ above the Fermi level. The indirect band gap in the G-X direction has a value of $6.04\,\text{eV}$ which is in agreement with the experimental value [12]. The valence band begins at -5 to $-1\,\text{eV}$ and is dominated by the p orbital associated with the oxygen atoms. The lower band structures reveal large effective masses, due to the high radius of curvature of the bands, thereby prohibiting the promotion of carriers from below the Fermi level. The valence band consists of a spdf hybrid with dominant p character. In addition, the density of states for these bands is relatively small at $5\,\text{eV}$ per volt. At levels below $-10\,\text{eV}$ the charges are bound, consisting of

CASTEP thermodynamic properties
Zero point energy = 0.1783406296 eV

CASTEP thermodynamic properties

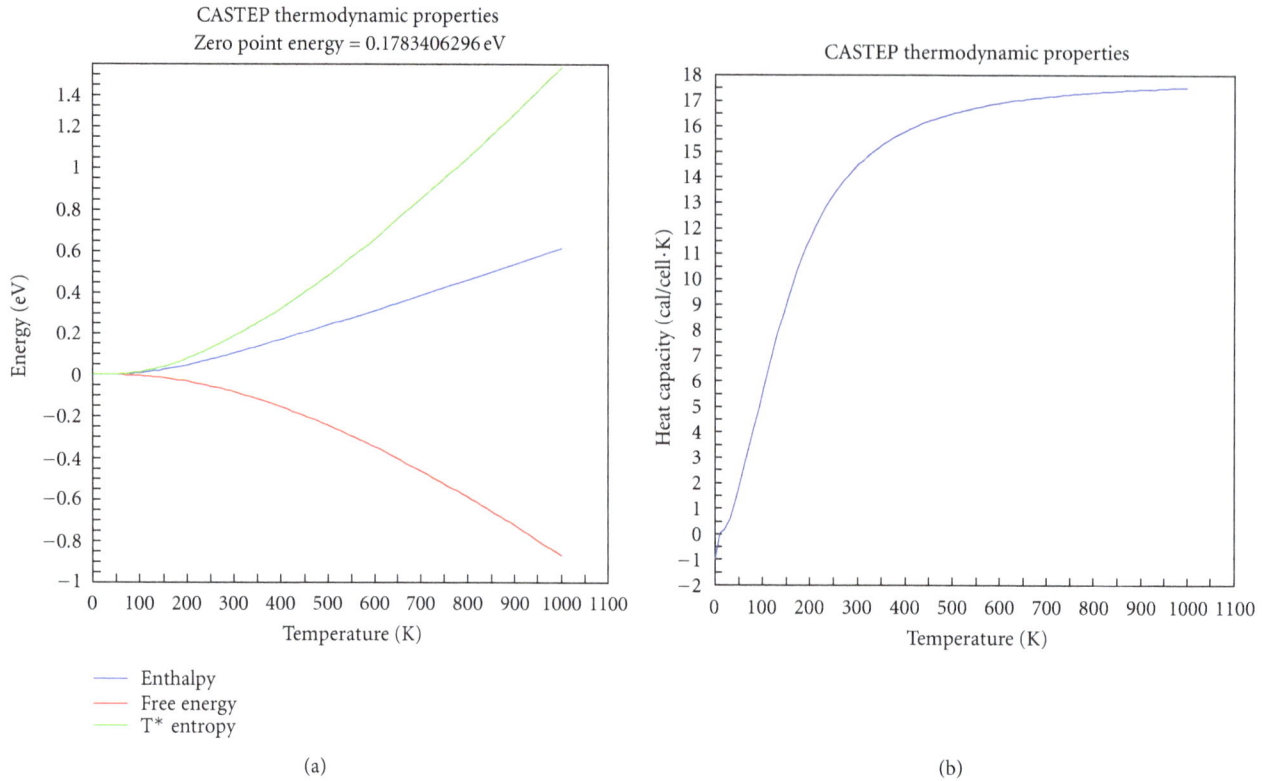

FIGURE 5: Thermodynamic properties enthalpy, free energy, and entropy (a) and heat capacity (b).

hybridized d, s, and p orbitals. The conduction band has a prominent set of bands which exists between 6 and 17 eV. The highest valence band occurs in the GX direction with a value of 17 eV. An oxygen p orbital beyond the Fermi level, centered at 10 eV, forms the bond with the cerium pdf hybrid. The cerium atom contributes to the majority of the conduction band dominated by an f orbital and to a lesser degree p as well as d, which forms the oxygen bonds. However, the prominent f orbital above the Fermi level is reported to be unoccupied [13] but nonetheless forms a hybrid with some minor p and d character. The work function was determined by constructing dual two-dimensional arrays of the CeO2 (110) surfaces separated by a vacuum gap of 30 Å. The electrostatic potential, determined midway of this gap, represents the energy required to remove one electron from the surface [14]. The (110) plane was selected as an intermediate between the (100) and (111) with regard to atomic population within a given plane. Figure 3 illustrates this potential profile between the two (110) planes in a vacuum. From this electrostatic potential versus fractional distance, the work function was calculated to be 5.287 eV.

3.3. Elastic Properties. The calculated elastic constants (Cij) elements are given in (1). From the three independent elastic constants c_{11}, c_{12}, and c_{14} and bulk modulus ($B = 277.521 \pm 0.348$ (GPa)), the mechanical stability criteria of the cubic crystal may be evaluated using various symmetry-based relationships [15, 16]. For this case, c_{11} and c_{44} must be greater than zero as well as the respective quantities ($c_{11} - c_{12}$)

and ($c_{11} + 2c_{12}$). In addition, the bulk modulus must be greater than c_{12} but less than c_{11}, thus for ceria the mechanical stability requirements are met:

$ci j$ [GPa]

$$
= \begin{bmatrix}
455.06 & 188.74 & 188.74 & 0.00 & 0.00 & 0.00 \\
188.74 & 455.06 & 188.74 & 0.00 & 0.00 & 0.00 \\
188.74 & 188.74 & 455.06 & 0.00 & 0.00 & 0.00 \\
0.00 & 188.74 & 0.00 & 81.48 & 0.00 & 0.00 \\
0.00 & 188.74 & 0.00 & 0.00 & 81.48 & 0.00 \\
0.00 & 188.74 & 0.00 & 0.00 & 0.00 & 81.48
\end{bmatrix}.
$$

(1)

For single-crystal ceria the predicted Young's modulus is 344.39 GPa along the three orthogonal axes. The polycrystalline bulk modulus was calculated as 277.52 GPa using the Voigt model which is comparable to the experimental value of 220 GPa [17]. The Poisson's ratio is 0.293, and the shear modulus was calculated to be 102.15 GPa. The elastic anisotropic factor ratio was calculated to be 0.773, which means polycrystalline ceria should be less susceptible to propagating microcracks at the grain boundary triple points [18], thereby facilitating its use as a suitable substrate for electronic devices.

3.4. Phonon Dispersion. The phonon is defined as the quantum of elastic strain energy. Therefore, the dispersion curves give insight into the material's thermal properties. For the cubic unit cell (N), the maximum number of branches, based upon 3 atoms per cell (z), yields 9 degrees of freedom through the relationship 3Nz [19]. The relationship (3z − 3) yields a maximum of six optical branches and hence the balance of three acoustic branches in any given direction. Figure 4 shows results of the phonon dispersion calculations. The lower acoustic branch in the G-X direction shows one acoustic longitudinal of higher energy than the transverse acoustic branch. The phonon dispersion curves become nondegenerate for all branches in the X-K direction giving the expected nine branches for this cubic structure. The higher frequency optical branches are split at the G symmetry point indicating some ionic bonding character [19]. The phonon dispersion calculation allows additional thermodynamic properties to be determined. The maximum Debye temperature was found to be 643 K. Figure 5(a) illustrates the enthalpy, free energy, and the temperature-scaled entropy as a function of temperature. The zero-point energy was found to be 0.178 eV. Figure 5(b) shows the temperature-dependent heat capacity per unit cell with a maximum value of 17 calories per Kelvin near the Debye temperature. The enthalpy was found to be −1914 eV.

4. Conclusions

It has been demonstrated that first-principles methods allow predictions of material properties and behavior of ceria. Judicious selection of a hybrid functional is necessary for the correct description of the electronic band structure for this insulating oxide-based material. Both static and dynamic lattice responses can be predicted, and the results can be used to guide processing and device fabrication. In future investigations, first-principles methods will be applied to obtain property predictions as a function of dopants and other processing variables.

References

[1] R. van de Krol and H. L. Tuller, "Electroceramics—the role of interfaces," *Solid State Ionics*, vol. 150, no. 1-2, pp. 167–179, 2002.

[2] R. Ramamoorthy, P. K. Dutta, and S. A. Akbar, "Oxygen sensors: materials, methods, designs and applications," *Journal of Materials Science*, vol. 38, no. 21, pp. 4271–4282, 2003.

[3] M. Ozawa, "Role of cerium-zirconium mixed oxides as catalysts for car pollution: a short review," *Journal of Alloys and Compounds*, vol. 277, pp. 886–890, 1998.

[4] G. Eranna, B. C. Joshi, D. P. Runthala, and R. P. Gupta, "Oxide materials for development of integrated gas sensors—a comprehensive review," *Critical Reviews in Solid State and Materials Sciences*, vol. 29, no. 3-4, pp. 111–188, 2004.

[5] M. D. Segall, P. J. D. Lindan, M. J. Probert et al., "First-principles simulation: ideas, illustrations and the CASTEP code," *Journal of Physics*, vol. 14, no. 11, pp. 2717–2744, 2002.

[6] J. P. Perdew, K. Burke, and M. Ernzerhof, "Generalized gradient approximation made simple," *Physical Review Letters*, vol. 77, no. 18, pp. 3865–3868, 1996.

[7] J. P. Perdew, A. Ruzsinszky, G. I. Csonka et al., "Restoring the density-gradient expansion for exchange in solids and surfaces," *Physical Review Letters*, vol. 100, no. 13, Article ID 136406, 2008.

[8] J. Paier, M. Marsman, and G. Kresse, "Why does the B3LYP hybrid functional fail for metals?" *Journal of Chemical Physics*, vol. 127, no. 2, Article ID 024103, 2007.

[9] J. Robertson, K. Xiong, and S. J. Clark, "Band structure of functional oxides by screened exchange and the weighted density approximation," *Physica Status Solidi B*, vol. 243, no. 9, pp. 2054–2070, 2006.

[10] G. J. Ackland, M. C. Warren, and S. J. Clark, "Practical methods in ab initio lattice dynamics," *Journal of Physics*, vol. 9, no. 37, pp. 7861–7872, 1997.

[11] E. A. Kümmerle and G. Heger, "The Structures of C-$Ce_2O_{3+\delta}$, Ce_7O_{12}, and $Ce_{11}O_{20}$," *Journal of Solid State Chemistry*, vol. 147, no. 2, pp. 485–500, 1999.

[12] E. Wuilloud, B. Delley, W. D. Schneider, and Y. Baer, "Spectroscopic evidence for localized and extended f-symmetry states in CeO_2," *Physical Review Letters*, vol. 53, no. 2, pp. 202–205, 1984.

[13] D. R. Mullins, S. H. Overbury, and D. R. Huntley, "Electron spectroscopy of single crystal and polycrystalline cerium oxide surfaces," *Surface Science*, vol. 409, no. 2, pp. 307–319, 1998.

[14] C. W. Chen and M. H. Lee, "Ab initio calculations of dimensional and adsorbate effects on the workfunction of single-walled carbon nanotube," *Diamond and Related Materials*, vol. 12, no. 3–7, pp. 565–571, 2003.

[15] D. M. Han, X. J. Liu, S. H. Lv, H. P. Li, and J. A. Meng, "Elastic properties of cubic perovskite $BaRuO_3$ from first-principles calculations," *Physica B*, vol. 405, no. 15, pp. 3117–3119, 2010.

[16] G. V. Sin'ko and N. A. Smirnov, "Ab initio calculations of elastic constants and thermodynamic properties of bcc, fcc, and hcp Al crystals under pressure," *Journal of Physics*, vol. 14, no. 29, pp. 6989–7005, 2002.

[17] L. Gerward, J. S. Olsen, L. Petit, G. Vaitheeswaran, V. Kanchana, and A. Svane, "Bulk modulus of CeO_2 and PrO_2—an experimental and theoretical study," *Journal of Alloys and Compounds*, vol. 400, no. 1-2, pp. 56–61, 2005.

[18] V. Tvergaard and J. W. Hutchinson, "Microcracking in ceramics induced by thermal-expansion or elastic-anisotropy," *Journal of the American Ceramic Society*, vol. 71, no. 3, pp. 157–166, 1988.

[19] G. Burns, *Solid State Physics*, Academic Press, San Diego, Calif, USA, 1985.

12

Structural and Ferroic Properties of La, Nd, and Dy Doped BiFeO₃ Ceramics

Ashwini Kumar, Poorva Sharma, and Dinesh Varshney

Materials Science Laboratory, School of Physics, Vigyan Bhawan, Devi Ahilya University, Khandwa Road Campus, Indore 452001, India

Correspondence should be addressed to Ashwini Kumar; ashu1220@gmail.com

Academic Editor: Shaomin Liu

Polycrystalline samples of $Bi_{0.8}RE_{0.2}FeO_3$ (RE = La, Nd, and Dy) have been synthesized by solid-state reaction route. X-ray diffraction (XRD) patterns of $Bi_{0.8}La_{0.2}FeO_3$ and $Bi_{0.8}Nd_{0.2}FeO_3$ were indexed in rhombohedral ($R3c$) and triclinic ($P1$) structure, respectively. Rietveld refined XRD pattern of $Bi_{0.8}Dy_{0.2}FeO_3$ confirms the biphasic ($Pnma + R3c$ space groups) nature. Raman spectroscopy reveals the change in BiFeO₃ mode positions and supplements structural change with RE ion substitution. Ferroelectric and ferromagnetic loops have been observed in the $Bi_{0.8}RE_{0.2}FeO_3$ ceramics at room temperature, indicating that ferroelectric and ferromagnetic ordering coexist in the ceramics at room temperature. The magnetic measurements at room temperature indicate that rare-earth substitution induces ferromagnetism and discerns large and nonzero remnant magnetization as compared to pristine BiFeO₃.

1. Introduction

Multiferroic materials, such as BiFeO₃ (BFO), have been a subject of unprecedented interest due to coexistence of simultaneous ferroelectric and antiferromagnetic/ferromagnetic ordering in the same phase [1]. These are studied extensively due to their wide range of potential applications, including information-storage device, spintronics, and sensors [2, 3]. The common exclusive nature of magnetism and electric polarization makes natural multiferroic materials rare [3, 4]. Needless to say, majority of compounds have low ordering temperatures; however room temperature achievement has yet to be noticed. BFO has a rhombohedrally distorted perovskite structure (space group $R3c$) [5] with high Curie temperature ($T_C \sim 1100$ K) and antiferromagnetic Neel temperature ($T_N \sim 675$ K) with a spatially modulated spiral spin structure [6, 7].

Efforts have been made to improve the ferroelectric and magnetic properties in antiferromagnetic BiFeO₃ ceramic. Enhancing the magnetic moments by reducing particle size of BFO has been one of the important tasks [8]. Ion substitution in BFO is believed to be effective and the most convenient way to enhance the ferroelectric and magnetic properties.

From the existing literature, it has been earlier observed that partial substitution of rare-earth (RE) elements like La [9], Pr [10], Nd [11, 12], Gd [13], Dy [14], and Ho [15] at Bi site can eliminate the impurity phase along with a structural phase transformation and improve ferroelectric properties and induced ferromagnetism in BFO ceramic. The relationship between the structural, evolution, and magnetic properties among these doped BFO ceramics still needs further investigations. With these priorities, we have synthesized RE ions doped (Bi site) BFO ceramics.

A structural phase transition from rhombohedral to orthorhombic structure is observed in $Bi_{1-x}La_xFeO_3$ near $x = 0.3$ and enhances the magnetoelectric interaction [9]. For Nd doped BiFeO₃ a rhombohedral structure at $x = 0$, a triclinic structure between $x = 0.05$ and 0.15, and a pseudotetragonal structure between $x = 0.175$ and 0.2 have been reported [11]. A structural transformation from rhombohedral to monoclinic structure for $Bi_{1-x}Nd_xFeO_3$ ($x = 0.0-0.15$) ceramic prepared by an improved rapid liquid phase sintering method is also reported [12]. On the other hand, 20% Dy substitution confirms the orthorhombic structure with $Pnma$ structural model [14]. Substitutional effect of Ho on BFO bulk ceramic infers that the remnant polarization

(P_r) and switching characteristic were improved at low field by reducing the leakage current apart from enhanced ferromagnetic properties [15]. These improved properties obtained by RE doping demonstrate the possibility of enhancing multiferroic applicability of BFO.

From this viewpoint, it is necessary to be aware of the crystal structure of compounds. Keeping the important features in mind and taking interesting crystallography of $BiFeO_3$ into consideration, it is worth studying the properties of rare-earth doped Bi-ferrite. We have thus synthesized rare-earth substituted $BiFeO_3$ abbreviated as $Bi_{0.8}RE_{0.2}FeO_3$ (RE = La, Nd, and Dy) multiferroic samples via solid-state reaction route and reported the consistency of evolution of crystal structure, dielectric, electrical, and magnetic properties. The synthesized samples of $Bi_{0.8}RE_{0.2}FeO_3$ (RE = La, Nd, and Dy) are further designated as BLFO, BNFO, and BDFO, respectively.

2. Experimental Details

Polycrystalline samples with the composition $Bi_{0.8}RE_{0.2}FeO_3$ (RE = La, Nd, and Dy) were prepared by conventional solid-state route and the typical synthesized process is described as follows. For the synthesis of $Bi_{0.8}RE_{0.2}FeO_3$ (RE = La, Nd, and Dy) ceramics Bi_2O_3 (Loba Chemie, 99.9% purity), Fe_2O_3 (Sigma-Aldrich, 99.99% purity), La_2O_3 (Sigma-Aldrich, 99.9% purity), Nd_2O_3 (Loba Chemie, 99.9% purity), and Dy_2O_3 (Loba Chemie, 99.9% purity) reagents were used as starting materials. All the chemicals were of GR grade and were used without any further purification. All the starting materials were weighed, mixed, and grounded thoroughly in an agate mortar for 6 hours using acetone and calcined for 6 hours at 650°C for the desired composition of $Bi_{0.8}RE_{0.2}FeO_3$. All the calcined compositions were uniaxially dye-pressed into pellets of size 10 mm in diameter and 2 mm in thickness. Sintering was performed at 820°C for 3 hours, with intermediate grinding.

X-ray diffraction was carried out with $CuK\alpha_1$ (1.5406 Å) radiation using Bruker D8 Advance X-ray diffractometer over the angular range 2θ (20°–80°) with a scanning rate of 2° per minute at room temperature working at 40 kV voltage and 40 mA current. The lattice parameters and other detailed structural information were obtained by the Rietveld refinement FullPROF program [16]. Raman measurements were carried out using LabRam HR800 micro-Raman spectrometer equipped with a 50x objective, an appropriate edge filter, and a Peltier-cooled charge coupled device detector. The spectra were excited with 488 nm radiations (2.53 eV) from an air-cooled Argon laser. Ferroelectric hysteresis (P-E) loop of the samples was measured using RT6000 (Radiant Technologies, USA) at frequency 50 Hz. The magnetization as a function of applied field (M-H) curve for BNFO sample at room temperatures was performed using a superconducting quantum interference device (SQUID) magnetometer (MPMS, Quantum Design). However, for BLFO and BDFO samples magnetic measurements were performed using vibrating sample magnetometer (VSM-lakeshore).

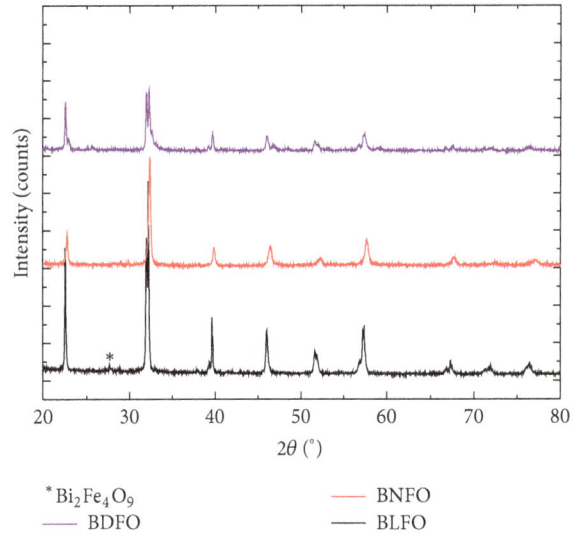

FIGURE 1: Room temperature XRD pattern of the $Bi_{0.8}RE_{0.2}FeO_3$ (RE = La, Nd, and Dy) samples designated as BLFO, BNFO, and BDFO, respectively.

3. Results and Discussion

3.1. Structural Analysis. Figure 1 shows the XRD pattern of $Bi_{0.8}RE_{0.2}FeO_3$ (RE = La, Nd, and Dy) ceramic samples. From the measured XRD pattern it has been observed that all the samples exhibit different crystal structures. BLFO sample was indexed in rhombohedral structure with space group $R3c$ where all the diffraction peaks match closely with the JCPDS file number 86-1518. A minor low level impurity phase (marked with ∗) was detected around $2\theta \approx 28°$ associated with $Bi_2Fe_4O_9$ [17]. This impurity peak matches well with the JCPDS file number 72-1832. The diffraction peaks change in both intensity and 2θ values with different dopant ionic radii as a result of change in crystal structure [18]. The values of lattice constants were manually calculated using different forms of crystal structures as defined earlier [19].

In order to further analyze the crystal structure the Rietveld refinement of measured XRD pattern was performed for all the samples as shown in Figures 2(a), 2(b), and 2(c). La doped BFO compound holds the polar rhombohedral $R3c$ structure similar to pure bismuth ferrite (BFO). The XRD pattern of BLFO sample was indexed in rhombohedral ($R3c$) system with lattice parameters $a = b = 5.5604(5)$ Å and $c = 13.7596(6)$ Å as shown in Figure 2(a). Earlier study reports a substitutional induced structural phase transition ($R3c \rightarrow C222$) for BLFO compound [9, 21]. It is worth noting that our attempt to fit the XRD pattern of present BLFO sample with $C222$ structural model was completely failed. Indeed, the reflection conditions derived from indexed reflection for BLFO cell were $l = 2n$ for hhl, $k = 2n$ for hkh, $h = 2n$ for hkk, $h = 2n$ for $h00$, $k = 2n$ for $0k0$, and $l = 2n$ for $00l$ which are compatible with the space group $R3c$.

The XRD pattern of BNFO sample was indexed in triclinic structure ($P1$ space group) with cell parameters $a = 3.9074(5)$ Å, $b = 3.9112(6)$ Å, and $c = 3.9002(5)$ Å. The X-ray diffraction pattern shows that all the X-ray peaks of

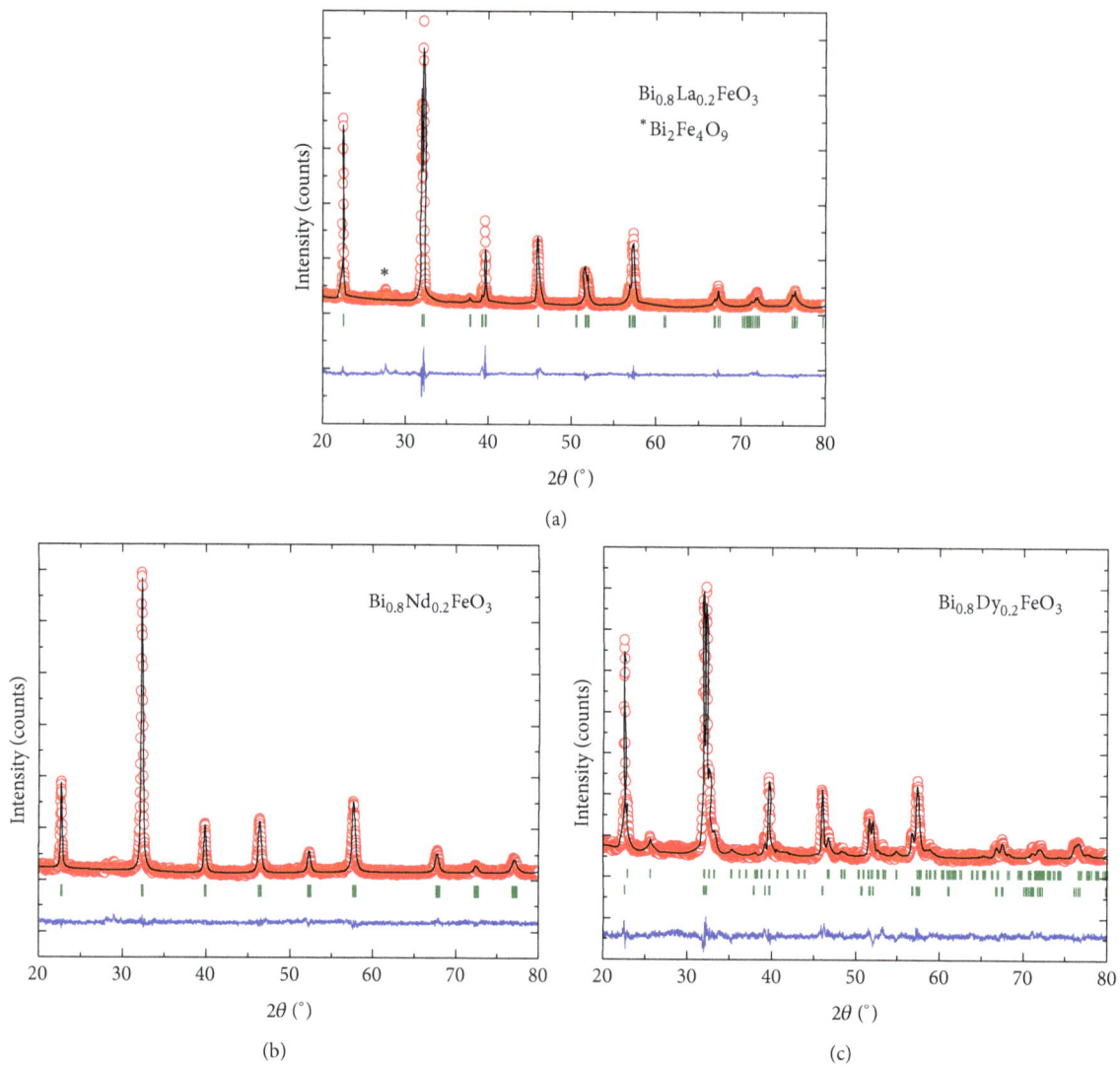

FIGURE 2: Rietveld refined X-ray diffraction pattern of the $Bi_{0.8}RE_{0.2}FeO_3$ (RE = La, Nd, and Dy) samples. The figure shows observed intensity (Y_{obs}), calculated intensity (Y_{calc}), and difference in observed and calculated intensities ($Y_{obs} - Y_{calc}$) and peak positions of different phases are shown at the base line as small ticks (|).

the BLFO and BNFO samples are well matching with the earlier reported data [17]. Nd addition at Bi site is helpful in suppressing the secondary phase in BFO. Therefore present samples have a single phase triclinic perovskite structure with all constituent components forming a solid solution rather than a mixture of Bi_2O_3, Fe_2O_3, Nd_2O_3, or any other impurity phase except $Bi_2Fe_4O_9$ observed in pure BFO.

Similarly, for BDFO compound, the refinement was performed with $Pnma + R3c$ structural model. The Rietveld refined XRD pattern of BDFO is shown in Figure 2(c). The dominant contribution is related to orthorhombic phase ($Pnma$, 80.62%) with lattice parameters a = 5.4014(5) Å, b = 7.7842(4) Å, and c = 5.5904(5) Å. The other component is related to rhombohedral phase ($R3c$, 19.38%). Reflection conditions obtained for $Pnma$ model are found to be almost similar for $Pn2_1a$ model except h = $2n$ for $h0l$. The best iteration gives $\chi^2 \approx 4.32$ for $R3c$, $\chi^2 \approx 10.08$ for $Pnma$, and $\chi^2 \approx 1.84$ for $R3c + Pnma$ model attributing to the fact

that crystal structure of BDFO compound is characterized by coexistence of two phases with a minimum χ^2 value. The obtained result is consistent with the earlier reported work [14]. Rietveld refined structural parameters of the $Bi_{0.8}RE_{0.2}FeO_3$ (La, Nd and Dy) samples simulated based on the measured XRD patterns are documented in Table 1.

It is well known that ferroic order and spontaneous polarization in BFO mainly result from the Bi^{3+} stereochemical $6s^2$ lone pair electron. Thus it is expected that systematic doping of rare-earth ions will distort the cation spacing between the oxygen octahedra and alter the long-range ferroelectric order. The ferroelectric properties have a close relation with the Fe–O bond length. The interatomic bond lengths of all the four samples were calculated by using Bond_Str program and are tabulated in Table 2. In BLFO compound with rhombohedral ($R3c$) crystal structure the octahedra bond environment is composed of three long degenerate Fe–O bond lengths and three short degenerate Fe–O bond lengths. The FeO_6

TABLE 1: Rietveld refined structural parameters of the $Bi_{0.8}RE_{0.2}FeO_3$ samples simulated based on the measured XRD patterns. The error values are presented in the parentheses.

Structure	Cell	Atoms	x	y	z	R-factors (%)
			$Bi_{0.8}La_{0.2}FeO_3$			
	$a = 5.5604\,(5)$	Bi/La	0.0000	0.0000	0.2724	$R_{Bragg} = 4.89$
$R3c$	$b = 5.5604\,(5)$	Fe	0.0000	0.0000	0.0000	$R_p = 5.36$
(100%)	$c = 13.7596\,(6)$	O	0.6679	0.7647	0.5489	$R_{wp} = 7.15$
	$V = 368.43\,(2)$					$\chi^2 = 2.91$
						GoF = 1.7
			$Bi_{0.8}Nd_{0.2}FeO_3$			
	$a = 3.9074\,(5)$	Bi/Nd	0.0000	0.0000	0.0000	$R_{Bragg} = 10.0$
$P1$	$b = 3.9112\,(6)$	Fe	0.5689	0.4362	0.5467	$R_p = 7.79$
(100%)	$c = 3.9002\,(5)$	O1	-0.0815	0.4542	0.6774	$R_{wp} = 11.2$
	$V = 59.60\,(2)$	O2	0.4538	-0.0722	0.6835	$\chi^2 = 1.85$
		O3	0.4541	0.4756	0.0176	GOF = 1.16
			$Bi_{0.8}Dy_{0.2}FeO_3$			
	$a = 5.4014\,(5)$	Bi/Dy	0.0472	0.2500	0.9933	$R_{B1} = 10.7$
$Pnma$	$b = 7.7822\,(4)$	Fe	0.0000	0.0000	0.5000	$R_{B2} = 6.96$
(80.62%)	$c = 5.5904\,(5)$	O1	0.3832	0.2500	0.0818	$R_p = 5.62$
	$V = 234.99\,(1)$	O2	0.2076	0.5414	0.2044	$R_{wp} = 7.2$
						$\chi^2 = 1.85$
$R3c$	$a = 5.5504\,(2)$	Bi/Dy	0.0000	0.0000	0.2676	GOF = 1.4
(19.38%)	$b = 5.5504\,(2)$	Fe	0.0000	0.0000	0.0000	
	$c = 13.7888\,(5)$	O	0.6794	0.7801	0.5544	
	$V = 367.88\,(2)$					

TABLE 2: Important bond lengths of $Bi_{0.8}RE_{0.2}FeO_3$ (RE = La, Nd, and Dy) samples. The error values are presented in the parentheses.

Compounds	Bond type	Bond length (Å)
BLFO	Fe–O (3)	1.7775
	Fe–O (3)	2.3165
BNFO	Fe–O (1)	1.4653
	Fe–O (1)	1.9015
	Fe–O (1)	2.0505
	Fe–O (1)	2.1114
	Fe–O (1)	2.1248
	Fe–O (1)	2.5882
BDFO	Fe–O (2)	1.9760
	Fe–O (2)	2.0226
	Fe–O (2)	2.0958

octahedron gets distorted due to Nd and Dy substitution, resulting in change in bond lengths as mentioned in Table 2.

3.2. Raman Analysis. Raman spectroscopy is a powerful tool to probe the structural and vibrational property of a material and also provide valuable information about ionic substitution and electric polarization. Raman spectra of BLFO, BNFO, and BDFO samples with excitation wavelength of 488 nm at room temperature are shown in Figure 3.

It has been reported that undoped $BiFeO_3$ with distorted rhombohedral structure, $R3c$ space group, and ten atoms in a unit cell of this structure yields 18 optical phonon modes that can be summarized using the following irreducible representation: $\Gamma_{opt} = 4A_1 + 5A_2 + 9E$; according to group theory 13 observed modes ($\Gamma_{Raman,R3c} = 4A_1 + 9E$) are Raman active, whereas $5A_2$ are Raman inactive modes [22, 23]. The dependence of mode positions on BLFO, BNFO, and BDFO samples is summarized in Table 3.

In the present study, for BLFO ceramic A_1 and E-symmetry normal modes for $R3c$ symmetry including A_1-1, A_1-2, A_1-3, A_1-4, E-1, E-4, E-6, E-7, E-8, and E-9 at around 138, 174, 191, 434, 67, 273, 373, 475, 527, and 627 cm^{-1} are clearly observed. These results are matched well with the Raman active vibration modes identified for BFO ($R3c$) [20, 24]. The ferroelectricity of BFO generally originates from the stereochemical activity of the Bi^{3+} $6s^2$ lone pair electron that is mainly responsible for the change in both Bi–O covalent bonds. The low frequency characteristic modes below 200 cm^{-1} may be responsible for the ferroelectric nature of the bismuth ferrite samples. As evident from the X-ray diffraction the changes in crystal symmetries are attributed to the A-site disorder created by Nd and Dy substitution, which leads to the shifting of Raman modes with sudden disappearance of some modes. The A_1-2, A_1-3, and A_1-4 were decomposed completely in BNFO and BDFO samples.

TABLE 3: Raman modes (cm^{-1}) for Bi$_{0.8}$RE$_{0.2}$FeO$_3$ (RE = La, Nd, and Dy) samples and the bulk BiFeO$_3$ (Kothari et al. [20]).

Raman modes (cm^{-1})	BFO bulk [20]	BLFO	BNFO	BDFO
A_1-1	135.15	135.95	139.06	142.33
A_1-2	167.08	174.52	—	176.88
A_1-3	218.11	—	—	—
A_1-4	430.95	434.14	—	—
E-1	71.39	67.44	68.24	—
E-2	98.36	—	97.91	—
E-3	255.38	—	—	234.32
E-4	283.0	273.93	298.01	278.77
E-5	321.47	—	—	326.52
E-6	351.55	373.35	—	386.31
E-7	467.60	475.53	488.53	475.47
E-8	526.22	527.78	—	531.89
E-9	598.84	627.47	619.53	585.45

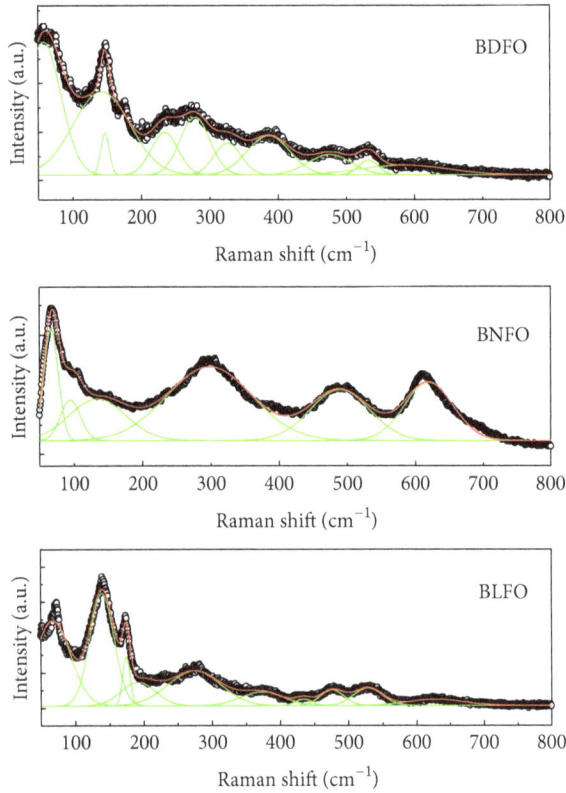

FIGURE 3: Raman spectra for Bi$_{0.8}$RE$_{0.2}$FeO$_3$ (RE = La, Nd, and Dy) samples at room temperature with excitation wavelength of 488 nm.

The normal modes related to the Bi–O covalent bonds (i.e., E-1, A_1-1, A_1-2, A_1-3, and E-2 modes) shift gradually toward higher frequencies and are attributed to the substitution of relatively light (mass) La^{3+} (138.90 g), Nd^{3+} (144.24 g), and Dy^{3+} (162.50 g) ion for Bi^{3+} (208.98 g) ion in the BiFeO$_3$.

3.3. Ferroelectric (P-E) Measurement. We have also made attempts to measure ferroelectric hysteresis loop at room

temperature for BLFO, BNFO, and BDFO compounds (Figure 4). It has been observed from the graph that BLFO and BDFO sample represent ferroelectric (FE) behaviour whereas BNFO sample represents paraelectric (PE) nature. The BLFO compound represents ferroelectric loop without any saturation value. This is attributed to the fact that BLFO sample is highly conductive at room temperature which results in the partial reversal of the polarization. For BDFO the spontaneous polarization ($2P_s$), remnant polarization ($2P_r$), and coercive field (E_c) are about ~0.23 μc/cm^2, 0.16 μc/cm^2, and 1.50 kV/cm, respectively, under the electric field of ~3 kV/cm, whereas for BNFO compound the obtained values of $2P_s$ and $2P_r$ are found to be ~2.6 μc/cm^2 and ~0.0 μc/cm^2, respectively, under the electric field of ~100 kV/cm.

It has been reported that Dy substitution is very much supportive in decreasing the leakage behavior of BiFeO$_3$. Moreover, too much Dy substitution will degrade ferroelectric nature [25]. This is attributed to the fact that higher Dy substitution will transform the crystal structure from rhombohedral to orthorhombic symmetry. The orthorhombic structure is more centrosymmetric, which in turn suppresses the ferroelectricity [26]. Furthermore, the origin of ferroelectricity in BiFeO$_3$ is generally due to Bi^{3+} ($6s^2$) lone pair electron. The substitution of Bi^{3+} with Dy^{3+} ion will weaken the stereochemical activity of lone pair and weaken the ferroelectricity. In present study, despite 20% Dy substitution, BDFO compound is ferroelectric in nature with a small leakage current. This might be because of the coexistence of two phases ($Pnma + R3c$) in present BDFO compound.

Furthermore, for paraelectric BNFO sample the obtained values of $2P_s$ (~2.6 μc/cm^2) and $2P_r$ (~0 μc/cm^2) are found to be much lower even at a high field of 100 kV/cm. The lower P_r value of BNFO despite its higher applied electric field indicates that Nd doping degrades the ferroelectric nature. In the present case of BNFO, the Bi^{3+} lone pair electron hybridizes with empty p orbital of Bi^{3+} or an O^{2-} ion to form Bi–O covalent bonds ensuing the noncentrosymmetric distortion

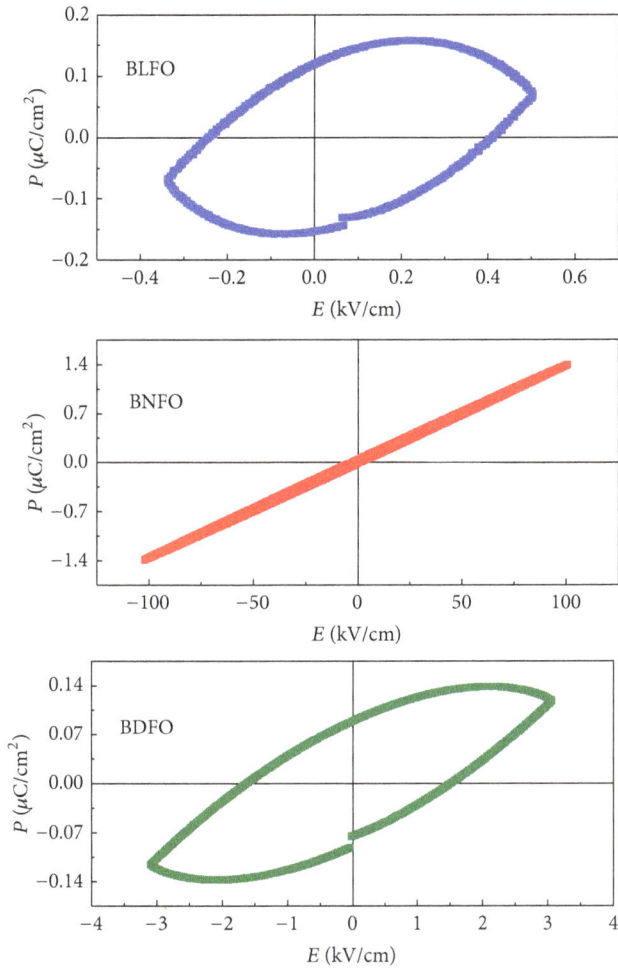

FIGURE 4: Room temperature ferroelectric (*P-E*) loop measurement of $Bi_{0.8}RE_{0.2}FeO_3$ (*RE* = La, Nd, and Dy) samples.

and ferroelectricity. Substitution of Nd^{3+} ion weakens the stereochemical activity and results in FE-PE transition.

3.4. Magnetic Hysteresis Analysis. The magnetization-magnetic field (*M-H*) curve recorded at room temperature for BLFO and BDFO samples with a maximum applied field 10 kOe has been illustrated in Figure 5. Similarly, for BNFO sample, the maximum applied magnetic field (H_m) up to 50 kOe has been also shown in Figure 5. It is well known that pristine BFO has *G*-type antiferromagnetic order with a long period of ~62 nm of canted spins between two successive ferromagnetically coupled (111) planes and zero net magnetization [27]. There are sharp and noticeable changes in magnetization observed for $Bi_{0.8}RE_{0.2}FeO_3$ (*RE* = Nd and Dy) samples at room temperature. Magnetization increases presentably with doping content as shown in the inset of Figure 5.

In Figure 5(a) *M-H* curve of BLFO sample shows a very narrow magnetic hysteresis loop, no saturation magnetization, with a small but nonzero remnant magnetization (M_r) of ~0.005 emu/g and coercive field (H_c) of ~454.6 Oe. From Figure 5(b), it is clearly seen that BDFO sample exhibits

typical ferromagnetic hysteresis loop. The saturation magnetization (M_s), M_r, and H_c values were found to be 1.36 emu/g, 0.05 emu/g, and 4.71 Oe, respectively. A weak ferromagnetic behaviour was observed for BLFO and BDFO sample due to the suppression of space modulated spin structure [28, 29]. Doping by La at Bi site in BFO leads to distortion in rhombohedral structure. However, doping of Dy further enhances the distortion in rhombohedral structure leading to coexistence of orthorhombic phase (*Pnma*, 80.62%) + rhombohedral phase (*R3c*, 19.38%). We thus make a note that apart from the suppression of space modulated spin structure mismatch of ionic radii of La and Dy ion leading to different structures is also plausible for weak ferromagnetic behaviour. This can be probed by neutron diffraction and we shall study this problem in near future.

Figure 5(c) represents the room temperature *M-H* curve for BNFO sample with a maximum applied magnetic field (H_m) 50 kOe. The magnetic field dependent magnetization of BNFO showed similar behaviour. In case of ceramic BNFO, it exhibits a deviation from linear loop at 300 K as documented in Figure 5(c). Interestingly, the peculiar double hysteresis loop like *M-H* curve in both BNFO samples with a low coercive field (H_c) has been observed; however this implies that additional factors contribute to enhanced magnetization. Similar behaviour was earlier observed in Ho doped BFO and is attributed to ferromagnetic interaction between Fe^{3+} and Ho^{3+} [15].

The magnetization appears in all the samples due to suppression of spiral magnetic spin structures in $Bi_{0.8}RE_{0.2}FeO_3$ (*RE* = La, Nd, and Dy). As the substituting ions effectively disturb the crystal structure differently reasons for each rare-earth dopant in this report are different for destruction of spiral magnetic spin structure [29]. In other words, the *RE* ion as (*RE* = La, Nd, and Dy) doping in BFO can only suppress but cannot destroy the spin cycloid structure completely at 20% substitution in BFO. Similar behavior is also observed in previously rare-earth doped $BiFeO_3$ [10–15, 30]. Henceforth, partial substitution should be considered as one of the most effective ways for improving the multiferroic characteristic of antiferromagnetic BFO.

4. Conclusion

Conclusively, polycrystalline $Bi_{0.8}RE_{0.2}FeO_3$ (*RE* = La, Nd, and Dy) samples were successfully prepared through solid-state reaction route and were further investigated by XRD, Raman spectroscopy, and ferroelectric and magnetic measurements. All the samples fitted with Rietveld refinement using FullPROF program revealed the existence of rhombohedral structure with space group *R3c* for BLFO (La^{3+} doped $BiFeO_3$) and also revealed the substitutional induced structural transformation *R3c* → *P*1 (BNFO) and *R3c* → *Pnma* (BDFO) systems. The changes in crystal structure are attributed to the *A*-site disorder created by Nd and Dy substitution, which leads to the shifting of Raman modes with sudden disappearance of some modes as A_1-2, A_1-3, and A_1-4 decomposed completely in BNFO and BDFO samples.

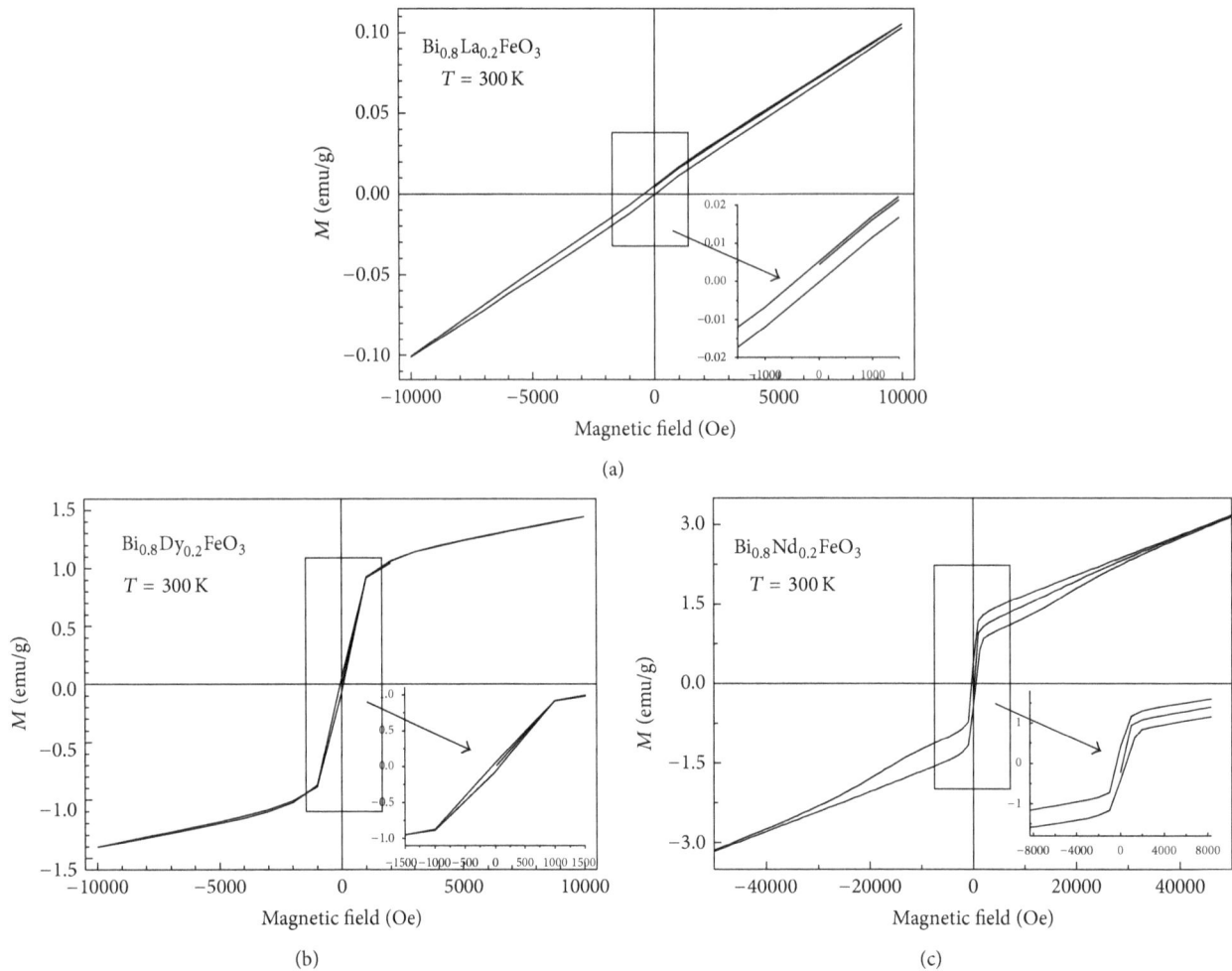

FIGURE 5: Magnetization versus magnetic field (*M-H*) loops of $Bi_{0.8}RE_{0.2}FeO_3$ (RE = La, Nd, and Dy) samples recorded at room temperature. The inset shows the enlarged data.

Ferroelectric and ferromagnetic loops have been observed in the $Bi_{0.8}RE_{0.2}FeO_3$ ceramics at room temperature, indicating that ferroelectric and ferromagnetic ordering coexist in the ceramics at room temperature. A significant enhancement in magnetization with enhanced remnant magnetization and coercive field is inferred in rare-earth doped $Bi_{0.8}RE_{0.2}FeO_3$ (RE = La, Nd, and Dy) samples. These results validate the doping induced destruction of the cycloidal structure in $Bi_{0.8}RE_{0.2}FeO_3$ (RE = La, Nd, and Dy) samples.

Conflict of Interests

The authors declare that there is no conflict of interests regarding the publication of this paper.

Acknowledgments

Financial support from UGC-DAE Consortium for Scientific Research, Indore, is gratefully acknowledged. XRD, Raman, and *P-E* measurements were performed at UGC-DAE Consortium for Scientific Research, Indore. The authors wish to thank Dr. Mukul Gupta, Dr. V. G. Sathe, and Dr. V. R. Reddy, UGC-DAE CSR, Indore, for useful discussions. They are also grateful to Dr. V. P. S. Awana, National Physical Laboratory, New Delhi, for his timely support in magnetic measurements.

References

[1] M. Fiebig, "Revival of the magnetoelectric effect," *Journal of Physics D: Applied Physics*, vol. 38, no. 8, pp. R123–R152, 2005.

[2] G. Catalan and J. F. Scott, "Physics and applications of bismuth ferrite," *Advanced Materials*, vol. 21, no. 24, pp. 2463–2485, 2009.

[3] N. A. Hill, "Why are there so few magnetic ferroelectrics?" *The Journal of Physical Chemistry B*, vol. 104, no. 29, pp. 6694–6709, 2000.

[4] D. I. Khomskii, "Multiferroics: different ways to combine magnetism and ferroelectricity," *Journal of Magnetism and Magnetic Materials*, vol. 306, no. 1, pp. 1–8, 2006.

[5] J. B. Neaton, C. Ederer, U. V. Waghmare, N. A. Spaldin, and K. M. Rabe, "First-principles study of spontaneous polarization in multiferroic BiFeO3," *Physical Review B: Condensed Matter and Materials Physics*, vol. 71, no. 1, Article ID 014113, 2005.

[6] I. Sosnowska, T. P. Neumaier, and E. Steichele, "Spiral magnetic ordering in bismuth ferrite," *Journal of Physics C: Solid State Physics*, vol. 15, no. 23, pp. 4835–4846, 1982.

[7] W. Kaczmarek, Z. Pająk, and M. Połomska, "Differential thermal analysis of phase transitions in $(Bi_{1-x}La_x)FeO_3$ solid solution," *Solid State Communications*, vol. 17, no. 7, pp. 807–810, 1975.

[8] S. Vijayanand, M. B. Mahajan, H. S. Potdar, and P. A. Joy, "Magnetic characteristics of nanocrystalline multiferroic $BiFeO_3$ at low temperatures," *Physical Review B—Condensed Matter and Materials Physics*, vol. 80, no. 6, Article ID 064423, 2009.

[9] S.-T. Zhang, Y. Zhang, M.-H. Lu et al., "Substitution-induced phase transition and enhanced multiferroic properties of $Bi_{1-x}La_xFeO_3$ ceramics," *Applied Physics Letters*, vol. 88, no. 16, Article ID 162901, 3 pages, 2006.

[10] D. Varshney, P. Sharma, S. Satapathy, and P. K. Gupta, "Structural, magnetic and dielectric properties of Pr-modified $BiFeO_3$ multiferroic," *Journal of Alloys and Compounds*, vol. 584, pp. 232–239, 2014.

[11] G. L. Yuan, S. W. Or, J. M. Liu, and Z. G. Liu, "Structural transformation and ferroelectromagnetic behavior in single-phase $Bi_{1-x}Nd_xFeO_3$ multiferroic ceramics," *Applied Physics Letters*, vol. 89, no. 5, Article ID 052905, 2006.

[12] G. L. Yuan and S. W. Or, "Enhanced piezoelectric and pyroelectric effects in single-phase multiferroic $Bi_{1-x}Nd_xFeO_3$ ($x = 0$–0.15) ceramics," *Applied Physics Letters*, vol. 88, p. 062905, 2006.

[13] V. A. Khomchenko, D. A. Kiselev, I. K. Bdikin et al., "Crystal structure and multiferroic properties of Gd-substituted $BiFeO_3$," *Applied Physics Letters*, vol. 93, no. 26, Article ID 262905, 2008.

[14] V. A. Khomchenko, D. V. Karpinsky, A. L. Kholkin et al., "Rhombohedral-to-orthorhombic transition and multiferroic properties of Dy-substituted $BiFeO_3$," *Journal of Applied Physics*, vol. 108, no. 7, Article ID 074109, 2010.

[15] N. Jeon, D. Rout, W. Kim, and S. L. Kang, "Enhanced multiferroic properties of single-phase $BiFeO_3$ bulk ceramics by Ho doping," *Applied Physics Letters*, vol. 98, Article ID 072901, 2011.

[16] J. Rodríguez-Carvajal, "Recent advances in magnetic structure determination by neutron powder diffraction," *Physica B: Physics of Condensed Matter*, vol. 192, no. 1-2, pp. 55–69, 1993.

[17] A. Kumar and D. Varshney, "Crystal structure refinement of $Bi_{1-x}Nd_xFeO_3$ multiferroic by the Rietveld method," *Ceramics International*, vol. 38, pp. 3935–3942, 2012.

[18] Z. Tao, Y. Huang, and H. J. Seo, "Blue luminescence and structural properties of Ce^{3+}-activated phosphosilicate apatite $Sr_5(PO_4)2(SiO_4)$," *Dalton Transactions*, vol. 42, no. 6, pp. 2121–2129, 2013.

[19] B. D. Cullity, *Introduction to Magnetic Materials*, Addison-Wesley, Reading, Mass, USA, 1972.

[20] D. Kothari, V. R. Reddy, V. G. Sathe, A. Gupta, A. Banerjee, and A. M. Awasthi, "Raman scattering study of polycrystalline magnetoelectric $BiFeO_3$," *Journal of Magnetism and Magnetic Materials*, vol. 320, no. 3-4, pp. 548–552, 2008.

[21] Q.-H. Jiang, C.-W. Nan, and Z.-J. Shen, "Synthesis and properties of multiferroic La-modified $BiFeO_3$ ceramics," *Journal of the American Ceramic Society*, vol. 89, no. 7, pp. 2123–2127, 2006.

[22] M. K. Singh, H. M. Jang, S. Ryu, and M. H. Jo, "Polarized Raman scattering of multiferroic $BiFeO_3$ epitaxial films with rhombohedral *R3c* symmetry," *Applied Physics Letters*, vol. 88, p. 42907, 2006.

[23] R. Haumont, J. Kreisel, P. Bouvier, and F. Hippert, "Phonon anomalies and the ferroelectric phase transition in multiferroic $BiFeO_3$," *Physical Review B*, vol. 73, no. 13, Article ID 132101, 2006.

[24] J. Wu and J. Wang, "$BiFeO_3$ thin films of (1 1 1)-orientation deposited on $SrRuO_3$ buffered $Pt/TiO_2/SiO_2/Si(1\ 0\ 0)$ substrates," *Acta Materialia*, vol. 58, no. 5, pp. 1688–1697, 2010.

[25] S. Zhang, L. Wang, Y. Chen, D. Wang, Y. Yao, and Y. Ma, "Observation of room temperature saturated ferroelectric polarization in Dy substituted $BiFeO_3$ ceramics," *Journal of Applied Physics*, vol. 111, no. 7, Article ID 074105, 5 pages, 2012.

[26] W.-M. Zhu, L. W. Su, Z.-G. Ye, and W. Ren, "Enhanced magnetization and polarization in chemically modified multiferroic $(1 − x)BiFeO_3$–$xDyFeO_3$ solid solution," *Applied Physics Letters*, vol. 94, no. 14, Article ID 142908, 2009.

[27] M. M. Kumar, S. Srunath, G. S. Kumar, and S. V. Suryanarayana, "Spontaneous magnetic moment in $BiFeO_3$–$BaTiO_3$ solid solutions at low temperatures," *Journal of Magnetism and Magnetic Materials*, vol. 188, no. 1-2, pp. 203–212, 1998.

[28] D. Maurya, H. Thota, A. Garg, B. Pandey, P. Chand, and H. C. Verma, "Magnetic studies of multiferroic Bi1-xSmxFeO3 ceramics synthesized by mechanical activation assisted processes," *Journal of Physics Condensed Matter*, vol. 21, no. 2, Article ID 026007, 2009.

[29] V. A. Khomchenko, D. A. Kiselev, J. M. Vieira et al., "Effect of diamagnetic Ca, Sr, Pb, and Ba substitution on the crystal structure and multiferroic properties of the $BiFeO_3$ perovskite," *Journal of Applied Physics*, vol. 103, no. 2, Article ID 024105, 2008.

[30] Y.-J. Zhang, H.-G. Zhang, J.-H. Yin et al., "Structural and magnetic properties in $Bi_{1-x}R_xFeO_3$ ($x = 0$–1, $R =$ La, Nd, Sm, Eu and Tb) polycrystalline ceramics," *Journal of Magnetism and Magnetic Materials*, vol. 322, no. 15, pp. 2251–2255, 2010.

13

Simple and Rapid Fabrication of $Na_{0.5}K_{0.5}NbO_3$ Thin Films by a Chelate Route

A. Fernández Solarte, N. Pellegri, O. de Sanctis, and M. G. Stachiotti

Laboratorio de Materiales Cerámicos, Universidad Nacional de Rosario, IFIR, CONICET, Avenida Pellegrini 250, S2000BTP Rosario, Argentina

Correspondence should be addressed to N. Pellegri; pellegri@fceia.unr.edu.ar

Academic Editor: Shaomin Liu

$Na_{0.5}K_{0.5}NbO_3$ (NKN) thin films were prepared by a chelate route which offers the advantage of a simple and rapid solution synthesis. The route is based on the use of acetoin as a chelating agent. The process was optimized by investigating the effects of alkaline volatilization on film properties. While we observed no evidence of stoichiometry problems due to potassium volatilization loss during the heat treatments, thin films synthesized with insufficient sodium excess presented a potassium-rich secondary phase, which has a significant influence on the ferroelectric properties. We show that the amount of spurious phase decreases with increasing Na^+ concentration, in such a way that a 20 mol% Na^+ excess is necessary to fully compensate the volatilization loss that occurred during the heat treatment. In this way, NKN thin films annealed at 650°C presented a well-crystallized perovskite structure, no secondary phases, well-defined ferroelectric hysteresis loops ($P_r \sim 9\,\mu C/cm^2$, $E_C \sim 45\,kV/cm$), and low leakage current density ($2 \times 10^{-7}\,A/cm^2$ at $80\,kV/cm$).

1. Introduction

Lead-free $K_{0.5}Na_{0.5}NbO_3$ (NKN) based ceramics are the most promising candidates for the replacement of lead zirconate titanate (PZT), due to their relatively good piezoelectric properties and high Curie temperature [1]. The miniaturization and integration of devices demand layered structures, which are supported on thin films processing. The fabrication of NKN-based thin films has been investigated by several methods such as pulsed laser deposition (PLD) [2], aerosol-deposited films [3], and mainly by sol-gel techniques [4–6]. Sol-gel techniques are based on the 2-methoxyethanol process. Although this route offers an excellent control and reproducibility of process chemistry, the preparation of the precursor solutions involves distillation and refluxing strategies which makes the solution preparation process difficult.

NKN films produced by chemical solution deposition methods typically have large leakage currents, and it is always difficult to obtain well-saturated polarization hysteresis loops [6]. This drawback comes out from the loss of stoichiometry because Na^+ and K^+ are easy to volatilize during the thermal treatment of the films. The evaporation of the alkali ions produces the appearance of spurious phases and causes the formation of oxygen vacancies in the films that lead to a large leakage current and thus poor ferroelectric and piezoelectric properties. The main strategy for overcoming the problem caused by the loss of alkali ions has been addition an excess of Na^+ and/or K^+ in the initial compositions. How alkaline compositions are lost is an important area of research, including issues such as what the amount of loss is. For example, Tanaka et al. [7] obtained single perovskite phase in 300 nm thick NKN films prepared by sol-gel with the adding of 10 mol% excess of Na^+ in the precursor solution. On the contrary, Kupec et al. [8] used a 5 mol% K^+ excess solution to synthesize 250 nm thick films. Other authors prepared NKN films from precursor solutions with excess of both K^+ and Na^+ [9, 10]. Evidently, there are no identical conclusions in the literature as regards which cation volatilizes more easily, K^+ or Na^+. We note, however, that the results mentioned previously were obtained by using different synthesis parameters, such as thermal sequences, drying, and annealing temperatures, and in some cases different starting materials.

The aim of this work is twofold. Firstly, to develop a chelate route for the synthesis of NKN thin films which offers the advantage of simple and rapid solution synthesis. In comparison with the 2-methoxyethanol process, distillation and refluxing strategies are not required. Secondly, to optimize the route by investigating the effects of different excess amounts of Na^+ and K^+ in the precursor solutions for the growth of NKN thin films of submicron thickness with good ferroelectric properties. Phase structure and microstructure, ferroelectric P-E hysteresis loops, and J-E characteristics, as well as the thermal evolution of partner powders, are compared. The influence of the annealing temperature on the ferroelectric properties of the thin films is also investigated.

2. Experimental Procedure

The chelate route relies on the molecular modification of the alkoxide compounds through reactions with other reagents, namely, chelating agents which provide stability to the precursor solutions. Regarding the chelating agent, acetic acid, acetylacetone, or amine compounds are the most commonly used. α-Hydroxyketones, on the other hand, have emerged as good chelating agents due to their stabilization effects on metal alkoxides solutions. In particular, it was found that acetoin (3-hydroxy 2-butanone) has the highest stabilization effect on Ti, Zr, Ta, and Nb metal alkoxides [11, 12]. In the present case, the NKN solutions were prepared starting from sodium ethoxide, potassium ethoxide, and niobium ethoxide as metal precursors, using acetoin and ethanol as chelating agent and solvent, respectively. Firstly, niobium pentaethoxide ($Nb(OCH_2CH_3)_5$, 99.95% Aldrich) was dissolved in a solution of acetoin (3-hydroxy-2butanone, Aldrich) and ethanol with a ratio $[Nb(OCH_2CH_3)_5]/[acetoin] = 1/4$. Afterwards, one solution of potassium ethoxide ($KOCH_2CH_3$, 24% wt. solution in ethanol, Aldrich) and subsequently another one of sodium ethoxide ($NaOCH_2CH_3$, 21% wt solution in ethanol, Aldrich) were added to the niobium precursor solution. Each step of the procedure was performed under a nitrogen atmosphere and a continuous stirring. The nitrogen atmosphere was maintained during the whole solution preparation process through the use of a controlled atmosphere chamber filled with dry nitrogen. Finally, water (dissolved in ethanol) was added up to reach a molar ratio $[H_2O]/[NKN]$ = 2. The NKN molar concentration of the solution was equal to 0.125. Precursor solutions were prepared with a 0 mol% alkaline excess, a 10 and 20 mol% excess amount of sodium ethoxide, and a 15 mol% amount of potassium ethoxide that correspond to the compositions (labeled as): $Na_{0.5}K_{0.5}NbO_3$ (NK-0), $Na_{0.55}K_{0.5}NbO_3$ (N-10), $Na_{0.6}K_{0.5}NbO_3$ (N-20), and $Na_{0.5}K_{0.575}NbO_3$ (K-15), respectively.

To grow the films, the precursor solutions were spin-coated on $Pt/TiO_2/SiO_2/Si$ substrates at 3000 rpm for 30 s in a clean bench. The wet films were dried at 200°C for 7 min. Subsequently, the burning of residual groups in the films was performed at 400°C for 10 min in air. Finally, the samples were annealed in air at temperatures between 600°C and 750°C for 5 min. For thicker films, a multilayer process was used, repeating the coating/heat-treatment cycle three times. The thickness of each sample was determined by ellipsometry using a Rudolph ellipsometer with a wavelength of 634 nm.

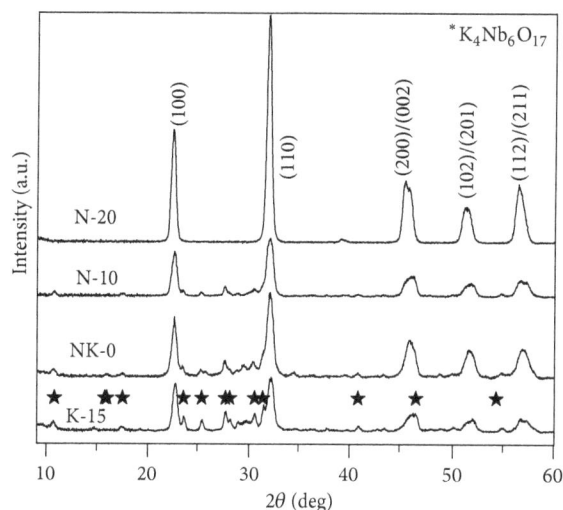

FIGURE 1: XRD spectra of the powders derived from 20 mol % (N-20), 10 mol% Na^+ excess (N-10), 0 mol% alkaline excess (NK-0), and 15 mol% K^+ excess in precursor solutions (K-15) after annealing at 600°C.

For instance, the thickness of a 0 mol% excess NKN three-layer film annealed at 600°C was 235 ± 3 nm, and its refractive index was 2.074 ± 0.008. As all films were grown in similar conditions, their thicknesses are similar to the previous value.

Gel powders were obtained from the precursor solution by solvent evaporation under a pressure of 40 torr, and, then, the gels were dried at 200°C for 24 h. Thermal analyses (DTA-TG) of the gel powders were performed using a Shimadzu DTG 60H equipment with a heating rate of 10°C/min from room temperature to 600°C in normal atmosphere. Crystal structures of thin films and ceramic powders were analyzed at room temperature using a Philips X'Pert Pro X-ray diffractometer with Cu K_α radiation of wavelength 1.5406 A, at a scan rate of 0.02^0/s. For thin films, the measurements were made with grazing incident X-ray diffraction (GIXRD), using a Soller slit to get parallel beam for diffraction. Surface morphology was observed with scanning electron microscopy using an FE-SEM Philips. To measure the electrical properties of the films, 0.3 mm diameter Pt top electrodes were deposited by DC sputtering on the films and then annealed at 400°C for 60 min. The Pt layer of the substrates was used as bottom electrodes. The dielectric constant and the dielectric loss were measured by complex impedance spectroscopy using an HP 4192 A LF impedances analyzer at frequencies between 1 KHz and 1 MHz. Ferroelectric properties were evaluated using a ferroelectric test system (a conventional Sawyer-Tower circuit) applying AC signals at 1 kHz and at room temperature. The current density (J)-electric field (E) characteristics of the NKN thin films were measured at room temperature using an electrometer/high-resistance meter.

3. Results and Discussions

Figure 1 shows the XRD patterns of partner powders treated at 600°C which are related to each film composition, that is, using 0 mol% alkaline excess, 10 mol%, 20 mol% Na^+, and

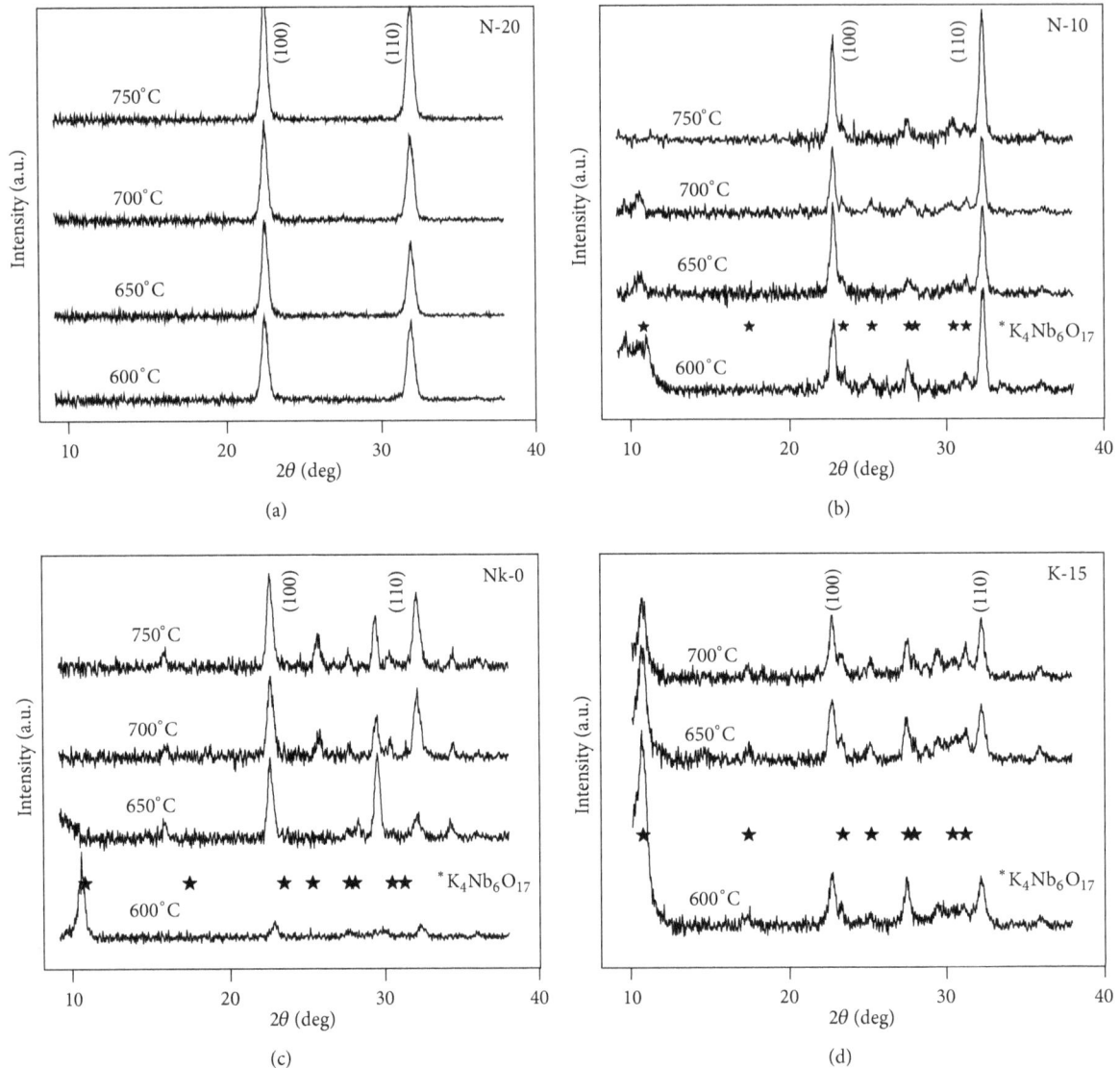

FIGURE 2: XRD spectra of the thin films grown using (a) 20 mol %, (b) 10 mol% Na$^+$ excess, (c) 0 mol% alkaline excess, and (d) 15 mol% K$^+$ excess in precursor solutions on Pt/TiO$_2$/SiO$_2$/Si substrate after annealing at 600–750°C.

15 mol% K$^+$ excess precursor solutions, and the names are abbreviated as KN-0, N-10, N-20, and K-15, respectively. We anticipate that the spectra of the thin films are similar to that obtained by the powders. In the case of the powders, however, the peaks corresponding to spurious phases are more intense, clarifying the spectra. While the N-20 powder shows a well-crystallized NKN perovskite as unique phase, the spectrum of the K-15 powder shows additional peaks corresponding to the K$_4$Nb$_6$O$_{17}$ tungsten bronze structure, which is the only secondary phase present in the system. It is clear that the increase of the Na$^+$/K$^+$ ratio in the precursor solution (see NK-0 and N-10) produces a decrease in the intensity of the peaks corresponding to the spurious phase. Thus, the excess of Na reduces the content of the nonperovskite phase, indicating that the volatilization of Na$^+$ is more important than K$^+$ in powders annealed at 600°C.

Figure 2 shows the XRD patterns of the thin films annealed at different temperatures from 600 to 750°C. We

show results for 2θ ranged between 10 and 40 degrees in the region where the peaks of the spurious phase are more intensive. All grown films exhibit diffraction peaks corresponding to the NKN perovskite phase. However, only the films derived from the 20 mol% Na$^+$ excess precursor solution (N-20) (Figure 2(a)) exhibited a spectrum free of spurious phases, with a well-crystallized perovskite structure after being annealed at 600°C. NK-0 thin films annealed at 600°C (Figure 2(c)) showed weak peaks corresponding to the perovskite phase together with K$_4$Nb$_6$O$_{17}$ peaks. The intensity of perovskite peaks as well as the one corresponding to spurious phases increase as the annealing temperatures increase. N-10 thin films (Figure 2(b)) developed a better crystallinity; however, their XRD spectra also show nonperovskite peaks although their intensities are much lower than the ones of the NK-0 films. For NKN thin films grown from 15 mol% K$^+$ excess (Figure 2(d)), the XRD spectra show also peaks corresponding to the spurious phase.

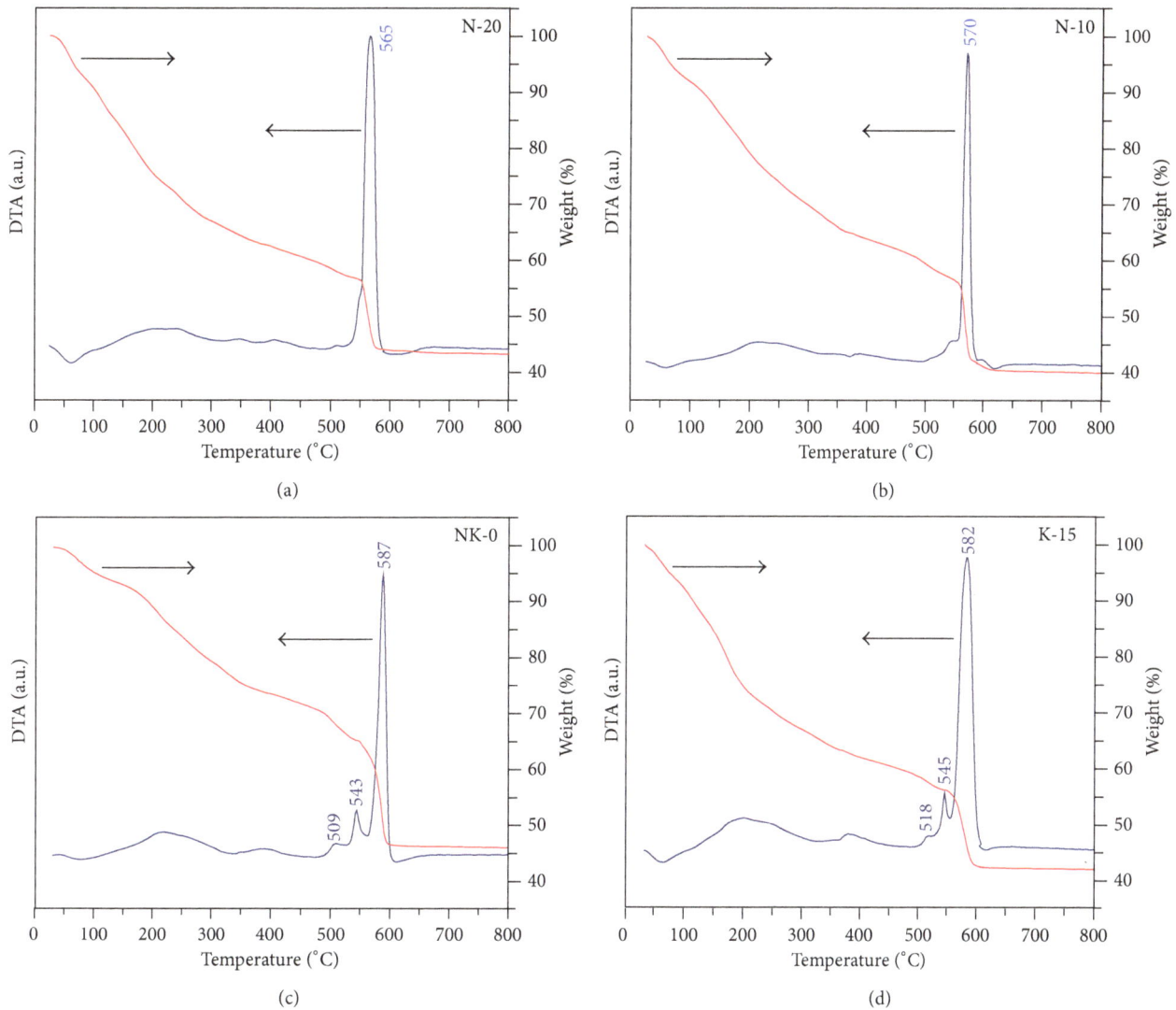

FIGURE 3: DTA-TG curves of the gel powders derived from precursor solutions with (a) 20 mol %, (b) 10 mol% Na^+ excess, (c) 0 mol% alkaline excess, and (d) 15 mol% K^+ excess.

Figure 3 shows the DTA-TG curves of partner powders obtained from precursor solutions after the solvent has been removed. No significant differences between them are observed at temperatures below 450°C. At higher temperatures, the DTA curves of all powders show exothermic peaks centered between 565, and 587°C, coinciding each one with a strong mass loss. Those peaks can be attributed to a massive crystallization of NKN that occurred during the amorphous-crystalline transition. With the increase of Na^+ excess in the precursor solutions, NKN crystallization temperature decreases, the width of the DTA exothermic peak decreases, and the temperature range in which that mass loss occurs is narrower. Other two exothermic peaks are observed in the temperature range between 500°C and 550°C. One is a small peak at around 509°C that tends to disappear when the Na^+ content increases. This peak could be attributed to the crystallization of the $K_4Nb_6O_{17}$ phase, since it is known that the $K_4Nb_6O_{17}$ phase appeared as the first crystalline phase during the heating of the amorphous

precursor gel at temperatures around 500°C [13, 14]. The second exothermic peak around 543°C is probably due to the early formation of part of the NKN perovskite phase induced by reaction of the preexisting $K_4Nb_6O_{17}$ phase with the surrounding amorphous material [15]. This peak decreases as the Na content increases and almost disappears for the N-10 sample (Figure 3(b)), and it is completely covered by the peak corresponding to the massive crystallization of NKN in the powder derived from the precursor solution with the highest Na content (Figure 3(a)). Both exothermic peaks at around 509°C and 543°C are accompanied by the loss of mass, whose slope decreases as the Na content increases, indicating that the fraction of amorphous material consumed in each of those reactions decreases with Na^+ content. These results reveal that the increase of Na^+ content in the precursor solutions partially inhibits the early crystallization of the $K_4Nb_6O_{17}$ phase and lowers the crystallization temperature of the NKN perovskite phase, while a Na^+ deficient structure favors the crystallization of the rich potassium phase

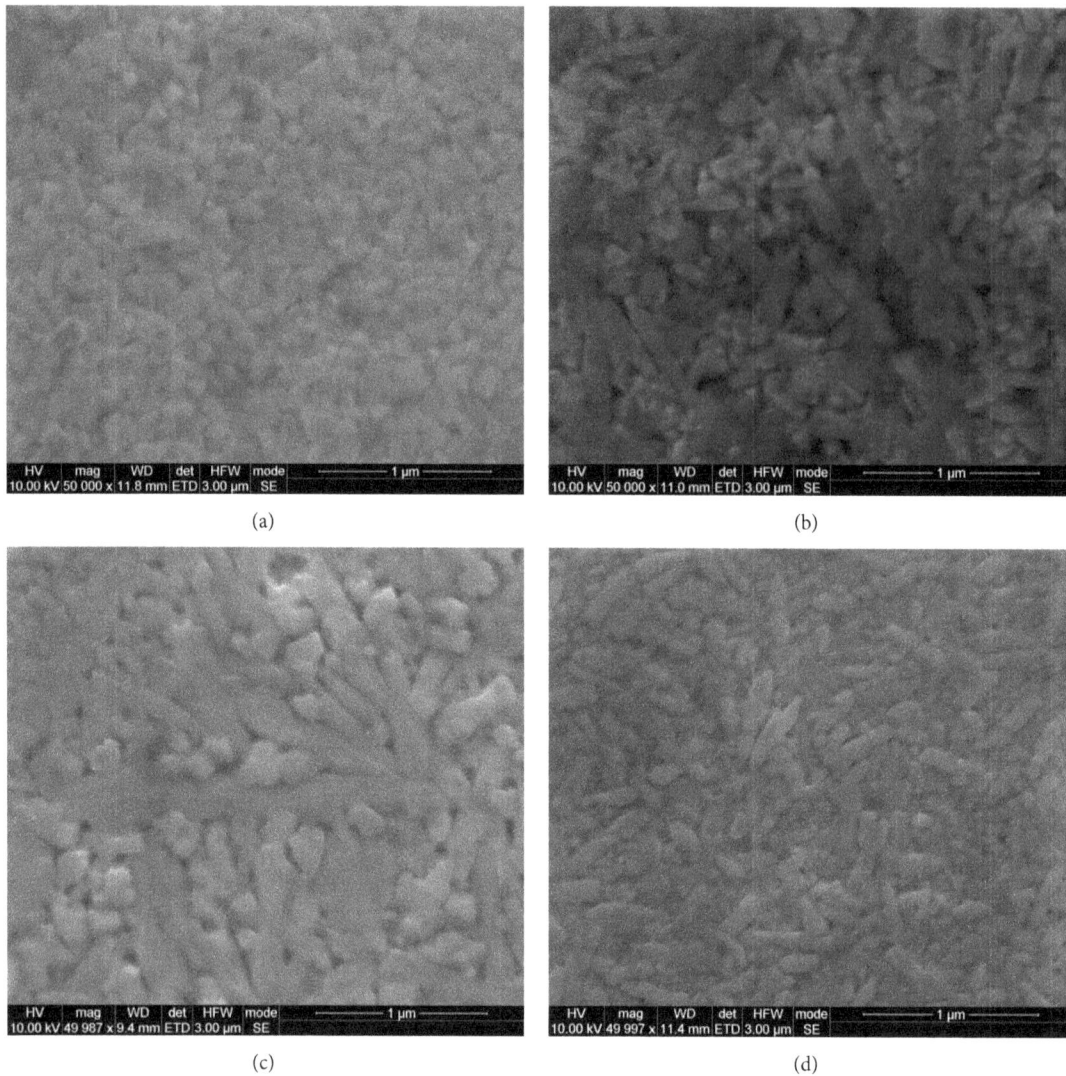

(a)

(b)

(c)

(d)

FIGURE 4: Scanning electron microscopy micrographs of the thin films grown using (a) 20 mol %, (b) 10 mol% Na$^+$ excess, (c) 0 mol% alkaline excess, and (d) 15 mol% K$^+$ excess annealed at 650°C.

K$_4$Nb$_6$O$_{17}$ [16]. We can thus conclude that a 20 mol% Na$^+$ excess must be added to the precursor solution to mitigate the Na$^+$ volatilization. The compensation of the volatilized Na$^+$ content allows for the achievement of stoichiometry in order to obtain a unique fully crystallized NKN perovskite phase in powder as well as in thin films annealed at 600°C.

Figure 4 shows the microstructure of the NKN thin films that were grown using 0 mol% alkaline excess, 10 mol%, 20 mol% Na$^+$ excess and 15 mol% K$^+$ excess, precursor solutions and after being annealed at 650°C, obtained by SEM. Thin films grown with the highest Na$^+$ content (Figure 4(a)) exhibit a homogeneous and dense microstructure with grains of about 100 nm in size and with almost a single-modal distribution of grain size, indicating a normal grain growth. On the contrary, the microstructure of K-15 thin film is formed by elongated-shaped grains (Figure 4(d)). NK-0 and N-10 films present polycrystalline microstructures that consist of two different grain types. We can observe in Figures 4(b) and 4(c) the presence of highly anisotropic

grains immersed into a fine-grained dense material. This highly anisotropic grain growth can be related to the fact that the crystallization of these grains does not occur directly from the amorphous precursor but through an asymmetric intermediate crystalline phase; that is, during the heating of the Na-deficient amorphous films, this asymmetric phase is the first to be formed.

Figure 5 shows the P-E hysteresis loops for NK-0, N-10, N-20, and K-15 thin films annealed at 650°C. The measurements were conducted at a switching frequency of 50 Hz and a maximum applied field of 150 kV/cm. The NK-0 and K-15 thin films almost do not display ferroelectric behavior, while the N-10 thin film exhibited a very slim ferroelectric hysteresis loop. Only the film grown from the 20 mol% Na$^+$ excess precursor solution exhibited a well-saturated hysteresis loop, with a relatively high remnant polarization of $P_r = 8.2\,\mu$C/cm^2 and a coercive field of $E_C = 55$ kV/cm. Moreover, the N-20 thin films tolerate a very high applied electric field, over 300 kV/cm. Figure 6 shows the P-E

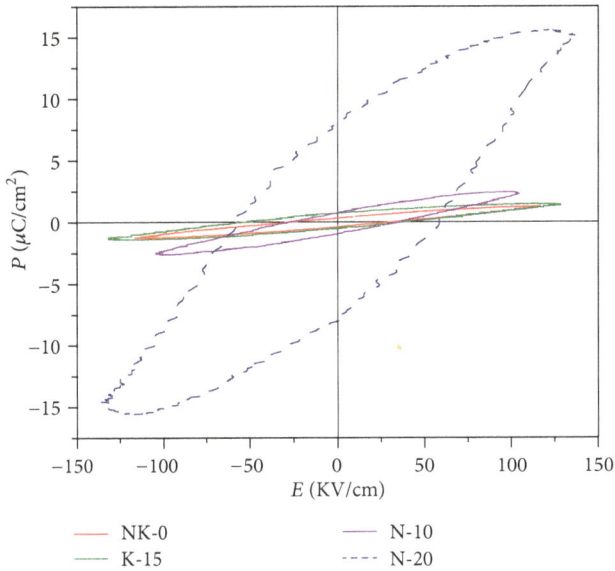

FIGURE 5: Ferroelectric P-E hysteresis loops for NKN thin films grown using 20 mol % (N-20), 10 mol% Na^+ excess (N-10), 0 mol% alkaline excess (NK-0), and 15 mol% K^+ excess (K-15) in precursor solutions on $Pt/TiO_2/SiO_2/Si$ substrate after annealing at 650°C.

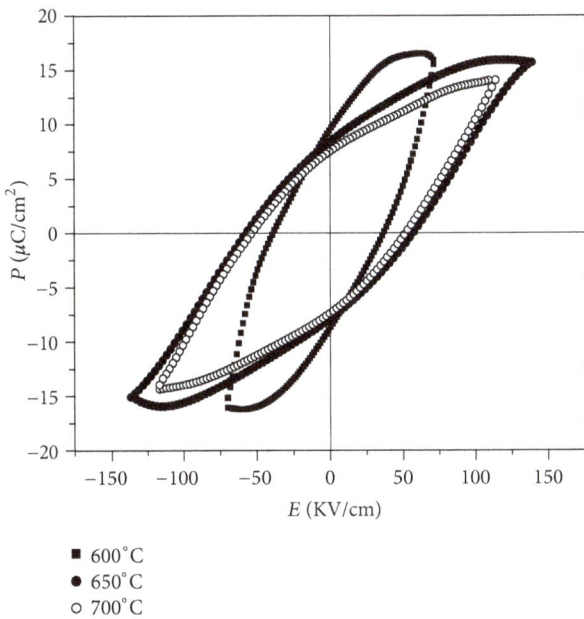

FIGURE 7: J-E curves of 20%mol Na excess$^+$ NKN thin films after annealing at 600°C (■), 650°C (●), and 700°C (○).

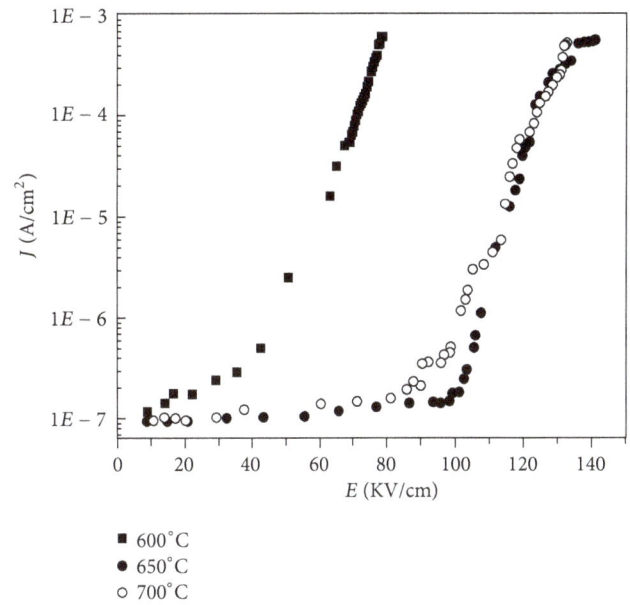

FIGURE 6: The P-E hysteresis loops of the 20%mol Na^+ excess NKN thin films after annealing at 600°C (■), 650°C (●), and 700°C (○).

ferroelectric hysteresis loops for N-20 thin films annealed at 600, 650 and 700°C. All films show typical ferroelectric behavior. Remnant polarizations of 10, 8.2, and 7.5 $\mu C/cm^2$ and the coercive fields of 33, 45, and 55 kV/cm were obtained for the films annealed at 600, 650, and 700°C, respectively. Although the film annealed at 600°C presents the highest P_r and the lowest E_C, its hysteresis loop is rounded in shape, indicating large leakage components. The films annealed

at higher temperatures exhibited better ferroelectric loops. Figure 7 shows room-temperature J-E characteristics of the N-20 films annealed at 600, 650, and 700°C. The leakage current densities were around 10^{-7} A/cm^2 at electric fields lower than 30 kV/cm. For electric fields ranged between 40 and 80 kV/cm, the leakage current density of the film annealed at 600°C rapidly increased from 10^{-6} to 10^{-3} A/cm^2. On the contrary, the leakage current densities of the films annealed at 650 and 700°C remain lower than 2×10^{-7} A/cm^2 for electric fields up to 100 kV/cm. Once this value is achieved, the leakage current increases abruptly.

The leaky ferroelectric P-E hysteresis loops as well as the high-leakage current at low field of the N-20 thin film annealed at 600°C would be caused by residual defects. The amorphous gel-crystalline transition left residual defects of synthesis, such as dangling bonds and "noncharge compensation vacancies of oxygen" [17, 18]. These defects remain in the film structure until the diffusion mechanism becomes an operative process, which allows the oxygen absorption into the structure and consequently produces structural healing and annihilation of the oxygen vacancies that occurs at a temperature range between 600 and 650°C [19]. "Noncharge compensation vacancies" are missing atoms that behave as neutral vacancies when the structure has crystallized. Neutral oxygen vacancies act as donor defects. First ionization, $V_O^{x} \rightarrow V_O^{\bullet} + e^-$, has a very low ionization energy ($\Delta h < 0.2$ eV); on the contrary, the ionization energy is higher for double ionization: $V_O^{x} \rightarrow V_O^{\bullet\bullet} + 2e^-$ ($\Delta h < 1.5$ eV) [20]. First-ionization vacancy would be responsible for the high-leakage current at low field and the leaky hysteresis loop display in the N-20 thin film annealed at 600°C. The incorporation of oxygen in the structure produces the annihilation of oxygen vacancies, but the annihilation of the "noncharge compensation vacancies" does not produce electrical carriers, $(1/2)O_2 + V_O^x \rightarrow (1/2)O_2 + V_O^\bullet + e^- \rightarrow O_O$.

4. Conclusions

We developed a chelate route for the synthesis of NKN thin films using acetoin as a chelating agent. The route offers the advantage of a simple and rapid solution synthesis; in comparison with the 2-methoxyethanol method, distillation and refluxing strategies are not required. The process was optimized by investigating the effects of alkaline volatilization on film properties. While no evidence of stoichiometry problems was detected due to potassium volatilization loss, the volatility of Na^+ complicates the growth of phase-pure films. Na-deficient compositions favor an early crystallization of a $K_4Nb_6O_{17}$ secondary phase. We showed that the addition of a 20 mol% Na^+ excess to the NKN precursor solution was effective in compensating for the volatilization of Na at moderate temperatures, lowering at the same time the crystallization temperature of the perovskite phase. In this way, NKN films annealed at 650°C present a well-crystallized perovskite structure, good ferroelectric properties, and low leakage current.

Conflict of Interests

The authors declare no conflict of interests regarding commercial entities mentioned in this paper.

Acknowledgments

The authors wish to acknowledge financial support from Consejo Nacional de Investigaciones Científicas y Técnicas (CONICET) and Universidad Nacional de Rosario. M. G. Stachiotti thanks CIUNR support.

References

[1] Y. Saito, H. Takao, T. Tani et al., "Lead-free piezoceramics," *Nature*, vol. 432, no. 7013, pp. 84–87, 2004.

[2] T. Saito, T. Wada, H. Adachi, and I. Kanno, "Pulsed laser deposition of high-quality (K,Na)NbO₃ thin films on SrTiO₃ substrate using high-density ceramic targets," *Japanese Journal of Applied Physics B*, vol. 43, no. 9, pp. 6627–6631, 2004.

[3] J. Ryu, J.-J. Choi, B.-D. Hahn, D.-S. Park, W.-H. Yoon, and K.-H. Kim, "Fabrication and ferroelectric properties of highly dense lead-free piezoelectric (K₀.₅Na₀.₅)NbO₃ thick films by aerosol deposition," *Applied Physics Letters*, vol. 90, no. 15, Article ID 152901, 2007.

[4] F. Lai and J.-F. Li, "Sol-gel processing of lead-free (Na,K)NbO₃ ferroelectric films," *Journal of Sol-Gel Science and Technology*, vol. 42, no. 3, pp. 287–292, 2007.

[5] K. Tanaka, K.-I. Kakimoto, and H. Ohsato, "Fabrication of highly oriented lead-free (Na, K)NbO₃ thin films at low temperature by Sol-Gel process," *Journal of Crystal Growth*, vol. 294, no. 2, pp. 209–213, 2006.

[6] Y. Nakashima, W. Sakamoto, H. Maiwa, T. Shimura, and T. Yogo, "Lead-free piezoelectric (K,Na)NbO₃ thin films derived from metal alkoxide precursors," *Japanese Journal of Applied Physics*, vol. 46, no. 12–16, pp. L311–L313, 2007.

[7] K. Tanaka, H. Hayashi, K.-I. Kakimoto, H. Ohsato, and T. Iijima, "Effect of (Na,K)-excess precursor solutions on alkoxy-derived (Na,K)NbO₃ powders and thin films," *Japanese Journal of Applied Physics B*, vol. 46, no. 10, pp. 6964–6970, 2007.

[8] A. Kupec, B. Malic, J. Tellier, E. Tchernychova, S. Glinsek, and M. Kosec, "Lead-free ferroelectric potassium sodium niobate thin films from solution: composition and structure," *Journal of the American Ceramic Society*, vol. 95, no. 2, pp. 515–523, 2012.

[9] Y. Nakashima, W. Sakamoto, T. Shimura, and T. Yogo, "Chemical processing and characterization of ferroelectric (K,Na)NbO₃ thin films," *Japanese Journal of Applied Physics B*, vol. 46, no. 10, pp. 6971–6975, 2007.

[10] C. W. Ahn, S. Y. Lee, H. J. Lee et al., "The effect of K and Na excess on the ferroelectric and piezoelectric properties of K₀.₅Na₀.₅NbO₃ thin films," *Journal of Physics D*, vol. 42, no. 21, Article ID 215304, 2009.

[11] T. Ohya, M. Kabata, T. Ban, Y. Ohya, and Y. Takahashi, "Effect of α-hydroxyketones as chelate ligands on dip-coating of zirconia thin films," *Journal of Sol-Gel Science and Technology*, vol. 25, no. 1, pp. 43–50, 2002.

[12] Y. Takahashi, A. Ohsugi, T. Arafuka, T. Ohya, T. Ban, and Y. Ohya, "Development of new modifiers for titanium alkoxide-based sol-gel process," *Journal of Sol-Gel Science and Technology*, vol. 17, no. 3, pp. 227–238, 2000.

[13] K. Tanaka, K.-I. Kakimoto, H. Ohsato, and T. Iijima, "Effects of Pt bottom electrode layers and thermal process on crystallinity of alkoxy-derived (Na,K)NbO₃ thin films," *Japanese Journal of Applied Physics A*, vol. 46, no. 3, pp. 1094–1099, 2007.

[14] K. Tanaka, K.-I. Kakimoto, and H. Ohsato, "Morphology and crystallinity of KNbO₃-based nano powder fabricated by sol-gel process," *Journal of the European Ceramic Society*, vol. 27, no. 13–15, pp. 3591–3595, 2007.

[15] I. Pribošič, D. Makovec, and M. Drofenik, "Formation of nanoneedles and nanoplatelets of KNbO₃ perovskite during templated crystallization of the precursor gel," *Chemistry of Materials*, vol. 17, no. 11, pp. 2953–2958, 2005.

[16] L. Y. Wang, K. Yao, P. C. Goh, and W. Ren, "Volatilization of alkali ions and effects of molecular weight of polyvinylpyrrolidone introduced in solution-derived ferroelectric K₀.₅Na₀.₅NbO₃ films," *Journal of Materials Research*, vol. 24, no. 12, pp. 3516–3522, 2009.

[17] R. Caruso, E. Benavídez, O. de Sanctis et al., "Phase structure and thermal evolution in coating films and powders obtained by sol-gel process: part II. ZrO₂ -2.5 mole % Y₂O₃," *Journal of Materials Research*, vol. 12, no. 10, pp. 2594–2601, 1997.

[18] M. C. Caracoche, P. C. Rivas, M. M. Cervera et al., "Zirconium oxide structure prepared by the sol-gel route: I, the role of the alcoholic solvent," *Journal of the American Ceramic Society*, vol. 83, no. 2, pp. 377–384, 2000.

[19] M. G. Stachiotti, R. MacHado, A. Frattini, N. Pellegri, and O. de Sanctis, "Effects of the chemical modifier on the thermal evolution of SrBi₂Ta₂O₉ precursor powders," *Journal of Sol-Gel Science and Technology*, vol. 36, no. 1, pp. 53–60, 2005.

[20] I. Tanaka, F. Oba, K. Tatsumi, M. Kunisu, M. Nakano, and H. Adachi, "Theoretical formation energy of oxygen-vacancies in oxides," *Materials Transactions*, vol. 43, no. 7, pp. 1426–1429, 2002.

14

Preparation, Structural, Electrical, and Ferroelectric Properties of Lead Niobate–Lead Zirconate–Lead Titanate Ternary System

Rashmi Gupta, Seema Verma, Vishal Singh, and K. K. Bamzai

Crystal Growth & Materials Research Laboratory, Department of Physics and Electronics, University of Jammu, Jammu 180006, India

Correspondence should be addressed to K. K. Bamzai; kkbamz@yahoo.com

Academic Editor: Shaomin Liu

A ternary system of lead niobate–lead zirconate–lead titanate with composition xPN–yPZ–$(x-y)$PT where $x = 0.5$ and $y = 0.15$, 0.25, and 0.35 known as PNZT has been prepared by conventional mixed oxide route at a temperature of 1100°C. The formation of the perovskite phase was established by X-ray diffraction analysis. The surface morphology studied by scanning electron microscopy shows the formation of fairly dense grains and elemental composition was confirmed by energy dispersive X-ray analysis. Dielectric properties like dielectric constant and dielectric loss (ε' and $\tan\delta$) indicate poly-dispersive nature of the material. The temperature dependent dielectric constant (ε') curve indicates relaxor behaviour with two dielectric anomalies. The poly-dispersive nature of the material was analysed by Cole-Cole plots. The activation energy follows the Arrhenius law and is found to decrease with increasing frequency for each composition. The frequency dependence of ac conductivity follows the universal power law. The ac conductivity analysis suggests that hopping of charge carriers among the localized sites is responsible for electrical conduction. The ferroelectric studies reveal that these ternary systems are soft ferroelectric.

1. Introduction

Materials with high Curie temperature and piezoelectric properties are required for high temperature applications in industries. Barium titanate (BT) or lead zirconate titanate (PZT) is used for many commercial applications like medicine, industry, and research but due to their low Curie temperature, that is, ~ -120°C for BT [1] and ~ 390°C for PZT [2], they are not suitable for high temperature applications. Recently, a great deal of interest has been focused on relaxor ferroelectric materials due to their excellent piezoelectric as well as dielectric properties [3, 4]. Lead metaniobate ($PbNb_2O_6$ or PN) reported by Goodman [5] is a member of tungsten bronze family and is often used in nondestructive testing and medical diagnostic imaging and for deep submergence hydrophones. However, problems such as high level of porosity and relatively low mechanical strength are often encountered in its use and since its discovery no much work has been done on it in spite of the high quality pioneering work [6, 7]. Niobium based perovskite oxides have attracted much attention because they show combined electrical and mechanical properties for use in the electronic industries [8–10]. In spite of the reports of an attractive high Curie temperature (517–570°C) [11, 12], fewer studies [13–15] on PN have been carried out which can be attributed to the preparing of piezoelectric form (the metastable orthorhombic structure) in pure phase. On the other hand, one of the most widely studied ferroelectric materials due to its various potential technological applications is lead zirconate titanate, that is, $PbZr_{1-x}Ti_xO_3$ (PZT) [16]. Aliovalent modifications to PZT which can be either with higher valence substitutions (donors) or with lower valence ions (acceptors) can alter its properties and have been studied by many authors [16, 17]. One of the most used additives is niobium which substitutes either Zr^{4+} or Ti^{4+}, promotes the lead vacancies, and is considered as a donor dopant. Most of the studies have reported on the effect of niobium modification on the dielectric, piezoelectric, pyroelectric, and electrooptic properties of PZT in bulk as well as thin film form [18–25]. However, there are so many reports that due to lead evaporation during processing, both doped and undoped PZT, are seldom observed to be perfect [26–28]. In the present investigation,

(a) $x = 50$, $y = 15\%$

(b) $x = 50$, $y = 25\%$

(c) $x = 50$, $y = 35\%$

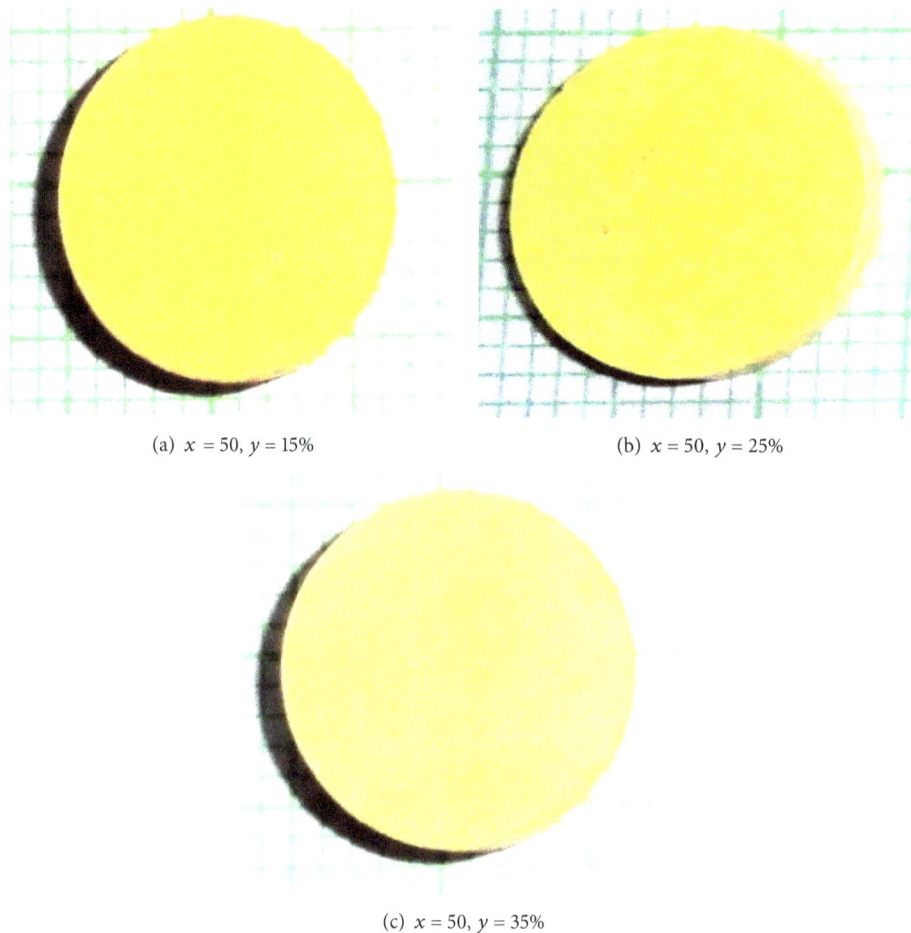

FIGURE 1: Photographs of the sintered pellets for xPN–yPZ–$(x\text{-}y)$PT (PNZT) ($x = 50$, $y = 15$, 25, and 35%).

we have investigated the preparation and properties of lead niobate–lead zirconate–lead titanate (PN–PZ–PT) ternary system. The effects on the electrical properties of the system have been investigated in detail by varying the PZ and PT concentration while keeping the PN concentration steady. To the best of author's knowledge, there is no such report on preparation and structural and electrical properties of such a ternary compound having general formula (x)lead niobate (PN)–(y)lead zirconate (PZ)–$(x\text{-}y)$lead titanate (where $x = 0.5$, $y = 0.15$, 0.25, 0.35).

2. Materials and Methods

Polycrystalline ceramic of xPN–yPZ–$(x\text{-}y)$PT (where $x = 0.5$, $y = 0.15$, 0.25, and 0.35) popularly known as lead niobate zirconate titanate (PNZT) was prepared using the conventional mixed oxide route. Oxide powders of PbO, ZrO_2, TiO_2, and Nb_2O_5 from S D Fine-Chem Limited (99% pure) were used as raw materials. Stoichiometric amount as per the molecular formula was weighed and mixed together in a mortar and pestle first and then in a high energy ball mill, as well as homogenized with triple distilled water for 48 h using zirconia balls as the milling media. The mixture was then dried in an oven at 100°C for 2 h. After drying,

the reaction of the uncalcined powders taking place during heat treatment was investigated by thermogravimetric and differential thermal analysis (TGA–DTA, Shimadzu), using a heating rate of 10°C/min in air from room temperature up to 1000°C. Based on the TGA–DTA results, the mixture was calcined at a temperature of 900°C for a dwell time 2 h and heating/cooling rates ranging 4°C/min in closed platinum crucible in order to ensure the proper chemical homogeneity of the constituent elements. The calcined powder was then grinded and mixed with 5 wt% of polyvinyl alcohol as binder and uniaxially pressed into pellets of 13 mm diameter by applying a pressure of 200 MPa in a hydraulic press. These pellets were then sintered at 1100°C using a heating/cooling rate of 4°C/min with dwell time of 4 h in a sealed platinum crucible. Figure 1 shows the photographs of the sintered ceramic pellets for all the three compositions considered in the present investigation. X-ray diffraction pattern was collected using a Rigaku X-ray diffractometer with CuKα radiation ($\lambda = 1.5405$ Å) and nickel filter in a wide 2θ range of 20–80°C using step scanning with a step size of 0.02° and step time of 65.6 s. The surface morphology and microstructures were examined using scanning electron microscope (SEM) of JEOL model number JSM–6390 LV and the presence of all the major elements was confirmed by energy dispersive

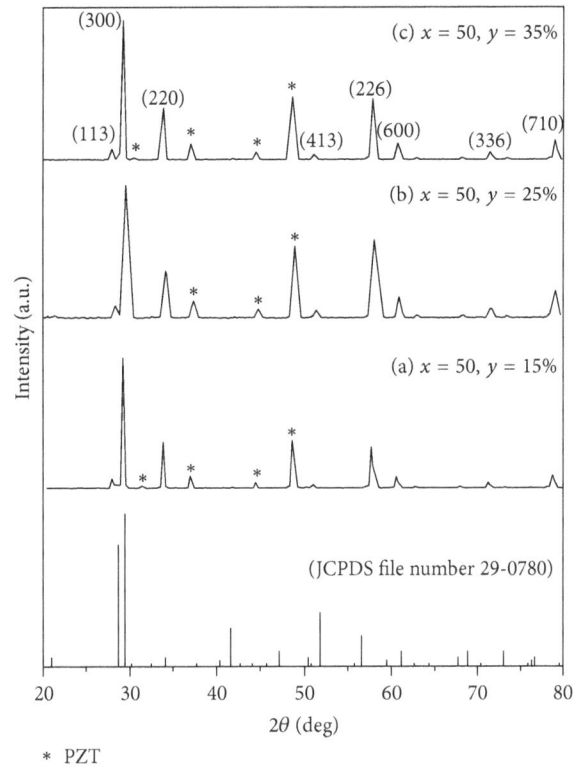

FIGURE 2: Room temperature X-ray diffraction pattern of xPN–yPZ–(x-y)PT ceramics sintered at 1100°C matched with JCPDS file number 29–0780 for the rhombohedral $PbNb_2O_6$; (*) indicates the PZT phase.

X-ray analysis (EDAX) of model number JEOL JED-2300. The dielectric measurements have been recorded with the help of automated impedance analyzer (4192A LF model) interfaced with USB–GPIB converter 82357 (Agilent) and further automated by using a computer for data recording, storage, and analysis. The polarization versus electric field (P–E) hysteresis loop at room temperature was traced using an automatic PE loop tracer (Marine India Electrocom Ltd., New Delhi, India).

3. Results and Discussion

3.1. Structural Characterization. Figure 2 shows the X-ray diffraction (XRD) patterns of three different compositions of PNZT, namely, xPN–yPZ–(x-y)PT (where x = 0.5, y = 0.15, 0.25, and 0.35) with a well-crystallized structure without the presence of pyrochlore or unwanted phases. The major diffraction peaks observed correspond to rhombo-hedral phase of $PbNb_2O_6$ (JCPDS 29 - 0780) with lattice parameters a = 10.501, $\alpha = \beta = 90°$, $\gamma = 120°$, and space group R3m (160) [15, 29]. The peaks marked with star (*) in the figure correspond to PZT, thereby suggesting the presence of mixed phases in PNZT. The peaks were indexed according to the JCPDS card. The peaks with maximum intensity and maximum crystallite size were observed for the composition having 35% zirconium which shows the improved crys-tallinity of the material. The backscattered scanning electron micrograph (SEM) images are shown in Figures 3(a)–3(c).

The micrographs clearly show the formation of fairly dense grains and the average grain size calculated using linear intercept method was found to be 1.40, 1.46, and 1.18 μm for 15, 25, and 35% of zirconium content, respectively. The elemental composition was analyzed by energy dispersive X-ray analysis (EDAX) (Figures 4(a)–4(c)) which confirms the presence of almost all the constituent elements, that is, Pb, Nb, Ti, Zr, and O in the composition. Table 1 shows the grain size, crystalline size calculated by Debye Scherrer formula [30] as well as Williamson-Hall plot [31], and strain and crystallinity index for all the three compositions. Crystallinity is evaluated from the crystallinity index equation [32]:

$$I_{cry} = \frac{D_p \, (\text{SEM, TEM})}{D_{cry} \, (\text{XRD})}, \quad I_{cry} \, (\geq 1.00), \quad (1)$$

where "I_{cry}" is the crystallinity index; "D_p" is the particle size (obtained from either TEM or SEM morphological analysis); "D_{cry}" is the particle size (calculated from the Scherrer equation). If "I_{cry}" value is close to 1, then it is assumed that the crystallite size represents monocrystalline whereas a polycrystalline has a much larger crystallinity index.

3.2. Thermal Analysis. The result of thermogravimetric anal-ysis (TGA) and differential thermal analysis (DTA) of a milled powder having the composition 0.50PN–0.25PZ–0.25PT is shown in Figure 5. Three endothermic peaks followed by small weight loss were observed in the DTA curve in the temperature range 250–460°C. This weight loss

TABLE 1: Variation of grain size, particle size, and strain and crystallinity index for PNZT system.

Composition	Grain size (μm)	Particle size (nm)		Strain (ε)	Crystallnity index, I_{cry}
		Debye Scherrer method	Williamson-hall method		
0.5PN–0.15PZ–0.35PT	1.40	28.41	34.07	0.00229	49.27
0.5PN–0.25PZ–0.25PT	1.46	19.19	20.85	0.00035	76.08
0.5PN–0.35PZ–0.15PT	1.18	38.47	66.34	0.00108	30.67

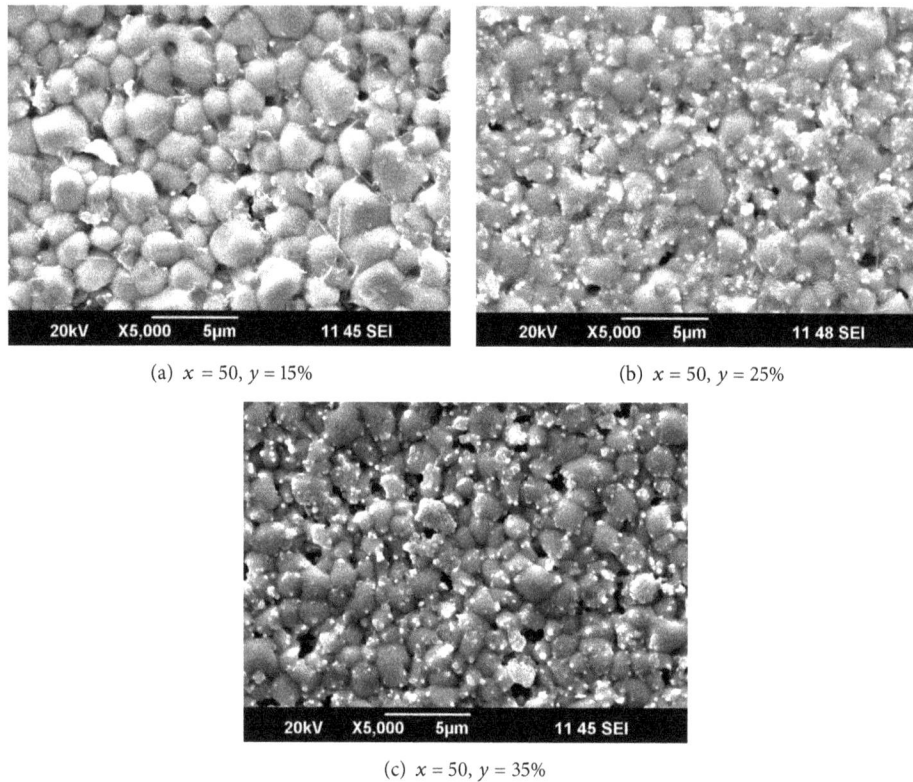

(a) $x = 50$, $y = 15\%$

(b) $x = 50$, $y = 25\%$

(c) $x = 50$, $y = 35\%$

FIGURE 3: Scanning electron micrographs of xPN–yPZ–$(x$-$y)$PT (PNZT) composite ($x = 50$, $y = 15, 25$, and 35%).

may be attributed due to the presence of adsorbed and trapped solvent. Also, there is a sharp increase of weight loss between 325 and 375°C which is due to the reason that the solvent adsorbed on the outer surface of the particles requires relatively lower temperature to evaporate than those entrapped between the particles during milling. The weight loss at 578°C may be assigned due to loss of lead at high temperature suggesting that the precursor materials should be heated well at 600°C [33]. No significant weight loss was observed for temperature higher than 600°C. The weight loss and the associated endothermic peak in DTA may be due to transformations into metastable phases, which formed prior to the crystallization of the perovskite phase. The last endothermic peak over the temperature range of 800–1000°C is not associated with weight loss in the TGA curve which corresponds to the crystallization of perovskite phase of lead niobate zirconate titanate (PNZT) [34].

3.3. Electrical Studies. Figures 6 and 7 show the dependence of real part of dielectric permittivity (ε') and dielectric loss ($\tan\delta$) on frequency at different temperatures. From the figures, the value of both ε' and $\tan\delta$ is found to decrease with increasing frequency which is a normal dielectric behaviour followed by all dielectric materials. In the lower frequency region, decrease in the value of ε' and $\tan\delta$ is observed which is due to the dominance of space charge polarization and interface effects at lower frequencies. However, in the high frequency region, frequency independent behaviour of these parameters is observed. At very low frequencies ($\omega \ll 1/\tau$), dipoles are able to follow the applied field and we have the value of dielectric constant at quasistatic field ($\varepsilon' \sim \varepsilon_s$). As the frequency increases, the dipoles began to lag behind the field resulting in a small increase in dielectric constant and as the characteristic frequency is reached, the value of dielectric constant drops which is the

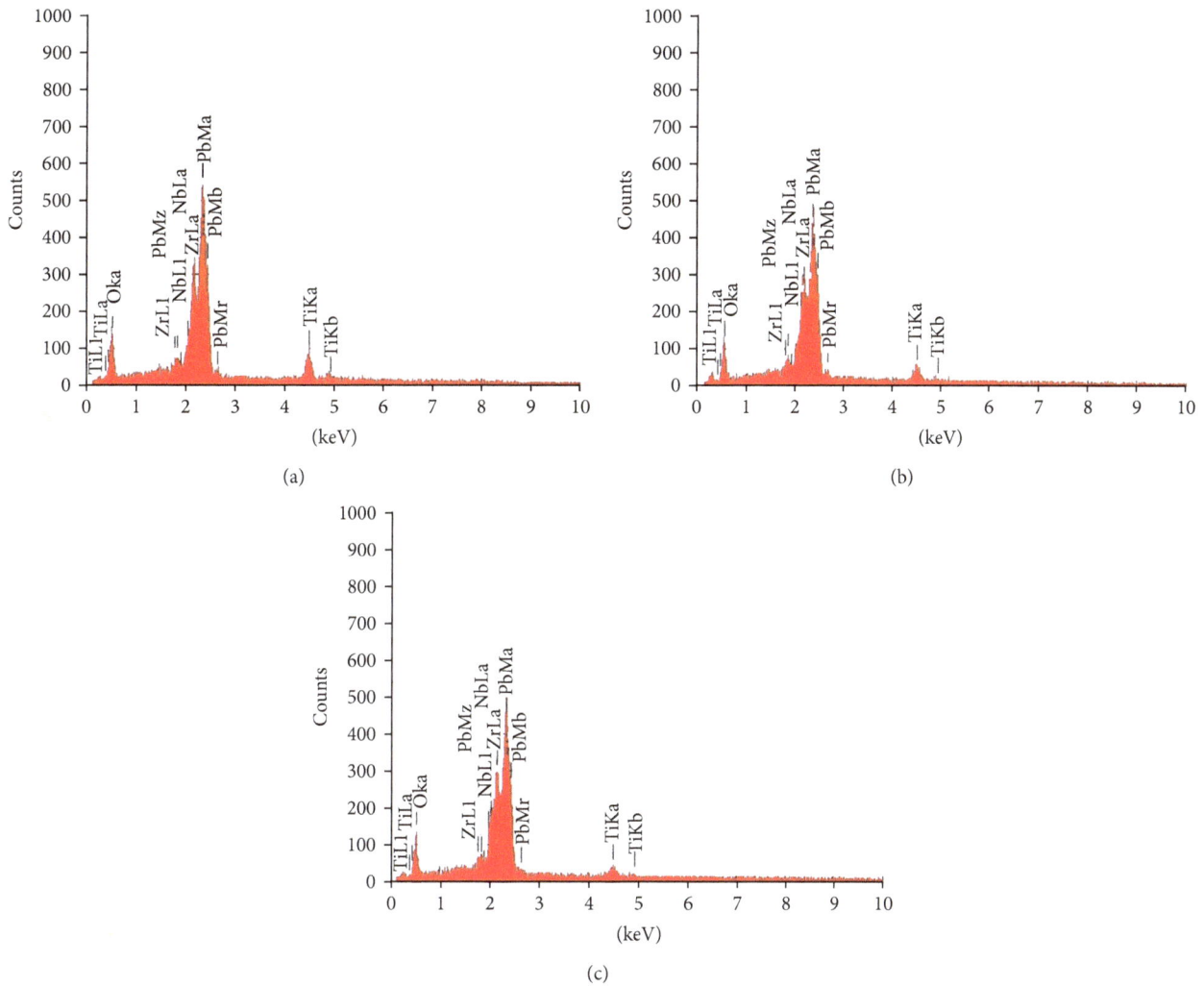

FIGURE 4: Energy dispersive X-ray analysis results for xPN–yPZ–$(x$-$y)$PT (PNZT) composite: (a) $x = 50$, $y = 15\%$; (b) $x = 50$, $y = 25\%$; and (c) $x = 50$, $y = 35\%$.

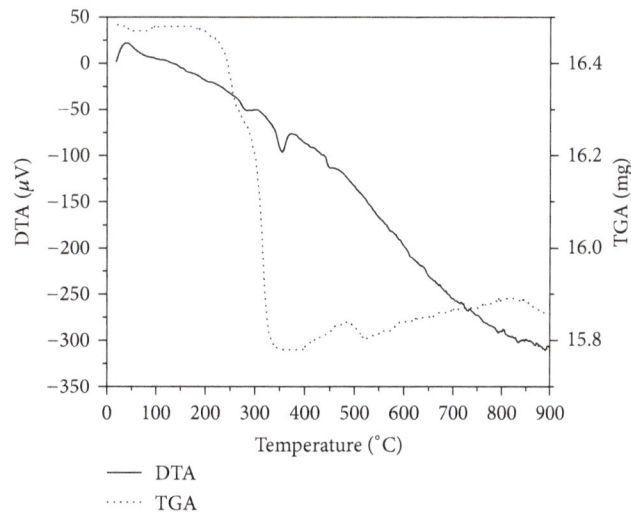

FIGURE 5: Thermogravimetric analysis along with differential thermal analysis (TGA/DTA) results for milled powder having composition 0.5PN–0.15PZ–0.35PT.

(a)

(b)

(c)

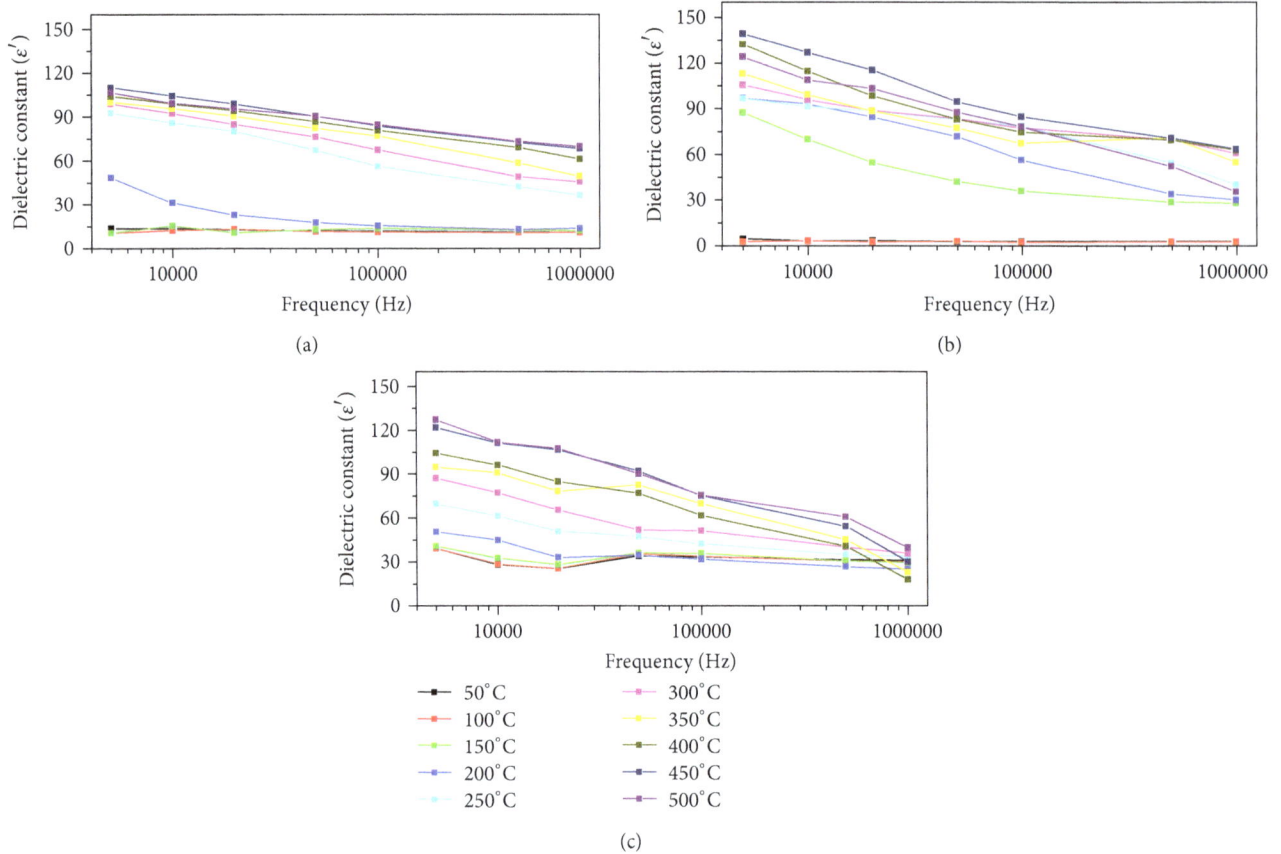

FIGURE 6: Variation of real part of dielectric constant with frequency for xPN–yPZ–$(x$-$y)$PT (a) $x = 50$, $y = 15\%$; (b) $x = 50$, $y = 25\%$; and (c) $x = 50$, $y = 35\%$.

relaxation process. At very high frequencies ($\omega \gg 1/\tau$), the dipoles are not able to align with the applied field and hence we have ($\varepsilon' \sim \varepsilon_\infty$). The maximum and minimum value of dielectric constant was found to be 111, 145, 128 and 11.7, 0.65, 37.4 for $x = 50$, $y = 15$, 25, and 35% in xPN–yPZ–$(x$-$y)$PT, respectively. The temperature dependence of dielectric constant (ε', the real part of dielectric permittivity) at different frequencies is shown in Figure 8. Two types of dielectric anomalies were observed, one below 350°C, which may be due to ferroelectric to antiferroelectric phase transition and another above 350°C, which may be due to antiferroelectric to paraelectric phase transition. Thus, for these three compositions, two dielectric anomalies were seen which is of relaxor type. This relaxor-like anomaly was widely reported in various materials [35] and the anomaly in the temperature range of 400–900°C in oxide materials especially for those containing titanium is related to oxygen vacancies [36]. However, in present case, the formation of two much diffused peak at two different dielectric anomalies may be suggested to be due to the formation of mixed phase. Hence, in these three compositions because of poor peak formation, we are not able to give exact Curie temperature. At a given temperature other than the anomaly regions, the dielectric constant decreases with an increase in frequency and the dependence of dielectric constant with frequency increasingly varies with temperature. With increase in temperature,

the peak position of dielectric constant in the curve shifts towards higher frequency side and this indicates frequency dispersion, thereby suggesting the thermally activated nature of dielectric relaxation. Further, all the plots show broad maxima, that is, diffuse phase transitions. In solid solutions, this broadening of peak or diffuse phase transition is a common feature due to the presence of more than one cation in the sublattice which produces some kind of heterogeneity. The poly-dispersive nature of dielectric relaxation can be analysed quantitatively through complex Argand plane plots of imaginary part of dielectric constant (ε'') versus real part of dielectric constant (ε') commonly called Cole-Cole plots. For a pure nondispersive Debye process, the plot is a perfect semicircle with its center located on the real axis. However, for poly-dispersive relaxation, the plot is a distorted semicircle with end points on the real axis and whose center lies below the real axis. Mathematically, the Cole-Cole plot obeys the following empirical relation [37]:

$$\varepsilon^* = \varepsilon' - j\varepsilon'' = \varepsilon_\infty + \frac{(\varepsilon_s - \varepsilon_\infty)}{\left[1 + (i\omega\tau)^{1-\alpha}\right]}. \qquad (2)$$

Here, "ε_∞" is the high frequency limit of the permittivity, "ε_s" is the static dielectric constant, "$\varepsilon_s - \varepsilon_\infty$" is the dielectric strength, "ω" is the angular frequency, "τ" is the mean relaxation time, and the parameter "α" represents the distribution

(a)

(b)

■ 220°C	▲ 400°C
● 250°C	◄ 450°C
▲ 300°C	● 500°C
▼ 350°C	

(c)

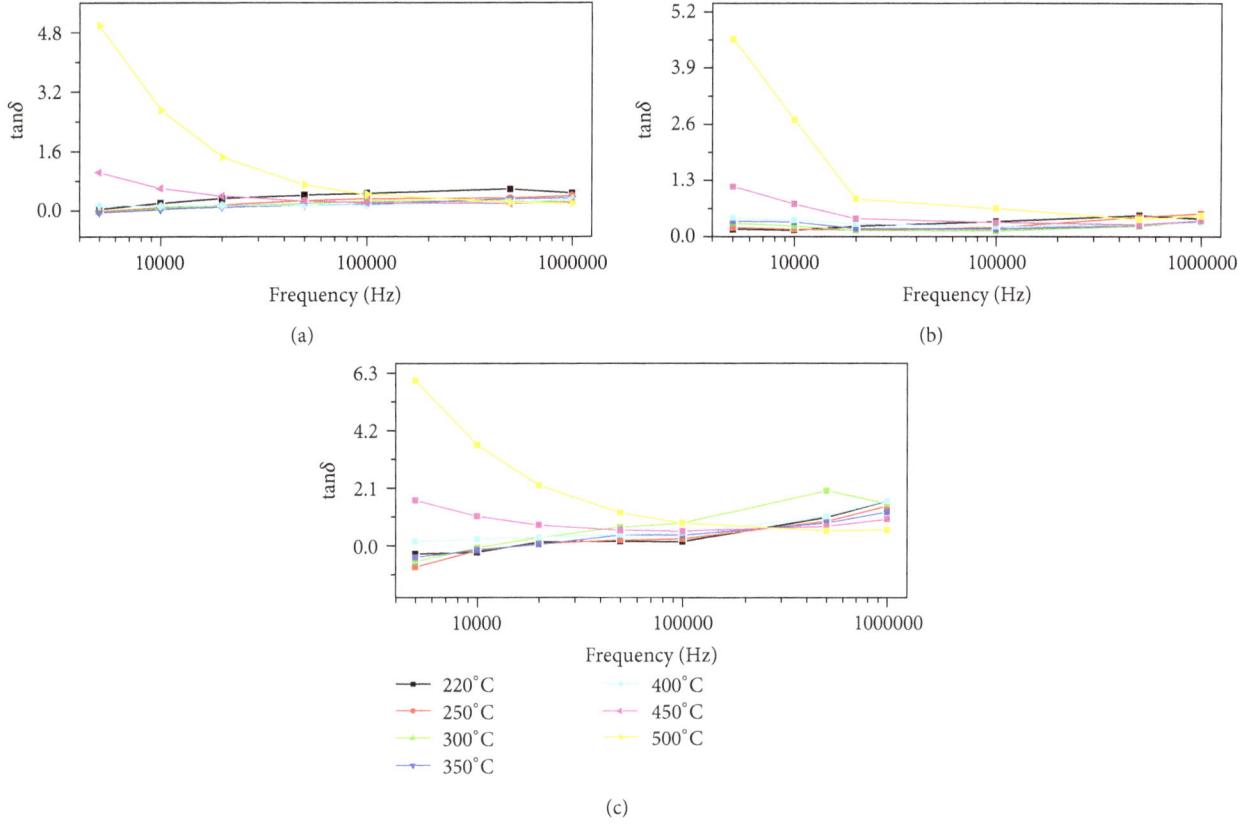

FIGURE 7: Variation of dielectric loss with frequency for xPN–yPZ–$(x$-$y)$PT (a) $x = 50$, $y = 15\%$; (b) $x = 50$, $y = 25\%$; and (c) $x = 50$, $y = 35\%$.

of relaxation time and can be determined from the angle subtended by the radius of Cole-Cole plot with the real axis passing through the origin of the imaginary axis. At "ε_∞" and "ε_s", there will be no dielectric loss and maximum loss occurs at the midpoint between the two dielectric values. Figures 9(a)–9(c) depict the Cole-Cole plot representations at $T = 200°C$ for all the three compositions. At low temperatures, that is, less than 200°C, the Cole-Cole plot resembles almost straight lines (not shown here) with large slopes, suggesting the insulating behaviour at low temperatures. However, with increasing temperature, the slopes of these curves decrease and they bend towards the real axis forming semicircles suggesting an increase in conductivity of the material with temperature. The formation of two semicircular arcs in the Cole-Cole plot can be explained on the basis of two parallel RC elements connected in series, one associated with grain (higher frequency range) and the other with grain boundary (lower frequency range) as the relaxation time for the grain boundary is much larger than that of the bulk crystal, thereby indicating negative temperature coefficient of resistance (NCTR) behaviour of PNZT like that of a semiconducting material [38]. As "ε'" is directly related to capacitance, that is, $\varepsilon' = Ct/\varepsilon_0 A$, where "$A$" is area in mm^2 and "$t$" is thickness of sample in mm, the capacitances C_g (associated with grain) and C_{gb} (associated with grain boundary) have been calculated from the low frequency intercept of the semicircles on the real ε'-axis. As relaxation

TABLE 2: Values of grain resistance (R_g), grain capacitance (C_g), grain boundary resistance (R_{gb}), and grain boundary capacitance (C_{gb}) obtained from Cole-Cole plot for PNZT system.

Composition	$R_g \times 10^4$ (Ω)	C_g (pF)	$R_{gb} \times 10^5$ (Ω)	C_{gb} (pF)
0.5PN–0.15PZ–0.35PT	42.3	23.61	24.8	40.26
0.5PN–0.25PZ–0.25PT	30.0	59.54	68.5	29.21
0.5PN–0.35PZ–0.15PT	65.1	5.12	44.5	14.97

time, $\tau_g = R_g C_g$ and $\tau_{gb} = R_{gb} C_{gb}$, the resistances R_g and R_{gb} were obtained from the maximum condition of the semicircles, that is, $\omega_{mac} = 1/\tau = 2\pi f_{max}$, where "$f_{max}$" is the frequency at the maximum of semicircle. The values of R_g, R_{gb}, C_g, and C_{gb} obtained from the Cole-Cole plot are given in Table 2. As seen from table, bulk resistance decreases with increase in grain size. Also, "α" lies below the real axis and its value was found to be 0.42, 0.37, and 0.16 radians for the three compositions, thereby confirming the non-Debye type relaxation in PNZT.

3.4. ac Conductivity Analysis. ac conductivity (σ_{ac}) has been calculated from the dielectric loss (tan δ) data using the empirical formula [39]:

$$\sigma_{ac} = \omega\varepsilon_0\varepsilon' \tan \delta = 2\pi f \varepsilon_0 \varepsilon' \tan \delta, \qquad (3)$$

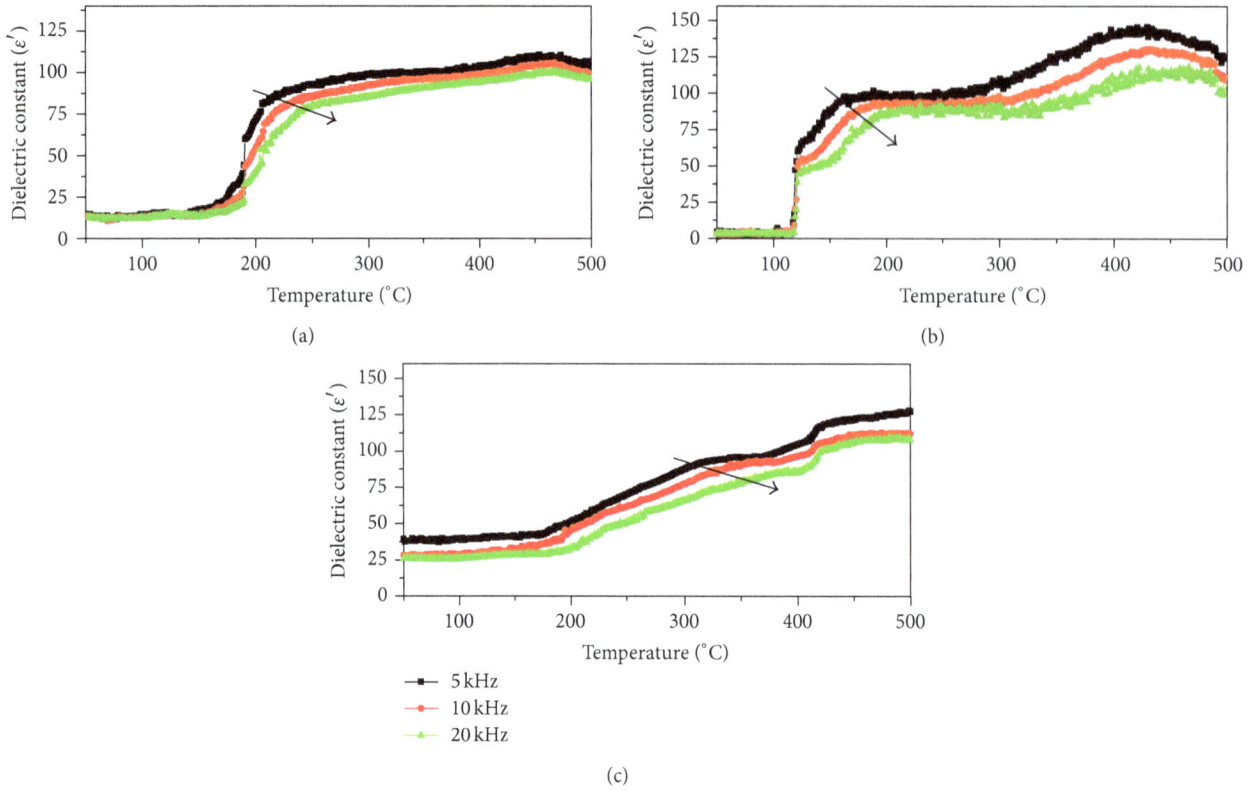

FIGURE 8: Variation of real part of dielectric constant with temperature at 5, 10, and 20 kHz frequency for xPN–yPZ–$(x$-$y)$PT (a) $x = 50$, $y =$ 15%; (b) $x = 50$, $y = 25$%; and (c) $x = 50$, $y = 35$%.

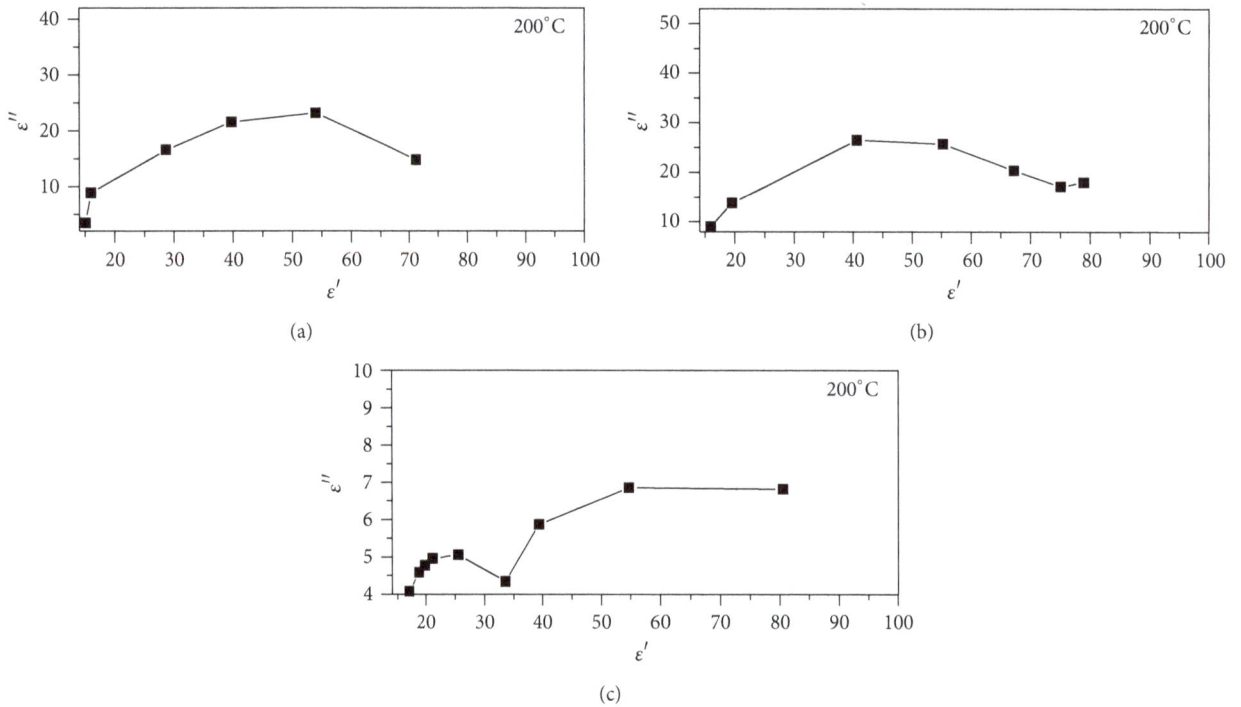

FIGURE 9: Complex plane Argand plot between ε'' and ε' at 200°C for xPN–yPZ–$(x$-$y)$PT (a) $x = 50$, $y = 15$%; (b) $x = 50$, $y = 25$%; and (c) $x = 50$, $y = 35$%.

TABLE 3: Variation of activation energy (E_a) with frequency and exponent (s) with temperature for three different compositions of PNZT system.

Composition	Activation energy (E_a) for different frequencies (eV)			Exponent (s) for different temperatures (°C)					
	5 kHz	10 kHz	20 kHz	250	300	350	400	450	500
0.5PN–0.15PZ–0.35PT	0.66	0.46	0.27	1.0	1.1	1.3	1.0	0.7	0.3
0.5PN–0.25PZ–0.25PT	0.24	0.16	0.13	1.0	0.9	0.8	0.8	0.6	0.3
0.5PN–0.35PZ–0.15PT	0.99	0.50	0.23	0.8	1.2	1.0	0.8	0.4	0.1

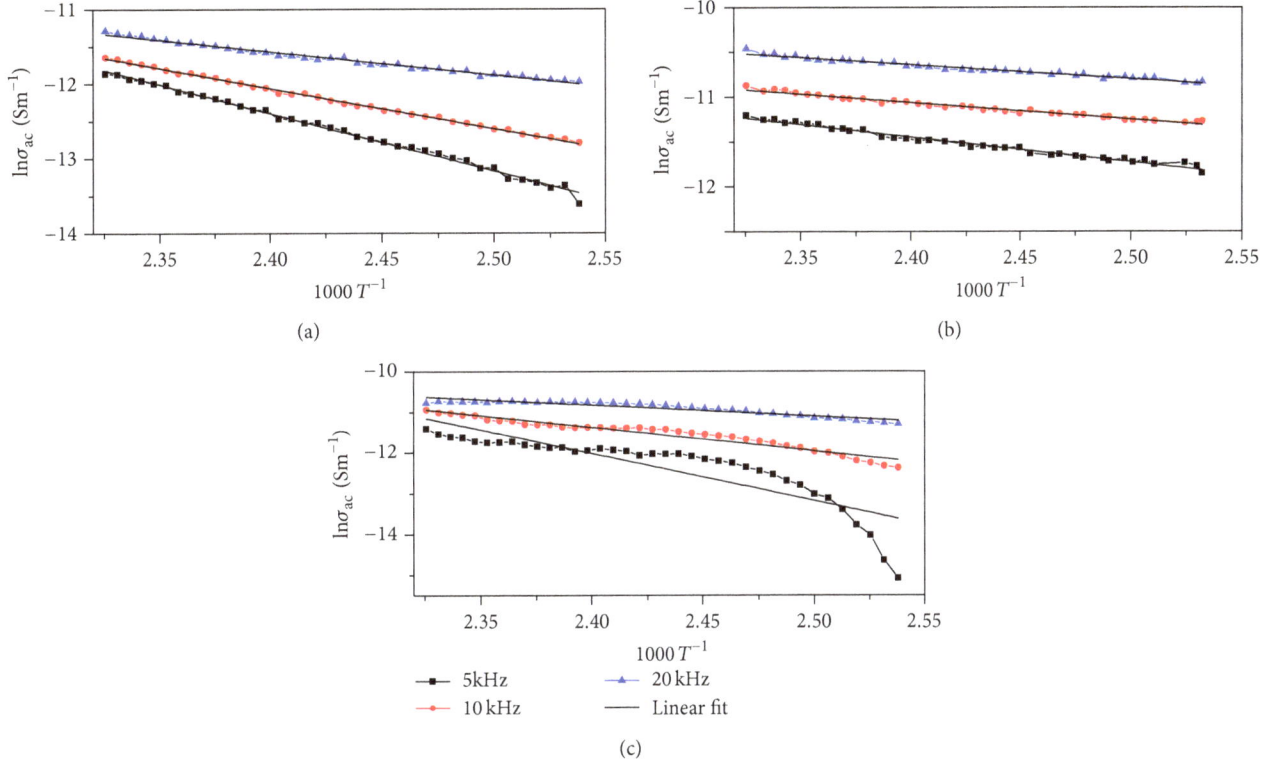

FIGURE 10: Temperature dependence of the ac conductivity at 5, 10, and 20 kHz frequency for xPN–yPZ–$(x$-$y)$PT (a) $x = 50$, $y = 15\%$; (b) $x = 50$, $y = 25\%$; and (c) $x = 50$, $y = 35\%$.

where $\varepsilon_0 = 8.845 \times 10^{-12}$ Fm^{-1} is the permittivity of free space. Figure 10 shows the variation of ac conductivity as a function of $1000/T$. The results indicate an increase in conductivity with rise in temperature, thereby indicating the negative temperature coefficient of resistance (NTCR) character. This type of temperature dependence indicates that the electrical conduction in this material is the thermally activated transport process governed by Arrhenius equation [29]:

$$\sigma_{ac} = \sigma_0 \exp\left(-\frac{E_a}{kT}\right), \tag{4}$$

where "σ_0" is the ac conductivity preexponential factor and "E_a," the ac conductivity activation energy. The activation energy of these three compositions in the temperature range of 394–430°C was calculated from the slope of $\ln(\sigma_{ac})$ versus $1000/T$ graph and is given in Table 3. As seen from the table, the activation energy shows a decreasing trend with increase in frequency. This is because of the reason that the increase in

the frequency of the applied field results in the enhancement of the charge carriers to jump between localized states which therefore results in the decrease in electrical activation energy with increasing frequency [40, 41]. The low value of activation energy may be due to the carrier transport through hopping between localized states in a disordered manner [42, 43]. Also, the activation energy is found to be minimum for the 0.5PN–0.25PZ–0.25PT composition thereby suggesting the decrease in resistivity. The frequency dependence of ac conductivity follows the universal power law [44, 45], that is,

$$\sigma(\omega) = A\omega^s, \tag{5}$$

where "A" is temperature dependent constant, "ω" is the angular frequency of the applied ac field, and the exponent "s" is temperature as well as frequency dependent constant and its value approaches unity at low temperature and decreases with increase in temperature. The values of "s" deduced from the slopes of $\ln(\sigma_{ac})$ versus $\ln(\omega)$ curves are given in Table 3, from which one can see that "s" is approaching

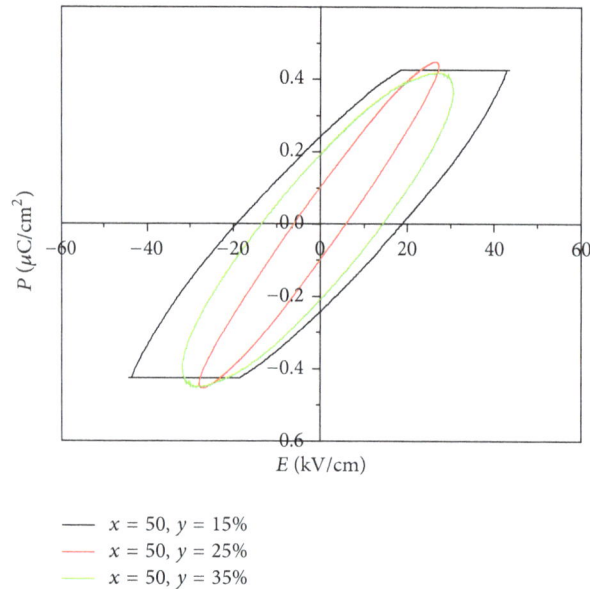

FIGURE 11: Variation of polarization with electric field for xPN–yPZ–$(x-y)$PT ($x = 50$, $y = 15$, 25, and 35%) at room temperature.

unity at low temperature which confirms assumption that ac conductivity is predominant due to the relaxation dipole moment. The power law dependence of the ac conductivity on frequency corresponds to the short range hopping of carriers through trap sites separated by an energy barrier of various heights. Since the value of "s" actually scales the extent of localized charge carriers [45], for example, in the case of $s = 0$, the above equation shows the usual reciprocal frequency behaviour, and the system is nondispersive transport of free charge carriers process; for $s = 1$, the system has the feature of nearly constant loss relating to strictly localized carriers, while for $0 < s < 1$, the system obeys the universal power law with confined hopping carriers. It, therefore, follows that the relaxing species for the low-temperature anomaly are confined carriers and for the high-temperature anomaly are free ones.

3.5. Ferroelectric Studies. The polarization versus electric field (P-E) hysteresis loops with different Zr concentrations are shown in Figure 11 whereas the values of remanent polarization and coercivity are given in Table 4. The minimum value of remnant polarization, $P_r \sim 0.10 \, \mu C/cm^2$, and the coercive field, $E_c \sim 6.33 \, kV/cm$, was observed for the composition 0.5PN–0.25PZ–0.25PT. The saturation polarization was obtained only in the case of 0.5PN–0.15PZ–0.35PT with a value of $P_s = 0.43 \, \mu C/cm^2$ at a maximum applied field of 20 kV/cm. However, in the other two cases, no saturation in polarization in P-E curve was achieved up to the maximum applied field which is due to the low resistivity of the composition [38]. The compositions with 25 and 35% Zr content could not withstand the electric field beyond 14 and 16 kV/cm and further increase in electric field led to the electrical breakdown and thus unsaturated P-E loops were obtained in these two compositions. Low value of coercivity and remanent polarization suggests that these ternary systems are soft ferroelectric.

TABLE 4: Variation of remnant polarization (P_r) and coercivity (E_c) for PNZT ternary system.

Composition	Remnant polarization (P_r) ($\mu C/cm^2$)	Coercivity (E_c) (kV/cm)
0.5PN–0.15PZ–0.35PT	0.24	19.15
0.5PN–0.25PZ–0.25PT	0.10	6.33
0.5PN–0.35PZ–0.15PT	0.20	13.61

4. Conclusions

Polycrystalline compositions of xPN–yPZ–$(x-y)$PT (where $x = 0.5$, $y = 0.15$, 0.25, and 0.35) known as PNZT ceramic were prepared by conventional mixed oxide route. Room temperature X-ray diffraction analysis of PNZT confirms the well-crystallized rhombohedral phase of the prepared compositions with the presence of both PN and PZT phase. Scanning electron micrographs depict formation of fairly dense grains with average grain size in the submicrometric range whereas presence of major elements was confirmed by energy dispersive X-ray analysis (EDAX). Thermal analysis of the milled powder suggests the optimum temperature for calcination is 900°C. Electrical properties carried by dielectric spectroscopy on PNZT indicate that the material exhibits two types of dielectric anomalies, one below 350°C and another above 350°C and shifting of maximum value of dielectric constant towards higher temperature side with increase in frequency indicating the relaxor behaviour. The temperature dependence of ac conductivity indicates increase in conductivity with rise in temperature thereby suggesting negative temperature coefficient of resistance (NCTR) character. The poly-dispersive nature of dielectric relaxation has been studied by complex plane Argand plot and its shape indicates the presence of both grain and grain boundary effect in the relaxation process in PNZT. The frequency dependence of ac conductivity obeys the universal power law. The ferroelectric studies suggest these ternary systems to be soft ferroelectric.

Conflict of Interests

The authors declare that there is no conflict of interests regarding publication of this paper.

Acknowledgments

The authors express their gratitude to Sophisticated Test and Instrumentation Centre (STIC), Cochin University, for providing XRD and SEM–EDAX facilities. They are also thankful to Dr. Binay Kumar, Department of Physics, University of Delhi, for providing the facility of P–E loop tracer.

References

[1] M. M. Vijatović, J. D. Bobić, and B. D. Stojanović, "History and challenges of barium titanate: part II," *Science of Sintering*, vol. 40, no. 3, pp. 235–244, 2008.

[2] N. Vittayakorn, G. Rujijanagul, X. Tan, M. A. Marquardt, and D. P. Cann, "The morphotropic phase boundary and dielectric properties of the $xPb(Zr_{1/2}Ti_{1/2})O_3$-(1-x)$Pb(Ni_{1/3}Nb_{2/3})O_3$ perovskite solid solution," *Journal of Applied Physics*, vol. 96, no. 9, pp. 5103–5109, 2004.

[3] L. E. Cross, "Relaxor ferroelectrics: an overview," *Ferroelectrics*, vol. 151, no. 1, pp. 305–320, 1994.

[4] S.-E. Park, "Relaxor based ferroelectric single crystals for electro-mechanical actuators," *Materials Research Innovations*, vol. 1, no. 1, pp. 20–25, 1997.

[5] G. Goodman, "Ferroelectric properties of lead metaniobate," *Journal of the American Ceramic Society*, vol. 36, no. 11, pp. 368–372, 1953.

[6] M. H. Francombe and B. Lewis, "Structural, dielectric and optical properties of ferroelectric leadmetaniobate," *Acta Crystallographica*, vol. 11, no. 10, pp. 696–703, 1958.

[7] E. C. Subbarao and G. Shirane, "Nonstoichiometry and ferroelectric properties of $PbNb_2O_6$-type compounds," *The Journal of Chemical Physics*, vol. 32, no. 6, pp. 1846–1851, 1960.

[8] L. A. Reznichenko, G. A. Geguzina, and N. V. Dergunova, "Piezoelectric solid solutions based on alkali niobates," *Inorganic Materials*, vol. 34, no. 2, pp. 167–173, 1998.

[9] I. P. Raevski and S. A. Prosandeev, "A new, lead free, family of perovskites with a diffuse phase transition: $NaNbO_3$-based solid solutions," *Journal of Physics and Chemistry of Solids*, vol. 63, no. 10, pp. 1939–1950, 2002.

[10] J. M. Haussonne, G. Desgardin, A. Herve, and B. Boufrou, "Dielectric ceramics with relaxors and a tetragonal tungsten bronze," *Journal of the European Ceramic Society*, vol. 10, no. 6, pp. 437–452, 1992.

[11] M. Venet, A. Vendramini, F. L. Zabotto, F. Guerrero, D. Garcia, and J. A. Eiras, "Piezoelectric properties of undoped and titanium or barium-doped lead metaniobate ceramics," *Journal of the European Ceramic Society*, vol. 25, no. 12, pp. 2443–2446, 2005.

[12] K. R. Sahu and U. De, "Thermal characterization of piezoelectric and non-piezoelectric Lead Meta-Niobate," *Thermochimica Acta*, vol. 490, no. 1-2, pp. 75–77, 2009.

[13] U. De, K. R. Sahu, K. R. Chakraborty, and S. K. Pratihar, "Dielectric and thermal investigations on $PbNb_2O_6$ in pure piezoelectric phase and pure non-piezoelectric phase," *Integrated Ferroelectrics*, vol. 119, no. 1, pp. 96–109, 2010.

[14] K. R. Sahu and U. De, "Dielectric properties of $PbNb_2O_6$ up to 700°C from impedance spectroscopy," *Journal of Materials*, vol. 2013, Article ID 702946, 15 pages, 2013.

[15] K. R. Sahu and U. De, "Dielectric properties of rhombohedral $PbNb_2O_6$," *Journal of Solid State Physics*, vol. 2013, Article ID 451563, 9 pages, 2013.

[16] Y. Xu, *Ferroelectric Materials and Their Applications*, Elsevier Science, Amsterdam, The Netherlands, 1991.

[17] G. H. Haertling, "Ferroelectric ceramics: history and technology," *Journal of the American Ceramic Society*, vol. 82, no. 4, pp. 797–818, 1999.

[18] X. Dai and Y. Wang, "Multistage heat-electric energy conversion working on $F_{R(LT)}$-$F_{R(HT)}$ phase transition in NB doped PZT97/3 ceramics," *Ferroelectrics*, vol. 109, no. 1, pp. 253–258, 1990.

[19] N. Duan, N. Cereceda, B. Noheda, and J. A. Gonzalo, "Dielectric characterization of the phase transitions in $Pb_{1-y/2}(Zr_{1-x}Ti_x)_{1-y}Nb_yO_3(0.03 < x < 0.04, 0.02 < y < 0.05)$," *Journal of Applied Physics*, vol. 82, article 779, 1997.

[20] Z. Ujma, J. Handerek, and G. E. Kugel, "Phase transitions in Nb-doped $Pb(Zr_{0.95}Ti_{0.05})O_3$ ceramics investigated by dielectric, pyroelectric and Raman scattering measurements," *Ferroelectrics*, vol. 198, no. 1, pp. 77–97, 1997.

[21] H. C. Nie, X. L. Dong, N. B. Feng et al., "Quantitative dependence of the properties of $Pb_{0.99}(Zr_{0.95}Ti_{0.05})_{0.98}Nb_{0.02}O_3$ ferroelectric ceramics on porosity," *Materials Research Bulletin*, vol. 45, no. 5, pp. 564–567, 2010.

[22] R. D. Klissurska, K. G. Brooks, I. M. Reaney, C. Pawlaczyk, M. Kosec, and N. Setter, "Effect of Nb doping on the microstructure of sol-gel-derived PZT thin films," *Journal of the American Ceramic Society*, vol. 78, no. 6, pp. 1513–1520, 1995.

[23] V. Kayasu and M. Ozenbas, "The effect of Nb doping on dielectric and ferroelectric properties of PZT thin films prepared by solution deposition," *Journal of the European Ceramic Society*, vol. 29, no. 6, pp. 1157–1163, 2009.

[24] H. Han, S. Kotru, J. Zhong, and R. K. Pandey, "Effect of Nb doping on pyroelectric property of lead zirconate titanate films prepared by chemical solution deposition," *Infrared Physics and Technology*, vol. 51, no. 3, pp. 216–220, 2008.

[25] D. F. Ryder and N. K. Raman, "Sol-Gel processing of Nb-doped $Pb(Zr, Ti)O_3$ thin films for ferroelectric memory applications," *Journal of Electronic Materials*, vol. 21, no. 10, pp. 971–975, 1992.

[26] R. Gerson, "Variation in ferroelectric characteristics of lead zirconate titanate ceramics due to minor chemical modifications," *Journal of Applied Physics*, vol. 31, no. 1, pp. 188–194, 1960.

[27] R. Gerson and H. Jaffe, "Electrical conductivity in lead titanate zirconate ceramics," *Journal of Physics and Chemistry of Solids*, vol. 24, no. 8, pp. 979–984, 1963.

[28] S. K. Hau, K. H. Wong, P. W. Chan, and C. L. Choy, "Intrinsic resputtering in pulsed-laser deposition of lead-zirconate-titanate thin films," *Applied Physics Letters*, vol. 66, pp. 245–247, 1995.

[29] K. R. Chakraborty, K. R. Sahu, A. De, and U. De, "Structural characterization of orthorhombic and rhombohedral lead meta-niobate samples," *Integrated Ferroelectrics*, vol. 120, no. 1, pp. 102–113, 2010.

[30] B. D. Cullity and S. R. Stock, *Elements of X-Ray Diffraction*, Prentice Hall, Englewood Cliffs, NJ, USA, 3rd edition, 2001.

[31] V. D. Mote, Y. Purushotham, and B. N. Dole, "Williamson-Hall analysis in estimation of lattice strain in nanometer-sized ZnO particles," *Journal of Theoretical and Applied Physics*, vol. 6, article 6, 2012.

[32] M. N. Rifaya, T. Theivasanthi, and M. Alagar, "Chemical cappingsynthesis of nickel oxide nanoparticles and their characterization studies," *Nanoscience and Nanotechnology*, vol. 2, no. 5, pp. 134–138, 2012.

[33] A. Shukla and R. N. P. Choudhary, "Ferroelectric phase-transition and conductivity analysis of La^{3+}/Mn^{4+} modified $PbTiO_3$ nanoceramics," *Physica B: Condensed Matter*, vol. 405, no. 11, pp. 2508–2515, 2010.

[34] B. Praveenkumar, H. H. Kumar, D. K. Kharat, and B. S. Murty, "Investigation and characterization of La-doped PZT nanocrystalline ceramic prepared by mechanical activation route," *Materials Chemistry and Physics*, vol. 112, no. 1, pp. 31–34, 2008.

[35] C. C. Wang and S. X. Dou, "Pseudo-relaxor behaviour induced by Maxwell-Wagner relaxation," *Solid State Communications*, vol. 149, no. 45-46, pp. 2017–2020, 2009.

[36] C. Ang, Z. Yu, and L. E. Cross, "Oxygen-vacancy-related low-frequency dielectric relaxation and electrical conduction in $Bi:SrTiO_3$," *Physical Review B—Condensed Matter and Materials Physics*, vol. 62, no. 1, pp. 228–236, 2000.

[37] K. S. Cole and R. H. Cole, "Dispersion and absorption in dielectrics I. Alternating current characteristics," *The Journal of Chemical Physics*, vol. 9, no. 4, pp. 341–351, 1941.

[38] A. K. Roy, A. Singh, K. Kumari, K. A. Nath, A. Prasad, and K. Prasad, "Electrical properties and AC conductivity of $(Bi_{0.5}Na_{0.5})_{0.94}Ba_{0.06}TiO_3$ ceramic," *ISRN Ceramics*, vol. 2012, Article ID 854831, 10 pages, 2012.

[39] R. N. P. Choudhary, D. K. Pradhan, G. E. Bonilla, and R. S. Katiyar, "Effect of La-substitution on structural and dielectric properties of $Bi(Sc_{1/2}Fe_{1/2})O_3$ ceramics," *Journal of Alloys and Compounds*, vol. 437, no. 1-2, pp. 220–224, 2007.

[40] A. A. Ebnalwaled, "On the conduction mechanism of p-type GaSb bulk crystal," *Materials Science and Engineering B: Solid-State Materials for Advanced Technology*, vol. 174, no. 1–3, pp. 285–289, 2010.

[41] N. A. Hegab, M. A. Afifi, H. E. Atyia, and A. S. Farid, "AC conductivity and dielectric properties of amorphous $Se_{80}Te_{20-x}Ge_x$ chalcogenide glass film compositions," *Journal of Alloys and Compounds*, vol. 477, no. 1-2, pp. 925–930, 2009.

[42] S. Upadhyay, A. K. Sahu, D. Kumar, and O. Parkash, "Probing electrical conduction behavior of $BaSnO_3$," *Journal of Applied Physics*, vol. 84, no. 2, pp. 828–832, 1998.

[43] K. Prasad, C. K. Suman, and R. N. P. Choudhary, "Electrical characterisation of $Pb_2Bi_3SmTi_5O_{18}$ ceramic using impedance spectroscopy," *Advances in Applied Ceramics*, vol. 105, no. 5, pp. 258–264, 2006.

[44] A. G. Hunt, "Ac hopping conduction: perspective from percolation theory," *Philosophical Magazine B*, vol. 81, no. 9, pp. 875–913, 2001.

[45] A. R. Long, *Hopping Transport in Solid*, North-Holland, Amsterdam, The Netherlands, 1991.

15

Coefficient of Thermal Diffusivity of Insulation Brick Developed from Sawdust and Clays

E. Bwayo and S. K. Obwoya

Department of Physics, Kyambogo University, P.O. Box 1, Kyambogo, Kampala, Uganda

Correspondence should be addressed to S. K. Obwoya; ksobwoya@yahoo.co.uk

Academic Editor: Jim Low

This paper presents an experimental result on the effect of particle size of a mixture of ball clay, kaolin, and sawdust on thermal diffusivity of ceramic bricks. A mixture of dry powders of ball clay, kaolin of the same particle size, and sawdust of different particle sizes was mixed in different proportions and then compacted to high pressures before being fired to 950°C. The thermal diffusivity was then determined by an indirect method involving measurement of thermal conductivity, density, and specific heat capacity. The study reveals that coefficient of thermal diffusivity increases with decrease in particle size of kaolin and ball clay but decreases with increase in particle size of sawdust.

1. Introduction

In a recent study by Manukaji [1], thermal diffusivity is very important in all nonequilibrium heat conduction problems in solid objects. The time rate of change of temperature depends on the numerical value of thermal diffusivity. The physical significance of thermal diffusivity is associated with the diffusion of heat into the medium during changes of temperature with time. Nonequilibrium heat transfer is important because of the large number of heating and cooling problems occurring industrially [2]. In metallurgical processes it is necessary to predict cooling and heating rates for various geometries of conductors in order to predict the time required to reach certain temperatures. Materials with high thermal mass will take longer for heat to travel from the hot face of the brick to the cold face and will also take long to release their heat once the heat source is removed [3, 4]. A paper by Aramide, [5], points out that when brick samples made with sawdust are fired, the sawdust admixture burns off between 450–550°C, [6] leaving pores (air voids) in the brick, which retards heat flow.

One of the problems facing the building industry in Uganda is the high consumption of electric energy caused by poor ventilation and air conditioning systems. This is mainly due to lack of thermal insulation techniques in buildings, [7, 8]. Nevertheless, there are no classified thermal insulators produced in Uganda. The country depends on imported insulation materials which are very expensive and are not easily accessed by the local industry and yet there are abundant mineral deposits in different parts of the country, which can provide potential raw materials for the production of different ceramic products like thermal insulating bricks. Thus, this paper presents the results of an experimental study of the effect of particle size on thermal diffusivity of clay bricks of composition as shown in Table 1, which have been fabricated from a combination of kaolin, ball clay, and wood sawdust of different particle sizes.

2. Experimental Procedures

2.1. Materials Processing. The raw materials used in this study were kaolin, ball clay, and hard wood sawdust. The sawdust was obtained from mahogany. The hard wood was preferred because when incorporated in clay bricks, it forms uniform pores, has high calorific values, and does not cause bloating, [9]. The kaolin was collected from Mutaka in South Western Uganda while ball clay was collected from Ntawo (Mukono), 25 km east of the capital city, Kampala. The ball clay and kaolin were separately soaked in water for seven days to allow them to dissolve completely in order to separate the colloids from heavy particles like stones, sand, and roots. The clay was then dried and milled into powder form in electric ball mill. The powders were sieved through test sieves stuck together on a mechanical test sieve shaker. The particle size ranges 0–45 μm, 45–53 μm, 53–63 μm, 63–90 μm, 90–125 μm, and 125–154 μm were obtained separately for kaolin and ball clay. Similarly, powders of sawdust of particle size ranges 0–125 μm, 125–154 μm, 154–180 μm, 180–355 μm, and 355–425 μm were also prepared.

The study was carried out using two sets of batch formulations. In the first part, the batch formulations A_1–A_5 had compositions of kaolin and ball clay of the same particle size ranges, which were mixed with equal masses of sawdust of three different particle size ranges in the ratios 9 : 7 : 4 by weight as shown in Table 1. The mixture of these powders was first sun dried and then compacted to pressure of 50 MPa into rectangular specimens with dimensions of 10.51 cm × 5.25 cm × 1.98 cm. The test samples were fired to 950°C in an electric furnace in two stages. In the first stage they were dried at a heating rate of 2.33°C min^{-1} to 110°C and this temperature was maintained for four hours in order to remove any water in the sample. In the second stage, the samples were fired at a rate of 6°C min^{-1} to 950°C. At this temperature, the holding time was one hour before the furnace was switched off to allow the samples to cool naturally to room temperature.

In the second part of the study, batch formulations B_1–B_5 had each of the particle size ranges 0–125 μm, 125–154 μm, 154–180 μm, 180–355 μm, and 355–425 μm of sawdust mixed with kaolin and ball clay of the same particle size ranges in the ratio 4 : 9 : 7 as shown in (Table 1) before compacting them at a pressure of 50 MPa into rectangular specimens of dimensions 10.51 cm × 5.25 cm × 1.98 cm. A similar firing process as for the first batch formulation was followed. Each of the sample formulations had a total mass of 200 g (90 g of kaolin, 70 g of ball clay, and 40 g of sawdust).

2.2. Determination of the Coefficient of Thermal Diffusivity. The coefficient of thermal diffusivity was determined from measured values of the specific heat capacity, thermal conductivity, and density using the following equation derived from Fourier's law of heat conduction through solid:

$$\alpha = \frac{k}{\ell c_p},$$ (1)

where α is the thermal diffusivity, k is the thermal conductivity, ℓ is density, and c_p is specific heat capacity [10].

The thermal conductivity was measured by the Quick Thermal Conductivity Meter (QTM-500) with sensor probe (PD-11) which uses transient technique (nonsteady state) to study the heat conduction of the samples [11, 12]. Specific heat capacity was determined by method of mixtures [13] and the density was determined by measuring the dimensions and mass of the sample. Measurements of thermal conductivity, density, and specific heat capacity were performed at room temperature.

TABLE 1: Formulation of the brick samples.

Sample	Kaolin (90 g) + ball clay (70 g)	Sawdust addition (40 g)		
	Particle size (μm)			
A_1	90–125	0–125	125–154	154–180
A_2	63–90	0–125	125–154	154–180
A_3	53–63	0–125	125–154	154–180
A_4	45–53	0–125	125–154	154–180
A_5	0–45	0–125	125–154	154–180

Sample	Sawdust (40 g)	Kaolin (90 g) + ball clay (70 g) addition		
	Particle size (μm)			
B_1	0–125	63–90	90–125	125–154
B_2	125–154	63–90	90–125	125–154
B_3	154–180	63–90	90–125	125–154
B_4	180–355	63–90	90–125	125–154
B_5	355–425	63–90	90–125	125–154

2.3. Chemical Composition. The chemical composition of the fired samples was determined by the X-ray fluorescence (XRF), spectrometer, model X′ Unique ll [14], to establish the chemical composition of major compounds that influence the thermal properties of insulation clay bricks Table 2.

3. Results and Discussions

3.1. Effect of Particle Size on the Coefficient of Thermal Diffusivity. The coefficient of thermal diffusivity was determined by an indirect method involving the measurement of thermal conductivity, specific heat capacity, and density of fired samples [2, 10]. The effect of particle size on thermal conductivity, density, and specific heat capacity and thermal diffusivity is discussed below.

3.1.1. Effect of Particle Size on Thermal Conductivity. The results (Figure 1) show that thermal conductivity increases with decrease in particle size of kaolin and ball clay for fixed particle size of sawdust. This is because larger particles create large pores due to poor filling of the voids that contain air after firing compared to small particle sizes [15, 16]. Thermal conduction of a ceramic material depends on thermal conductive pathways which are affected by the microstructure, particle size distribution, and the amount of air space or voids created during the firing of a body [17]. Figure 2 shows that thermal conductivity decreases when the particle size of sawdust incorporated into the clay mix increases. This is because particle size of combustible organic waste determines the amount of air spaces that are created in the insulation clay brick [18–20]. In addition, thermal conductivity decreases further when particle size of the mixture of kaolin and ball clay increases due to less contact between particles [21]. Interlocking of clay particles depends on particle size distribution and particle size range of small

TABLE 2: Chemical composition of fired samples.

Compound	SiO_2	Al_2O_3	Fe_2O_3	CaO	TiO_2	Na_2O	MgO	K_2O	MnO_2	P_2O_5
Weight (%)	68.98	22.29	1.87	1.15	0.48	2.04	1.04	2.54	0.05	0.57

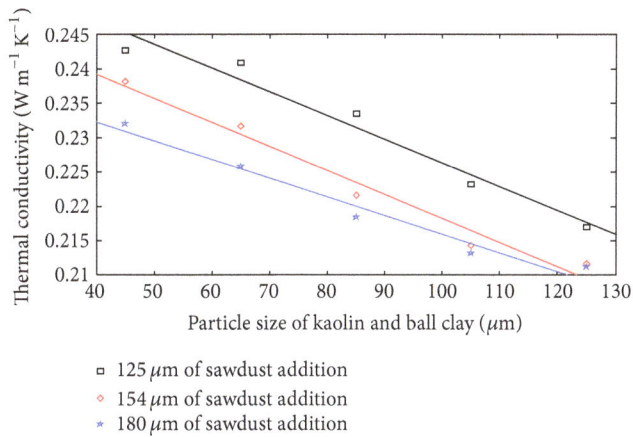

□ 125 μm of sawdust addition
◇ 154 μm of sawdust addition
✳ 180 μm of sawdust addition

FIGURE 1: Effect of particle size of kaolin and ball clay on thermal conductivity with different particle size of sawdust.

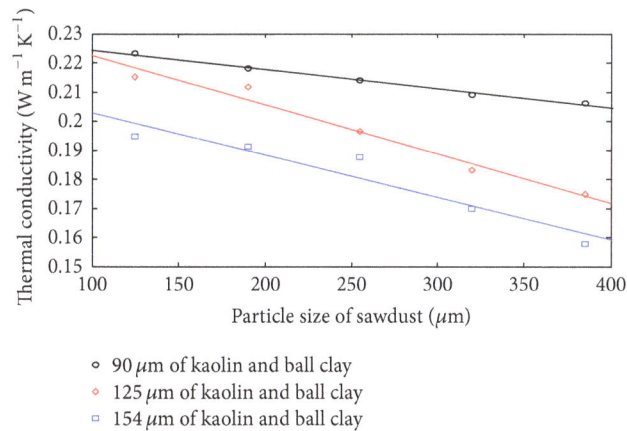

▽ 125 μm of sawdust addition
△ 154 μm of sawdust addition
▷ 180 μm of sawdust addition

FIGURE 3: Variation of density with particle size of kaolin and ball clay for different particle size of sawdust.

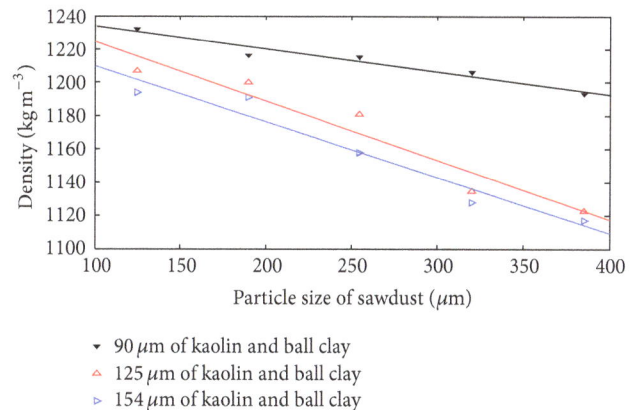

○ 90 μm of kaolin and ball clay
◇ 125 μm of kaolin and ball clay
□ 154 μm of kaolin and ball clay

FIGURE 2: Effect of particle size of sawdust on thermal conductivity at different particle size of a mixture of kaolin and ball clay.

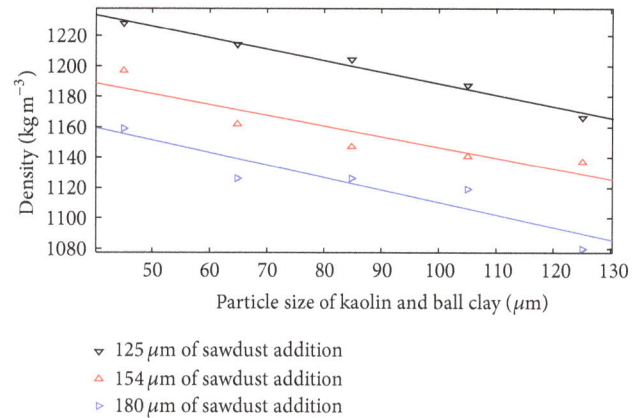

▼ 90 μm of kaolin and ball clay
△ 125 μm of kaolin and ball clay
▷ 154 μm of kaolin and ball clay

FIGURE 4: Variation of density with particle size of sawdust at different particle size of kaolin and ball clay.

and large particles and also on whether the body is made of monosized particles or multiple size particles.

3.1.2. Effect of Particle Size on Density. The density of the samples increases with decreasing particle size of the mixture of kaolin and ball clay at fixed particle size of sawdust (Figure 3). Smaller particle sizes have more contact points that allow for more cohesion and lubrication of kaolin by ball clays. Multiple particle sizes in a ceramic body increase packing of particles and produce high density body because finer grains go into the interparticle voids of the coarser particles and thus increase the packing density. This study also shows that there is further decrease in density with increase in particle size of sawdust at fixed particle size of kaolin and ball clay [20].

In Figure 4, density of the samples decreases with increase in particle size of sawdust for fixed particle size of kaolin and ball clay. Small pores that are created by small particle size of sawdust tend to be closed during densification as a result of formation of intergranular contact areas, while large pores will remain in the clay matrix during firing and maturing [18]. This is attributed to the sufficient length of sawdust that improves the bond at the sawdust-clay interface to oppose the deformation and clay contraction during drying and firing [9].

3.1.3. Variation of Specific Heat Capacity with Particle Size. The specific heat capacities for samples A_1 to A_5 are generally lower than those of B_1 to B_5 (Figures 5 and 6). This implies that lower thermal diffusivity can be achieved by use of bigger particle sizes of sawdust [9]. The specific heat capacity increases with increase in particle size of clay materials

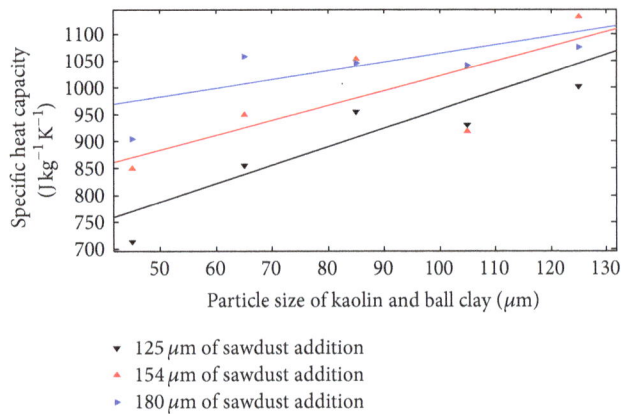

FIGURE 5: Effect of particle size of a mixture of kaolin and ball clay of specific heat capacity at different particle sizes of sawdust addition.

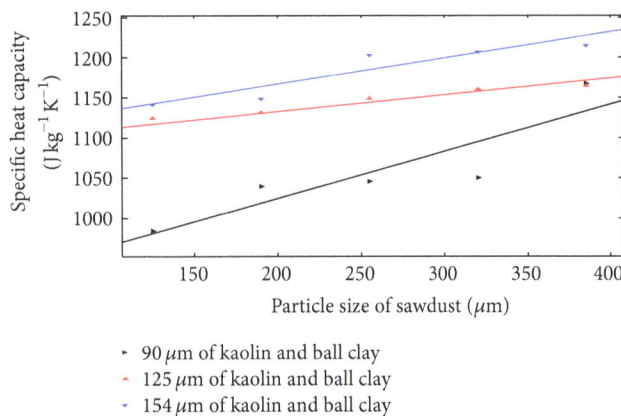

FIGURE 6: Effect of particle size of sawdust on specific heat capacity at different particle size of a mixture of kaolin and ball clay.

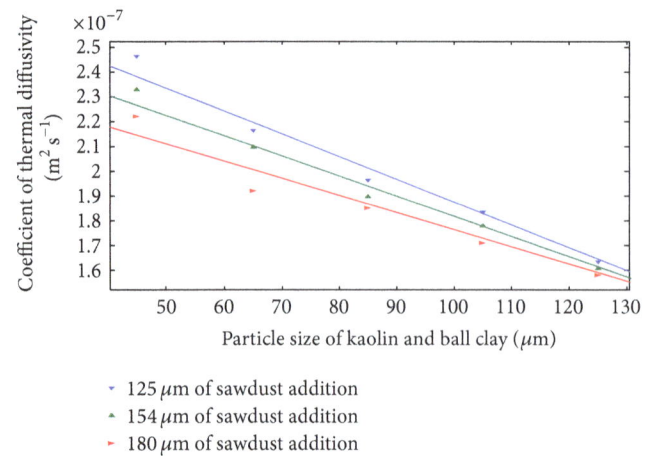

FIGURE 7: Effect of particle size of a mixture of kaolin and ball clay on thermal diffusivity at fixed particle size of sawdust.

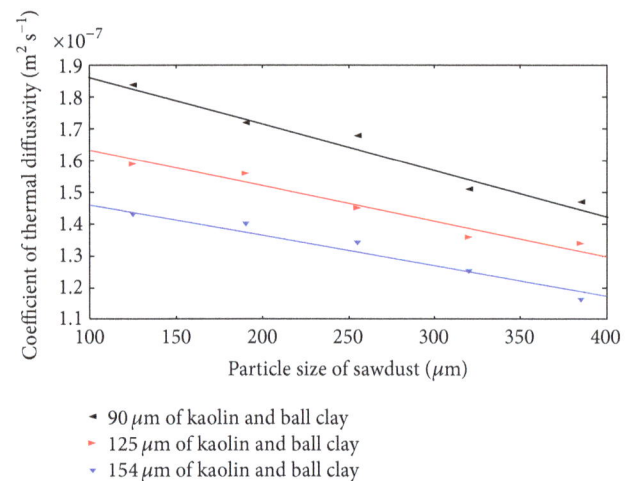

FIGURE 8: Variation of thermal diffusivity with particle size of sawdust at different particle size of a mixture of kaolin and ball clay.

(Figure 5) used and increase in particle size of sawdust addition (Figure 6).

3.1.4. Coefficient of Thermal Diffusivity. The coefficient of thermal diffusivity increases as the particle size of the mixture of kaolin and ball clay decreases at fixed particle size of sawdust addition (Figure 7). The principal effect of particle size on the thermal diffusivity of the solid material is related to the amount of solid and air space which the heat has to transverse in passing through the material. This is attributed to large particle size which results in high porosity levels due to poor filling of voids between large size particles compared to small sizes thus creating large air spaces [21]. The large proportion of air produces low value of the coefficient of thermal diffusivity because of its low thermal conductivity. The decrease in particle size increases particle content per unit volume which decreases the average interparticle distance of the clay matrix. This results in close packing of particles that leads to densification of clay bricks which increases thermal diffusivity [16, 20]. Hence, a fine-grained, closed-textured material (small particle size) has a much greater thermal diffusivity than one with a coarser open texture

(large particle size). Small particle sizes enhance low thermal resistance because the contact points for thermal conduction are very closely packed. Large grain size of kaolin and ball clay produces bricks that are more porous and thus more resistant to sudden temperature changes across the specimen [1, 22]. The low thermal diffusivity values are suitable for minimizing heat conduction. It is observed (Figure 7) that increasing particle size of sawdust addition further decreases the thermal diffusivity.

Thermal diffusivity decreases with increase in particle size of sawdust at fixed particle size of a combination of kaolin and ball clay (Figure 8). This is because particles of sawdust burn out between 450-550°C, [6] leaving pores or voids in the samples. During drying and firing, densification takes place and small pores created by small particle size of sawdust tend to be closed by clay minerals as a result of formation of intergranular contact areas, while large pores will persist in the clay matrix [18].

Incorporation of sawdust into the ceramic body which is removed during the firing step leaves pores whose sizes are related to the organic particle sizes. Finer sawdust forms smaller pores most of which may be eliminated during densification while large particle sizes form large pores. Sawdust of large particle size improves the bond at the sawdust-clay interface that opposes the deformation and clay contraction. This yields high porosity, low density, low thermal conductivity, and low rate of change of temperature across the specimen. Hence thermal diffusivity decreases as particle size of sawdust increases. Generally, the values of thermal diffusivity of B_1 to B_5 are lower than those of A_1 to A_5. This is a result of the multiplicative porosity created by clay and sawdust addition.

3.2. Chemical Composition. The percentage composition of SiO_2 is 68.0% while that of Al_2O_3 is 22.0%. According to the Bureau of Energy Efficiency [23] report on fireclay refractories, low density fireclay refractories consist of aluminum silicates with varying silica content between 67 and 77% and Al_2O_3 content between 23 and 33%. The chemical composition of alumina in the developed samples can be improved either by beneficiating the raw materials (kaolin and ball clay) or by increasing the percentage composition of kaolin in the samples. The clay samples contain less than 9.0% of the fluxing components (K_2O, Na_2O, and CaO).

3.3. Implications. The physical significance of low values of thermal diffusivity is associated with the low rate of change of temperature through the material during the heating process. The samples, therefore, have low values of the coefficient of thermal diffusivity and are suitable for use as thermal insulators. The suitable thermal insulator is the sample containing a combination of kaolin and ball clay of particle size range of 125–154 μm with sawdust of particle size range of 355–425 μm. This combination was characterized by the lowest value thermal diffusivity of 1.16×10^{-7} m^2 s^{-1} and can easily be prepared for commercial production of thermal insulation bricks.

4. Conclusions

The results of the study show that all the samples analyzed are good thermal insulators and the coefficient of thermal diffusivity is directly affected by particle size of a combination of kaolin and ball clay minerals as well as particle size of sawdust addition. Thus from the overall experimental analysis carried out, the following was observed.

(1) The coefficient of thermal diffusivity increases with decrease in particle size of a mixture of kaolin and ball clay at fixed particle size of sawdust addition. Sawdust addition of larger particle size decreases thermal diffusivity even at very small particle size of kaolin and ball clay.

(2) The coefficient of thermal diffusivity decreases with increase in particle size of sawdust addition to a fixed particle size of kaolin and ball clay. Incorporation of kaolin and ball clay of a much higher particle size further decreases the coefficient of thermal diffusivity due to the multiplicative effect of higher porosity produced by sawdust and clay minerals.

(3) The samples contain suitable compositions of silica and alumina which are suitable for lightweight high temperature insulation bricks.

(4) The samples, therefore, have low values of the coefficient of thermal diffusivity and are suitable for use as thermal insulators.

Conflict of Interests

The authors declare that there is no conflict of interests regarding the publication of this paper.

Acknowledgments

The authors would like to thank the staff at Kyambogo University for their guidance and support during the course of the study and research. Further thanks go to the Management and staff of Uganda Industrial Research Institute, UIRI (Department of Ceramics), for availing their labs and equipment to be used for the research and the Department of Physics, Makerere University. In a special way the authors wish to appreciate the financial support they received from Ms. Nanyama Christine, Dr. Mayeku Robert, and his wife Mrs. Kate Mayeku.

References

[1] J. U. Manukaji, "The effects of sawdust addition on the insulating characteristics of clays from the Federal Capital Territory of Abuja," *International Journal of Engineering Research and Applications*, vol. 3, no. 2, pp. 6–9, 2013.

[2] F. Colangelo, G. de Luca, and C. Ferone, *Experimental and Numerical Analysis of Thermal and Hygrometric Characteristics of Building Structures Employing Recycled Plastic Aggregates and Geopolymer Concrete*, University of Napoli Parthenope, Napoli, Italy, 2013.

[3] R. T. Faria Jr., V. P. Souza, C. M. F. Vieira, and R. Toledo, "Characterization of clay ceramics based on the recycling of industrial residues," in *Characterization, Raw Materials, Processing, Properties, Degradation and Healing*, C. Sikalidis, Ed., 2011.

[4] A. B. E. Bhatia, *Overview of Refractory Materials*, PDH Online, Fairfax, Va, USA, 2012.

[5] O. Aramide, Journal of Minerals and Materials Characterization and Engineering, 2012, http://www.SciRP.org/journal/jmmce.

[6] Government of Uganda, *State of Uganda Population Report: Planned Urbanization for Uganda's Growing Population*, Government of Uganda, Kampala, Uganda, 2007.

[7] S. Mukiibi, *The Effect of Urbanisation on the Housing Conditions of the Urban Poor in Kampala*, Department of Architecture, Makerere University, Kampala, Uganda, 2008.

[8] H. Chemani and B. Chemani, "Valorization of wood sawdust in making porous clay brick," *Scientific Research and Essays*, vol. 8, no. 15, pp. 609–614, 2013.

[9] R. E. B. Sreenivasula, *Basic Mechanisms of Heat Transfer—Fourier's Law of Heat Conduction*, College of Food Science and Technology, Ranga Agricultural Unuversity, Pulivendula, India, 2013.

[10] ASTM D 5334-92, D 5930-97 and IEEE 442-1981 standards.

[11] A.-B. Cherki, B. Remy, A. Khabbazi, Y. Jannot, and D. Baillis, "Experimental thermal properties characterization of insulating cork-gypsum composite," *Construction and Building Materials*, vol. 54, pp. 202–209, 2014.

[12] ASTM D411-08, *Standard Test Method for Specific Heat Capacity of Rocks and Soils*, ASTM International, West Conshohocken, Pa, USA, 2008.

[13] M. S. Shackley, *X-Ray Fluorescence Spectrometry (XRF) in Geoarchaeology*, Department of Anthropology, University of California, Berkeley, Calif, USA, 2011.

[14] A. A. Kadir, A. Mohajerani, F. Roddick, and J. Buck-eridge, "Density, strength, thermal conductivity and leachate characteristics of light-weight fired clay bricks incorporating cigarette butts," *International Journal of Civil and Environmental Engineering*, vol. 2, no. 4, pp. 1035–1040, 2010.

[15] G. Viruthagiri, S. Nithya Nareshananda, and N. Shanmugam, "Analysis of insulating fire bricks from mixtures of clays with sawdust addition," *Indian Journal of Applied Research (Physics)*, vol. 3, no. 6, 2013.

[16] W. Ling, A. Gu, G. Liang, and L. Yuan, "New composites with high thermal conductivity and low dielectric constant for microelectronic packaging," *Polymer Composites*, vol. 31, no. 2, pp. 307–313, 2010.

[17] H. Binici, O. Aksogan, M. N. Bodur, E. Akca, and S. Kapur, "Thermal isolation and mechanical properties of fibre reinforced mud bricks as wall materials," *Construction and Building Materials*, vol. 21, no. 4, pp. 901–906, 2007.

[18] R. Saiah, B. Perrin, and L. Rigal, "Improvement of thermal properties of fired clays by introduction of vegetable matter," *Journal of Building Physics*, vol. 34, no. 2, pp. 124–142, 2010.

[19] S. Zhang, X. Y. Cao, Y. M. Ma, Y. C. Ke, J. K. Zhang, and F. S. Wang, "The effects of particle size and content on the thermal conductivity and mechanical properties of Al_2O_3/high density polyethylene (HDPE) composites," *Express Polymer Letters*, vol. 5, no. 7, pp. 581–590, 2011.

[20] N. Meena Seema, *Effects of particle size distribution on the properties of alumina refractories [M.S. thesis]*, Department of Institute of Technology, Rourkela, India, 2011.

[21] A. G. E. Marwa, F. M. Mohamed, S. A. H. El-Bohy, C. M. Sharaby, C. M. El-Menshawi, and H. Shalabi, "Factors that affected the performance of fire clay refractory bricks," in *Gornictwo i Geoinzynieria, Rok 33, Zeszyt 4*, Central Metallurgical Research and Development Institute, Helwan, Egypt, 2009.

[22] Bureau of Energy Efficiency, *Energy Efficiency in Thermal Utilities*, Energy Efficiency Guide for Industry in Asia, Ministry of Power, UNEP, 2005.

[23] F. Colangelo, G. De Luca, and C. Ferone, *Experimental and Numerical Analysis of Thermal and Hygrometric Characteristics of Building Structures Employing Recycled Plastic Aggregates and Geopolymer Concrete*, University of Napoli Parthenope, Centro Direzionale, Naples, Italy, 2013.

16

Modeling of Thermal and Mechanical Behavior of ZrB$_2$-SiC Ceramics after High Temperature Oxidation

Jun Wei,[1] **Lokeswarappa R. Dharani,**[1] **K. Chandrashekhara,**[1]
Gregory E. Hilmas,[2] **and William G. Fahrenholtz**[2]

[1] *Department of Mechanical and Aerospace Engineering, Missouri University of Science and Technology, Rolla, MO 65409-0050, USA*
[2] *Department of Materials Science and Engineering, Missouri University of Science and Technology, Rolla, MO 65409-0340, USA*

Correspondence should be addressed to Lokeswarappa R. Dharani; dharani@mst.edu

Academic Editor: Guillaume Bernard-Granger

The effects of oxidation on heat transfer and mechanical behavior of ZrB$_2$-SiC ceramics at high temperature are modeled using a micromechanics based finite element model. The model recognizes that when exposed to high temperature in air ZrB$_2$-SiC oxidizes into ZrO$_2$, SiO$_2$, and SiC-depleted ZrB$_2$ layer. A steady-state heat transfer analysis was conducted at first and that is followed by a thermal stress analysis. A "global-local modeling" technique is used combining finite element with infinite element for thermal stress analysis. A theoretical formulation is developed for calculating the thermal conductivity of liquid phase SiO$_2$. All other temperature dependent thermal and mechanical properties were obtained from published literature. Thermal stress concentrations occur near the pore due to the geometric discontinuity and material properties mismatch between the ceramic matrix and the new products. The predicted results indicate the development of thermal stresses in the SiO$_2$ and ZrO$_2$ layers and high residual stresses in the SiC-depleted ZrB$_2$ layer.

1. Introduction

Ultrahigh temperature ceramics (UHTCs) such as zirconium diboride and hafnium diboride (ZrB$_2$ and HfB$_2$) have been proposed for thermal protection of hypersonic aerospace vehicles, which may be exposed to temperatures above 1500°C in oxidizing environments. These materials are chemically and physically stable above 1600°C and have melting points above 3000°C [1]. In particular, ZrB$_2$ because of its lower theoretical density is attractive for aerospace applications [2]. Exposure of solid zirconium boride (ZrB$_2$ (s)) to air at elevated temperatures results in its oxidation to solid zirconia (ZrO$_2$ (s)) and liquid boria (B$_2$O$_3$ (l)). The oxidation resistance of ZrB$_2$ (s) can be improved by adding SiC (s) to promote the formation of a silica-rich scale. At high temperature, above 1100°C, SiC (s) oxidizes by reaction to form SiO$_2$ (l) which has a lower volatility and a higher melting point and viscosity compared with B$_2$O$_3$ (l) [3–5]. Based on the experimental observations, Fahrenholtz [3] proposed a reaction sequence for the formation of the SiC-depleted layer

during the oxidation of ZrB$_2$-SiC at 1500°C in air. The oxide scales that form on ZrB$_2$-SiC consist of an outer layer of SiO$_2$, a middle layer of porous ZrO$_2$, sometimes filled with SiO$_2$, and a layer of SiC-depleted ZrB$_2$ adjoining the unoxidized ZrB$_2$-SiC at around 1500°C [2–6]. Parthasarathy et al. [7] developed a chemical reaction model for the oxidation of ZrB$_2$-SiC ceramics to predict the thicknesses of the above three new productions. For temperatures below ~1600°C, an external glassy SiO$_2$ layer forms and completely fills in pores of the porous ZrO$_2$ scale whereas at higher temperatures, the glassy scale recedes due to evaporation of SiO$_2$ (l) so that it only partially fills the pores in the ZrO$_2$ layer.

The region of particular interest, from a mechanical perspective, is the interface between the pores and the corner of the pores in the ZrO$_2$ scale. The pore itself may or may not be filled with liquid SiO$_2$ (l). The interface therefore consists of three materials (ZrO$_2$ scale, solid/liquid SiO$_2$, and SiC-depleted ZrB$_2$ layer) of significantly different thermal and mechanical properties. This thermomechanical mismatch and geometric discontinuity would lead to residual stresses,

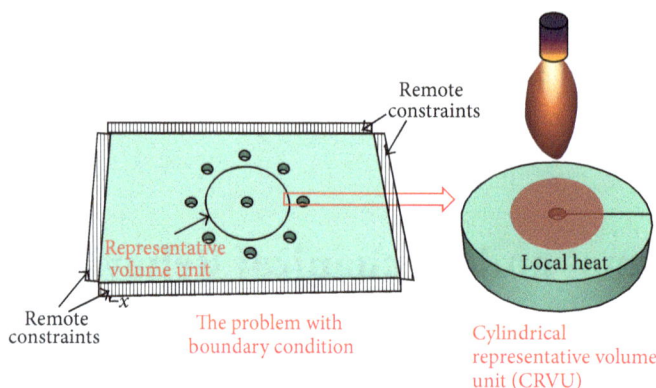

FIGURE 1: A schematic of a model with cylindrical representative volume unit (CRVU) subjected to local heating.

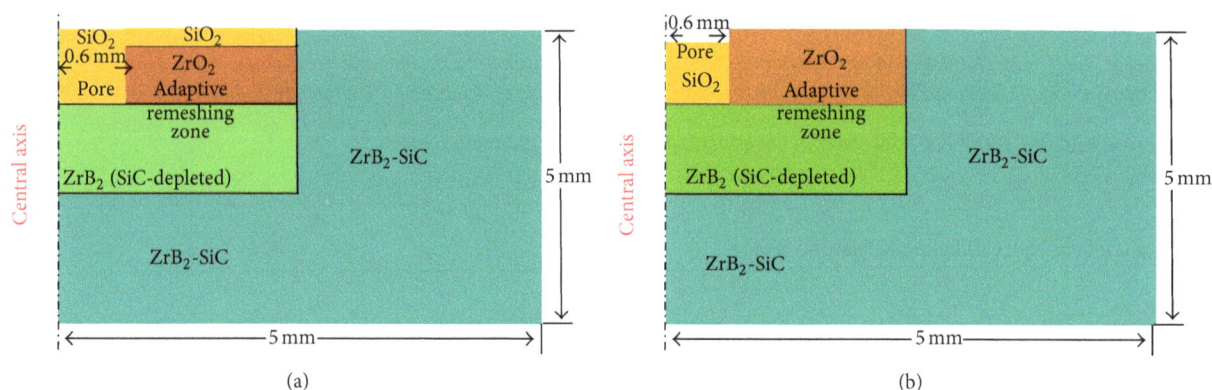

FIGURE 2: FEA models for ZrB$_2$-SiC ceramic after oxidation. (a) Lower temperature (external layer of SiO$_2$), and (b) higher temperature (partial evaporation of the glassy SiO$_2$).

and additional stress concentrations during the cool-down process from the processing temperature, thereby leading to potential cracking. Some researchers [8–12], having performed furnace oxidation and high velocity thermal shock tests on ZrB$_2$-SiC, have indeed shown cracking in the ZrO$_2$ scale.

The purpose of this study is to develop a thermal and mechanical simulation model for ZrB$_2$-SiC ceramics after oxidation. A steady-state heat transfer analysis was conducted using finite element analysis (FEA) modeling. An adpative remeshing technique is employed in both heat transfer and thermal stress analysis. A "global-local modeling" technique is used to combine finite element with infinite element for the thermal stress and the stress concentration analysis near a pore. Temperature, thermal, and residual stress distributions will be presented.

2. FEA Model and Simulation Procedure for ZrB$_2$-SiC after Oxidation

To simplify the problem, the ZrO$_2$ scale was assumed to be of uniform thickness with regularly distributed pores. The pores were assumed to be straight, columnar in structure without tortuosity. A cylindrical representative volume unit (CRVU) was constructed and further treated as a two-dimensional

(2D, pseudo-3D) axisymmetric problem subjected to local heating as shown in Figure 1. It is assumed that the body is stress free prior to heating.

The oxide scale that forms on ZrB$_2$-SiC consists of an outer layer of SiO$_2$, a middle layer of porous ZrO$_2$, and a layer of SiC-depleted ZrB$_2$ next to the unoxidized ZrB$_2$-SiC [3]. Based on the chemical oxidation models [7] and the experimental observation [3], the FEA models for ZrB$_2$-SiC ceramic after oxidation at high temperature were created as shown in Figure 2. An adaptive remeshing zone was created to cover the SiO$_2$, ZrO$_2$, ZrB$_2$ (SiC-depleted) and part of the ZrB$_2$-SiC base near the pore. The temperature dependent dimensions of the ZrO$_2$ scale (crystalline oxide), glassy SiO$_2$, and SiC-depleted ZrB$_2$ layer in ZrB$_2$-20 vol% SiC were obtained from the chemical oxidation model [7]. The unoxidized ZrB$_2$-SiC ceramic is treated as a macroscale continuous solid with properties of a predetermined ratio of 4:1 of ZrB$_2$ to SiC (ZrB$_2$-20 vol% SiC).

The heat conduction equation for an axisymmetric problem can be expressed as

$$\rho c \frac{\partial T}{\partial t} = \frac{k}{r}\frac{\partial}{\partial r}\left(r\frac{\partial T}{\partial r}\right) + k\frac{\partial^2 T}{\partial z^2}, \quad (1)$$

where t is the time, T is the temperature, r and z are polar axis and longitudinal axis, ρ is the mass density, c is the specific

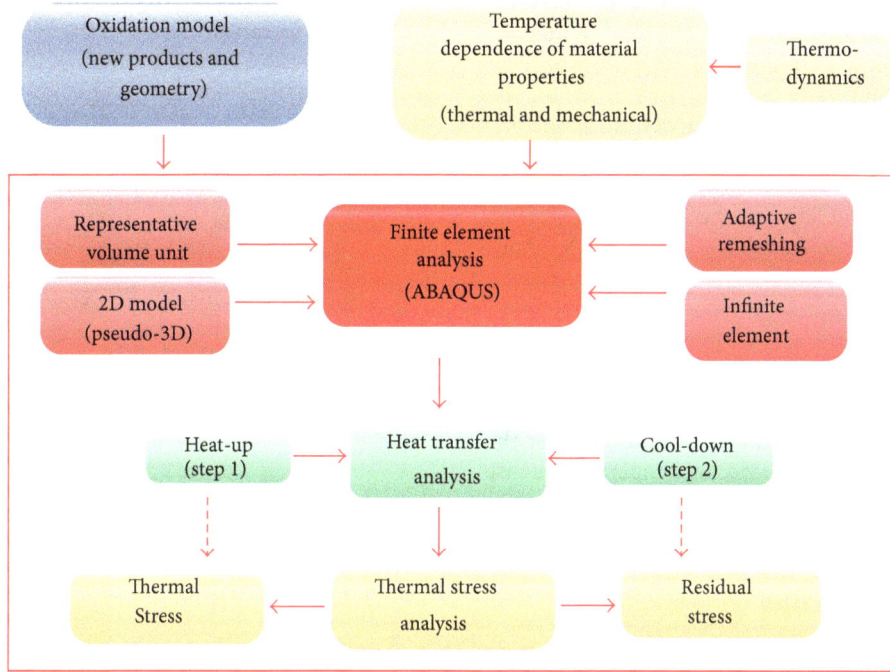

FIGURE 3: Flow diagram showing the procedure and steps for thermal and mechanical analyses.

heat, and k is the thermal conductivity. The thermoelastic model is given by

$$\{\sigma\} = [D]\left(\{\varepsilon\} - \{\alpha\}\,\Delta T\right), \tag{2}$$

where $[D]$ is the elasticity matrix, α the coefficient of thermal expansion, ΔT the temperature increment, σ the stress vector, and ε the strain vector.

The heat flux condition is given by

$$q = -\beta\left(T - T_0\right), \tag{3}$$

where, q is the heat flux, β is the surface film coefficient, and T_0 is the sink temperature. All simulations were conducted using ABAQUS finite element code.

The temperature dependent thermal and mechanical properties of the solid phases, needed for the heat transfer and mechanical analyses, can be found in the literature or databases [13, 14]. However, the thermal conductivity of liquid phase of SiO_2 (l) and the elastic constants cannot be found. As such, the temperature dependence of the thermal conductivity of liquid SiO_2 (l) and the elastic constants have to be predicted based on thermodynamics and some available test data. The predictive methods used for calculating the above properties are outlined in the next section. The cylindrical representative volume unit with equivalent pore diameter was treated as a 2D axisymmetric model (pseudo-3D). The modeling involves a steady-state heat transfer analysis representing local heat-up to calculate the temperature distribution and then a transient heat transfer analysis for 30 minutes representing a cool-down event to calculate the residual temperature distribution. The resulting temperature distributions were then applied to a thermomechanical finite element model to calculate the thermal stress distribution in the material. Adaptive remeshing technique was employed for the heat transfer analysis to improve accuracy. A "global-local modeling", along with the adaptive remeshing technique, is used to combine finite element with infinite element for thermal stress analysis. The procedure is summarized in Figure 3.

3. Thermal and Mechanical Properties

As mentioned earlier, the thermal conductivity and elastic constants of liquid phases of SiO_2 are not readily available. In an earlier work [15], the authors developed a method for calculating the thermal conductivity of liquid SiO_2 at a given temperature. The following thermal conductivity equation for a liquid by Hirschfelder et al. [16] is used in the method [15]:

$$k = 2.8 k_B \left(\frac{A\rho(T)}{M}\right)^{2/3}\left(\frac{c_p}{c_V}\right)^{1/2} c_l. \tag{4}$$

In the above equation, k_B is Boltzmann's constant, n is molecules per unit volume for the liquid, A is Avogadro's number, M is molar mass, and $\rho(T)$ is the temperature dependent bulk density of the liquid, c_p and c_V are the specific heats at constant pressure and at constant volume, respectively, and c_l is speed of sound in the liquid. The temperature dependent specific heat at constant pressure, c_p, for liquid SiO_2 was reported in [14]. Then, the specific heat at constant volume, c_V, is calculated using the following relationship [17]:

$$c_V = \frac{c_p^2}{c_p + \alpha^2 T c_l^2}, \tag{5}$$

where α is the coefficient of thermal expansion. The temperature dependence of density and coefficient of thermal expansion of liquid SiO$_2$ were given in [18]. The speed of sound in liquids of SiO$_2$ was found in [19]. With all the needed parameters, the thermal conductivity of liquid SiO$_2$ was calculated using (4).

Using the temperature dependent values for density and speed of sound in liquid of SiO$_2$, the bulk and shear moduli (K, G) of liquid SiO$_2$ were calculated using the Newton-Laplace equation [20–24]:

$$K = \rho C_L^2; \qquad G = \rho C_T^2, \qquad (6)$$

where C_L and C_T are the sound velocity of longitudinal and transverse wave, respectively. To simplify the problem, the temperature dependent elastic properties were used for the liquid phase of SiO$_2$ instead of the viscous properties because the stress state in liquid phase of SiO$_2$ was not of interest in the present study.

4. Results and Discussions

A 2D (pseudo-3D) 4-node linear axisymmetric heat transfer quadrilateral element was used in the thermal analysis. Heat flux was used as an error indicator variable to control the adaptive remeshing rule [25]. The dimensions of the new products after oxidation were taken from the chemical oxidation model [7].

Two steps were used in the heat transfer analyses. The first step in the heat transfer analysis was a steady-state analysis representing local heating at the top surface to calculate the temperature distribution. The second step was a transient heat transfer analysis for 30 minutes representing a cooling event to room temperature to predict residual temperature distribution. The surface heating temperature was set as T_{heat} (1780 K or 2240 K) during heating and 293 K during cooling. Outside the local heating area, the sink temperature was set at 293 K. The initial temperature of the material was 293 K. The heating, cooling and sink temperature conditions are summarized in Figure 4. The surface film coefficient, β, was set as 2500 W/(m^2·K) during the heating representing a high speed fluid flow and 100 W/(m^2·K) during cooling assuming a cooler fluid flow next to a solid boundary in air. The surface film coefficient was set as 100 W/(m^2·K) at all other boundaries during both heating and cooling.

4.1. Results of Heat Transfer Analysis. The calculated temperature distributions in the body after surface heating temperatures of 1780 K and 2240 K are shown in Figures 5 and 6, respectively. In the following results, the temperatures shown in parenthesis correspond to the case of 2240 K. The maximum temperature at the top surface of the outer SiO$_2$ layer is 1168 K (1492 K) which is less than the applied heating temperature of 1780 K (2240 K). This is due to the effect of the surface film coefficient on heat transfer between a fluid and a solid and the thermal conduction at the boundaries. The temperatures at the interface between the outer SiO$_2$ layer and the ZrO$_2$, and at the interface between the oxide scale and ZrB$_2$, are 1160 K (1432 K) and 1148 K (1404 K), respectively.

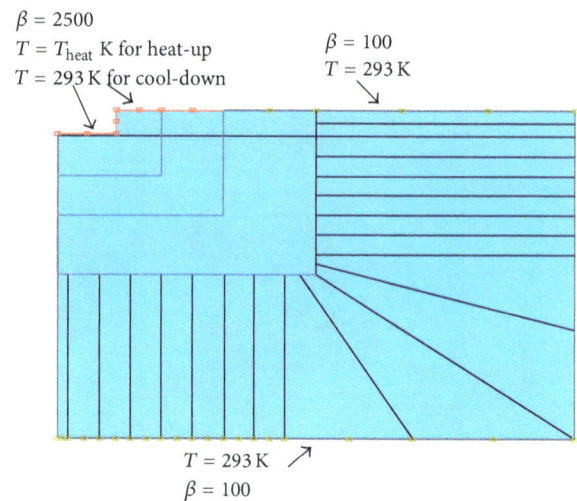

FIGURE 4: The heating, cooling, and sink temperature conditions used in the analysis.

The temperature at the bottom surface is 1124 K (1370 K) which is much less than the heating temperature applied at the top surface. The temperatures at locations shown in Figures 5 and 6 were also calculated for different heating temperatures. The predicted temperatures in various materials and at the interfaces are linearly dependent on heating temperature, except in ZrO$_2$ layer. This deviation could be due to the increase in ZrO$_2$ layer thickness accompanied by a decrease in SiO$_2$ layer thickness. Figure 7 shows the predicted heat flux distribution of ZrB$_2$-SiC after steady-state analysis for heating to 2240 K. It is seen that a heat flux concentration occurs at the pore corner due to the geometric discontinuity and thermal conductivity mismatch.

4.2. Results of Thermal Stress Analysis. In the thermal stress analysis, the layout of infinite elements and finite elements, as well as the displacement constraints for the stress analysis shown in Figure 8, are used. The distribution of maximum principal stresses for the steady-state heating at 1780 K is shown in Figure 9(a). The maximum principal stress distribution after cooling from 1780 K is shown in Figure 9(b). The maximum value of the maximum principal stresses of 946 MPa occurs at the top surface of SiO$_2$ layer. The temperature at this location is about 1166 K (Figure 5), which is below the glass melting point. The brittle glassy SiO$_2$ is sensitive to tensile stress with an average tensile strength of 364 ± 57 MPa [26]. Therefore, a tensile stress of 946 MPa may induce cracking in the SiO$_2$ layer. The maximum value of the maximum principal stresses of 568 MPa occurs at the upper corner of the pore in the ZrO$_2$ layer and is less than the flexural strength (900 MPa) of ZrO$_2$ [27]. The maximum stress in the ZrB$_2$ is 451 MPa and occurs near the lower corner of the pore. This is higher than the measured bend strength of ZrB$_2$ [28]. The largest principal stress in the ZrB$_2$-SiC is 191 MPa, located near the lower corner of the pore. The flexural strength of ZrB$_2$-SiC is 1000 MPa [29–31]. For

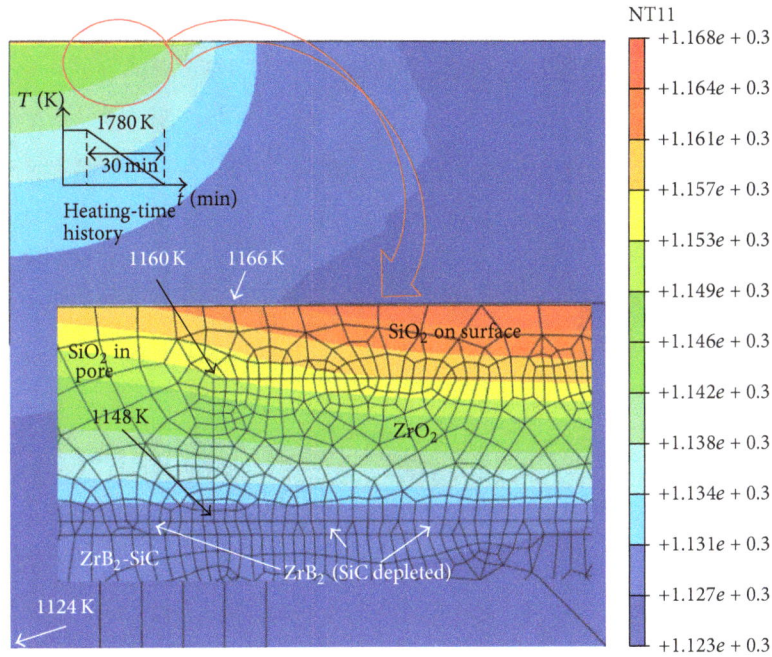

FIGURE 5: Predicted temperature distribution of ZrB_2-SiC after steady-state analysis at heating to 1780 K.

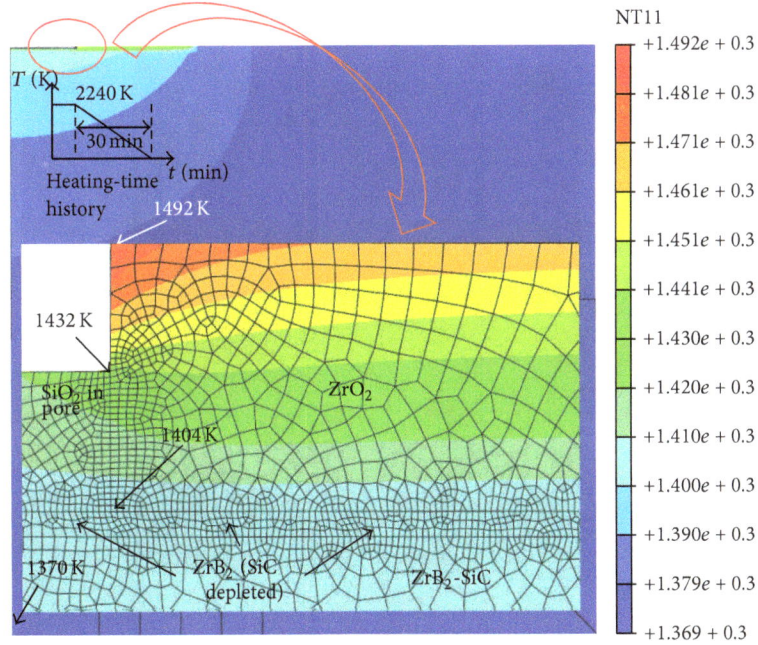

FIGURE 6: Predicted temperature distribution of ZrB_2-SiC after steady-state analysis for heating to 2240 K.

the cool-down case, the residual stresses in Figure 9(b) vary between 153 and 360 MPa.

The distribution in the maximum principal stresses near the pore for steady-state heating to 2240 K and cool-down from 2240 K to 293 K are shown in Figures 10(a) and 10(b), respectively. The stresses for heating vary between 182 and 2702 MPa while the corresponding stresses for cool-down vary between 281 and 529 MPa. Once again, the maximum values occur in the SiO_2 (2702 MPa) and ZrO_2 (2224 MPa) near the pore. This may initiate tensile cracking. These results are consistent with the experimental observations by Levine et al. [8] who performed furnace oxidation and high velocity thermal shock of ZrB_2 + 20 vol.% SiC ceramic tests. Their results show that both pores and cracks appeared in the ZrO_2 when oxidized in air at 1927°C for ten 10-min cycles. The highest maximum principal stress near the lower

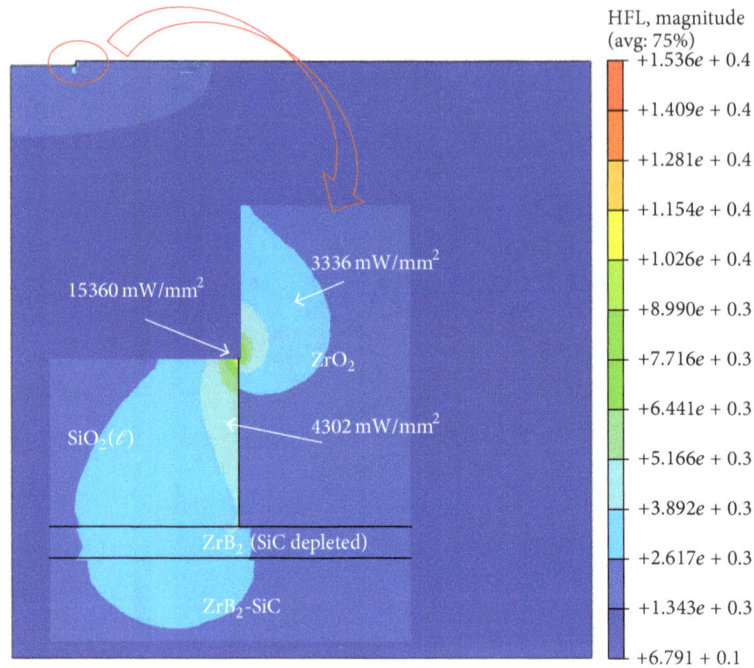

FIGURE 7: Predicted heat flux distribution of ZrB_2-SiC after steady-state analysis for heating to 2240 K.

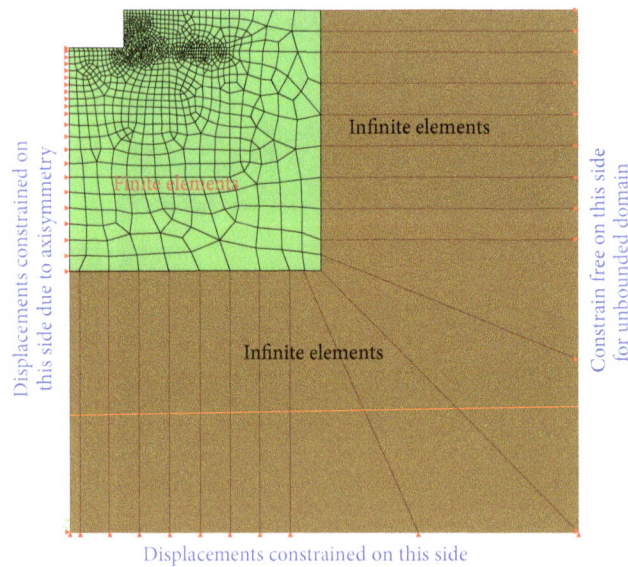

FIGURE 8: The infinite element, the finite element, and the boundary conditions for the thermal stress analysis.

corner of the pore in the ZrB_2 layer is 657 MPa shown in Figure 10(a) which is higher than the bending strength of the ZrB_2 as indicated above. For the cool-down case, the maximum principal stresses shown in Figure 10(b) are less than the respective material strengths except in the ZrB_2 layer.

The results were also obtained for additional temperatures. The variation of the maximum principal stress at indicated locations in Figures 9 and 10 with temperature T is shown in Figure 11(a) for heating up to T and in Figure 11(b) for cool-down from T to 293 K. The thermal stress at heating temperature T in the ZrB_2-SiC matrix is relatively small and does not vary much with heating temperature T (Figure 11(a)). The residual stress at 293 K (Figure 11(b)) in the ZrB_2-SiC decreases with the increasing heating temperature T. Watts et al. [32] measured thermal residual stresses in ZrB_2 -30 vol% SiC composites using neutron diffraction. Their results indicated that stresses begin to accumulate at about 1673 K during cool-down from the processing temperature of 2172 K. The stress increased to an

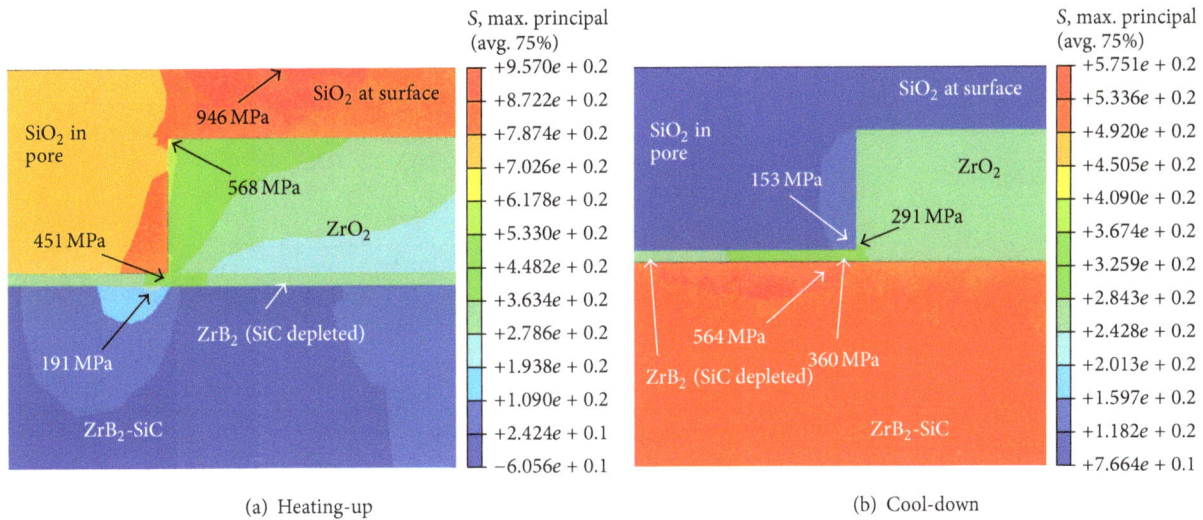

(a) Heating-up

(b) Cool-down

FIGURE 9: Maximum principal stresses distribution in the enlarged area near pore after steady-state thermal analysis at 1780 K.

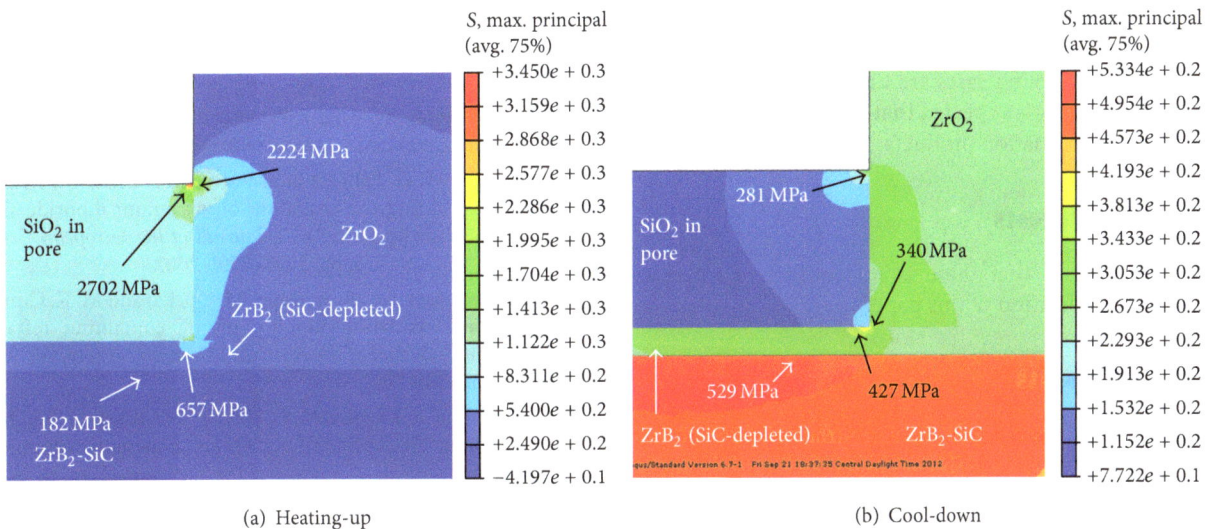

(a) Heating-up

(b) Cool-down

FIGURE 10: Maximum principal stresses distribution in the enlarged area near pore after steady-state thermal analysis at 2240 K.

average compressive stress of 880 MPa in the SiC phase and to an average tensile stress of 450 MPa in the ZrB_2 phase. By using the rule of mixtures for 34 vol% SiC, the stress in the SiC (880 MPa) converts to an equivalent stress of 453 MPa which is very close to the measured stress of 450 MPa [32].

5. Conclusion

A "global-local modeling" technique is used combining finite element with infinite element for thermal stress analysis for the oxidation effects on heat transfer and mechanical behavior of ZrB_2-SiC ceramics at high temperature. Thermal conductivity was calculated for the liquid phase of SiO_2 based on a theoretical formulation. The predicted temperature at

the top surface of the outer SiO_2 layer is less than the applied heating temperature due to the surface film coefficient effect on the heat transfer between a fluid and a solid and the thermal conduction at the boundaries. An increase in ZrO_2 layer thickness, accompanied by a decrease in SiO_2 layer thickness, during oxidation will affect heat transfer in the body. Heat flux concentration occurs at the pore corner due to the geometric discontinuity and the material property mismatch. Thermal and residual stress concentrations occur near the pore due to geometric discontinuity and the material properties mismatch between the ceramic matrix and the new products. Thermal stresses in the surface oxide layers consisting of SiO_2 and ZrO_2, are higher than their respective materials strengths. Thermal and residual stresses in the layer

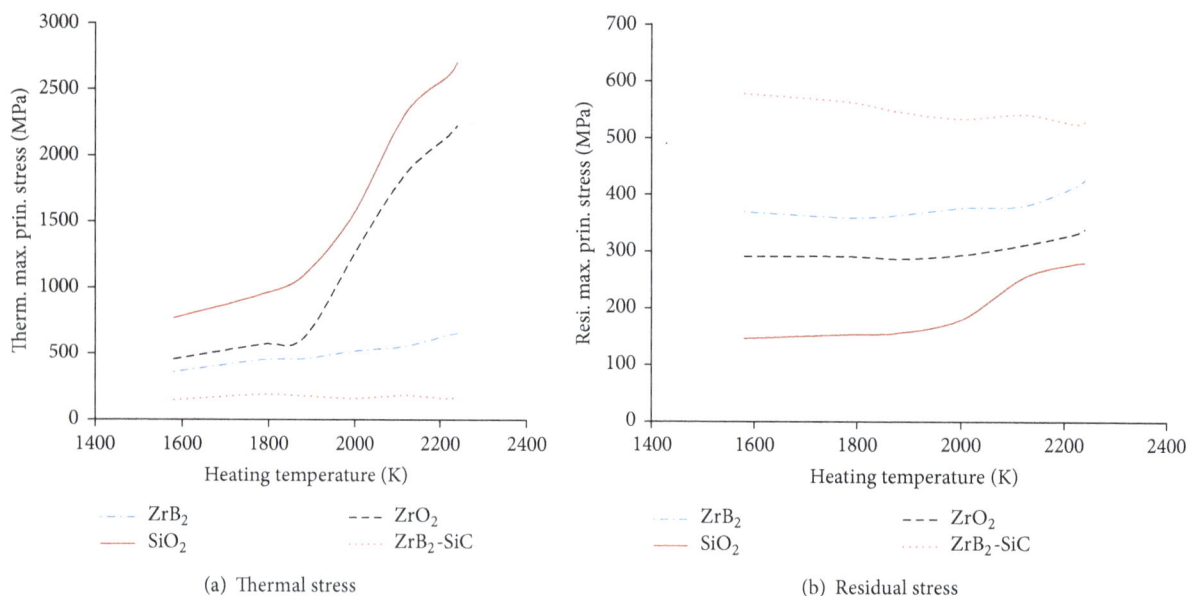

(a) Thermal stress

(b) Residual stress

FIGURE 11: Thermal and residual maximum principal stress at indicated locations for different heating temperatures.

of a new oxidation product of SiC-depleted ZrB_2 layer for both heating and cooling cases are higher than the material strength. Therefore, it is expected that damage may initiate in the layers of new oxidation products.

Conflict of Interests

The authors declare that there is no conflict of interests regarding the publication of this paper.

Acknowledgments

This project was funded under subcontract 10-S568-0094-01-C1 through the Universal Technology Corporation under prime contract number FA8650-05-D-5807. The authors are grateful to the technical support on the program by the Air Force Research Laboratory and specifically to Dr. Mike Cinibulk at AFRL for both his collaboration and guidance.

References

[1] M. M. Opeka, I. G. Talmy, and J. A. Zaykoski, "Oxidation-based materials selection for 2000°C + hypersonic aerosurfaces: theoretical considerations and historical experience," *Journal of Materials Science*, vol. 39, no. 19, pp. 5887–5904, 2004.

[2] W. G. Fahrenholtz, G. E. Hilmas, A. L. Chamberlain, and J. W. Zimmermann, "Processing and characterization of ZrB_2-based ultra-high temperature monolithic and fibrous mono-lithic ceramics," *Journal of Materials Science*, vol. 39, no. 19, pp. 5951–5957, 2004.

[3] W. G. Fahrenholtz, "Thermodynamic analysis of ZrB_2-SiC oxidation: formation of a SiC-depleted region," *Journal of the American Ceramic Society*, vol. 90, no. 1, pp. 143–148, 2007.

[4] C. M. Carney, P. Mogilvesky, and T. A. Parthasarathy, "Oxidation behavior of zirconium diboride silicon carbide produced

by the spark plasma sintering method," *Journal of the American Ceramic Society*, vol. 92, no. 9, pp. 2046–2052, 2009.

[5] W. G. Fahrenholtz, "The ZrB_2 volatility diagram," *Journal of the American Ceramic Society*, vol. 88, no. 12, pp. 3509–3512, 2005.

[6] A. Rezaie, W. G. Fahrenholtz, and G. E. Hilmas, "Evolution of structure during the oxidation of zirconium diboride-silicon carbide in air up to 1500∘C," *Journal of the European Ceramic Society*, vol. 27, no. 6, pp. 2495–2501, 2007.

[7] T. A. Parthasarathy, R. A. Rapp, M. Opeka, and M. K. Cinibulk, "Modeling oxidation kinetics of SiC-containing refractory diborides," *Journal of the American Ceramic Society*, vol. 95, no. 1, pp. 338–349, 2012.

[8] S. R. Levine, E. J. Opila, M. C. Halbig, J. D. Kiser, M. Singh, and J. A. Salem, "Evaluation of ultra-high temperature ceramics for aeropropulsion use," *Journal of the European Ceramic Society*, vol. 22, no. 14-15, pp. 2757–2767, 2002.

[9] J. C. Han, P. Hu, X. Zhang, S. Meng, and W. Han, "Oxidation-resistant ZrB2-SiC composites at 2200°C," *Composites Science and Technology*, vol. 68, no. 3-4, pp. 799–806, 2008.

[10] X.-H. Zhang, P. Hu, and J.-C. Han, "Structure evolution of ZrB_2-SiC during the oxidation in air," *Journal of Materials Research*, vol. 23, no. 7, pp. 1961–1972, 2008.

[11] P. Hu, W. Guolin, and Z. Wang, "Oxidation mechanism and resistance of ZrB_2-SiC composites," *Corrosion Science*, vol. 51, no. 11, pp. 2724–2732, 2009.

[12] P. Sarin, P. E. Driemeyer, R. P. Haggerty et al., "In situ studies of oxidation of ZrB2 and ZrB2–SiC composites at high temperatures," *Journal of the European Ceramic Society*, vol. 30, no. 11, pp. 2375–2386, 2010.

[13] "Material Property Database (MPDB)," JAHM Software, Inc., http://www.jahm.com.

[14] I. Barin, *Thermochemical Data of Pure Substances*, Wiley-VCH, New York, NY, USA, 3rd edition, 1995.

[15] J. Wei, L. R. Dharani, K. Chandrashekhara, G. E. Hilmas, and W. G. Fahrenholtz, "Modeling of oxidation effects on heat transfer behavior of ZrB_2 and ZrB_2-SiC Ceramics at high temperature,"

in *Proceedings of the 53rd AIAA/ASME/ASCE/AHS/ASC Structures, Structural Dynamics and Materials Conference*, AIAA, Honolulu, Hawaii, USA, April 2012.

[16] J. O. Hirschfelder, C. F. Curtiss, and R. B. Bird, *Molecular Theory of Gases and Liquid*, John Wiley & Sons, New York, NY, USA, 1954.

[17] R. S. Hixson, M. A. Winkler, and J. W. Shaner, "Sound speed measurements in liquid lead at high temperature and pressure," in *Conference: 10. High Pressure Conference on Research in High Pressure Science and Technology*, Amsterdam, The Netherlands, 1985, LA-UR-85-2363.

[18] M. S. Ghiorso, "An equation of state for silicate melts. I. Formulation of a general model," *American Journal of Science*, vol. 304, no. 8-9, pp. 637–678, 2004.

[19] A. Polian, V.-T. Dung, and P. Richet, "Elastic properties of a-SiO_2 up to 2300 K from Brillouin scattering measurements," *Europhysics Letters*, vol. 57, no. 3, pp. 375–381, 2002.

[20] 2013, https://en.wikipedia.org/wiki/Speed_of_sound.

[21] D. D. Joseph, A. Narain, and O. Riccius, "Shear-wave speeds and elastic moduli for different liquids. Part 1. Theory," *Journal of Fluid Mechanics*, vol. 171, pp. 289–308, 1986.

[22] D. D. Joseph, O. Riccius, and M. Arney, "Shear-wave speeds and elastic moduli for different liquids. Part 2: experiments," *Journal of Fluid Mechanics*, vol. 171, pp. 309–338, 1986.

[23] O. Riccius, D. D. Joseph, and M. Arney, "Shear-wave speeds and elastic moduli for different liquids Part 3. Experiments-update," *Rheologica Acta*, vol. 26, no. 1, pp. 96–99, 1987.

[24] M. S. Greenwood and J. A. Bamberger, "Measurement of viscosity and shear wave velocity of a liquid or slurry for on-line process control," *Ultrasonics*, vol. 39, no. 9, pp. 623–630, 2002.

[25] "ABAQUS 6.9 documentation," http://www.simulia.com/support/documentation.html.

[26] W. N. Sharpe Jr., J. Pulskamp, D. S. Gianola, C. Eberl, R. G. Polcawich, and R. J. Thompson, "Strain measurements of silicon dioxide microspecimens by digital imaging processing," *Experimental Mechanics*, vol. 47, no. 5, pp. 649–658, 2007.

[27] http://www.ferroceramic.com/zirconia.htm.

[28] D. Kalish, E. V. Clougherty, and K. Kreder, "Strength, fracture mode, and thermal stress resistance of HfB_2 and ZrB_2," *Journal of the American Ceramic Society*, vol. 52, no. 1, pp. 30–36, 1969.

[29] A. L. Chamberlain, W. G. Fahrenholtz, G. E. Hilmas, and D. T. Ellerby, "High-strength zirconium diboride-based ceramics," *Journal of the American Ceramic Society*, vol. 87, no. 6, pp. 1170–1172, 2004.

[30] A. Rezaie, W. G. Fahrenholtz, and G. E. Hilmas, "Effect of hot pressing time and temperature on the microstructure and mechanical properties of ZrB_2-SiC," *Journal of Materials Science*, vol. 42, no. 8, pp. 2735–2744, 2007.

[31] S. Zhu, W. G. Fahrenholtz, and G. E. Hilmas, "Influence of silicon carbide particle size on the microstructure and mechanical properties of zirconium diboride-silicon carbide ceramics," *Journal of the European Ceramic Society*, vol. 27, no. 4, pp. 2077–2083, 2007.

[32] J. Watts, G. Hilmas, W. G. Fahrenholtz, D. Brown, and B. Clausen, "Measurement of thermal residual stresses in ZrB_2-SiC composites," *Journal of the European Ceramic Society*, vol. 31, no. 9, pp. 1811–1820, 2011.

Synthesis of β-SiC Fine Fibers by the Forcespinning Method with Microwave Irradiation

Alfonso Salinas,[1] Maricela Lizcano,[2] and Karen Lozano[1]

[1]*Mechanical Engineering Department, The University of Texas-Pan American, Edinburg, TX 78539, USA*
[2]*National Aeronautics and Space Administration, Materials and Structures Division, Glenn Research Center at Lewis Field, Cleveland, OH 44135, USA*

Correspondence should be addressed to Karen Lozano; lozanok@utpa.edu

Academic Editor: Yuan-hua Lin

A rapid method for synthesizing β-silicon carbide (β-SiC) fine fiber composite has been achieved by combining forcespinning technology with microwave energy processing. β-SiC has applications as composite reinforcements, refractory filtration systems, and other high temperature applications given their properties such as low density, oxidation resistance, thermal stability, and wear resistance. Nonwoven fine fiber mats were prepared through a solution based method using polystyrene (PS) and polycarbomethylsilane (PCmS) as the precursor materials. The fiber spinning was performed under different parameters to obtain high yield, fiber homogeneity, and small diameters. The spinning was carried out under controlled nitrogen environment to control and reduce oxygen content. Characterization was conducted using scanning electron microscopy (SEM), X-ray diffraction (XRD), and Fourier transform infrared spectroscopy (FTIR). The results show high yield, long continuous bead-free nonwoven fine fibers with diameters ranging from 270 nm to 2 μm depending on the selected processing parameters. The fine fiber mats show formation of highly crystalline β-SiC fine fiber after microwave irradiation.

1. Introduction

In the past 50 years, microwave irradiation has been utilized to process various materials such as semiconductors and inorganic and polymeric materials. More recently, microwave energy has been used to sinter powdered metals as well as ceramic systems [1]. Microwaves are electromagnetic radiation with a wavelength from 1 mm to 1 m with frequencies in the range of 1 to 300 GHz [2]. Most common microwaves furnaces used for industrial and scientific applications operate at a frequency of 2.45 GHz [3]. The most effective way to produce microwaves is from a magnetron source, but they can also be produced from klystrons, power grid tubes, traveling wave tubes, and gyrotrons [4].

Microwave processing of ceramics was first reported in 1968 by Tinga and Voss [5]. However, it was not until the 1980s that high temperature processing with microwave energy started gaining much ground [6]. Although microwave processing of advanced ceramic materials is still developing, it offers many advantages over conventional ceramic processing

methods such as reduced heating times and lower power consumption [7].

In traditional thermal material processing, energy is transferred by convection and radiant heat onto the surface of the material and then through the material by conduction heating. Materials requiring long processing times via traditional methods undergo thermal gradients within the material, wherein the surface of the material is exposed to more heat than the core of the material, resulting in surface damage [8]. However, in microwave thermal processing, heat is directly transferred to the material volumetrically by molecular interactions with the electromagnetic field. Since the diffusion of heat through the surface, as in traditional thermal processing, is bypassed by volumetric heating, uniform heating and fast processing times can be achieved with heating rates as high as 1000°C/min [8–10].

The major advantages of microwave processing over conventional heating methods are reduced thermal gradients within the material, faster reaction times, lower processing temperatures, high density microstructures, and improved mechanical properties [11, 12]. These advantages strongly support the use of microwave thermal processing for advanced materials development.

The forcespinning (FS) process is a rapid method to produce nanosize to micron size fibrous materials. Unlike electrospinning, yields as high as 1 g/min in a lab scale unit are easily achieved. The combination of this fiber making technology with the fast heating rates of MW irradiation provides a rapid route for producing ceramic materials. FS utilizes centrifugal forces to overcome shear forces promoting fiber elongation. Process, theory, and schematics have been reported elsewhere [13–15]. In this research, the development and optimization of the parameters involved in the production of β-SiC fine fibers were carefully analyzed and developed materials characterized. The prepared green fine fibers were spun from polymeric precursors. β-SiC nanomaterials have been intensively studied given their unique properties such as high mechanical strength, high thermal conductivity, low thermal expansion coefficient, and chemical inertness when compared to those of their bulk counterparts [16]. In this case, the utilization of a fibrous morphology also provides a significant increase in surface to volume ratio. Many studies have shown the potential applications of β-SiC fine fibers and the lab scale results have shown promising applications; therefore, scientists are researching new and easier methods to develop β-SiC nanostructures.

The preceramic fine fibers were developed utilizing a solution of polystyrene and polycarbomethylsilane. The developed nonwoven fine fiber mats were characterized by FESEM (field emission scanning electron microscope), XRD (X-ray diffraction), EDS (energy dispersive spectroscopy), and FTIR (Fourier transform infrared spectroscopy).

2. Experimental

2.1. Materials. Polystyrene (PS) with a molecular weight of 280,000 g/mol and polycarbomethylsilane (PCmS) with a molecular weight of 800 g/mol were purchased from Sigma-Aldrich (Milwaukee, WI, USA) and used as received. Toluene was purchased from Fisher Scientific (Waltham, MA, USA) and used as received. The PS/PCmS/Toluene (15, 20, and 25 wt% of PS with a 2 : 1 ratio of PS : PCmS) solutions were prepared inside a MBRAUN (Stratham, NH) glovebox under nitrogen atmosphere in order to prevent oxidation. The solutions were prepared in 20 mL scintillation vials and sealed with parafilm to prevent solvent evaporation. Solutions were magnetically stirred for a period of 4 hours.

2.2. Fine Fiber Development. A FS system was placed inside a glovebox under nitrogen environment. Approximately 2 mL of solution was inserted into a cylindrical type spinneret using a 10 mL syringe. The spinneret was outfitted with 30 gage needles. The angular velocity at which the fibers were spun was varied from 5,000 rpm to 9,000 rpm. The solution was depleted in less than 30 sec. The fibers were deposited on a circular collector having 16 equally spaced polytetrafluoroethylene (PTFE) bars. The developed fibers were stored in a glovebox to prevent fiber oxidation.

Fibers were cross-linked to maintain geometric integrity of the precursor fine fibers because PS reaches the glass transition temperature before the preceramic polymer (PCmS) is converted to ceramic SiC during heat treatment. The collected fine fibers were placed under a 254 nm wavelength UV light source for a period of 24 hours. The cross-linking was performed in a glovebox under nitrogen environment.

2.3. Microwave Pyrolysis. A microwave furnace, Hi-Tech single mode microwave applicator, was used. The system is fitted with a MH 2.0 W-S water cooled magnetron head assembly. It supplies 2 kW of adjustable microwave energy at 2.45 GHz. The magnetron outputs into a WR340 waveguide. An Omega iSeries iR2 infrared pyrometer was used to record the temperature. The samples were placed between small SiC susceptor plates which absorb electromagnetic energy and convert it to heat. The sample chamber was fitted with a quartz tube attached to a turbo pump. The microwave heating was carried out under nitrogen gas after evacuation of air with the turbo pump. Power was increased 100 W every 4 minutes up to 600 W. The total processing time at 600 W was 3 minutes. The temperature was observed to be approximately 1140°C at 600 W.

2.4. Fiber Characterization. Fiber morphology was analyzed using the Carl Zeiss Sigma VP scanning electron microscope. Fiber diameters were measured using the Carl Zeiss Axio-Vision software. For the X-ray diffraction analysis, a Bruker AXS D8 diffractometer was utilized. The fine fibers were scanned from 20 to 80° (2θ angles) using a 2D detector. FTIR-ATR with a diamond tip was carried out using an Agilent Technologies Cary 600 Series FTIR spectrometer.

3. Results and Discussion

The optimization of the SiC fiber precursors was conducted in a previous study [17]. This study focused on developing highly crystalline β-SiC fine fibers through conventional pyrolysis methods. Several parameters were evaluated that resulted in fibers with average diameters ranging from 270 nm to 2 μm. It was concluded that the parameters that synergistically contributed to the development of homogeneous, high yield, bead-free continuous green SiC fibers were of a polymer concentration of 20 wt% in the developed solution which was then processed at an angular velocity of 7000 rpm. Figure 1 shows a nonwoven fine fiber mat with its observable corresponding fiber diameter distribution. Figure 2 shows an SEM micrograph of the precursor fibers showing micron and submicron size fibers. Fiber-fiber adhesion can be observed as indicated by the red boxes. The above mentioned fiber spinning parameters were selected to prepare samples utilized in this work to further analyze the effect of heat treatment via microwave energy processing.

The materials in this study consisted of PS and PCmS with molecular formulas $(C_8H_8)_n$ and $(C_2H_6Si)_n$, respectively. At temperatures between 550°C and 800°C, the precursor becomes an inorganic material as it begins to decompose Si–H, Si–CH_3, Si–CH_2–Si, H_2, CH_4, CO, and CO_2 species which

TABLE 1: Corresponding XRD peak list from Figure 3.

Pos.[°2Th.]	Height [cts]	d-spacing [Å]	Rel. Int. [%]	Matched by
26.7407	197.39	3.33386	9.92	04-014-0337
35.6177	1989.65	2.5207	100	00-029-1129
41.4258	111.8	2.17973	5.62	00-029-1129
59.9538	501.03	1.54295	25.18	00-029-1129
71.7441	328.15	1.31564	16.49	00-029-1129

Pos = position of the peak on the 2θ axis in the XRD spectra.

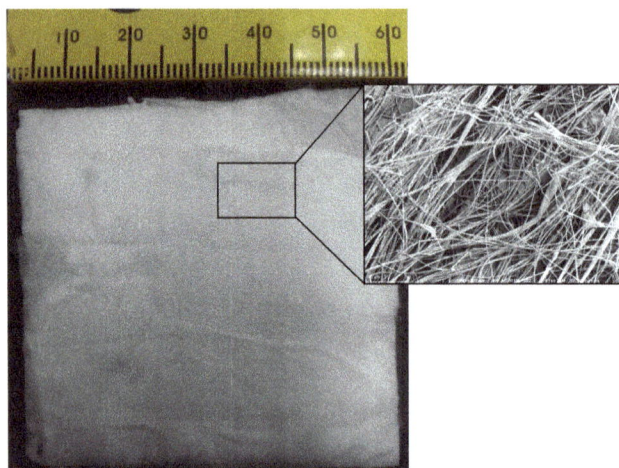

FIGURE 1: 6 cm × 6 cm SiC precursor mat obtained after 30 sec.

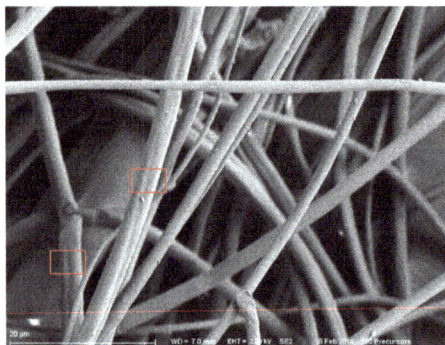

FIGURE 2: SEM micrograph of SiC precursors. Red boxes show fiber surface roughness.

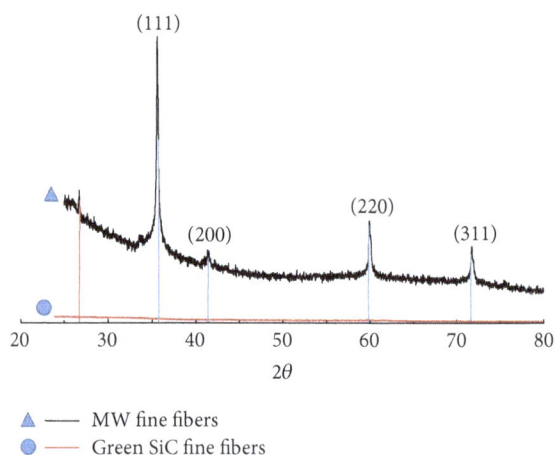

FIGURE 3: XRD spectra of fine fibers before and after microwave assisted heating.

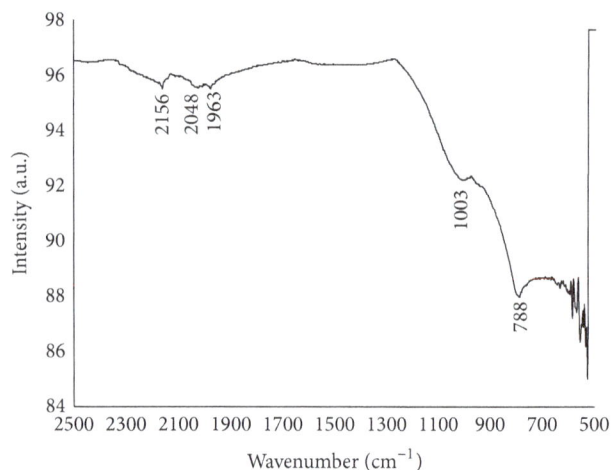

FIGURE 4: FTIR of SiC fine fibers after microwave irradiation.

are eliminated from the precursor [18–20]. Increasing the temperature from 800°C up to 1000°C results in amorphous SiC while crystalline SiC begins to form at 1000°C with the evolution of H_2 [18].

XRD results are shown in Figure 3 with a corresponding peak list given in Table 1. The peaks indicate conversion of precursor to β-SiC with $2\theta = 35.62°$, $41.42°$, $59.95°$, and $71.74°$ (reference code: 00-029-1129) ascribed to lattice planes (111), (200), (220), and (311), respectively [21]. Additionally, the slight broadening base of the peaks is indicative of either unreacted material or some amorphous SiC remaining in the sample. A small carbon peak is noted at $2\theta = 26.74°$ (reference code: 04-014-0337).

SiC FTIR reflectance bands can be seen between wave numbers 740 and 970 cm^{-1} [22–24]. The FTIR scan shown in Figure 4 depicts a reflectance band at wave number 788 cm^{-1}, supporting ceramic conversion of precursor fine fibers shown in Figure 3. The peak at wave number 1003 cm^{-1} corresponds to Si–O bonds indicating the presence of SiO_2 in the sample [25, 26]. Although the SiO_2 was not detected in XRD results, it may be due to the nanosized geometries of its content. If the uncured precursor is exposed to oxygen during handling,

TABLE 2: EDS of SiC precursor and microwaved fine fibers.

Element	SiC precursor NFs		Microwaved NFs	
	wt%	at%	wt%	at%
Si	17.63	8.71	37.19	21.32
C	68.99	79.69	46.29	62.06
O	13.38	11.6	16.52	16.62

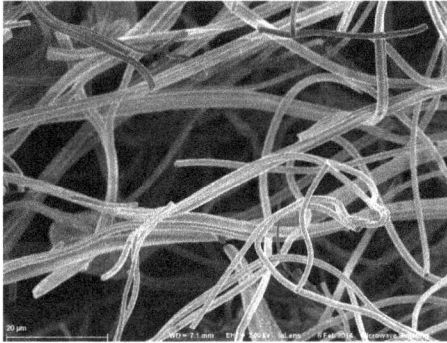

FIGURE 5: SEM micrograph of ceramic converted fine fibers.

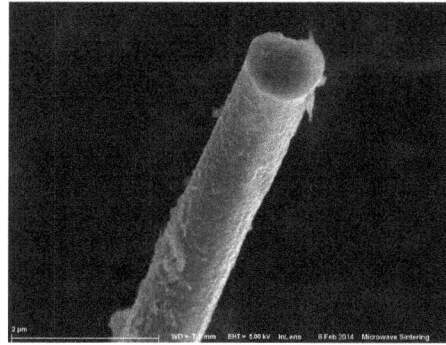

FIGURE 6: High magnification SEM micrograph of a β-SiC fine fiber.

oxidation of Si–H and Si–CH$_3$ can occur resulting in the formation of Si–O–Si bonds [27]. The available oxygen species in the precursor can then form amorphous SiO$_2$ at temperatures below 1200°C [19]. EDS was performed in the precursor fiber as well as in the ceramic converted fibers. The small presence of oxygen in both samples is given in Table 2. Reflectance band at 2156 cm^{-1} is consistent with MeSiH$_3$ [28] while bands seen at 2048 cm^{-1} and 1963 cm^{-1} indicate the presence of CO species [29].

These types of composite materials have applications in nanoelectronics, nanomechanics, reinforced composite materials, and nanosensors to mention some [30]. More importantly, these materials may be used to improve SiC/SiO$_2$ bonding interfaces for improved performance in electronic devices [31, 32]. SiC/SiO$_x$ materials have been previously synthesized including the use of microwave irradiation [21, 33–36]. However, in these studies Si, SiO$_2$, and graphite powders were used as the precursor materials rather than a preceramic polymer. An SEM micrograph of the converted ceramic fine fiber is shown in Figure 5. Micron and submicron size fibers can be seen in this figure. Also, noted is a rough fiber structure in some areas which may be due to adhesion of two or more fibers forming ribbon-like fibers or shrinkage that a fiber undergoes as a result of volatilization of solvent during microwave processing as previously observed in annealed fibers [37]. The micrograph also shows many short fibers, although fibers are expected to be up to 6 feet in length [14]. The precursor fine fibers are removed with tweezers resulting in broken fibers before being placed on the susceptor for MW processing. A continuous deposition method with subsequent MW treatment will preserve fiber length. Figure 6 shows a SEM micrograph of a single SiC nanofiber.

4. Conclusion

In this study, we successfully show ceramic conversion of spun PS/PCmS fine fibers to β-SiC fine fibers utilizing forcespinning technology for rapid fiber development and microwave energy processing for rapid ceramic conversion of spun fibers. The combination of these two technologies illustrates a processing route that can be utilized to produce rapid novel ceramic nanomaterials. XRD and FTIR results confirmed ceramic conversion of the fine fibers. Both micron and submicron fibers were observed in SEM images.

Conflict of Interests

Dr. K. Lozano and The University of Texas-Pan American have research-related interest in FibeRio Technology Corporation.

References

[1] A. Chang, H. Zhang, Q. Zhao, and B. Zhang, *Microwave Sintering of Thermistor Ceramics*, InTech, 1980.

[2] J. D. Katz, "Microwave sintering of ceramics," *Annual Review of Materials Science*, vol. 22, no. 1, pp. 153–170, 1992.

[3] D. K. Agrawal, "Microwave processing of ceramics," *Current Opinion in Solid State and Materials Science*, vol. 3, no. 5, pp. 480–485, 1998.

[4] S. Chandrasekaran, S. Ramanathan, and T. Basak, "Microwave material processing—a review," *AIChE Journal*, vol. 58, no. 2, pp. 330–363, 2012.

[5] W. R. Tinga and W. A. G. Voss, *Microwave Power Engineering*, edited by E. C. Okress, Academic Press, New York, NY, USA, 1968.

[6] D. Agrawal, "Microwave sintering of ceramics, composites and metallic materials, and melting of glasses," *Transactions of the Indian Ceramic Society*, vol. 65, no. 3, pp. 129–144, 2006.

[7] D. E. CLark, D. C. Folz, C. E. Folgar, and M. M. Mahmoud, *Microwave Solutions for Ceramic Engineers*, The American Ceramic Society, Westerville, Ohio, USA, 2005.

[8] E. T. Thostenson and T.-W. Chou, "Microwave processing: fundamentals and applications," *Composites Part A: Applied Science and Manufacturing*, vol. 30, no. 9, pp. 1055–1071, 1999.

[9] Y. C. Kim, C. H. Kim, and D. K. Kim, "Effect of microwave heating on densification and $\alpha \rightarrow \beta$ phase transformation of

silicon nitride," *Journal of the European Ceramic Society*, vol. 17, no. 13, pp. 1625–1630, 1997.

[10] S. Somiya, F. Aldinger, R. M. Spriggs, K. Uchino, K. Koumoto, and M. Kaneno, *Handbook of Advanced Ceramics: Materials, Applications, Processing, and Properties*, Academic Press, 2003, http://books.google.com.eg/books?id=SMCRJi52OcQC&redir_esc=y.

[11] D. D. Upadhyaya, A. Ghosh, G. K. Dey, R. Prasad, and A. K. Suri, "Microwave sintering of zirconia ceramics," *Journal of Materials Science*, vol. 36, no. 19, pp. 4707–4710, 2001.

[12] M. Oghbaei and O. Mirzaee, "Microwave versus conventional sintering: a review of fundamentals, advantages and applications," *Journal of Alloys and Compounds*, vol. 494, no. 1-2, pp. 175–189, 2010.

[13] B. Vazquez, H. Vasquez, and K. Lozano, "Preparation and characterization of polyvinylidene fluoride nanofibrous membranes by forcespinning," *Polymer Engineering and Science*, vol. 52, no. 10, pp. 2260–2265, 2012.

[14] S. Padron, A. Fuentes, D. Caruntu, and K. Lozano, "Experimental study of nanofiber production through forcespinning," *Journal of Applied Physics*, vol. 113, no. 2, Article ID 024318, 2013.

[15] K. Sarkar, C. Gomez, S. Zambrano et al., "Electrospinning to forcespinning," *Materials Today*, vol. 13, no. 11, pp. 12–14, 2010.

[16] Z.-M. Huang, Y.-Z. Zhang, M. Kotaki, and S. Ramakrishna, "A review on polymer nanofibers by electrospinning and their applications in nanocomposites," *Composites Science and Technology*, vol. 63, no. 15, pp. 2223–2253, 2003.

[17] A. Salinas, *Mass production of β-silicon carbide nanofibers by the novel [M.S. thesis]*, The University of Texas Pan American, 2013.

[18] P. Colombo, R. Riedel, G. D. Soraru, and H.-J. Kleebe, Eds., *Polymer Derived Ceramics: From Nano-Structure to Applications*, DEStech Publications, 2010.

[19] K. Okamura, "Ceramic fibres from polymer precursors," *Composites*, vol. 18, no. 2, pp. 107–120, 1987.

[20] S. Somiya, F. Aldinger, R. M. Spriggs, K. Uchino, K. Koumoto, and M. Kaneno, Eds., *Handbook of Advanced Ceramics: Materials, Applications, Processing and Properties*, vol. 2, Academic Press, New York, NY, USA, 2013.

[21] J. Wang, S. Liu, T. Ding, S. Huang, and C. Qian, "Synthesis, characterization, and photoluminescence properties of bulk-quantity β-SiC/SiO$_x$ coaxial nanowires," *Materials Chemistry and Physics*, vol. 135, no. 2-3, pp. 1005–1011, 2012.

[22] S. B. Qadri, A. W. Fliflet, A. Imam, B. B. Rath, and E. P. Gorzkowski, "Silicon carbide synthesis from agricultural waste," US 20130272947Al, 2013.

[23] J. P. Li, A. J. Steckl, I. Golecki et al., "Structural characterization of nanometer SiC films grown on Si," *Applied Physics Letters*, vol. 62, no. 24, pp. 3135–3137, 1993.

[24] M. Perný, V. Šály, M. Váry, and J. Huran, "Electrical and structural properties of amorphous silicon carbide and its application for photovoltaic heterostructures," *Elektroenergetika*, vol. 4, no. 3, 2011.

[25] J. Bullot and M. P. Schmidt, "Physics of amorphous silicon-carbon alloys," *Physica Status Solidi (B) Basic Research*, vol. 143, no. 2, pp. 345–418, 1987.

[26] J. Aguilar, L. Urueta, and Z. Valdez, "Polymeric synthesis of silicon carbide with microwaves," *Journal of Microwave Power and Electromagnetic Energy*, vol. 40, no. 3, pp. 145–154, 2007.

[27] B. Shokri, M. A. Firouzjah, and S. Hosseini, "FTIR analysis of silicon dioxide thin film deposited by metal organic-based PECVD," in *Proceedings of the International Plasma Chemistry Society (IPCS '09)*, 32009 ISPC19, Shahid Beheshti University, Bochum, Germany, 2009.

[28] W. J. Miller, *High temperature oxidation of silicon carbide [M.S. thesis]*, Air Force Institute of Technology, Wright Paterson Air Force Base, Ohio, USA, 1972.

[29] A. L. Smith and N. C. Angelotti, "Correlation of the SiH stretching frequency with molecular structure," *Spectrochimica Acta*, vol. 15, pp. 412–420, 1959.

[30] N. Koizumi, K. Murai, S. Tamayama, H. Kato, T. Ozaki, and M. Yamada, "Diffuse reflectance IR spectroscopic study on the role of promoters in the reactivity of carbon monoxide with hydrogen over novel Pd sulfide catalyst," *Fuel Chemistry Division Preprints*, vol. 47, no. 2, p. 520, 2002.

[31] H. F. Zhang, C. M. Wang, and L. S. Wang, "Helical crystalline SiC/SiO$_2$ core-shell nanowires," *Nano Letters*, vol. 2, no. 9, 2002.

[32] J. G. Wang, S. Liu, T. Ding, S. Huang, and C. Qian, "Synthesis, characterization, and photoluminescence properties of bulk-quantity of β-SiC/SiO$_x$ coaxail nanowires," *Materials Chemistry and Physics*, vol. 135, pp. 1005–1011, 2012.

[33] D. M. Wolfe, B. J. Hinds, F. Wang et al., "Thermochemical stability of silicon-oxygen-carbon alloy thin films: a model system for chemical and structural relaxation at SiC-SiO$_2$ interfaces," *Journal of Vacuum Science and Technology A: Vacuum, Surfaces and Films*, vol. 17, no. 4, pp. 2170–2177, 1999.

[34] O.-S. Kwon, S.-H. Hong, and H. Kim, "The improvement in oxidation resistance of carbon by a graded SiC/SiO$_2$ coating," *Journal of the European Ceramic Society*, vol. 23, no. 16, pp. 3119–3124, 2003.

[35] H. Zhao, Z. Fu, C. Tang, X. Liu, Z. Li, and K. Zhang, "Study of SiC/SiO$_2$ oxidation-resistant coatings on matrix graphite for HTR fuel element," *Nuclear Engineering and Design*, vol. 271, pp. 217–220, 2014.

[36] Z. He, R. Tu, H. Katsui, and T. Goto, "Synthesis of SiC/SiO$_2$ core-shell powder by rotary chemical vapor deposition and its consolidation by spark plasma sintering," *Ceramics International*, vol. 39, no. 3, pp. 2605–2610, 2013.

[37] Y. Rane, A. Altecor, N. S. Bell, and K. Lozano, "Preparation of superhydrophobic Teflon AF 1600 sub-micron fibers and yarns using the Forcespinning technique," *Journal of Engineered Fibers and Fabrics*, vol. 8, no. 4, pp. 88–95, 2013.

Increasing Bending Strength of Porcelain Stoneware via Pseudoboehmite Additions

Omar Aguilar-García,[1] Rafael Lara-Hernández,[1] Azucena Arellano-Lara,[2] José L. Gil-Vázquez,[1] and Jaime Aguilar-García[1]

[1] *Departamento de Ingeniería Industrial, Instituto Tecnológico de Morelia, Avenida Tecnológico 1500, Col. Lomas de Santiaguito, 58120 Morelia, MICH, Mexico*
[2] *Departamento de Cerámica, Instituto de Investigaciones Metalúrgicas, Universidad Michoacana de San Nicolás de Hidalgo, Apdo. Postal 888, 58000 Morelia, MICH, Mexico*

Correspondence should be addressed to Omar Aguilar-García; omarag@mail.com

Academic Editor: Keizo Uematsu

Pseudoboehmite nanoparticles synthesized through the desulfation of $Al_2(SO_4)_3$ were used to investigate the reinforcement of commercial porcelain stoneware. Fractured specimens investigated by SEM suggest that the added pseudoboehmite precursor generated a nanometric primary mullite phase dispersed in the porcelain glassy phase that limited and stopped the intergranular crack propagation. The porcelain modulus of rupture increased twice the value of the modulus of rupture (108 MPa) as compared with that samples without pseudoboehmite additions. Pseudoboehmite also led to increased densification of porcelain stoneware bodies up to 1250°C as shown by thermodilatometry data.

1. Introduction

A high volume of porcelain stoneware factories installed around the world have a typical 40–50 wt% kaolinitic clay, 35–45 wt% feldspar, and 10–15 wt% quartz sand composition. Such material is characterized by its high technological properties, like low water absorption (<0.5%) and high bending strength (>35 MPa). After firing, porcelain stoneware shows a typical matrix that consisted of mullite crystals embedded in the glassy phase that holds the coarse quartz particles. Mullitization, a controversial issue, has been studied in the literature for diverse ceramic systems with a view to strengthen the ceramic body. In fact, Zoellner's [1] theory on the strength of porcelain defines the nature of the mullite as a key phase responsible for the mechanical strength. It appears that generating the correct amount of properly sized mullite is vital in achieving the desired strength. Furthermore, the dispersion-strengthening hypothesis of porcelains proposes that the dispersed particles limit the size of Griffith flaws, leading to increased strength [2]. Strength is a function of the volume fraction of the dispersed phase at low volume fractions, while at high volume fractions the strength is dependent on both the volume fraction and the particle size of the dispersed phase. It has also been extensively reported in the literature [3] that the improvement in mechanical strength of porcelains arises from a prestressing effect whereby the quartz is under the tensile stress, and consequently, the glassy matrix surrounding the quartz grains is under a compressive stress.

Primary mullite first occurs as derived from kaolinite reaction series in porcelain compositions heated to around 1000°C. Secondary mullite appears at higher temperature after molten feldspar dissolves clayey phases. In some porcelain systems, tertiary mullite has also been reported [1] as a result of a solution precipitation process from an alumina rich phase. Kaolin and gibbsite as well as synthetic mullite were added by Zanelli et al. [4] to porcelain stoneware to enhance mechanical properties. They found that though the mullite content in the porcelain body increased significantly, the physical and technological properties did not always increase proportional to the mullite content. Boehmite can be a source of nanosized alumina and presumably will make porcelain more reactive and strong in both green and fired states. The influence of boehmite gel additions on the green and sintered properties of alumina porcelain systems was investigated by Belnou et al. [5]. They found that boehmite gel additions increased both bending strength and thermal shock resistance of quartz-free porcelain by shifting the size of residual pores towards lower diameters and enhancing mullitization. Pseudoboehmite may be used in a porcelain paste in order to enhance the mechanical properties that may be expected from the increase in mullite content. A source for mullite in porcelain may come from the dissolution of alumina or alumina precursors saturating the porcelain glassy phase. The dissolution of alumina in alkali silicate glasses (Na_2O and K_2O) has been studied in the literature [6] by measuring the reduction of the magnitude of the enthalpy of mixing alkali silicates with alumina addition and has been attributed to the reduction of the glass nonbridging oxygen content. The aim of the present work was to study the role of the mullite phase on the mechanical properties of commercial porcelain stoneware and analyzing the resulting microstructure after systematically adding 2, 5, and 10 wt% pseudoboehmite.

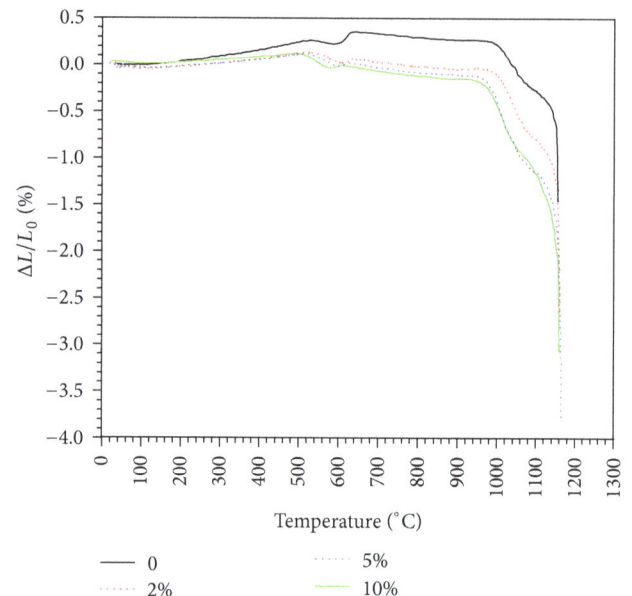

FIGURE 1: Thermodilatometry data showing higher shrinkage within the whole range of densification as pseudoboehmite additions increase.

2. Experimental Procedure

Raw materials used included nepheline syenite (48 wt%), kaolin (30 wt%), ball clay (10 wt%), and quartz sand (12 wt%). The pseudoboehmite was obtained using the U. G. process of the alunite [7]. This process is associated with the alkaline desulfation of the basic compounds [8]. Also, this compound can been obtained with the desulfation of the $Al_2(SO_4)_3$ using an ammonia solution. The obtained pseudoboehmite has a chemical composition of $Al_4O_3(OH)_6$. Furthermore, the compound was derived from the commercial sulphate with a chemical composition (wt%) given by Al_2O_3 (98.9010), SiO_2 (0.5560), CaO (0.4960), Fe_2O_3 (0.0259), ZnO (0.0061), CuO (0.0053), SO_3 (0.0048), NiO (0.0032), and K_2O (0.0015) [9]. The powders were oven-dried at 100°C for 24 h, and this material was ball-milled to pass a 200 mesh sieve and finally added to the slurry of porcelain by the 2, 5, and 10 wt%. Specimens were formed by slip casting at 69 wt% solids loading, adjusting dispersion conditions by the use of Darvan 7, Darvan 811, and sodium silicate deflocculants [10]. Disc specimens with dimensions of 100 × 10 mm were cast from properly aged slips and dried up at 110°C for 24 h. Firing was done by an electric furnace up to 1250°C, 10°C/min heating rate and 2 h soaking time. Bulk density was determined by water immersion procedures based on ASTM C20. Fired specimens polished and HF etched were characterized by SEM using a JEOL JSM-6300 and X-ray diffraction by a Siemens 400, CuKα 30 kV 25 mA.

The hardness (H) and the fracture toughness (K_{IC}) were measured (Mitutoyo MVK-E3 tester) by Vickers indentation on the polished surfaces of the sintered samples (diamond pastes of 6, 3, and 1 μm). Vickers microhardness measurements $HV_{0.2}$ were made by ISO 6507 using indentation load of 1.961 N [11], and a series of ten measurements were performed to made the statistical analysis. For fracture toughness, the samples were submitted to 10 loads of 9.8 N for 15 s on each indentation. The cracks were measured using the microscope attachment on the microhardness tester immediately after indentation. Crack measurements were only made on indents that were well defined without chipping and for which the cracks did not terminate at pores. The indentation fracture toughness of the material was evaluated selecting a model included by the Palmqvist crack system [11].

The Young's modulus of the samples was measured ultrasonically using an impulse-excitation of vibration technique (Grindo-Sonic, J. W. Lemmens Inc.) according to ASTM standards C 1259-94. This method covers a dynamic determination of the elastic properties of materials at ambient temperature. Young's modulus was computed using the resonant frequency in the flexural and torsion modes of vibration. The bending strength of fired test bars was measured under a three-point bending test by a universal instron machine model 3366 according to ASTM 1161-90. The final results were taken of the average of four replications for Young's modulus, the bending strength, and the physical properties.

3. Results and Discussion

The synthesized pseudoboehmite with a $Al_4O_3(OH)_6$ composition that consisted of nanometric whiskers is known to undergo several structural thermal transformations including the γ-alumina formation at 500°C; however, at porcelain vitrification temperatures, it is expected to dissolve in the glass and contribute to mullite formation. Figure 1 shows shrinkage behavior of porcelain samples with increasing pseudoboehmite additions. It is seen that such additions increase the shrinkage level within the whole range of

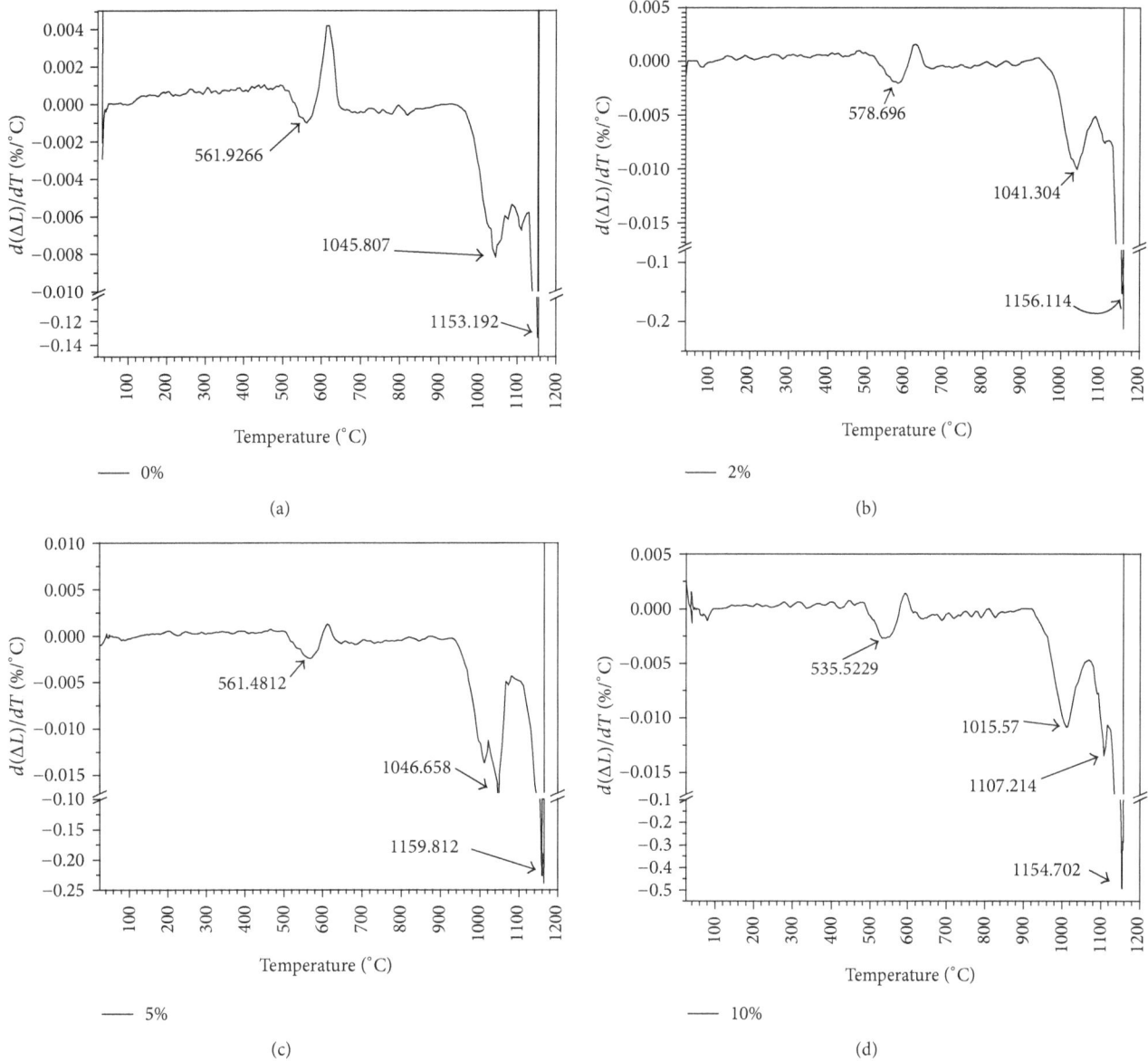

FIGURE 2: TDD curves corresponding to TD curves shown in Figure 1.

densification, implying a drastic reduction in porosity, particularly from 1100°C to 1160°C with obvious implications on strengthening. This is in agreement with the work of Belnou et al. [5] who found that pseudoboehmite additions in alumina porcelain shifted the size of residual pores towards lower diameters. It is also seen by Figure 1 that the rate of shrinkage in pseudoboehmite (PB) added samples becomes more gradual which may reduce faults and defects brought about by sudden structural rearrangements within the body.

The dilatometric curves of the compositions have been studied (Figure 2). The composition with 0 wt% of PB shows a typical behavior of the porcelain stoneware. As it can be noticed at 510–570°C, as a result of the kaolinite dehydration and metakaolinite formation, some shrinkage is observed, followed by a volume increase at ~580°C, due to the transformation of α-quartz to β-quartz. The compositions

show considerable shrinkage at ~1000°C. The shrinkage may be associated with the liquid development, mullitization, and densification. As pseudoboehmite is added, implying lower mullitization temperature was brought about by the highly reactive pseudoboehmite sol dissolved in the glass. It therefore appears that pseudoboehmite additions may ease mullite formation.

The derived dilatometric curves have been prepared (Figure 3). These curves show the differences between the compositions. As PB is added, decomposition of clays occurs at lower temperatures. Around 1045°C, a peak is observed in the sample without additions of PB; this change is associated with the nucleation of primary mullite from metakaolinite. As PB is added, it can be seen that peak appears at a lower temperature. Samples with additions of 5 wt% and 10 wt% of PB, the third peak appears associated with the formation

TABLE 1: Experimental physical and mechanical properties.

Pseudoboehmite addition (wt%)	Firing Temp. (°C)	Sample	Modulus of rupture (MPa)	Bulk density (g/cm³)	Young's modulus (GPa)	Vickers micro-hardness (GPa)	Fracture toughness K_{IC} (MPa·m$^{1/2}$)	Porosity (%)
2	1150	1	92	2.1	49.5	6.1	1.4	5.6
	1200	4	56.9	2.0	57.5	7.0	1.6	6.6
	1250	7	42.7	1.9	64.34	6.8	1.7	7.6
5	1150	2	81.2	2.2	40.5	6.8	1.5	5.1
	1200	5	94.5	2.2	60.2	6.8	1.7	7.8
	1250	8	107.4	2.1	62.2	6.9	1.5	7.5
10	1150	3	76.3	2.0	32.8	6.4	1.2	7.9
	1200	6	70.2	2.1	53	7.1	1.5	9.4
	1250	9	101.8	2.0	56.8	6.4	1.5	12.6
M. Dondi et al. [12]	1200		50	2.4-2.5	79	6.5	1.2	
Zanelli et al. [4]	1220		42	2.2–2.4			1.1	4–12.8

TABLE 2: Analysis of variance of bulk Density and porosity.

	Bulk density						Porosity				
Factor	SS	df	MS	F	P	Factor	SS	df	MS	F	P
Temperature L + Q	**0.02**	**2.00**	**0.01**	**5.87**	**0.06**	Temperature L + Q	**13.88**	**2.00**	**6.94**	**7.05**	**0.05**
PB (%) L + Q	**0.05**	**2.00**	**0.02**	**11.34**	**0.02**	PB (%) L + Q	**21.74**	**2.00**	**10.87**	**11.04**	**0.02**
Error	0.01	4.00	0.00			Error	3.94	4.00	0.98		
Total	0.08	8.00				Total	39.55	8.00			

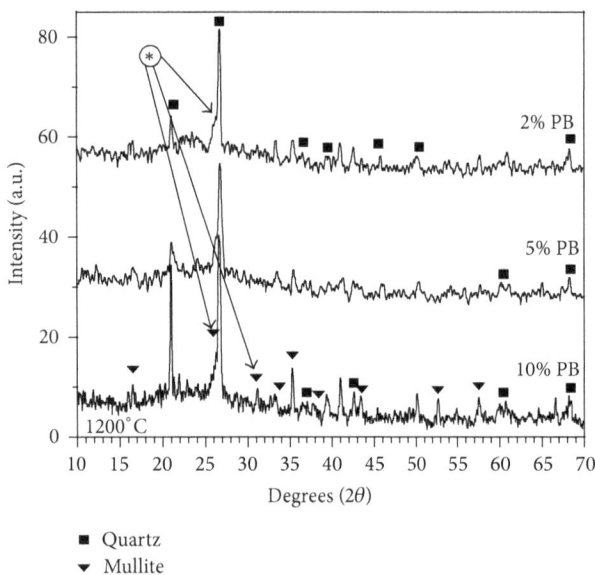

FIGURE 3: X-ray diffraction patterns of experimental porcelain compositions fired at 1200°C for boehmite additions (2, 5, 10 wt%). *Note moderate increase in the mullite peaks intensity.

of secondary mullite, attributable to the high reactivity of PB particles with the liquid phase. The importance of these curves is that the position of their minimums values indicates the reactivity of the various compositions. The sequence of reactivity by increasing temperatures is 10% > 5% > 2% > 0%. The reactivity of the composition is due to the corresponding additions of PB.

This assumption was confirmed by the XRD results of samples, sintered at 1200°C (Figure 3). Mullite phase increases moderately with added pseudoboehmite as shown in Figure 3 by X ray diffraction. The aspect ratio of the acicular mullite also increases with both pseudoboehmite additions and temperature as shown by SEM in Figure 4. In general it is seen that mullite needles coarsen, leading to a smaller number of larger needles. Figure 5 shows a fracture surface in a 10 wt% pseudoboehmite sintered at 1150°C specimen where an intergranular type of crack that propagates among primary mullite is being stopped by a cluster of nanometric mullite (2/3 Al_2O_3/SiO_2) crystals as determined by EDX.

The statistical treatment of physical and mechanical measurements has enabled the elaboration of mathematical models and permitted not only to simulate the mechanical behavior, but also to evaluate the contribution of the various experimental parameters involved in. By ANOVA analysis, the factors with statistical relevance have been defined, while by least squares method their coefficients in the mathematical model have been calculated. Table 1 lists average values of measured physical and mechanical properties obtained for three replications. The results of experiments designed in Table 1 were analyzed using ANOVA method and the results are in Tables 2 and 3. Based on the data of Table 2,

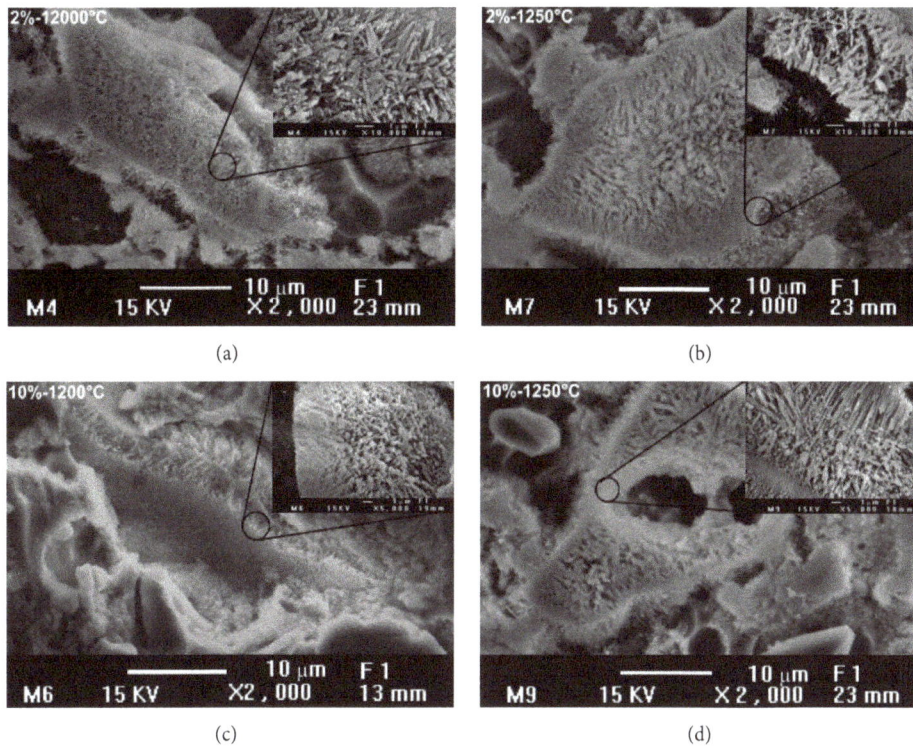

FIGURE 4: SEM micrographs of etched surfaces showing acicular mullite formation. Note increasing aspect ratio of mullite needles depending on both temperature (left to right) and composition (top to bottom).

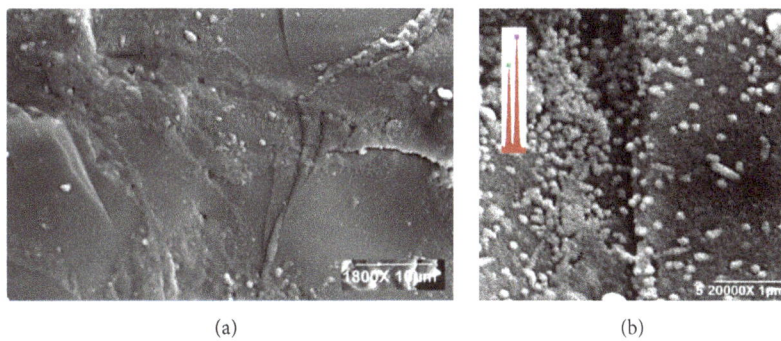

FIGURE 5: SEM micrographs porcelain stoneware fracture surfaces. (a) A crack stops at a spot of high mullite concentration. (b) Intergranular crack propagates among primary nanometric mullite (2/3 Al2O3/SiO2) crystals.

TABLE 3: Analysis of variance of Young's modulus and Vickers microhardness.

| | Young's modulus | | | | | | Vickers microhardness | | | | |
Factor	SS	df	MS	F	P	Factor	SS	df	MS	F	P
Temperature L + Q	679.92	2.00	339.96	26.71	0.00	Temperature L + Q	0.34	2.00	0.17	1.58	0.31
PB (%) L + Q	145.48	2.00	72.74	5.72	0.07	PB (%) L + Q	0.09	2.00	0.04	0.42	0.69
Error	50.90	4.00	12.73			Error	0.43	4.00	0.11		
Total	876.30	8.00				Total	0.86	8.00			

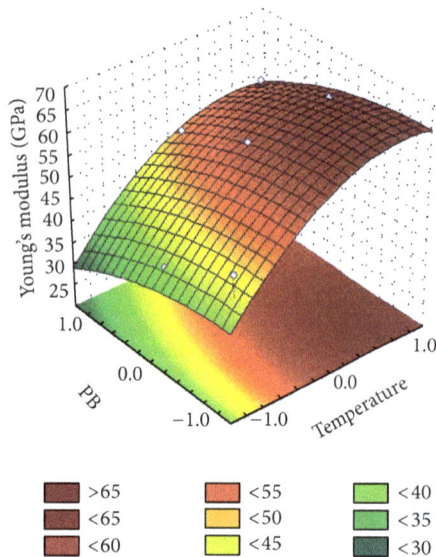

FIGURE 6: 3D surface response of Young's modulus versus pseudoboehmite additions and temperature.

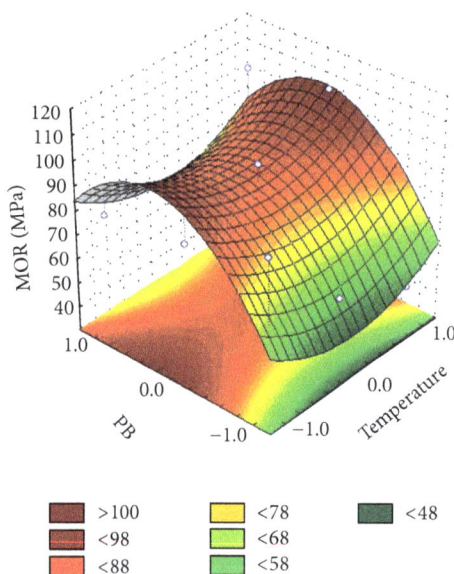

FIGURE 8: 3D surface response of fracture toughness versus pseudoboehmite additions and temperature.

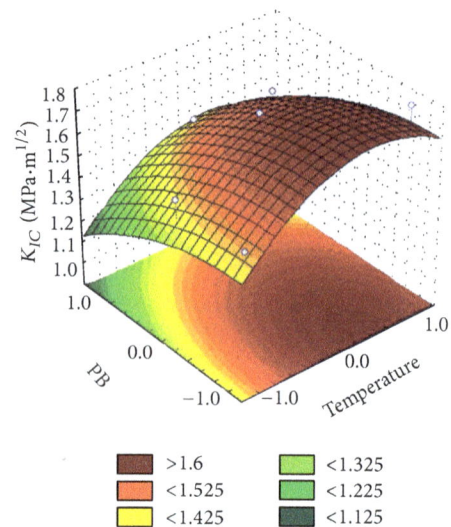

FIGURE 7: 3D surface response of modulus of rupture versus pseudoboehmite additions and temperature.

the pseudoboehmite and the temperature have statistical significance, the significance of the factors were obtained by comparing the F calculated value versus the F tabulated value in the Fisher-Snedecor distribution; if the F calculated values are higher than the F tabulated values (at the required level of significance) is considered to exist statistical significance, all the F values presented in Table 2 are higher than the tabulated. Both the temperature and the additions of PB have a positive effect on the bulk density and porosity. This is consistent with the stated above in dilatometric curves, where it was observed that the formation of glassy phase was obtained at higher temperatures and additions of PB, enabling that the liquid phase can surround the clay relicts and quartz,

and the achieving of a higher density. Increased porosity was consistent with the published results for Belnau et al. [5], limiting PB additions to the mixture of porcelain.

Table 3 shows that the independent variables have a statistical significance for the Young's modulus, but for the Vickers microhardness they do not have a statistical significance. The temperature and the additions of PB have a positive effect on Young's modulus. At higher temperature and additions of PB, the properties of modulus of elasticity increase (Figure 6). For Vickers microhardness, temperature and PB additions have no significant effect; this has important implications for improving the mechanical properties since it does not want to make more fragile materials. The modulus of rupture varies with both temperature and pseudoboehmite additions. The highest modulus (108 MPa) occurs for the 5% and 10% of pseudoboehmite additions, which is an important increase in strength as compared to samples without pseudoboehmite additions and those reported in the literature [12] that present about half of the latter value (50 Mpa). Regarding the effect of the temperature on the modulus of rupture, the higher modulus measurements (108 and 102 MPa) correspond to samples with higher levels of pseudoboehmite (5 and 10 wt%) heated up to 1250°C. On the contrary specimens that contain 2% pseudoboehmite and therefore less refractoriness present low strength (43 MPa) at the same temperature (1250°C). This entire phenomenon can be observed in Figures 6 and 7. Fracture toughness in pseudoboehmite containing samples almost doubles the value 1.2 presented by pseudoboehmite free specimens as shown in Table 1. The fracture toughness showed some variation both with sintering temperature and pseudoboehmite additions as shown in Figure 8. The highest value of K1C was obtained with 5% PB and sintering at 1200°C, and this is a consequence of the quantity of crystalline phase and lower porosity, limiting the propagation of Griffith cracks.

By analyzing the three graphs of surface, it can be seen that the higher areas are reached at a temperature of 1200°C and 5% PB additions. The combination of these factors achieves a synergistic effect between the amount of liquid present and PB particles, and this is a consequence of the highest density of the sample, the decrease in the porosity, and the extra mullitization promoted by the additions of PB that creates a barrier that stops the cracks.

4. Conclusions

The high reactivity of pseudoboehmite sol additions presumably increased the amount of alumina dissolved in the glass, thereby increasing moderately the precipitated mullite phase content. It appears that strengthening of the porcelain may be caused by a dispersion-strengthening mechanism rather than by a substantial increase in the mullite phase content. Primary mullite nanometric crystal clusters were shown to limit the intergranular crack propagation, in agreement with the dispersion-strengthening hypothesis whereby dispersed particles limit the size of Griffith flaws leading to increased strength. Also, pseudoboehmite additions allowed the reinforcement of green bodies producing less body defects and higher densification.

Conflict of Interests

The authors declare that there is no conflict of interests regarding the publication of this paper.

Acknowledgment

The authors acknowledge the financial support of the Investigación Científica y Desarrollo Tecnológico del Sistema Nacional de Institutos Tecnológicos, México.

References

[1] W. M. Carty and U. Senapati, "Porcelain—raw materials, processing, phase evolution, and mechanical behavior," *Journal of the American Ceramic Society*, vol. 81, no. 1, pp. 3–20, 1998.

[2] O. I. Ece and Z.-E. Nakagawa, "Bending strength of porcelains," *Ceramics International*, vol. 28, no. 2, pp. 131–140, 2002.

[3] H.-Y. Lu, W.-L. Wang, W.-H. Tuan, and M.-H. Lin, "Acicular mullite crystals in vitrified kaolin," *Journal of the American Ceramic Society*, vol. 87, no. 10, pp. 1843–1847, 2004.

[4] C. Zanelli, M. Dondi, G. Guarini, M. Raimondo, and I. Roncarati, "Influence of strengthening components on industrial mixture of porcelain stoneware tiles," *Key Engineering Materials*, vol. 264–268, pp. 1491–1494, 2004.

[5] F. Belnou, D. Goeuriot, P. Goeuriot, and F. Valdivieso, "Nanosized alumina from boehmite additions in alumina porcelain: 1. Effect on reactivity and mullitisation," *Ceramics International*, vol. 30, no. 6, pp. 883–892, 2004.

[6] P. J. Lezzi and M. Tomozawa, "Effect of alumina on enthalpy of mixing of mixed alkali silicate glasses," *Journal of Non-Crystalline Solids*, vol. 357, no. 10, pp. 2086–2092, 2011.

[7] W. X. López, "Three methods to produce alumina from alunite," in *Light Metals*, vol. 2, pp. 49–58, American Institute of Mining, Metallurgical and Petroleum Engineers, New York, NY, USA, 1977.

[8] H. Juárez-M, J. M. Martínez-R, and J. M. Ruvalcaba-L, "Obtención y caracterización de pseudoboehmita a partir de sulfato de aluminio comercial," in *IV Congreso Iberoamericano de Química Inorgánica y XI Congreso Mexicano de Química Inorgánica*, pp. 256–260, Guanajuato, México, 1993.

[9] S. D. de la Torre, A. Kakitsuji, H. Miyamoto et al., "Seeding with α-alumina for transformation and densification of boehmite-derived γ and δ-alumina by spark plasma sintering," *Ceramic Transactions*, vol. 94, pp. 83–89, 1999.

[10] O. Aguilar-García, S. Bribiesca-Vazquez, and J. Zarate-Medina, "Mixture design to optimize the amount of deflocculants in aqueous porcelain precursor suspensions," *Journal of Ceramic Processing Research*, vol. 10, no. 2, pp. 125–128, 2009.

[11] O. Aguilar-García, S. Bribiesca-Vazquez, and J. Zárate-Medina, "Evaluation of hardness and fracture toughness in a porcelain stoneware with pseudoboehmite additions," *Journal of Ceramic Processing Research*, vol. 10, no. 1, pp. 37–42, 2009.

[12] M. Dondi, M. Raimondo, C. Zanelli, and P. M. T. Cavalcante, "Sintering mechanisms of porcelain stoneware tiles," in *Proceedings of International Conference on the Science, Technology & Applications of Sintering (SINTERING '03)*, 2003.

Thermal Variation of Elastic Modulus on Nanocrystalline NiCuZn Ferrites

S. R. Murthy

Department of Physics, Osmania University, Hyderabad 500 007, India

Correspondence should be addressed to S. R. Murthy; ramanasarabu@gmail.com

Academic Editor: Matjaz Valant

The nanopowders of $Ni_{0.38}Cu_{0.12}Zn_{0.5}Fe_2O_4$ with particle size, 20 nm have been synthesised using Microwave-Hydrothermal method and characterized. Then the ferrite samples were microwave sintered at different temperatures in an air atmosphere and characterized. The magnetic properties were measured at room temperature. The dielectric constant (ε), initial permeability (μ_i) and quality factor (Q) has been measured on sintered samples at 1 MHz. Thermal variation of initial permeability has been measured over temperature range of 300 K–600 K. A detailed study of elastic behaviour of NiCuZn ferrites has been under taken using a composite piezoelectric oscillator method over a temperature of 300 K–600 K. The room temperature elastic moduli is found to be slightly sample dependent and decreases with increasing the temperature, except near the Curie temperature, T_C, where a small anomaly is observed. The internal friction at room temperature is also found to be more particle size dependent. The temperature variation of internal friction exhibits a broad maximum around 500 K, just below Curie temperature T_C 530 K. The above observations were carried on in the demagnetized state; on the application of a 400 mT magnetic field allowed us to reach the saturated state of the sample at any of the measuring temperature. The anomaly observed in the thermal variation of elastic moduli and internal friction is explained with the help of temperature variation of magneto-crystalline anisotropy constant.

1. Introduction

By the development of surface mount technology (SMT) and multilayer chip devices, NiCuZn ferrites have been extensively studied and widely used to fabricate chip inductors and EMI filters because of their relatively low sintering temperature and high resistivity with good performance at high frequencies [1–5]. However, in these ferrites, it was found that the mechanical and magnetic properties are easily changed or deteriorated by the stress caused at the internal electrode. These problems can be reduced by the preparation of NiCuZn ferrites under controlled experimental conditions and with the knowledge of mechanical properties. On the other hand, the mechanical properties of ferrites are often taken for granted during the manufacture and testing. Despite this, the strength of ferrites can be important particularly for parts, which are tightly clamped during their assembly and use to assure minimum changes in devices performance with mechanical shock, temperature, vibration, and so forth. A survey of the literature tells that no information is available for NiCuZn ferrites on the elastic properties.

There is an abundant literature on the propagation of ultrasonic waves in magnetic materials, but a few interpretations only refer to possible effects due to the domain wall (DW) [6]. It was known from the acoustic emission (AE) studies on magnetic materials that the space occupied by a DW is the location of magnetoelastic interactions and ultrasonic bursts are emitted when the DW are created or annihilated [7, 8]. It was also found that the specific DW/lattice magneto-elastic interactions might be one of the origin of the DW relaxation. In order to understand, an interaction of DW with elastic modulus, a detailed study of the thermal variation of elastic modulus, and internal friction on nanocrystalline $Ni_{0.38}Cu_{0.12}Zn_{0.5}Fe_2O_4$ have been undertaken over a wide temperature range.

2. Experimental Method

Pure (99.98%) nickel nitrate $[Ni(NO_3)2.6H_2O]$, (99.95%) copper nitrate $[Cu(NO_3)2.3H_2O]$, (99.98%) zinc nitrate

[Zn(NO$_3$)2.6H$_2$O], and (99.96%) iron nitrate [Fe(NO$_3$) 2.9H$_2$O] were dissolved in 50 mL of deionized water. The molar ratio of powders was adjusted to obtain the composition Ni$_{0.38}$Cu$_{0.12}$Zn$_{0.5}$Fe$_2$O$_4$. An aqueous NaOH solution was added to solution until the desired pH (~9.45) value was obtained. The mixture was then treated in Teflon lined vessel using a microwave digestion system (Model MDS-2000, CEM Corp., Mathews, NC). This system uses 2.45 GHz microwaves and can operate at 0–100% full power (630 ± 50 W). In the present investigation, the samples were synthesized at 160°C/30 min. The time, pressure, and power were computer controlled. The products obtained were filtered and then washed repeatedly with deionized water, followed by freeze drying overnight. The synthesized powders were weighed, and the percentage yields were calculated from the expected and the amount that was actually crystallized. The yield obtained in the present investigation was 95%. As synthesized powders were characterized by using powder X-ray diffraction (XRD, Philips analytical X-ray diffractometer with Cu-Kα radiation). Particle size and morphology were determined using transmission electron microscopy (TEM, Model JEM-2010, JEOL, Tokyo, Japan).

To nanopowders, 2 wt% of poly-vinyl-alcohol was added as a binder and pressed in a die under a pressure of 190 MPa for 10 min into pellets (10 mm diameter, 3 mm thickness), and (50 mm in length, 10 mm diameter), and toroids (12 mm outside diameter, 6 mm inside diameter, 3 mm thickness). After binder burnt out at 300°C, the samples were microwave sintered at 750°C/30 min (MW1), 800°C/30 min (MW2), 850°C/30 min (MW3), 900°C/30 min (MW4), and 930°C/30 min (MW5) in air. The microwave sintering process was carried out using a specially designed applicator which consists of a domestic microwave oven having an output power level tuneable up to a maximum of 800 W and operating frequency of 2.45 GHz.

All the sintered samples were characterized using the X-ray diffraction (XRD). The average grain size (Dm) of the samples was evaluated using the scanning electron microscopy (SEM; LEICA, S440i, UK) pictures. The thermal expansion of the sintered samples has been measured using dilatometer (DIL 402C) in air and at a heating rate of 20°C/min. The saturation magnetization (M_S) and coercive field (H_C) values of the present samples were obtained from the recorded hysteresis loops using the vibrating sample magnetometer (VSM, model DMS 1660) at room temperature. The initial permeability (μ_i) and quality factor has been measured on sintered samples at 1 MHz using a LCR meter. Thermal variation of initial permeability has been measured using the LCR meter at 1 kHz.

The elastic constants such as Young's modulus, Y, and shear modulus, G, were measured on sintered samples over the temperature range of 300 K–600 K using the composite piezoelectric oscillator method [9]. The appropriate quartz crystal transducer was cemented on the specimen with phenyl salicylate and two parts of sodium silicofluoride mixed with one part of barium sulphate in one drop of water glass for room and high temperature measurements, respectively. For Young's moduli measurements, the piezoelectric crystals were cut to oscillate at 100.245 ± 0.001 kHz in longitudinal direction. Shear moduli were measured on the same specimen sections, using shear crystals oscillating at approximately 165.25 ± 0.002 kHz. Young's and shear moduli were then determined in the usual way using the following expressions: $Y = (\lambda_l f_l)^2 \rho$, $G = (\lambda_s f_s)^2 \rho$, where ρ is density, λ is the wavelength, f is the resonant frequency, and subscripts "l" and "s" refer to longitudinal and shear waves, respectively. All the measurements were conducted keeping the composite oscillator in a vacuum (10^{-4} Torr). The vacuum chamber was placed at the centre of pole pieces of an electromagnet. The magnetic field was always applied parallel to the length of the specimen. The temperature of the sample is measured using Cr-Al thermocouple with an accuracy of ±0.1 K. The accuracy of Y, G, and internal friction measurements is about 0.1%, 0.1%, and 0.5% respectively.

3. Results and Discussion

Figure 1(a) gives the XRD patterns for as-synthesized powder of NiCuZn ferrite. It can be seen from the figure that the phase-pure ferrite was obtained. The broad peaks in the XRD pattern indicate the formation of nanopowders. All the peaks in the diffraction pattern have been indexed. The peaks (220), (311), (400), (511), and (440) have been deconvoluted to Lorentzian curves using Peak Fit software for the determination of the average diameter (DXRD) using full-width at half-maximum (FWHM) value. The average particle size (D_m) has been calculated using the Scherrer formula: $D_m = kl/b\cos\theta$, where k is a constant, b is the full width half maxima, l is the wavelength of X-rays used, and θ is the diffraction angle. The average particle size of NiCuZn ferrite powder was found to be ~25 nm.

Figure 1(b) shows TEM picture for the as-synthesized ferrite powder. The particles are nearly spherical in shape. The average particle size calculated from the different regions of the image was 20 nm. The selected area electron diffraction (SAED) pattern shows rings made up of discrete spots, indicating high crystallinity; the rings are consistent with the cubic spinel structure with an intense ring pattern form (hkl) planes. No secondary phases are found. The observed average size is consistent with that obtained from the X-ray analysis.

The powder XRD patterns for microwave sintered NiCuZn ferrites under investigation reveal a single-phase cubic spinel structure without any other impurity phases (Figure 2). The lines are board, which is a typical characteristic of nanocrystalline nature. The grain sizes were calculated from the XRD line broadening of the peak (311) using the Scherrer relationship, and the obtained values are presented in Table 1. Figure 3 shows the typical SEM pictures for microwave sintered samples. It can be seen from the figures that with an increase of sintering temperature, microstructure of the ferrite has become finer and finer. The microwave sintered samples exhibit fine and uniform microstructures due to the small soaking time. The grain sizes (D_m) as measured from SEM picture for the presently investigated ferrites were presented in Table 1. It can be observed from the table that the grain size of sintered ferrites increases with an increase of sintering temperature.

(a)

(b)

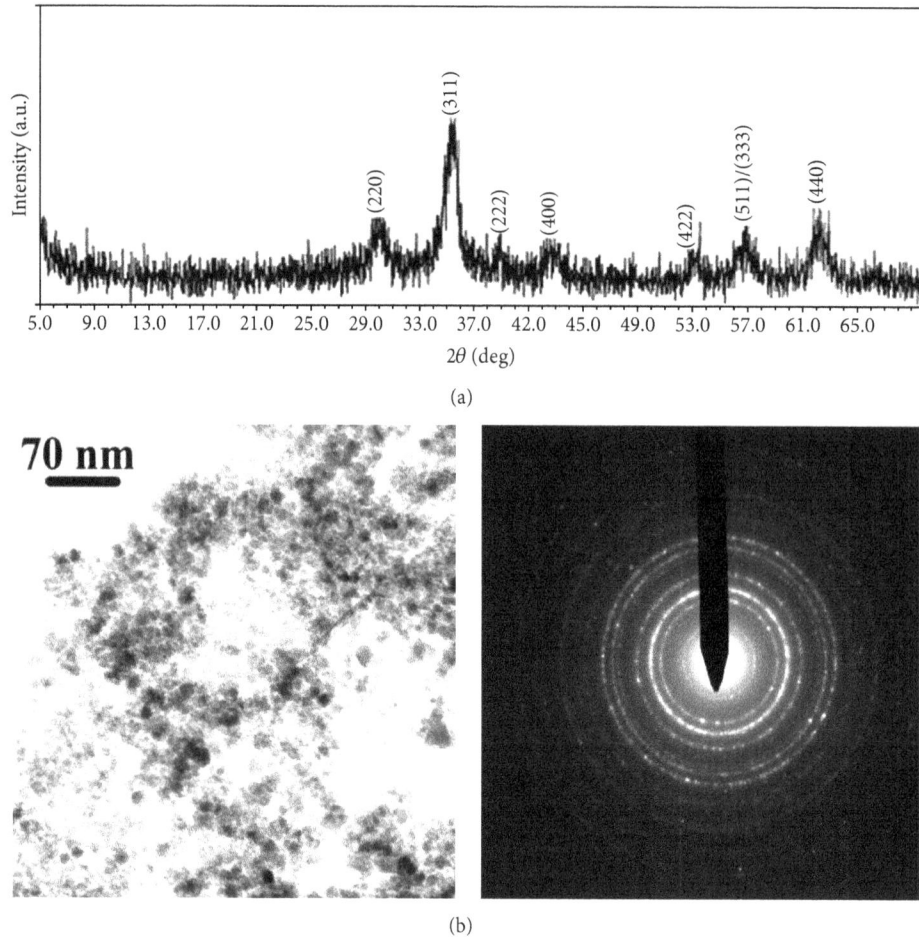

FIGURE 1: (a) XRD pattern and (b) TEM picture for as-synthesized ferrite powder.

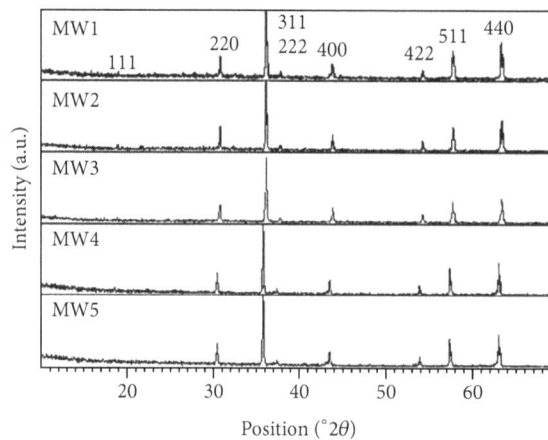

FIGURE 2: XRD patterns for microwave sintered NiCuZn ferrites.

The value of lattice constant (a) has been calculated using the XRD data, and the average value of lattice constant is 8.3374 ± 0.0001 Å. The bulk density (d_X) of the sintered samples has been measured using Archimedes's method, and the obtained results are presented in Table 1. It can be observed from the table that the value of bulk density increases with an increase of sintering temperature. It can also be seen from the table that the theoretical density (TD) samples have reached 97% at a sintering temperature of 750°C. Finally, a theoretical density of 99.5% has been achieved at a low sintering temperature of 930°C. The ultrafine particle size and relatively uniform grains are

FIGURE 3: SEM pictures for microwave sintered samples (a) MW3, (b) MW4, and (c) MW5.

TABLE 1: Room temperature data on NICuZn ferrites.

Sample	T_S (°C/min)	d_X (g/cc)	TD %	D_m (nm)	ε 1 MHz	μ_i 1 MHz	Q 1 MHz	M_S (emu/g)	H_C (Oe)
MW1	750/30	5.312	97	80	12	448	50	50	80
MW2	800/30	5.325	97.1	68	12	460	50	53	75
MW3	850/30	5.376	98	56	10	485	53	60	71
MW4	900/30	5.398	98.4	45	10	510	60	68	68
MW5	930/30	5.459	99.5	35	8	535	65	76	60

TABLE 2: Elastic modulus data for NiCuZn ferrites at room temperature.

Sample name	$Y \times 10^{-10}$ (N/m²)	$G \times 10^{-10}$ (N/m²)	$K \times 10^{-10}$ (N/m²)	Q^{-1}	V_l (m/s)	V_S (m/s)	α
MW1	17.82	7.28	10.76	0.71	6345	3345	0.22
MW2	18.03	7.34	11.12	0.65	6468	3675	0.22
MW3	18.10	7.45	8.42	0.45	6675	3890	0.22
MW4	18.25	7.56	10.39	0.26	6915	3962	0.21
MW5	18.54	7.67	9.38	0.20	6920	3974	0.21

responsible for achieving this very high densification at low temperature. We could not reach theoretical value (99.9% of TD) because of the ferrite particle being too fine to be completely dispersed in the liquid suspension and, therefore, flocculated to some extent by Van der Waals bonding. This slightly flocculated suspension resulted in some agglomerated regions distributed in the microstructure of samples. This was confirmed by microstructure. Rearrangement and differential micro-densification processes could not, thus, be wholly avoided during the sintering of these slightly inhomogeneous packed compacts, and therefore completely dense fired bodies were not achieved.

Magnetic properties such as saturation magnetization (M_S) and coercivity (H_C) obtained from the recorded hysteresis loops are presented in Table 1. The NiCuZn ferrite sintered at 930°C/30 min exhibits optimum magnetic property, a saturation magnetization of 76 emu/g, and $H_C = 60$ Oe. It is also observed from the table that value of M_S increases with increasing sintering temperature, as a result of increasing crystallinity. The values of H_C decrease with increasing sintering temperature. From the magnetic properties results,

one can notice that the maximum M_S corresponds to a smaller particle size of 35 nm; thus, control of crystal growth by varying sintering temperature can produce ultrafine magnetic powders.

Table 1 also gives initial permeability (μ_i), dielectric constant (ε), and quality factor (Q) values for sintered ferrites at 1 MHz. It can be seen from the table that the samples sintered at 930°C/30 min exhibit an optimum μ_i value of 535. It can be observed from the table that μ_i increases with increasing sintering temperature from 750°C to 930°C as a result of increasing densification. The ferrite sintered at 930°/30 min. also shows a maximum quality factor value of 65. The dielectric constant for present ferrites varies from 8 to 12. The smaller dielectric constant has been observed for MW5 sample.

The elastic constants such as Young's (Y), shear modulus (G), and internal friction (Q^{-1}) were measured at room temperature on sintered samples and presented in Table 2. The table also gives the computed values of Bulk modulus ($B = (YG)/3(3G - Y)$). It can be seen from table that the values of Y and G are found to increase with an increase of sintering temperature. This shows that an increase of sintering temperature increases the atomic binding force markedly. The average value of Poisson's ratio ($\sigma = (Y/2G) - 1$) for the samples under investigation is 0.3. The microwave sintered ferrites possess higher values of Y and G when compared to those of ferrites prepared by using solid state sintering method [10]. This is due to the high densification of the present samples.

It is clear from the table that the room temperature Y and G were found to increase from 17.82 N/m^2 to 18.54 N/m^2 and from 7.28 N/m^2 to 7.67 N/m^2 with an decreasing the grain size from 80 nm to 35 nm, respectively. Reciprocally, the internal friction was found to decrease from 0.71 to 0.20. In the present samples, the increase in density has decreased the grain size from 80 nm to 35 nm. As the porosity ($\sim$0.3) of samples remained nearly constant, the increase in the elastic modulus and decrease in the internal friction may be considered due to an decrease in the grain size. This is because, presently, microwave sintered samples have densities in the range from 5.312 g/cm^3 to 5.459 g/cm^3. This density variation is small compared to the variation of grain size (35 nm to 80 nm). Therefore grain size, not density, has the dominant effect on elastic properties.

No elastic modulus data is available in the literature on nanocrystalline NiCuZn ferrites. However, the present elastic modulus data is in fair agreement with that of previously investigated elastic behaviour of several polycrystalline ferrites [11–13]. Using the elastic modulus data, we have estimated the longitudinal velocity (V_l) and shear velocity (V_s) for all the samples, and results are presented in Table 2. The values of V_l and V_s are found to increase with an increase of sintering temperature. With the help of velocities, the Debye characteristic temperature (Θ) has been estimated using the Anderson method [14]. The average value of the Debye temperature for present ferrites is 546 K.

The elastic behaviour on NiCuZn ferrites has been measured in the temperature range of 300 K–600 K. For these experiments, a fused silica buffer rod was used to isolate the piezoelectric crystals from the heated specimen [15]. Fractional variation in frequency was measured and related to fractional modulus variations through the expansion coefficient.

Consider that Y_T or G_T/Y_{RT} or $G_{RT} = (1 + \alpha \Delta T)^{-1} (f_{sT}/f_s)^2$, where Y_{RT}, G_{RT}, and f_s are Young's modulus, shear modulus, and resonance of the specimen at the reference temperature (room temperature or 300 K), Y_T, G_T, and f_{sT} are the same quantities at any other temperature, ΔT is the difference between the measured and reference temperature, and α is the coefficient of linear expansion over the measured temperature range.

The values of the coefficient of thermal expansion (α) for the present specimens are measured using the dilatometer method in the temperature range of 300 K–600 K. It was found that the variation of "α" with temperature (T) for samples MW1, MW2, MW3, MW4, and MW5 are given by the following:

$$\alpha = 6.261 \times 10^{-6} + 0.65 \times 10^{-8}(T) + 0.532 \times 10^{-11}(T^2), \tag{1}$$

$$\alpha = 6.372 \times 10^{-6} + 0.75 \times 10^{-8}(T) + 0.58 \times 10^{-11}(T^2), \tag{2}$$

$$\alpha = 6.485 \times 10^{-6} + 0.92 \times 10^{-8}(T) + 0.542 \times 10^{-11}(T^2), \tag{3}$$

$$\alpha = 6.578 \times 10^{-6} + 0.71 \times 10^{-8}(T) + 0.645 \times 10^{-11}(T^2), \tag{4}$$

$$\alpha = 6.278 \times 10^{-6} + 0.25 \times 10^{-8}(T) + 0.141 \times 10^{-11}(T^2). \tag{5}$$

Figures 4 and 5 give the thermal variation of Young's and shear modulus measured in the demagnetized state, for all samples. It can be seen from the figures that the values of Y and G decrease smoothly with increasing temperature and attain a minimum at a certain temperature (T_1) below the Curie point. Beyond this temperature the elastic modulus shows a positive temperature coefficient and attains a maximum value at a temperature T_2. With further increase of temperature the both Y and, G are found to decrease. For samples, MW2, MW3, and MW4, the values of Y are very close to each other and are found to decrease by about 14% between 300 K and T_C. In all these samples, anomalous behaviour occurs in between 520 K and 530 K. The thermal variation of G for all the samples is found to be similar and the variation is about 5% between 300 K and T_C. A less (525 K to 535 K) pronounced increase of G with temperature is observed near the Curie point. No hysteresis has been observed during cooling. At a given temperature elastic modulus decreases when the mean grain size increases; this tendency is maintained all over the investigated temperature range. The decrease of Y and G with temperature is mainly due to the softening of the material. Murthy et al. [12, 13, 16, 17] also observed a similar variation of V_l and V_s with temperature in the vicinity of Curie temperature in case of polycrystalline YIG, and observed changes were used to understand the interaction of ultrasonic waves with annihilation/creation of domain walls.

The anomalous behaviour observed in the case of the presently investigated ferrites can be explained qualitatively with help of the temperature variation of the magneto-crystalline anisotropy constant (k_1). The temperature variation of k_1 was measured on the brother samples and found that k_1 becomes zero just below the Curie temperature, that is, at 522 ± 1 K. The measurements of initial permeability versus temperature on the brother sample have shown that the Curie temperature of the present samples is 530 ± 0.1 K.

As can be seen from Figures 4 and 5, the Y and G for present samples have reached a minimum value at temperature 520 K and 525 K (T_1), while k_1 becomes zero at 522 K. Similarly, the Y and G attain a maximum value at a temperatures 530 K and 535 K (T_2), respectively, while the Curie point of samples under investigation is 530 K. Thus, the minimum (T_1) and maximum (T_2) temperatures of Y and G verses temperature plots coincide with the temperatures of $k_1 = 0$ and the Curie temperature, respectively.

The magneto-crystalline anisotropy constant can be considered as a measure of the magnetic energy barrier to the movement of domain walls in the magnetic materials.

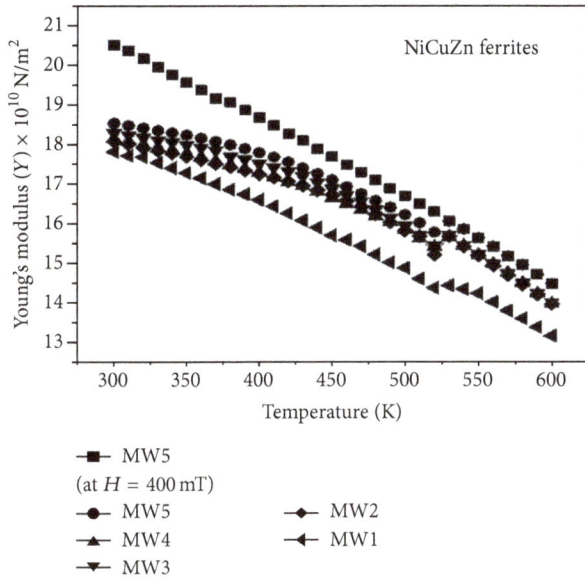

FIGURE 4: Thermal variation of Young's modulus (Y) with temperature for microwave sintered NiCuZn ferrites.

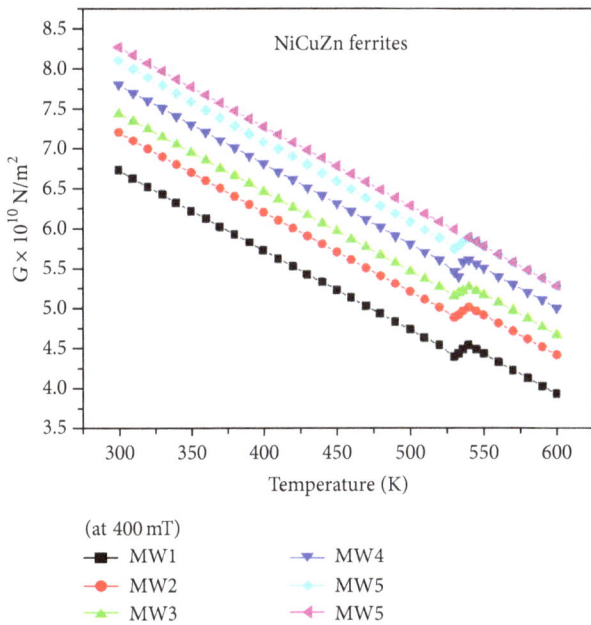

FIGURE 5: Thermal variation of shear modulus (G) with temperature for microwave sintered NiCuZn ferrites.

As such, domains will be free to move at a temperature at which $k_1 = 0$; the substance undergoes a maximum strain for a given stress; in other words, the Y and G increases till the Curie temperature is reached. Beyond the Curie temperature, both the elastic moduli have shown a decrease with an increase of temperature, since at T_C the ferrite losses its spontaneous magnetization and becomes paramagnetic.

In Figures 4 and 5 for the sample MW5, we present the temperature dependence of Y and G measured with $H = 400\,\text{mT}$; this field was checked to be high enough to saturate

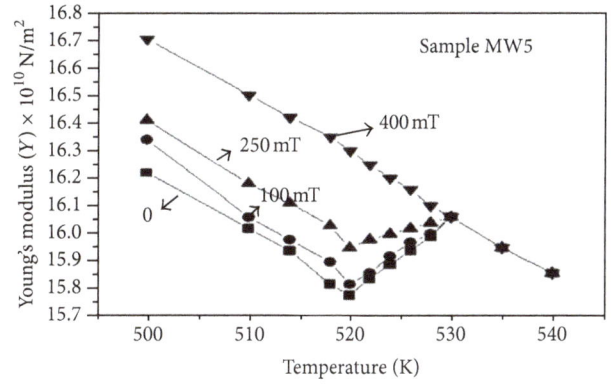

FIGURE 6: Thermal variation of Young's modulus (Y) with temperature for microwave sintered NiCuZn ferrites in the vicinity of the Curie temperature at applied magnetic fields.

the sample at any of the measuring temperatures. In saturated state Y (or G) is slightly higher than in the demagnetized state, expect above T_C, where both the values are identical, as expected. When the sample is subjected to magnetic field, it is observed that the minimum of elastic moduli-temperature curve is shifted to the low temperature side and becomes shallow. As the magnetic field increased further the minimum continuous to shift to the low temperature side becoming more and more shallow. Finally, the dip disappears when the magnetic field equal to saturation field of the sample (Figure 6).

The magnetic contribution to the velocity variations with temperature can be considered as a "second order effect," superimposed on the purely mechanical effects; it is more pronounced for the large grain size samples. The magnetic contribution dY defined as the difference between the elastic modulus in the saturated and demagnetized states is, respectively, given by $dY_S = Y_S(H = 400\,\text{mT}) - Y(H = 0)$ which is not more than 2% of the total elastic moduli and varies like the temperature variation of saturation magnetization (M_S). Similar variation of Y on temperature and magnetic field was observed for all the other samples.

Figure 7 gives the thermal variation of internal friction (Q^{-1}) measured at zero fields for all samples. It can be seen from the figure that the sample shows a small maximum around 400 K and broad maximum just below the Curie temperature. The rapid increase of Q^{-1} has been extended over 10 K below T_C: the very sharp decrease of Q^{-1} from about 0.97 down to 0.2 is accomplished in less than ±2 K. This abrupt change is the signature of the high chemical homogeneity of our samples.

In Figure 7, for the sample MW5, we present the temperature dependence of Q^{-1} measured with $H = 400\,\text{mT}$. The value of Q^{-1} in the saturated state is almost temperature independent (= 0.17) and is equal to the zero field value above T_C. Similar variation of Q^{-1} on temperature and magnetic field was also observed for all the other samples under investigation, it is observed that the temperature corresponding to the peak in the Q^{-1} versus temperature curve is near the temperature at which both the velocities

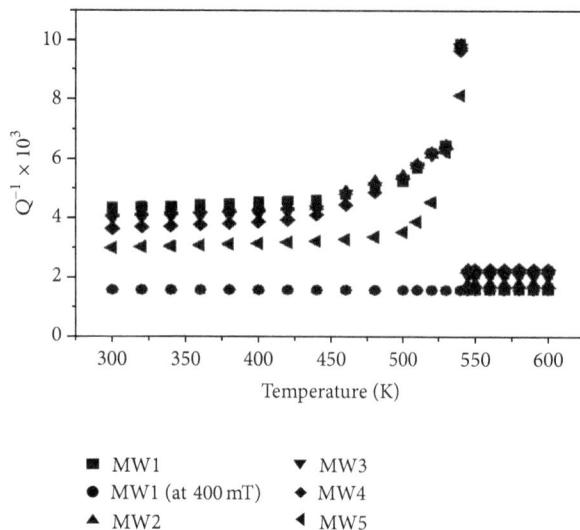

FIGURE 7: Thermal variation of internal friction (Q^{-1}) with temperature for microwave sintered NiCuZn ferrites.

show maximum value. As the internal friction is the complementary phenomena of the elastic behaviour, as such the peak observed in Q^{-1} versus T curve can be attributed to the $k_1 = 0$. As the internal friction is the complementary phenomena of elastic behaviour, the qualitative explanation given above can be extended for the occurrence of peaks in the Q^{-1} versus T curves.

The effects of microstructure are well marked on the internal friction, but they are not as dominant as for the elastic modulus. It can be seen from Figure 7 that the microstructure contribution to Q^{-1} is of the order of 95% of the total. The substantial magnetic contribution to Q^{-1} seems to result from two contributions: an electronic diffusion related magnetoelastic interaction in the 400 K regions and an interaction of ultrasonic waves with the domain walls in the vicinity of T_C. We know that the ferrites exhibit magnetic after-effects due to small deviations from the stoichiometry and the relaxation time of which falls in the microsecond range above room temperature; consequently, an ultrasonic wave can induce—through the magnetoelastic coupling—spin rotations or domain wall movements, which one leads to electronic migrations: this can explain the relative maximum of Q^{-1} observed around 400 K.

Acknowledgment

The author is thankful to UGC-BSR, New Delhi, for providing grants to carryout this work.

References

[1] R. Lebourgeois, J. Ageron, H. Vincent, and J. P. Ganne, "Low losses NiZnCu ferrites," in *Proceedings of the 8th International Conference on Ferrites (ICF '00)*, Tokyo, Japan, 2000.

[2] S. Yan, J. Geng, L. Yin, and E. Zhou, "Preparation of nanocrystalline NiZnCu ferrite particles by sol-gel method and their magnetic properties," *Journal of Magnetism and Magnetic Materials*, vol. 277, no. 1-2, pp. 84–89, 2004.

[3] Y.-P. Fu, C.-H. Lin, and C.-W. Liu, "Preparation and magnetic properties of $Ni_{0.25}Cu_{0.25}Zn_{0.5}$ ferrite from microwave-induced combustion," *Journal of Magnetism and Magnetic Materials*, vol. 283, no. 1, pp. 59–64, 2004.

[4] H. Su, H. Zhang, and X. Tang, "Effects of Bi_2O_3-WO_3 additives on sintering behaviors and magnetic properties of NiCuZn ferrites," *Materials Science and Engineering B*, vol. 117, no. 3, pp. 231–234, 2005.

[5] T. Krishnaveni, B. R. Kanth, V. S. R. Raju, and S. R. Murthy, "Fabrication of multilayer chip inductors using Ni-Cu-Zn ferrites," *Journal of Alloys and Compounds*, vol. 414, no. 1-2, pp. 282–286, 2006.

[6] R. C. Lecraw and R. L. Comstock, "Magnetoelastic interactions in ferromagnetic insulators," in *Physical Acoustics*, R. N. Thurston and A. D. Pierce, Eds., vol. 3B, chapter 4, pp. 127–199, Mason, 1965.

[7] S. R. Murthy, M. Guyot, and V. Cagan, "Ultrasonic velocity changes in Polycrystalline garnets," in *Proceedings of the International Conference on Magnetism (ICM '88)*, pp. 134–135, Paris, France, July 1988.

[8] M. Guyot and V. Cagan, "The acoustic emission along the hysteresis loop of various ferro and ferrimagnets," *Journal of Magnetism and Magnetic Materials*, vol. 101, no. 1-3, pp. 256–262, 1991.

[9] L. Balamuth, "A new method for measuring elastic moduli and the variation with temperature of the principal young's modulus of rocksalt between 78°K and 273°K," *Physical Review*, vol. 45, no. 10, pp. 715–720, 1934.

[10] S. R. Murthy, "Low temperature sintering of NiCuZn ferrite and its electrical, magnetic and elastic properties," *Journal of Materials Science Letters*, vol. 21, no. 8, pp. 657–660, 2002.

[11] S. R. Murthy, B. Revathi, and T. S. Rao, "Temperature and composition dependence of the elastic moduli of NiZn ferrites," *Journal of the Less-Common Metals*, vol. 57, no. 1, pp. 29–37, 1978.

[12] S. R. Murthy and T. S. Rao, "Effect of magnetic field and temperature on the elastic behaviour of cobalt-zinc ferrites," *Journal of the Less-Common Metals*, vol. 65, no. 1, pp. 19–26, 1979.

[13] Y. Kawai and T. Ogawa, "Acoustic losses in single crystals of Mn-Zn ferrite and titanium substituted Mn-Zn ferrites," *Journal of the Physical Society of Japan*, vol. 45, no. 6, pp. 1830–1834, 1978.

[14] O. L. Anderson, "Derivation of wachtman's equation for the temperature dependence of elastic moduli of oxide compounds," *Physical Review*, vol. 144, no. 2, pp. 553–557, 1966.

[15] B. Ramaiah and S. R. Murthy, "Measurement of Elastic moduli on MnZn ferrites at several Ultrasonic frequencies," *Acoustics Research Letters*, vol. 21, pp. 7–10, 1997.

[16] S. R. Murthy, "Elastic behaviour and internal friction studies on thin films of NiZn ferrites," *Physica Status Solidi*, vol. 91, p. 519, 2002.

[17] K. Praveena, K. Sadhana, and S. R. Murthy, "Elastic behaviour of microwave hydrothermally synthesized nanocrystalline Mn_{1-x}-Zn_x ferrites," *Materials Research Bulletin*, vol. 47, no. 4, pp. 1096–1103, 2012.

20

Effect of Processing on Synthesis and Dielectric Behavior of Bismuth Sodium Titanate Ceramics

Vijayeta Pal,[1] R. K. Dwivedi,[1] and O. P. Thakur[2]

[1] *Department of Physics and Material Science & Engineering, Jaypee Institute of Information Technology, Noida 201307, India*
[2] *Electroceramics Group, Solid State Physics Laboratory, Defence Research and Development Organization (DRDO), Timarpur, Delhi 110054, India*

Correspondence should be addressed to Vijayeta Pal; vijayetapal@yahoo.in

Academic Editor: Baolin Wang

An effort has been made to synthesize polycrystalline $(Bi_{1-x}La_x)_{0.5}Na_{0.5}TiO_3$ (abbreviated as BLNT) system with compositions $x = 0$, 0.02, and 0.04 by novel semiwet technique. Preparation of A-site oxides of BLNT for composition $x = 0$ was optimized using two precursor solutions such as ethylene glycol and citric acid. The XRD patterns revealed that the sample prepared by ethylene glycol precursor solution has single phase perovskite structure with a rhombohedral symmetry at RT as compared to the sample prepared by citric acid. Ethylene glycol precursor has been found to play a significant role in the crystallization, phase transitions, and electrical properties. The studies on structure, phase transitions, and dielectric properties for all the samples have been carried out over the temperature range from RT to 450°C at 100 kHz frequency. It has been observed that two phase transitions (i) ferroelectric to antiferroelectric and (ii) antiferroelectric to paraelectric occur in all the samples. All samples exhibit a modified Curie-Weiss law above Tc. A linear fitting of the modified Curie-Weiss law to the experimental data shows diffuse-type transition. The dielectric as well as ferroelectric properties of BLNT ceramics have been found to be improved with the substitution of La elements.

1. Introduction

Lead oxide-based ceramics with perovskite structure have been the subject of attraction for high-performance sensors, actuators, transducers, and other applications, owing to their superior dielectric, piezoelectric, and electromechanical coupling coefficients. Applications are restricted to temperature range −50 to 150°C [1]. However, in recent years many fields have expressed the need for actuation and sensing which can be used at higher temperatures (>400°C) such as automotive, aerospace, and related industrial applications. On the other hand, in most of the cases lead constitutes more than 60% of the composition of these piezoelectric devices. Lead, known to be highly toxic and volatile, is released to the atmosphere during sintering causing serious environmental and health problems. Another cause for concern is the disposal of these products at the end of the life cycle. Considering all these health concerns posed by lead, multinational governments

like the European Union have enacted laws that ban the use of lead in the manufacture of many industrial products [2]. This has led to the replacement of lead (Pb) in the field of piezoelectric ceramics. A lot of research has been carried out on lead-free piezoceramic products in the last fifty years but in the last few years, the momentum has tremendously increased, accounting for about 75% of all published works in this field. $Bi_{0.5}Na_{0.5}TiO_3$- (BNT-) and $K_{0.5}Na_{0.5}NbO_3$- (KNN-) based materials are two main material systems with perovskite structure, which have been studied to find the substitute of PZT for lead free piezoelectric applications. Pure BNT was discovered by Smolenskii et al. [3] and is a ferroelectric having Bi^{3+} and Na^+ complexes on the A-site of ABO_3-type perovskite structure with a rhombohedral symmetry. Because of a large remanent polarization ($P_r = 38 \, \mu C/cm^2$) at room temperature, BNT ceramic is considered as one of the promising candidates for lead free piezoelectric ceramics [4]. However, the poling of pure BNT ceramic is very difficult due to its high coercive field (E_c = 73 kV/cm). So, the pure BNT ceramic usually exhibits weak piezoelectric properties. Therefore, a number of BNT-based ceramics were prepared to improve the electrical properties of this material by the convectional solid state method [5, 6]. Recently, a lot of efforts have been made to prepare the material by various chemical methods, such as hydrothermal process [7], citrate method [8, 9], sol-gel, autocombustion [10], and stearic acid gel route [11]. In the present work, A-site oxides of lead-free $Bi_{0.5}Na_{0.5}TiO_3$ ceramics were optimized using two precursor solutions such as ethylene glycol and citric acid at low calcination temperature (750°C) and further some La doped BNTs (BLNTs) have been developed using ethylene glycols precursors by a novel semiwet technique and structural, dielectric, and ferroelectric properties have been studied for both systems. To the best of our knowledge, BLNT system has been synthesized for the first time using semiwet technique. This technique has been applied to make other systems enhance its properties [12].

2. Experimental Procedures

A novel semiwet technique was used to prepare lead-free ceramic $(Bi_{1-x}La_x)_{0.5}Na_{0.5}TiO_3$ (BLNT) system. These compositions were prepared using analytical-grade metal oxides or nitrate powders of sigma Aldrich as raw materials such as Bi_2O_3 (99%), La_2O_3 (99%), $NaNO_3$ (99%), TiO_2 (99.9%) citric acid, and ethylene glycol. In this method, A-site of BLNT with composition x = 0 was prepared by using two different precursor solutions; the first one is citric acid, and; the second is ethylene glycol. In the first process, an appropriate amount of citric acid solution was added to the solution of nitrates of A-site cations in a beaker (C/M ~ 1 : 1). Aqueous ammonia in the solution form was added drop by drop to adjust the pH value of the solution in the range of 6–8. The precursor solution was dehydrated by putting on heater with continuous stirring at 80°C for 3 hrs to form a viscous gel which was placed in an oven at 150°C for overnight to combust the gel into ash powder. In the second process, stoichiometry amount of the solution of nitrates of A-site cation was dissolved with the ethylene glycol (E/M ~ 1 : 1) with continuous stirring for 30 min to get homogeneously mixed solution. The precursor solution was dehydrated by the same conditions, which is discussed in the first method to form a gel into ash powder. Both precursor solutions are expected to distribute the cations atomically homogeneously throughout the polymeric structure forming a stable polymeric complex, which was combusted at appropriate temperature (T ~ 500°C) in the form of ash powder. The ash, highly fine, homogeneous, and highly reactive powder was mixed with appropriate amounts of TiO_2 powder thoroughly in ethanol using mortal pestle for 2 hrs followed by solid state route. These powders were dried and calcined at 750°C (2 hrs) for BNT-CA and BNT-EG and calcined at 850°C (2 hrs) for BLNT samples. The calcined powder was mixed thoroughly with a polyvinyl alcohol (PVA) binder solution and then pressed into the form of disk with 10 mm diameter. All samples are in pellet form, kept in alumina boat, and

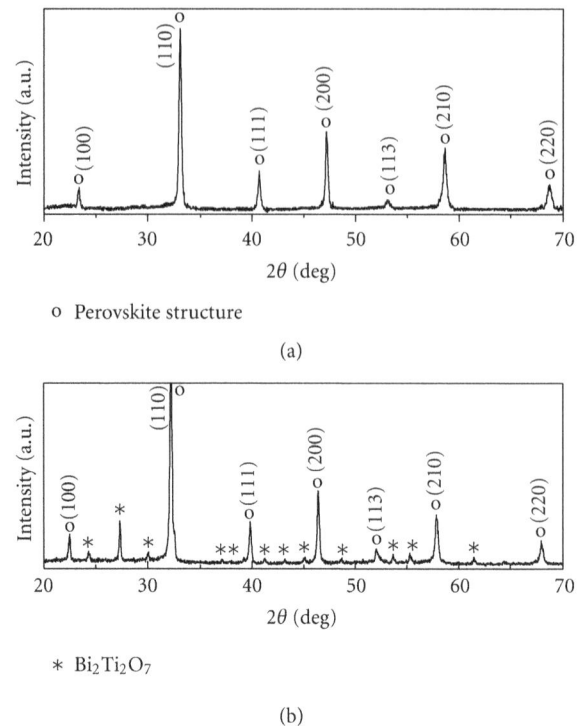

o Perovskite structure

(a)

* $Bi_2Ti_2O_7$

(b)

FIGURE 1: XRD patterns of (a) BNT-EG and (b) BNT-CA samples.

sintered at temperature 1150°C for 2 hours. Two pellets of each composition were electroded with silver paint on both the surfaces of the samples for the subsequent electrical measurements. The crystallite size of all the samples in BLNT system was calculated using Debye-Scherrer formula (D = $0.89\lambda/\beta\cos\theta_B$) where D is the average crystallite size, λ is the wavelength of X-ray radiation, β is the full width at half maximum (FWHM), and θ_B the diffraction angle. The corresponding values are reported in Table 1.

The crystalline structure of the sintered samples was examined using X-ray diffraction (XRD) analysis with Cu-$K\alpha$ radiation (DX-1000). The surface morphology of sintered ceramics was observed by scanning electron microscopy (SEM, model JEOL A 800). The dielectric constant ε_r and dielectric loss (tan δ) of the ceramic samples at 100 kHz were measured as a function of temperature over the temperature range from room temperature to 450°C using an LCR meter (Hioki 3522). A conventional P-E loop tracer (Marine India), which is based on Sawyer-Tower circuit, was used to measure the polarization-electrical field (P-E) hysteresis loop at 50 Hz.

3. Results and Discussion

In the present work, preparation of pure $Bi_{0.5}Na_{0.5}TiO_3$ was optimized using two chemical precursors, citric acid and ethylene glycol. These samples are abbreviated as BNT-CA and BNT-EG (Figures 1(a) and 1(b)). It is observed from the XRD patterns that the sample prepared by ethylene glycol precursor solution has shown better phase formation, whereas BNT-CA has formed partially along with

TABLE 1: Dielectric properties of all BLNT system, prepared using ethylene glycol precursor.

Composition (x)	Lattice parameter $a = b = c$ (Å)	Volume (m^3)	ε_r at RT	Tan δ at RT	ε_r at T_m	Tan δ at T_m	T_d °C	T_m °C	γ	D (nm)
BNT ($x = 0$)	3.868	$5.78 * 10^{-29}$	705	0.040	3200	0.08	180	353	1.26	34.98
BLNT ($x = 0.02$)	3.931	$6.07 * 10^{-29}$	1036	0.044	3630	0.03	200	355	1.49	39.33
BLNT ($x = 0.04$)	3.922	$6.03 * 10^{-29}$	3020	0.045	5630	0.05	210	380	1.80	34.81

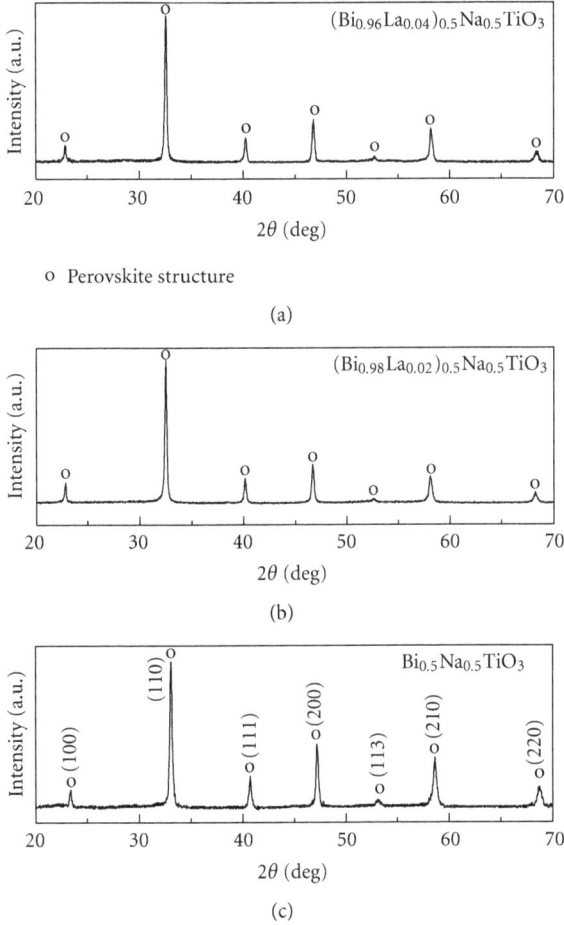

○ Perovskite structure

(a)

(b)

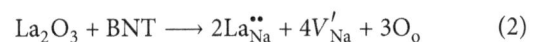

(c)

FIGURE 2: XRD patterns of BLNT system with compositions $x = 0, 0.02,$ and 0.04.

the presence of other phase identified as $Bi_2Ti_2O_7$ [13]. The experimental density observed in BNT-EG sample was $5.77 \, gm/cm^2$ which is 95% of theoretical density and the experimental density observed in BNT-CA sample was $5.56 \, gm/cm^2$ which is 91% of theoretical density. Therefore, a typical $(Bi_{1-x}La_x)_{0.5}Na_{0.5}TiO_3$ (BLNT) system with compositions $x = 0.02$ and 0.04 was prepared by semiwet technique using ethylene glycol precursor, calcined at $850°C$. The XRD patterns of all the samples in BLNT system have shown single phase formation with a rhombohedral symmetry in Figure 2.

Pure BNT and BLNT powders have rhombohedral symmetry at room temperature. However, rhombohedral structure is hard to distinguish due to the overlapping of peaks that could be due to nearly cubic lattice parameter. Owing to small degree of rhombohedral distortion, diffraction lines were indexed on the basis of pseudocubic unit cell [14] and lattice parameters of all the BLNT samples, prepared by ethylene glycol, are calculated by using the Unit Cell program package [15]. There is a variation in the lattice parameters because of the different sizes of ionic radius of La^{3+} (1.03 Å) which are close to Bi^{3+} ionic radius (0.96 Å) and Na^+ (0.99 Å). The values in the parentheses refer to the Shannon's effective ionic radius with the coordination number of six taken from [16].

The microstructure of the pure BNT ceramics, prepared by semiwet using ethylene glycol and citric acid, sintered at $1150°C$ is shown in Figure 3. Ethylene glycol precursor plays a significant role in the grain growth and densification. The microstructure of pure BNT-EG ceramics is denser, homogeneous, and uniform grains with grain size of $2.12 \, \mu m$ as compared to BNT-CA with grain size of $1.19 \, \mu m$.

The dielectric measurements of electroded samples of BNT-EG, BNT-CA, and BLNT samples were carried at $100 \, kHz$ frequency over temperature range from room temperature to $450°C$ and are shown in Figures 4(a), 4(b), 4(c), and 4(d), respectively.

Two dielectric anomalies in all the samples are shown at temperatures T_1 and T_2, termed as "T_d" and "T_m" respectively, which corresponds to dielectric transitions from ferroelectric (FE) to anti-ferroelectric (AFE) and anti-ferroelectric (AFE) to paraelectric (PE), respectively. The corresponding peaks also appear in tan δ versus T plots. However, the low value of dielectric constant of the sample, prepared using citric acid (BNT-CA), may be due to formation of other phases of $Bi_2Ti_2O_7$, which hampers the one dielectric anomaly and relative value of dielectric constant in this sample. It has been observed that the T_d and T_m shift to higher temperature with increasing the concentration of La^{3+}. The partial replacement of A-site cation, with larger ionic radius cations and larger amount, decreases the relative displacement of B-site cation with respect to the oxygen octahedral cage and hence the increase in transition temperature is observed with substitution of La at A-site. It is obvious that La^{3+} (1.03 Å) can occupy the A-site of Bi^{3+} (0.96 Å) or Na^+ (0.99 Å). When La^{3+} occupies A-site of Na^+, it will lead to charge imbalance, which creates defects on A-site. In general, A-site (Bi^{3+} and Na^+) substitution by La^{3+} in BNT ceramics can be formulated in the following way:

$$La_2O_3 + BNT \longrightarrow 2La_{Bi} + 3O_o \quad (1)$$

$$La_2O_3 + BNT \longrightarrow 2La_{Na}^{\bullet\bullet} + 4V_{Na}' + 3O_o \quad (2)$$

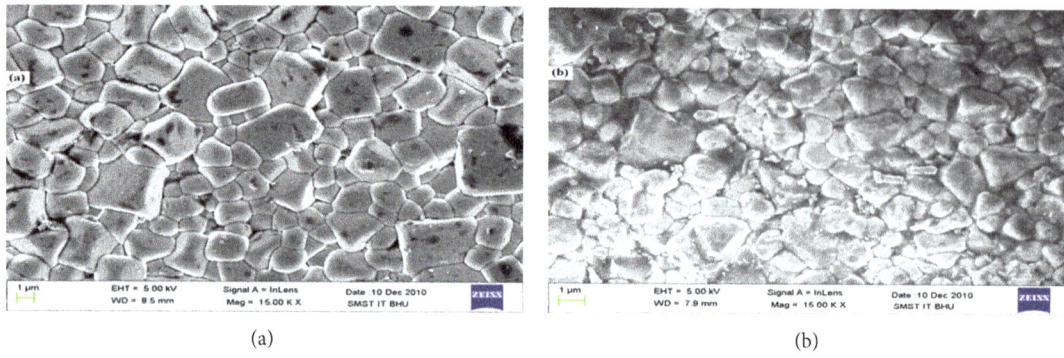

FIGURE 3: Microphotographs patterns of (a) BNT-EG and (b) BNT-CA samples.

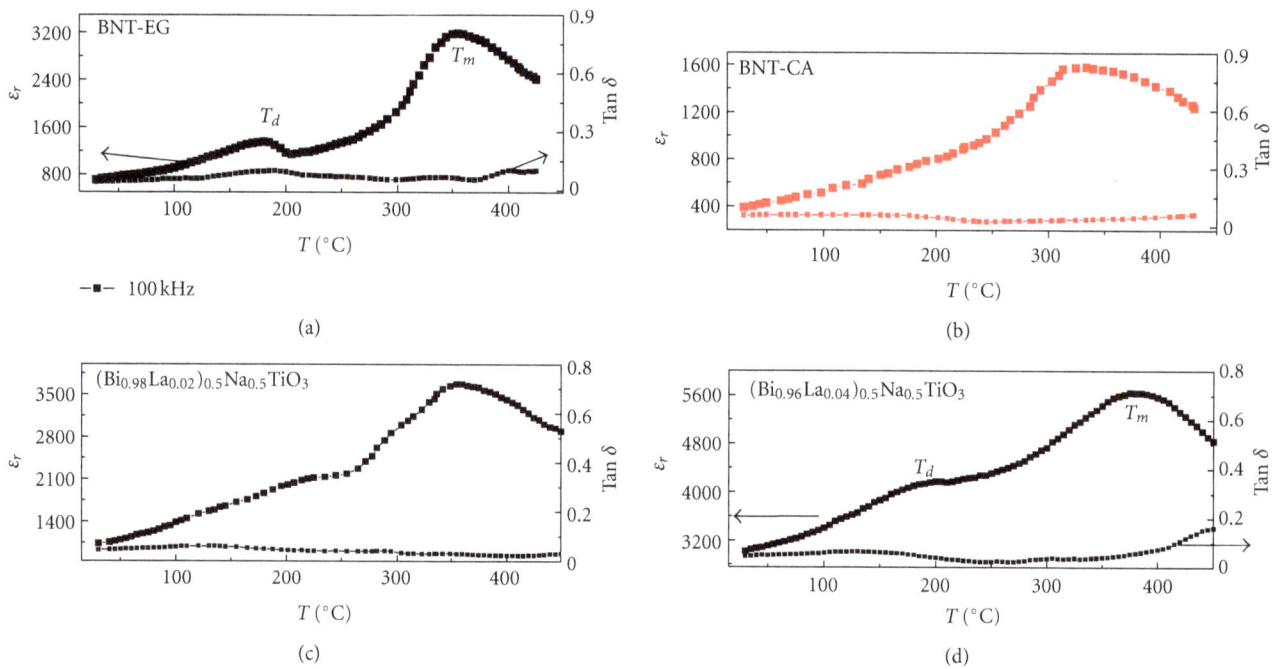

FIGURE 4: Variations of ε_r and tan δ with temperature for samples (a) BNT-EG, (b) BNT-CA, (c) BLNT with $x = 0.02$, and (d) $x = 0.04$.

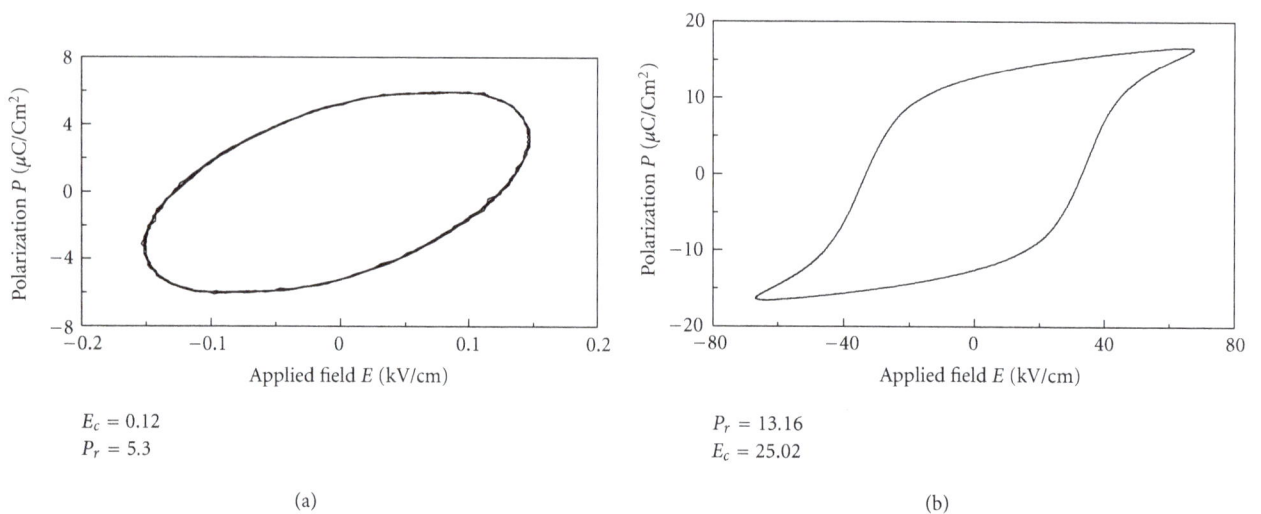

FIGURE 5: The P-E hysteresis loops at RT (50 Hz) (a) for BNT-CA and (b) BNT-EG.

TABLE 2: Ferroelectric properties of all saturated BLNT system, prepared using ethylene glycol precursor.

Composition (x)	P_r (μC/cm^2)	P_{max} (μC/cm^2)	E_c (kV/cm)
BNT-EG ($x = 0$)	13.16	16.90	25.02
BLNT ($x = 0.02$)	13.33	17.86	24.96
BLNT ($x = 0.04$)	17.70	20.60	36.73

When La^{3+} occupies Bi-site, as shown in (1), the substitution of Bi^{3+} by La^{3+} may cause the slack of BLNT lattice. The lattice deformation can make the ferroelectric domains reorientation more easily. It leads to the enhancement of dielectric as well as ferroelectric properties. Additionally, La^{3+} can also occupy the A-site of Na$^+$, as shown in (2). In this case, the valence of La^{3+} ion is higher than that of Na$^+$ ion. To maintain overall electrical neutrality, La^{3+} acts as a donor leading to some Na-site vacancies [V'_{Na}], which can relax the strain caused by reorientation of domains. Therefore, the movement of the domains becomes easier and thus the electrical properties of the BLNT ceramics are improved significantly. Thus, the substitution of La^{3+} in BNT system has significantly influenced the phase transition and dielectric behavior (Figures 4(c) and 4(d)). The highest value of ε_r (~3020) and lowest value of Tan δ (~0.045) are obtained with the composition $x = 0.04$ in the BLNT system at room temperature. The physical and dielectric properties of all the BLNT samples are tabulated at RT (100 kHz) in Table 1.

It shows the typical character of a ferroelectric behavior around transition temperature because of diffused phase transition. Dielectric constant exhibits strong frequency dependence above T_d and the maximum value of ε_r decreases as frequency increases in the BNT-EG samples suggesting that the ceramic is relaxor ferroelectric. The diffuseness in the phase transition can be described by $1/\varepsilon_r - 1/\varepsilon_{r\,max} = C^{-\gamma}(T - T_m)^\gamma$ in relaxor ferroelectrics [17], where $\varepsilon_{r\,max}$ is the maximum value of dielectric constant at T_m, γ is the degree of diffuseness, and C is the curie-like coefficient. γ can have a value ranging from 1 for normal ferroelectric to 2 for an ideal relaxor ferroelectric. This sample exhibits a linear relationship. The value of the exponent "γ" was determined by least-squared fitting experimental data to the equation which is in the range of 1.50 to 1.80. This confirms the diffuse phase transition in BLNT-EG system. The polarization-electrical field (P-E) hysteresis loop is shown in Figures 5(a) and 5(b). It has been found that the P-E loop of BNT using citric acid is round shaped, not well developed (saturated), Figure 4(a). BNT-CA sample gets breakdown with increasing electric field. This may be due to (I) large leakage current and (II) due to insufficient annealing process. P-E loop for BNT sample prepared using ethylene glycol is well developed or saturated, Figure 5(b).

This may be attributed to smaller grain size and relatively better density of the samples and confirm that all the samples are ferroelectric in nature.

The ferroelectric properties of all the BLNT system are tabulated at RT (50 Hz) in Table 2.

4. Conclusion

In summary, ethylene glycol prepared samples have revealed better crystallization of pure phase for BNT-EG sample. XRD patterns have revealed that the sample BNT-EG has single phase perovskite structure with a rhombohedral symmetry at RT. The structural, phase transition, and electrical properties of all the samples were investigated. The Bismuth sodium titanate, prepared by ethylene glycol precursor, has not only shown excellent dielectric but also ferroelectric behavior. The BNT sample has high value of dielectric constant ($\varepsilon_r = 705$), dielectric loss (Tan $\delta = 0.04$), remnant polarization ($P_r = 13.16\,\mu$C/Cm^2), and Coercive field ($E_c = 25.06$ kV/Cm) at room temperature. The relatively highest value of dielectric constant for BLNT with composition $x = 0.04$ may be attributed to the La doping. Composition with La substitution of $x = 0.04$ has shown the highest value of dielectric (3020) constant and low loss (0.045) as well as the highest value of remanent polarization ($P_r = 17.70\,\mu$C/Cm^2). The transition temperature Tm is also maximum (385°C) for this composition which reveals that the material can be useful for high-temperature device applications.

Acknowledgment

One of the authors Ms. V. Pal is thankful to JIIT for providing teaching assistance ship and other research facilities to carry out her research work at JIIT, Noida (India).

References

[1] B. Jaffe, W. R. Cook Jr., and H. Jaffe, *Piezoelectric Ceramics*, vol. 3, Academic Press, London, UK, 1971.

[2] M. Pecht, Y. Fukuda, and S. Rajagopal, "The impact of lead-free legislation exemptions on the electronics industry," *IEEE Transactions on Electronics Packaging Manufacturing*, vol. 27, no. 4, pp. 221–232, 2004.

[3] G. A. Smolenskii, V. A. Isupov, A. I. Agranovskaya, and N. N. Krainik, "New ferroelectrics of complex composition," *Soviet Physics, Solid State*, vol. 2, pp. 2651–2654, 1961.

[4] T. Takenaka, K. I. Maruyama, and K. Sakata, "(Bi$_{1/2}$Na$_{1/2}$)TiO$_3$-BaTiO$_3$ system for lead-free piezoelectric ceramics," *Japanese Journal of Applied Physics, Part 1*, vol. 30, no. 9, pp. 2236–2239, 1991.

[5] A. Sasaki, T. Chiba, Y. Mamiya, and E. Otsuki, "Dielectric and piezoelectric properties of (Bi$_{0.5}$Na$_{0.5}$)TiO$_3$-(Bi$_{0.5}$K$_{0.5}$)TiO$_3$ systems," *Japanese Journal of Applied Physics, Part 1*, vol. 38, no. 9, pp. 5564–5567, 1999.

[6] B. J. Chu, D. R. Chen, G. R. Li, and Q. R. Yin, "Electrical properties of Na$_{1/2}$Bi$_{1/2}$TiO$_3$-BaTiO$_3$ ceramics," *Journal of the European Ceramic Society*, vol. 22, no. 13, pp. 2115–2121, 2002.

[7] P. Pookmanee, G. Rujijanagul, S. Ananta, R. B. Heimann, and S. Phanichphant, "Effect of sintering temperature on microstructure of hydrothermally prepared bismuth sodium titanate ceramics," *Journal of the European Ceramic Society*, vol. 24, no. 2, pp. 517–520, 2004.

[8] Q. Xu, X. L. Chen, W. Chen, B. H. Kim, S. L. Xu, and M. Chen, "Structure and electrical properties of (Na$_{0.5}$Bi$_{0.5}$)$_{1-x}$Ba$_x$TiO$_3$ ceramics made by a citrate method," *Journal of Electroceramics*, vol. 21, no. 1–4, pp. 617–620, 2008.

[9] D. L. West and D. A. Payne, "Preparation of $0.95Bi_{1/2}Na_{1/2}TiO_3 \cdot 0.05BaTiO_3$ ceramics by an aqueous citrate-gel route," *Journal of the American Ceramic Society*, vol. 86, no. 1, pp. 192–194, 2003.

[10] J. G. Hou, Y. F. Qu, W. B. Ma, and D. Shan, "Synthesis and piezoelectric properties of $(Na_{0.5}Bi_{0.5})_{0.94}Ba_{0.06}TiO_3$ ceramics prepared by sol-gel auto-combustion method," *Journal of Materials Science*, vol. 42, no. 16, pp. 6787–6791, 2007.

[11] J. Hao, X. Wang, R. Chen, and L. Li, "Synthesis of $(Bi_{0.5}Na_{0.5})TiO_3$ nanocrystalline powders by stearic acid gel method," *Materials Chemistry and Physics*, vol. 90, no. 2-3, pp. 282–285, 2005.

[12] A. P. Singh, S. K. Mishra, D. Pandey, C. D. Prasad, and R. Lal, "Low-temperature synthesis of chemically homogeneous lead zirconate titanate (PZT) powders by a semi-wet method," *Journal of Materials Science*, vol. 28, no. 18, pp. 5050–5055, 1993.

[13] K. Kitagawa, T. Toyoda, K. Kitagawa, and T. Yamamoto, "$(Bi_{1/2}Na_{1/2})TiO_3$ additive effect for improved piezoelectric and mechanical properties in PZT ceramics," *Journal of Materials Science*, vol. 38, no. 10, pp. 2241–2245, 2003.

[14] E. Fukuchi, T. Kimura, T. Tani, T. Takeuch, and Y. Saito, "Effect of potassium concentration on the grain orientation in bismuth sodium potassium titanate," *Journal of the American Ceramic Society*, vol. 85, no. 6, pp. 1461–1466, 2002.

[15] N. Chaiyo, A. Ruangphanit, R. Muanghlua, S. Niemcharoen, B. Boonchom, and N. Vittayakorn, "Synthesis of potassium niobate (KNbO3) nano-powder by a modified solid-state reaction," *Journal of Materials Science*, vol. 46, no. 6, pp. 1585–1590, 2011.

[16] R. D. Shannon, "Revised effective ionic radii and systematic studies of interatomic distances in halides and chalcogenides," *Acta Crystallographica*, vol. 32, pp. 751–767, 1976.

[17] K. Uchino and S. Nomura, "Critical exponents of the dielectric constants in diffused-phase-transition crystals," *Ferroelectrics Letters Section*, vol. 44, no. 3, pp. 55–61, 1982.

21

Processing and Characterization of Yttria-Stabilized Zirconia Foams for High-Temperature Applications

**Ana María Herrera, Amir Antônio Martins de Oliveira Jr.,
Antonio Pedro Novaes de Oliveira, and Dachamir Hotza**

Graduate Program in Materials Science and Engineering (PGMAT), Departments of Chemical (EQA) and Mechanical Engineering (EMC), Federal University of Santa Catarina (UFSC), 88040-900 Florianópolis, SC, Brazil

Correspondence should be addressed to Dachamir Hotza; dhotza@gmail.com

Academic Editor: Young-Wook Kim

In this work ceramic foams of 3 and 8 mol% yttria-stabilized zirconia (3YSZ and 8YSZ) were manufactured by the replication method using polystyrene-polyurethane foams with pore sizes in the 7–10 ppi range. A second coating was carried out on presintered foams in order to thicken struts and hinder microstructural defects. The produced ceramic foams were structurally and thermomechanically characterized. Samples recoated with 3YSZ presented the highest relative densities (0.2 ± 0.1) which contributed to a better mechanical and thermal behavior.

1. Introduction

Porous radiant burners are devices used in applications benefit from radiant thermal heating. Numerous industries employ this technology for firing and drying, lower pollutant emission, and better product quality reached with these burners [1, 2].

Other functional advantages are a higher flame velocity within the porous structure when compared to a laminar free flame (due to the preheated of reagents), higher turn-down ration, higher efficiency in heat transfer by radiation, and the possibility of burning low calorific value fuels or very lean mixtures [3].

Ceramics foams for porous radiant burners are mainly produced by the replication method from polymeric templates. The method consists in the impregnation of a polymeric sponge with a ceramic slurry followed by a heat treatment which leads to the burning out of the organic body (sponge) and to the sintering of the ceramic skeleton [4]. Along the manufacture process, there are key parameters that must be taken into account to ensure the final product effectiveness. Among them, the most important are the linear expansion coefficient, maximum working temperature and thermal shock resistance of the ceramic material, thixotropy

and pseudoplasticity of the ceramic slurry, and, finally, the heating rate of the thermal treatment. According to the above, the most reliable ceramic materials to work with are cordierite, mullite, silicon carbide, alumina, partially stabilized zirconia (YSZ), and some composite systems [5–13].

Cellular ceramics are already commercially available for a wide range of technological applications including filters, membranes, catalytic substrates, thermal insulation, gas burner media, refractory materials, and lightweight structural panels [5–7]. In these cases, reticulated porous ceramics are used because of their functional properties, such as low thermal expansion coefficient and thermal conductivity, high permeability, and chemical inertness [7]. Considering the particularities associated with each specific application, reticulated porous ceramics are prepared and consolidated.

Variations of this method employ organic structures, such as wood or cellulose-based materials, as templates for the production of reticulated porous ceramics [3]. The use of a second coating applied to the presintered foams has been reported as a way of enhancing the mechanical resistance due to the thickening of the struts. This procedure hinders microstructural defects that are left in the walls of the ceramic foams during the burning out of the organic components [14–16].

Zirconia has been mentioned among possible candidates as material for high-temperature (for continuous operation above 1600°C) porous burners [17, 18]. Nevertheless, phase transformations that are dependent on temperature and composition are obstacles to a broader application of this material in radiant burners. Yttria-stabilized zirconia (YSZ) has been developed to hinder allotropic changes among the zirconia phases by replacing zirconium (Zr^{4+}) by yttrium ions (Y^{3+}) into the zirconia matrix [4]. Monoclinic is the most stable phase at room temperature, but YSZ allows tetragonal and cubic phases to be stable at lower temperatures with enhanced mechanical, thermal, and electrical properties.

The aim of this work was to evaluate the functionality, quality, and thermomechanical strength of 3 and 8 mol% yttria-stabilized zirconia (3YSZ or 8YSZ), as base materials for the porous radiant burners. From the mechanical point of view, 3YSZ would be the best option, but 8YSZ was chosen due its high oxygen ion conduction while blocking electronic conduction. This feature makes 8YSZ the standard electrolyte material for solid oxide fuel cells (SOFC), which must be operated at intermediate to high temperatures (600 to 1000°C). The motivation in this case is to associate a porous burner to a no-chamber SOFC, providing compatibility between both devices [19].

2. Materials and Methods

2.1. Materials. The starting materials were 3YSZ (TZ-3YB-E, Tosoh, Japan), with a mean particle size (d50) of 0.45 μm and density of 6.05 g/cm^3, and 8YSZ (YSZ8-U1, NexTech, USA) with a mean particle size (d50) of 0.32 μm and density of 7.80 g/cm^3. Ammonium polyacrylate (NH4PAA, Darvan, 821-A, Vanderbilt, USA) was used as dispersant and polyvinyl alcohol (PVA, $(CH_2CH(OH)_n)$, molecular weight: 70–100 g/mol Vetec, Brazil) as binder. To obtain a homogeneous slurry without bubbles, an antifoamer emulsion based on silicone was needed, (Sigma-Aldrich, Brazil).

Finally, for the manufacture of the ceramic foams by the replication method, a polystyrene-polyurethane foam (Crest Foam Industries, USA) with an average pore density of 10 ppi and nominal density of 0.03 g/cm^3 was employed.

2.2. Methods. Ceramic powders were characterized by zeta potential (Zetananosizer, Malvern, USA) and X-ray diffraction (X'pert, PW3710, Phillips, the Netherlands), based on 0.05° step, 1 s of step time, and a reading interval of 2θ from 0° to 118°. Rheological studies were performed for both 3YSZ and 8YSZ suspensions. The effect of the solids load, varied between 75 and 77.5 wt%, and binder amount, varied between 3.5 and 4.5 mass% suspension weight (base: 100 wt%), were evaluated. The ceramic foams were manufactured by the replication method. The impregnation step consisted in soaking the compressed sponge into the ceramic suspension and then releasing the tension so that the sponge could absorb as much as possible of the slurry. Subsequently, it is taken out to be wrung and remove all the slurry in excess, followed by a very soft air-spray to open the closed cells. Finally, it is dried under room temperature during 24 h and then presintered

according to the following thermal cycles: 1°C/min from 25 to 200°C; 0.5°C/min to 600°C; then 5°C/min to 1150°C with a holding time of 2 h. Subsequently, the samples were cooled (at 10–15°C/min) down to the room temperature.

A second coating was applied (in some cases) with a lighter suspension of the same material (70% solids, 3.5% binder). The sintering step was carried out with a heating rate of 7°C/min from room temperature to 900°C and 3°C/min to 1600°C, with a holding time of 2 h, and then samples were cooled (at 10°C/min) down to room temperature.

The sintered YSZ foams were characterized by its relative density and total porosity, wherein the geometrical density was calculated by measuring the mass and volume of the samples and the solid material density by helium picnometry.

Scanning electron microscopy (SEM, Phillips XL 30, the Netherlands) was used to access the microstructure of foams. Mechanical properties were measured according to usual procedures (compression strength [20], thermal shock resistance [21]). Finally, the foams were tested as porous radiant burners, at a constant fuel/air mixture ratio of 0.9 at different time cycles: 10 min^{-1} cycle, 10 min^{-5} cycles, and 360 min^{-1} cycle.

The mechanical testing was performed in 10 cylindrical samples (diameter equal to height) of sintered YSZ. The load velocity was 0.5 mm/s, based on the work of Elverum et al., [20] with the aim to compare results and materials. A universal mechanical testing machine was employed (DL2000, EMIC, Brazil).

The thermal shock test consisted in exposing the ceramic foams with a prismatic shape (170 × 30 × 25 mm) at high/low temperatures based on ASTM C1198-96 [22]. The samples were placed into a muffle furnace at 1000°C during 15 min and taken out and immediately cooled by forced air (hair dryer) to room temperature. This procedure was repeated 4, 8, and 12 times. The loss of the mechanical properties caused by the induced thermal shock was verified by measuring Young's modulus. This parameter was measured by a nondestructive method, submitting the ceramic foam to an impulse which excites the natural vibration frequencies of the structure. Young's modulus is related to the natural frequency, geometric dimensions, and mass of the ceramic foam sample by

$$E = 0.9465 \cdot \left(\frac{m \cdot f_f^2}{b} \right) \cdot \left(\frac{L^2}{t^2} \right) \cdot T_1, \quad (1)$$

where

E is the Young modulus,

m is the sample mass,

b, L, t are the width, the length, and the thickness of the sample, respectively,

f_f is the fundamental frequency, and

T_1 is the correction factor related to the fundamental flexional mode.

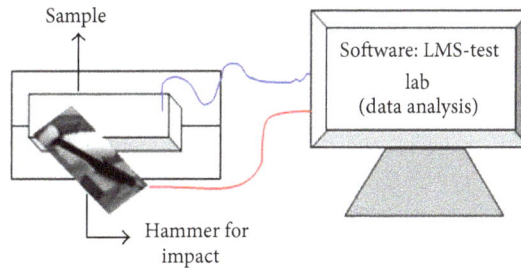

FIGURE 1: Dynamic test for identifying the structure's Young modulus.

The total damage in the ceramic structure due to the thermal shock may be estimated from

$$D_E = 1 - \frac{\sigma_c}{\sigma_0} \approx 1 - \frac{E_f}{E_0}, \qquad (2)$$

where

D_E is the total damage factor, from 0 to 1,

E_f is the last calculated Young modulus,

F_f is the last calculated mechanical strength,

E_0 is the initial Young modulus, and

σ_0 is the initial calculated mechanical strength.

The chart in Figure 1 represents the above-described thermal shock test.

An experimental test with the ceramic foams as porous burners was performed. The test employed disk shaped samples, with 35 mm diameter and 35 mm thickness, fitted inside a steel tube using a fiber mat to thermally insulate the ceramic foams from the tube wall. The burner was fed with a natural gas and air mixture with fuel equivalence ratio of 0.9. The flame, after approximately 2-3 min after ignition, was stabilized within the porous matrix. The outlet surface of the burner was allowed to radiate heat freely to the external ambient. The burner was submitted to periods of continuous operation of 10 min and 6 h. The adiabatic flame temperature corresponding to this equivalence ratio is approximately 2150 K [23]. This high operation temperature was specified with the aim to test the intrinsic properties of the YSZ samples.

The double-coated 3YSZ samples worked at the specified equivalence rate and with a flame velocity of 15 cm/s, as shown in Figure 2.

3. Results and Discussions

3.1. X-Ray Diffraction. Figure 3 shows the X-ray diffractograms obtained for the zirconium powders. As expected, 3YSZ (Figure 3(a)) corresponds mainly to the tetragonal phase, with a minor presence of monoclinic phase, which may be seen as little isles in the tetragonal matrix, which might contribute to a higher fracture toughness to the material [24]. For 8YSZ (Figure 3(b)), sharp peaks correspondent to the cubic phase were detected, as expected. This phase is known to present a characteristic grain growth, which reduces the mechanical resistance of the material [25].

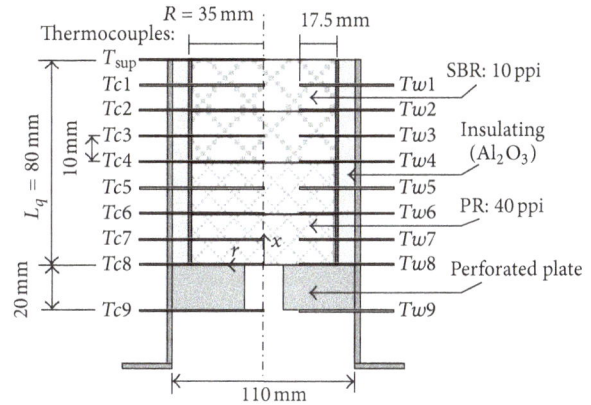

FIGURE 2: Setup of the porous burner test at a lab scale.

3.2. Rheological Characterization. The effect of the dispersant on the isoelectric point (IEP) of the materials was evaluated. In Figure 4 it is shown how the pH at the IEP is shifted from the basic to the acid range. The 3YSZ suspension has its original IEP at pH 6.9, the value that converges with the ones found in the literature, [26, 27], this IEP was fell in the range of 5.2 to 7.2. Adding one drop of dispersant to the solution, the IEP was then located at a pH 2.9. The same feature might be observed for the 8YSZ suspension. IEP is originally at pH 7.5, the value which converges with the IPE's of 8YSZ powder (Sigma-Aldrich) at 9.2 and the 8YSZ powder (Innovano) at 7.2. Adding a dispersant drop, the IEP moves to 3.2. This phenomenon is attributed to the change of the electrochemical properties of the particles surface when the dispersant particles are attached to them [26, 27].

Four formulations were analyzed for each material, as previously described. The guide parameters were the pseudo-plasticity and thixotropy, rheological properties that should be present in any suspension to be applied in the replica method. A high viscosity is desired at low shear stress and a low viscosity when at high shear stress. This ensures, respectively, that the slurry will flow into the reticulated structure by the impregnation, and, when at rest, the slurry will be strongly adhered to the polymeric walls. A high thixotropy (large area between the flow curves) corresponds to the facility that the slurry will recover its initial viscosity after the impregnation process [28, 29]. Figure 5 shows the thixotropy graphs of both systems, where those two properties can be observed. Figure 5(a) corresponds to the 3YSZ suspensions with 77.5%

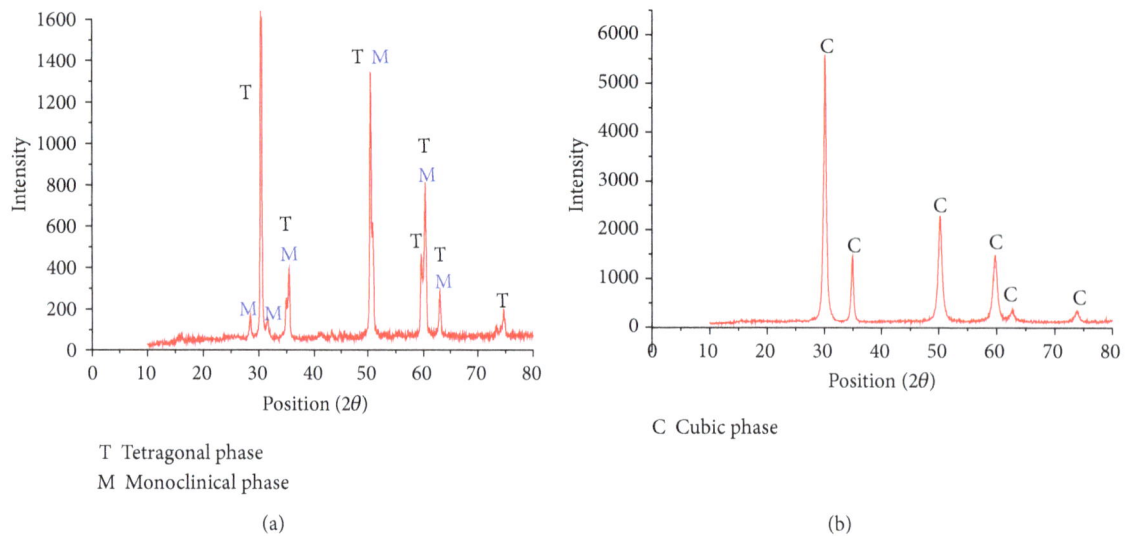

T Tetragonal phase
M Monoclinical phase

(a)

C Cubic phase

(b)

FIGURE 3: X-ray diffraction of the YSZ powders after sintering: (a) 3YSZ; (b) 8YSZ.

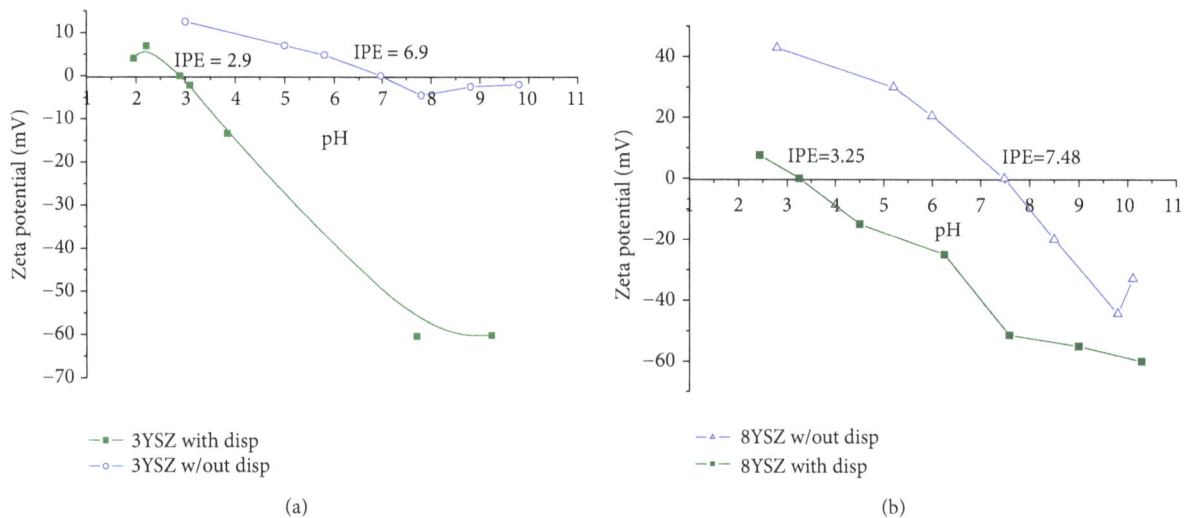

(a)

(b)

FIGURE 4: Zeta potential of zirconia suspensions as a function of pH and added dispersant. (a) 3YSZ; (b) 8YSZ.

solids and 3.5% binder. In the case of 8YSZ (Figure 5(b)), the formulation with 77.5% solids and 4.5% binder was selected. The maximum amount of solids is naturally desired [3, 14, 27, 30].

In Figure 5 the thixotropy curves present different behaviors. The steeper the curve, the higher the viscosity at high shear stress and consequently the higher the initial viscosity [15]. Those characteristics provide a thicker layer when coating and, consequently, higher homogeneity and better densification of the material [14].

3.3. Physical Characterization of Foams.
Relative density and porosity were determined for the three classes of yttria-stabilized zirconia foams produced, Table 1.

Higher relative density values were obtained for double coated 3YSZ in comparison to single-coated 3YSZ and 8YSZ.

This result is a consequence of the slightly larger amount of solids in the cellular structure, which also affects the porosity of the ceramic foam.

Images obtained by SEM allow for having an overall view of the foam structures obtained. Figure 6 presents the different types of foams manufactured. Figures 6(a) and 6(b) correspond to single-coated 3YSZ and 8YSZ, respectively. Homogeneous surfaces with few microstructural defects are observed, as micropores inside the struts and minicracks. Figure 6(c) corresponds to a double-coated 3YSZ foam showing how the internal surfaces have more micropores and microcracks than the external walls. Moreover, the vertices are rounded after the recoating, which might enhance the mechanical strength of the structure [14]. Figure 6(d) shows typical defects, characteristic of the replica method. Nevertheless, the thicker walls obtained with a recoating are seen, as well.

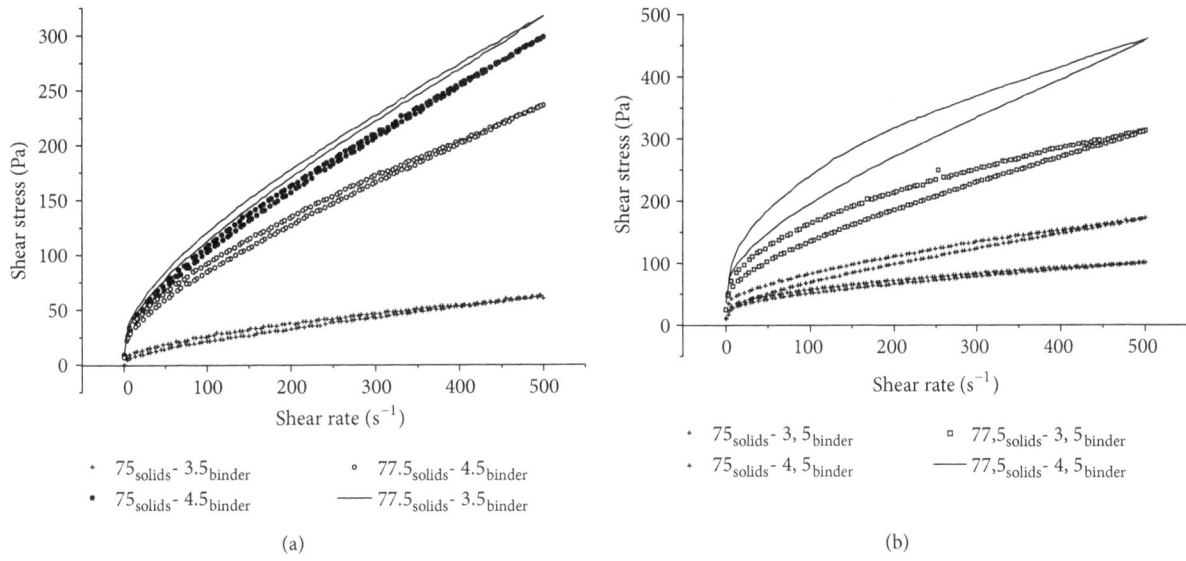

(a)

(b)

* $75_{solids}- 3.5_{binder}$ ○ $77.5_{solids}- 4.5_{binder}$
* $75_{solids}- 4.5_{binder}$ —— $77.5_{solids}- 3.5_{binder}$

* $75_{solids}- 3, 5_{binder}$ □ $77,5_{solids}- 3, 5_{binder}$
* $75_{solids}- 4, 5_{binder}$ —— $77,5_{solids}- 4, 5_{binder}$

FIGURE 5: Rheogram of the ceramic slurries: (a) 3YSZ; (b) 8YSZ, for different solid and binder concentrations.

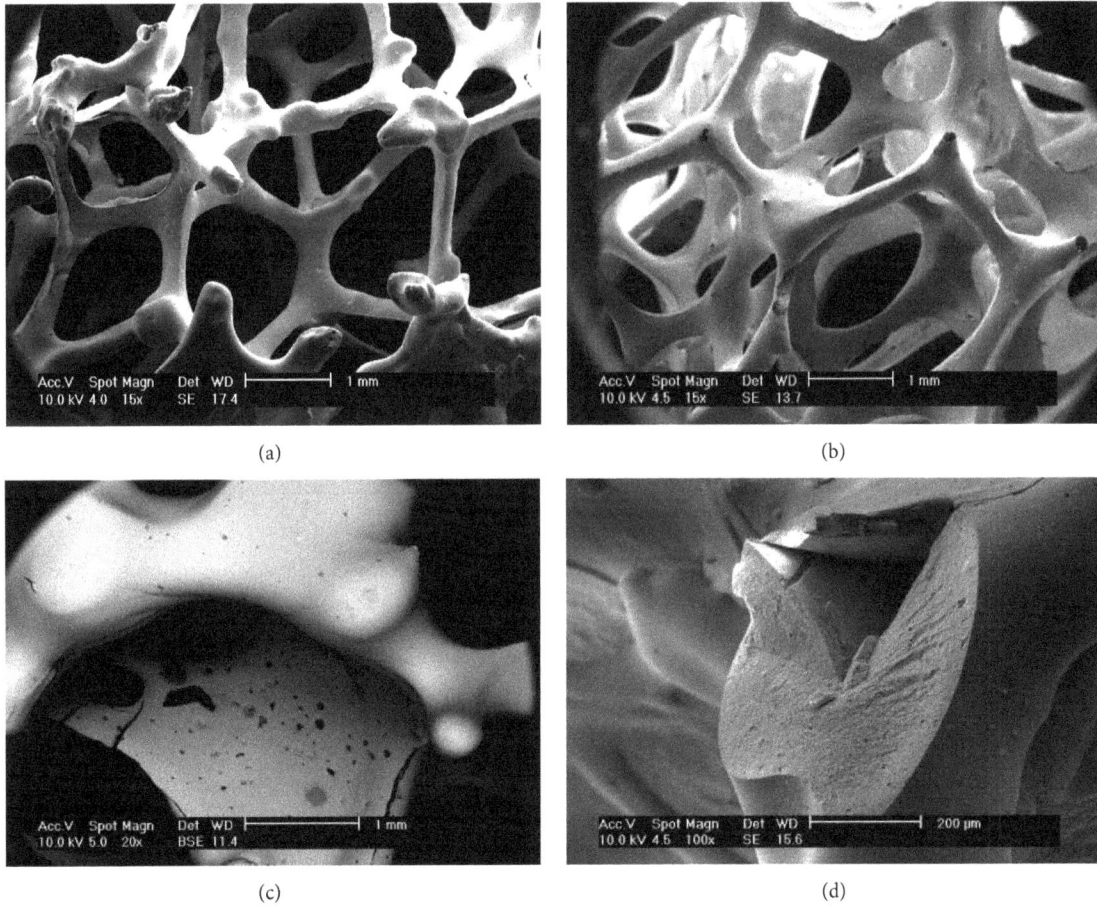

(a)

(b)

(c)

(d)

FIGURE 6: SEM images of YSZ foams: (a) single-coated 3YSZ; (b) single-coated 8YSZ; (c) double-coated 3YSZ; (d) conventional defect left by the organic compounds burning out in a double-coated 3YSZ foam.

TABLE 1: Relative density and total porosity values of zirconia foams.

YSZ foams	$\rho_{geometric}$ (g/cm^3)	$\rho_{solid\text{-}mat.}$ (g/cm^3)	Relative density, ρ_r	Porosity (%)
3YSZ, double coated	1.1 ± 0.1	5.7	0.20 ± 0.10	80.1 ± 0.1
3YSZ, single coated	0.5 ± 0.3	5.7	0.10 ± 0.30	90.0 ± 0.3
8YSZ	0.4 ± 0.3	5.8	0.08 ± 0.30	92.0 ± 0.3

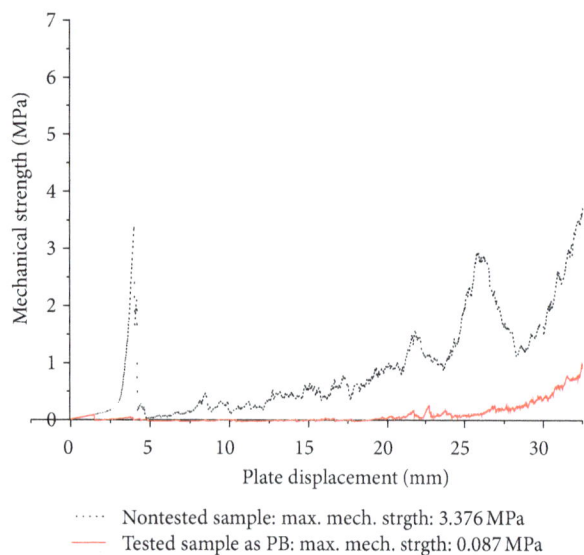

FIGURE 7: Compression strength of samples before/after the porous burner test.

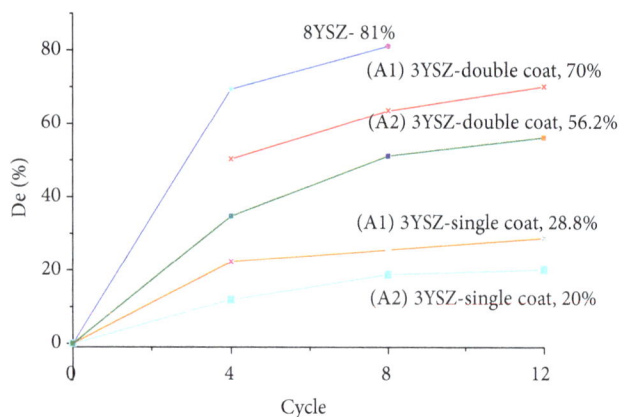

FIGURE 8: Damage done to the YSZ foams owed to a thermal shock test.

TABLE 2: Values of Young modulus for single/double YSZ samples after thermal shock test.

	f_f (Hz)	E (GPa)
8YSZ		
A1		
W/out cycles	597	0.10
4 cycles	330	0.03
8 cycles	259	0.02
3YSZ double coated		
A1		
W/out cycles	2373	5.07
4 cycles	1673	2.52
8 cycles	1430	1.84
12 cycles	1296	1.51
A2		
W/out cycles	1721	2.25
4 cycles	1390	1.47
8 cycles	1202	1.1
12 cycles	1139	0.98
3YSZ single coated		
A1		
W/out cycles	1257	0.42
4 cycles	1108	0.33
8 cycles	1084	0.32
12 cycles	1060	0.30
A2		
W/out cycles	1021	0.23
4 cycles	958	0.20
8 cycles	919	0.18
12 cycles	911	0.18

3.4. Mechanical Characterization of Foams. The mechanical strength of the double-coated 3YSZ samples (2.3 ± 0.7 MPa) is higher than that of the single-coated 3YSZ (0.8 ± 0.3 MPa) or 8YSZ samples (0.30 ± 0.04 MPa). This fact is owed to the lower density of microstructural defects and thicker walls for the 3YSZ double-coated samples. The average mechanical strength of samples was 2.3 ± 0.7 MPa which is comparable to the values found by Elverum et al. [20], in which average mechanical strength was of 1.8 MPa.

The compression strength test was also applied to samples which could overcome the burning test. Figure 7 presents the results obtained of a sample of double-coated 3YSZ that was tested as a porous burner in comparison to the best result obtained for a sample of the same group in Table 2. As can be seen, the mechanical properties of the tested sample were deeply reduced, with a maximum strength of 0.087 MPa, corresponding to 2.6% of the 3.37 MPa of the nontested sample.

Table 2 presents the fundamental frequencies and the corresponding Young moduli of the samples subjected up to 12 thermal shock cycles. As it can be seen, after 12 cycles, the highest Young modulus belonged to the double-coated 3YSZ samples. The 8YSZ samples could withstand just up to 8 cycles.

The total damage done to the structure due to the thermal shock test (DE, Figure 8) is lower for the single-coated 3YSZ sample (between 20–28%), while the double-coated 3YSZ

FIGURE 9: YSZ samples after tested as a porous burner after 6 h operation: (a) double-coated 3YSZ; (b) single-coated 3YSZ; (c) single-coated 8YSZ.

samples presented a loss between 55 and 70%, and the single-coated 8YSZ sample had a loss of 80% related to the original Young modulus.

The fact that the damage done to the single-coated 3YSZ samples was lower than that for the double-coated 3YSZ samples could be explained by the higher porosity that the former samples presented. Higher porosity contributes to relax the thermal stresses that are produced during the thermal shock and also to hinder the growth of microcracks [31, 32].

3.5. Porous Burners Performance. Double-coated 3YSZ samples maintained the structural integrity without any evidence of chemical reaction after contact with the flame. The maximum temperature within the burner reached 1600°C, smaller than the adiabatic flame temperature due to the surface heat transfer to the external ambient. Still, this is a very high temperature for the continuous operation of porous burners [33].

Figure 9 shows YSZ samples after 6 h operation as porous burners. It can be observed that the external reticulated structure kept its integrity; the internal struts, where the second coating was hindered, were broken. For the single-coated 3YSZ samples just part of the walls and the inferior base maintained the original structure, as it can be seen in Figure 9(b). The average operational temperature achieved was between 1450 and 1500°C. The samples of 8YSZ could not operate at the equivalence ratio of 0.9, which was then reduced to 0.8. They presented an average operational temperature between 1300 and 1350°C, for both standard periods. Figure 9(c) shows one of the tested samples, where it can be seen that the structure was severely destroyed when after contact with the flame.

4. Conclusions

The rheological properties of the 8YSZ suspension were more adequate than those of the 3YSZ suspension, due to the smaller particle size of the 8YSZ powder and the higher amount of binder.

Applying a second coating on to the presintered foams enhanced the structure morphology, hindering the formation of micropores, microcracks, and holes after burning organics. In this case, superior thermomechanical properties were achieved.

The highest thermal shock resistance corresponded to the samples of single-coated 3YSZ, presenting 25% damage, related to the Young modulus, due to the high porosity. Nevertheless, the samples of double-coated 3YSZ presented the highest Young modulus after the 12 thermal shock cycles.

Only the samples of double-coated 3YSZ withstood satisfactorily the porous burner test. Those samples achieved high work temperatures (1550–1600°C) during up to 6 h of operation at the highest equivalence ratio (0.9).

Acknowledgment

The authors gratefully acknowledge the financial support from the CNPq Foundation, Brazil.

References

[1] R. C. Catapan, A. A. M. Oliveira, and M. Costa, "Non-uniform velocity profile mechanism for flame stabilization in a porous radiant burner," *Experimental Thermal and Fluid Science*, vol. 35, no. 1, pp. 172–179, 2011.

[2] "Nitrogen Oxides (NOx), Why and How they are controlled," Clean Air Technology Center, 1999, http://www.epa.gov/ttn/catc/dir1/fnoxdoc.pdf.

[3] M. Scheffler and P. Colombo, *Cellular Ceramics: Structure, Manufacturing, Properties and Applications*, John Wiley & Sons, New York, NY, USA, 2005.

[4] P. Colombo and H. P. Degischer, "Highly porous metals and ceramics," *Materials Science and Technology*, vol. 26, no. 10, pp. 1145–1158, 2010.

[5] P. Greil, "Biomorphous ceramics from lignocellulosics," *Journal of the European Ceramic Society*, vol. 21, no. 2, pp. 105–118, 2001.

[6] T. Inui and T. Otowa, "Catalytic combustion of benzene-soot captured on ceramic foam matrix," *Applied Catalysis*, vol. 14, pp. 83–93, 1985.

[7] T. Mizrah, A. Maurer, L. Gauckler, and J. P. Gabathuler, SAE Paper 890172, 1989.

[8] L. Montanaro, Y. Jorand, G. Fantozzi, and A. Negro, "Ceramic foams by powder processing," *Journal of the European Ceramic Society*, vol. 18, no. 9, pp. 1339–1350, 1998.

[9] P. Colombo and E. Bernardo, "Macro- and micro-cellular porous ceramics from preceramic polymers," *Composites Science and Technology*, vol. 63, no. 16, pp. 2353–2359, 2003.

[10] M. F. Ashby, "The mechanical properties of cellular solids," *Metallurgical Transactions A*, vol. 14, no. 9, pp. 1755–1769, 1983.

[11] R. Brezny and J. D. Green, "Structure and properties of ceramics," *Materials Science and Technology*, vol. 11, pp. 467–516, 1992.

[12] F. Lange and T. K. Miller, "Open cell low density ceramics fabricated from reticulated polymer substrates," *Advanced Ceramics Materials*, vol. 2, no. 4, pp. 827–831, 1987.

[13] K. Lannguth, "Particle size range as a factor influencing compressibility of ceramic powder," *Ceramics International*, vol. 21, pp. 237–242, 1995.

[14] V. S. Stunican, R. C. Hink, and S. P. Ray, "Phase equilibriums and ordering in the system zirconia-yttria," *Journal of the American Ceramic Society*, vol. 61, pp. 17–21, 1988.

[15] X. W. Zhu, D. L. Jiang, S. H. Tan, and Z. Q. Zhang, "Improvement in the strut thickness of reticulated porous ceramics," *Journal of the American Ceramic Society*, vol. 84, no. 7, pp. 1654–1656, 2001.

[16] X. Yao, S. Tan, Z. Huang, and D. Jiang, "Effect of recoating slurry viscosity on the properties of reticulated porous silicon carbide ceramics," *Ceramics International*, vol. 32, no. 2, pp. 137–142, 2006.

[17] S. Y. Gómez, O. Álvarez, J. Escobar, J. B. Rodriguez, C. Rambo, and D. Hotza, "Relationship between rheological behaviour and final structure of Al_2O_3 and YSZ foams produced by replica," *Advances in Materials Science and Engineering*, vol. 2012, Article ID 549508, 9 pages, 2012.

[18] R. Stevens, *Zirconia and Zirconia Ceramics*, Magnesium Elektron Publication, 2nd edition, 1986.

[19] J. Aguilar and D. Hotza, "Configuraciones alternativas para celdas de combustible de óxido sólido," *Revista Latinoamericana de Metalurgia y Materiales*, vol. 33, pp. 172–185, 2013.

[20] P. J. Elverum, J. L. Ellzey, and D. Kovar, "Durability of YZA ceramic foams in a porous burner," *Journal of Materials Science*, vol. 40, no. 1, pp. 155–164, 2005.

[21] L. C. Cossolino and A. Pereira, "Módulos elásticos: visão geral e métodos de caraterização," ATCP Engenharia Física, 2010.

[22] ASTM C1198-96, "Standard test method for dynamic young's modulus, shear modulus and poisson's ratio for advanced ceramics," American Society for Testing Materials, 1996.

[23] F. M. Pereira, A. A. M. Oliveira, and F. F. Fachini, "Asymptotic analysis of stationary adiabatic premixed flames in porous inert media," *Combustion and Flame*, vol. 156, no. 1, pp. 152–165, 2009.

[24] J. Sun, L. Gao, and J. Guo, "Influence of the Initial pH on the adsorption behaviour of dispersant on nano zirconia powder," *Journal of the European Ceramic Society*, vol. 19, no. 9, pp. 1725–1730, 1999.

[25] J. Wang and L. Gao, "Surface and electrokinetic properties of Y-TZP suspensions stabilized by polyelectrolytes," *Ceramics International*, vol. 26, no. 2, pp. 187–191, 2000.

[26] J. Luo and R. Stevens, "Tetragonality of nanosized 3Y-TZP powders," *Journal of the American Ceramic Society*, vol. 82, no. 7, pp. 1922–1924, 1999.

[27] M. Filal, C. Petot, M. Mokchah, C. Chateau, and J. L. Carpentier, "Ionic conductivity of yttrium-doped zirconia and the 'composite effect'," *Solid State Ionics*, vol. 80, no. 1-2, pp. 27–35, 1995.

[28] B. R. Moreno, "Reología de suspensiones cerámicas," *Consejo Superior de Investigaciones Científicas*, vol. 17, p. 325, 2005.

[29] A. M. Herrera, O. Álvarez, J. Escobar, V. Moreno, and D. Hotza, "Fabrication and characterization of alumina foams for application in radiant porous burners," *Revista Matéria*, vol. 17, pp. 973–987, 2012.

[30] D. Hotza, "Colagem de folhas cerâmicas," *Cerâmica*, vol. 43, pp. 157–164, 1994.

[31] V. R. Vedula, D. J. Green, and J. R. Hellmann, "Thermal fatigue resistance of open cell ceramic foams," *Journal of the European Ceramic Society*, vol. 18, no. 14, pp. 2073–2081, 1998.

[32] L. Shen, M. Liu, X. Liu, and B. Li, "Thermal shock resistance of the porous Al_2O_3/ZrO_2 ceramics prepared by gelcasting," *Materials Research Bulletin*, vol. 42, no. 12, pp. 2048–2056, 2007.

[33] J. R. Howell, M. J. Hall, and J. L. Ellzey, "Combustion of hydrocarbon fuels within porous inert media," *Progress in Energy and Combustion Science*, vol. 22, no. 2, pp. 121–145, 1996.

22

Structural Evolution of Silicon Carbide Nanopowders during the Sintering Process

Galina Volkova, Oleksandr Doroshkevych, Artem Shylo, Tetyana Zelenyak, Valeriy Burkhovetskiy, Igor Danilenko, and Tetyana Konstantinova

Donetsk Institute for Physics and Engineering named after O.O. Galkin, NAS of Ukraine, R. Luxembourg Street, 72, Donetsk 83114, Ukraine

Correspondence should be addressed to Oleksandr Doroshkevych; nelya_dor@mail.ru

Academic Editor: Rajan Jose

Processes of sintering of silicon carbide nanopowder were investigated. Values of density (ρ = 3.17 g/cm^3) and strength (σ = 450 MPa) were obtained. Within the theory of dispersed systems, the temperature evolution of the materials structure was considered. The relationship between sintering temperature, characteristics of crystal structure and physical properties, in particular, density, and strength of aforementioned ceramics was established. It was concluded that it is necessary to suppress the anomalous diffusion at temperatures above 2080°C.

1. Introduction

High-density silicon carbide ceramics have a set of unique physical and mechanical properties such as high thermal conductivity, strength and wear resistance, corrosion resistance, resistance to high thermal and mechanical stresses, thermal conductivity, and semiconductor type of conductivity [1–3]. Products from SiC can operate in chemically corrosive environments at high mechanical and thermal loads. They are promising for applications in mechanical engineering, shipbuilding, metallurgy, gas and oil industry, and energetics (sealing rings for high-loaded friction units, heat-conducting substrate). Recently, a steady trend of expansion of field of use of SiC-materials in electronics, heat-conducting substrate, and semiconductor functional heterojunctions was outlined [4]. Sufficient frequent application SiC-ceramics are found in defense industry (armor ground of combat technical equipment, turbine blades for missiles, etc.) [5].

For a long time, obtaining products with complex shapes from SiC was hampered by extremely difficult conditions of sintering of micron powders. Using of hot-pressed methods (at pressures up to 1000 MPa) at operating temperatures of about 2100°C with subsequent polishing of products using diamond tools was required. Low technology and high cost of products manufacturing from SiC limited the applicability of this material in engineering.

With the advent of powder nanotechnologies [6], a unique opportunity of sintering of high-density silicon carbide ceramics within the traditional ceramic technology appeared. Prospects for improving in multiple time manufacturability and for reducing a cost of products are opened.

In particular, the possibility of sintering of products with complex shapes (geometric) and fine details, such as heat-removing substrate, offers the prospects of significant reduction in price of element base of electronics and solar energetic [7–9].

However, the process of sintering of SiC nanopowders within the ceramic technology currently is not fully understood.

The aim of this work was to study processes of structure formation during sintering of SiC nanopowder under laboratory conditions for establishment of relationship: temperature conditions of sintering—structure—physical properties.

2. Experimental

As starting material, a nanosized (d = 30 nm) silicon carbide powder α-SiC with addition up to 2% (by weight) of boron nitride (B_4N_3) as the dopant was used. Samples before sintering were prismatic blocks with size $4 \times 4 \times 30$ mm. It was obtained by successive technological operations of forming by uniaxial pressure ($P = 40$ MPa) and pressing by high hydrostatic pressure (HHP $= 400$ MPa). After forming and pressing, samples were sintered in the laboratory induction furnace type HFI-25 under argon atmosphere during 1 h with an average rate of increase of temperature 5°/min. A series of samples with different sintering temperatures (isothermal holding) was obtained. Sintering temperature for various samples was ranged from 2000 to 2200°C. Morphology of fractures of samples after sintering was analyzed by scanning electron microscopy (SEM, JSM640LV). The flexural strength was determined by four-point bending on device H50KT Tinus Olsen and density—by the method of hydrostatic weighing using electronic analytical weighing-machine type ADS50. Temperature was measured using pyrometer Raytek Marathon FR1CSF003H.

The crystalline structure of samples was investigated by X-ray diffraction (XRD) that was done using DRON-3 devices (Cu Kα-radiation, Ni-filter) with the possibility of obtaining X-ray spectra in digital form. Calculations of diffraction patterns were carried out using methods specified in [10–14].

3. Results and Discussion

XRD show, in all sintered samples presence of single phase—the hexagonal α-SiC 6H, with lattice parameters $a = 3,080$ Å, $c = 15,098$ Å, and $c/a = 4,90$.

Figure 1 shows diffractograms typical for the sample at temperature of sintering 2100°C and above (On diffractograms of samples sintered at temperatures above 2100°C, deviation of maximum peak intensities from the table values can be observed). High intensities of peaks indicate an abnormally large grain size in specified temperature range.

At lower sintering temperatures from 2000 to 2080°C, intensity of peaks similar is to intensities from the initial powder α-SiC (Figures 2(a), 2(b), and 2(c)).

Data of scanning electron microscope concerning grain structure is shown in Figure 3 and data concerning strength and density are in Figures 4(a) and 4(b), respectively.

Let us consider the processes of structure formation in samples that are realized during the thermal heating. As can be seen from Figure 3(a) at 2000°C, nanopowder dispersed system (NDS) based on SiC has a uniform topology. Structural elements are homogeneously distributed in volume of sample. Increase of temperature to 2050°C according to Figure 4 causes an increase of strength (3%) and density (0,65%) of samples at practically unchanged topology of fracture (Figure 3(b)). From the point of view of the theory of dispersed systems, this fact indicates beginning of condensation recrystallization processes (hardening contacts between particles) that are realized as a result of physical and chemical reaction of system on increase of internal energy by thermal heating. Further temperature increase, according to Figure 4, leads to a sharp increase of density (on 28%) and strength (in 1,5 times) of NDS. On the fracture (Figure 3(c))

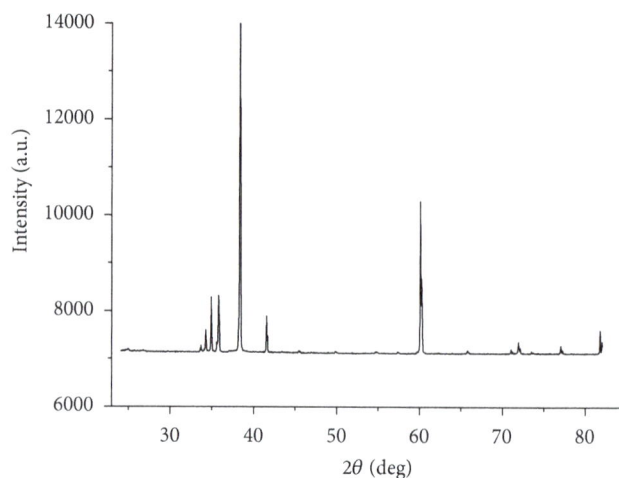

FIGURE 1: Diffractogram of SiC ceramics sintered at 2100°C, 1 h.

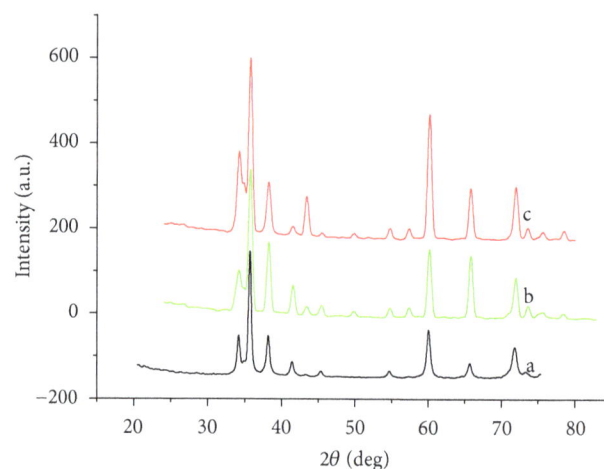

FIGURE 2: Diffractograms of starting powder SiC (a) and ceramic obtained at sintering temperatures 2000°C (b) and 2080°C, 1 h (c).

flattened areas that are typical for transgranular fracture are appearing. Topology becomes pronounced volume character. All of these facts are indicating the formation in the volume of so-called structure grid that determines the mechanical properties of samples. Subsequent rise of temperature to 2100°C (Figure 3(d)) initiates the formation of large anisometric crystals that formed the dendrite skeleton of formed ceramic structure. This reduces the strength of samples at practically constant density (Figure 4). Temperature rise above 2100°C is accompanied by increase in the degree of repletion of free space in the volume of sample (Figures 3(d)–3(f)) at monotonic decrease of strength (Figure 4(b)). It should be noted fact that at temperature above 2080°C ceramic sealing does not occur, which indicates presence of closed porosity almost at the whole range of investigated temperatures.

By comparing experimental data of X-ray analysis, SEM, density, and strength curves (Figures 4(a) and 4(b), resp.), we conclude that the optimal sintering temperature is 2080°C, 1 h: at this temperature, grains size is approximately equal (Figure 3(c)), which is indicated by standard value of intensity

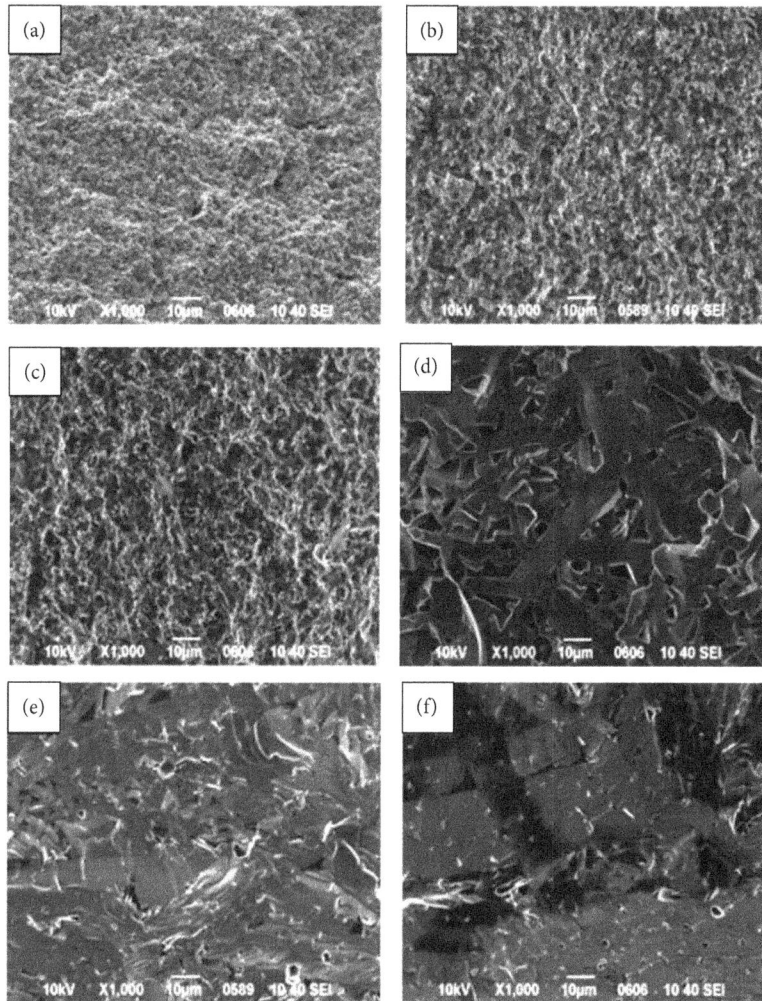

FIGURE 3: The topology of fractures of SiC samples obtained at different sintering temperatures: 2000°C (a), 2050°C (b), 2080°C (c), 2100°C (d), 2150°C (e), and 2200°C (f), 1 hour.

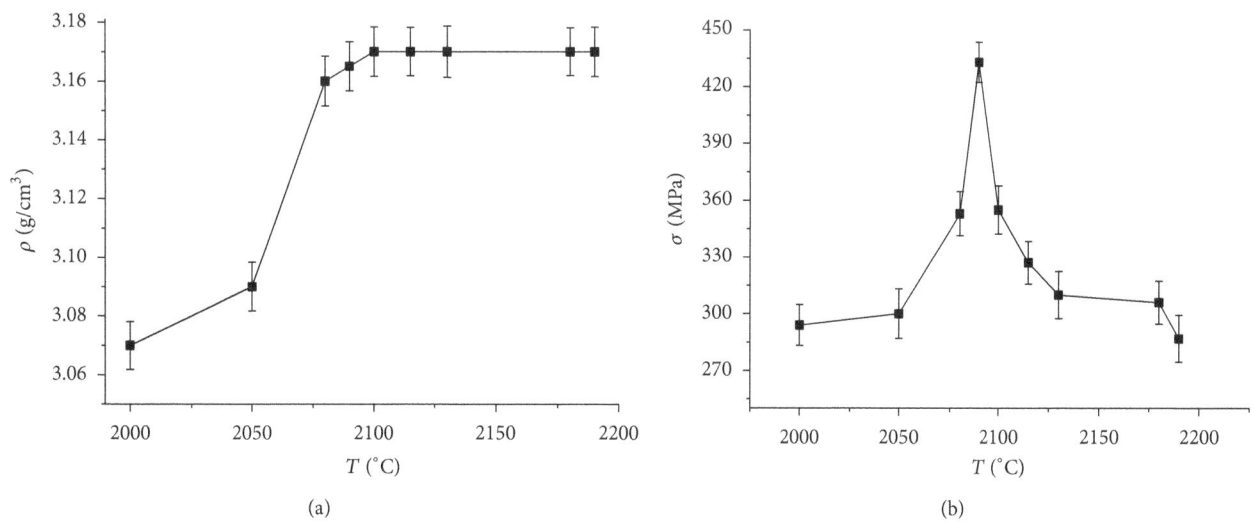

FIGURE 4: Dependence of the density (a) tensile strength and (b) from the sintering temperature of SiC ceramics.

of diffraction peaks (Figure 2(c)). This temperature corresponds to the maximum strength $\sigma = 435\,MPa$ in combination with a technically satisfactory density $\rho = 3{,}17\,g/cm^3$.

Let us consider processes of formation of grain structure. At temperatures below 2080°C, density and strength are low and grain structure has not been formed completely (Figures 3(a) and 3(b)); however, a small increase of sintering temperature to 2100°C leads to a sharp increase in grain size, which forms a specific grid with straight boundaries (Figure 3(d)), while the strength is reduced on ~50% at a constant density.

As can be seen from Figures 2 and 3, uncharacteristic for ceramic systems, behavior of physical and mechanical properties (Figure 4) is caused by threshold nature of grain growth process with temperature increasing. Temperature 2080°C is a critical point. At this temperature, maximum values of physical properties not only are observed, but also change the nature of nucleating process. Threshold character of SiC sintering, probably, is caused by the beginning of intense thermal diffusion.

Looking on the data in Figures 2 and 3 according to the theory of dispersed systems, it can concluded that in investigated system up to the temperature 2080°C a heterophase spatial structure with heterogeneous type of interfacial interactions (nearby dispersed particles have individual border), which, characteristic for dispersed systems, is saved [15, 16]. Rising of temperature to 2080°C leads to an increase of interaction energy between particles.

Coagulation with preservation of phase boundaries is take place. At exceeding of given temperature destruction of interfacial contacts is beginning. The process of coagulation of nanoparticles without saving phase boundaries, accompanied by intense grain growth, occurs. The system becomes monophasic, closer to ceramic system.

Thus, from the foregoing, it can be concluded that a level of physical and mechanical properties is caused by interfacial interaction. That is, surface component of energy provides a much larger contribution to physical and mechanical properties than bulk energy.

In case of SiC nanopowders at 2080°C, limit thermodynamic state is reached. At this temperature on borders of dispersed phase, a large maximum energy is concentrated at which the strength of the bond between particles is maximum, but border still retains individuality.

Thus, due to decrease of particles size of raw material powder, it is possible to reduce the activation energy of thermal diffusion processes and carry out synthesis of SiC-ceramics under ceramic technology, but their threshold nature is the problem. At standard mode of sintering with speed 5°/min at excess of energy threshold, an intensive grain growth is occurs. Relying on experimental results of this work, it is possible to make some conclusions about possible solutions of this problem. For this we consider the sintering process from the standpoint of classical crystallography.

From the viewpoint of crystallography, grain growth in the material occurs as a result of rapid formation of a "neck" by bulk diffusion mechanism [17]. Process of collective recrystallization of grains in silicon carbide is suppressed at

temperatures above 2080°C, probably, caused by the lack of curved boundaries. As a result, at temperatures above 2080°C, within a narrow range of temperatures, (10–20°C) anomalous recrystallization occurs with high intensity. Cause of straight direct borders should be sought, in our opinion, in highly elongated hexagonal cell where the axial ratio $c/a = 4{,}902$ and dihedral angle of the base is $\gamma = 120°$. To equilibrium configuration of borders is corresponds an orientation of three grains with apex angles 120° and it corresponds to the equilibrium surface tension forces and just such grids of boundaries with straight sides (inert boundary [17]) are formed during compacting the powder α-SiC 6H, and high temperature at sintering leads to immobilization of given configuration.

That is, threshold character of silicon carbide nanopowder sintering is associated with specificity of its hexagonal cell structure. It is possible to avoid these difficulties by applying a set of measures aimed at suppression of rapid growth, such as a significant decrease in the rate of temperature rise in the vicinity 2080°C or doping of grain boundaries by a small amount of corresponding impurity.

4. Conclusions

Given in the paper data and its interpretation allow to concluding that for the nanopowder system based on silicon carbide due to specific geometry of the elementary cell and high activation energy of diffusion processes (suppression of collective recrystallization) is characterized by

(1) presence of a relatively large proportion of heterophase boundaries,

(2) extreme nature of the formation of the grain structure.

Maximum strength of ceramics from SiC nanopowders is achieved with optimum grain size and strength (amount) of contacts between them. This ratio is determined by the heat treatment regime, in particular, by temperature increase rite and by temperature of isothermal holding.

The main scientific and technological task is creation of conditions for the implementation of heterogeneous crystallization. Possible solutions are to modify the thermal regime of sintering or chemical composition of solid solution of nanoparticles.

According to XRD-data (Figure 2(c)) and SEM (Figure 3(c)), at optimum temperature of sintering ($T = 2080°C$), nanostructured state of obtained material is saved. This fact has a scientific interest and physical properties of consolidated nanosized SiC require further investigation. In particular, taking into account small closed porosity (Figure 4), it is possible to expect a high resistance to high-speed thermal and mechanical stresses.

Conflict of Interests

The authors declare that there is no conflict of interests regarding the publication of this paper.

References

[1] A. Fialkov, *Carbon: Interlaminate and Composltes Based on It*, Aspect Press, Moscow, Russia, 1997.

[2] R. Wei, S. Song, K. Yang, Y. Cui, Y. Peng et al., "Thermal conductivity of 4H-SiC single crystals," *Journal of Applied Physics*, vol. 113, no. 5, Article ID 053503, 2013.

[3] B. A. Bilalov, M. K. Kurbanov, A. A. Gadzhiev, and S. M. Ramazanov, "Electric properties of $(SiC)_{1-x}(AlN)_x/SiC$ anisotropic heterostructures," *Russian Microelectronics*, vol. 40, no. 8, pp. 612–615, 2011.

[4] A. Lebedev and S. Sbruev, "SiC-electronics: past, present, future," *Electronics: Science, Technology, Business*, no. 5, pp. 28–41, 2006.

[5] L. I. Tuchinskii, *Composite Materials Obtained by Impregnation*, Metallurgy, Moscow, Russia, 1986.

[6] S. P. Bardakhanov, A. I. Korchagin, N. K. Kuksanov et al., "Nanopowder production based on technology of solid raw substances evaporation by electron beam accelerator," *Materials Science and Engineering B*, vol. 132, no. 1-2, pp. 204–208, 2006.

[7] B. Ozpineci, M. S. Chinthavali, L. M. Tolbert, A. Kashyap, and H. A. Mantooth, "A 55 kW three-phase inverter with Si IGBTs and SiC Schottky diodes," in *Proceedings of the 21st Annual IEEE Applied Power Electronics Conference and Exposition (APEC '06)*, pp. 541–546, March 2006.

[8] B. Luchinin and Yu. Tairov, "The domestic semiconductor silicon carbide: a step towards parity," *Modern Electronics*, no. 7, pp. 12–15, 2009.

[9] R. Singh, M. A. Green, and K. Rajkanan, "Review of conductor-insulator-semiconductor (CIS) solar cells," *Solar Cells*, vol. 3, no. 2, pp. 95–148, 1981.

[10] H. M. Rietveld, "A profile refinement method for nuclear and magnetic structures," *Journal of Applied Crystallography*, vol. 2, no. 2, pp. 65–71, 1969.

[11] A. Gusev, *Nanomaterials, Nanostructures, Nanotechnology*, Fizmatlit, Moscow, Russia, 2005.

[12] B. Warren, *X-Ray Diffraction*, Dover, New York, NY, USA, 1990.

[13] A. Gusev, A. Rempel, and A. Magerl, *Disorder and Order in Strongly Nonstoichiometric Compounds: Transition Metal Carbides, Nitrides and Oxides*, Springer, Berlin, Germany, 2001.

[14] C. Bernhard, *Particle Size Analysis: Classification and Sedimentation Methods*, Kluwer Academic, Dodrecht, The Netherlands, 1994.

[15] D. G. Knorre, F. V. Krylova, and V. S. Muzykantov, *Physical Chemistry*, Higher School, Moscow, Russia, 2nd edition, 1990.

[16] N. B. Ur'ev, "Structured disperse systems," *Soros Educational Journal*, no. 6, pp. 42–47, 1998.

[17] I. Kaynarskiy and E. Degtyareva, *Basic Refractories*, Metallurgy, Moscow, Russia, 1974.

23

Preparation and Characterization of Nano-Cadmium Ferrite

S. M. Ismail,[1] Sh. Labib,[2] and S. S. Attallah[1]

[1] Reactor Physics Department, Nuclear Research Center, Egyptian Atomic Energy Authority, P.O. Box 13759, Cairo, Egypt
[2] Nuclear Chemistry Department, Hot Laboratories Center, Egyptian Atomic Energy Authority, P.O. Box 13759, Cairo, Egypt

Correspondence should be addressed to Sh. Labib; shirazmd11@yahoo.com

Academic Editor: Young-Wook Kim

Nano-hematite (α-Fe_2O_3) and nano-cadmium ferrite ($CdFe_2O_4$) are prepared using template-assisted sol-gel method. The prepared samples are analyzed using X-ray diffraction (XRD), Fourier transform infrared spectroscopy (FTIR), scanning electron microscopy (SEM), and Mössbauer spectroscopy techniques for structural and microstructural studies. Nano-α-Fe_2O_3 with particle size ~60 nm is formed at 500°C, while nano-$CdFe_2O_4$ with smaller particle size (~40 nm) is formed at 600°C. It is found that with a simple sol-gel process we can prepare nano-$CdFe_2O_4$ with better conditions than other methods: pure phase at lower sintering temperature and time (economic point) and of course with a smaller particle size. So, based on the obtained experimental results, a proposed theoretical model is made to explain the link between the use of the sol-gel process and the formation of nano-$CdFe_2O_4$ as a pure phase at low temperature. This model is based on a simple magnetostatic interaction between the formed nuclei within the solution leading to the formation of the stable phase at low temperature.

1. Introduction

In recent years, the design and synthesis of nanomagnetic particles has the focus of the intense fundamental and applied research with special emphasis on their enhanced properties that are different from those of their bulk counterparts [1–6]. Nano-cadmium ferrite ($CdFe_2O_4$) is a normal spinel ferrite that can be applied in various fields [7]. Nano-$CdFe_2O_4$ exhibits ferromagnetism, and ~54% of Fe^{3+} ions occupy the A site in contrast to 0% for the bulk materials with normal spinel structures [8]. The enhanced occupancy of Fe^{3+} ions in $CdFe_2O_4$ is explained by a higher octahedral preferential energy of Cd^{2+} [8]. It is found that nano-$CdFe_2O_4$ can be obtained at low temperature by applying solution methods. Desai et al. [8] prepare nano-$CdFe_2O_4$ by sintering at 600°C for 4 h using the precipitation method, where the achieved crystallite size is 43 nm and the hematite (α-Fe_2O_3) as a minor phase is present. Otherwise, Sharma et al. [9] prepare nano-$CdFe_2O_4$, having crystallite sizes (100–200 nm), as a pure phase by sintering at 900°C for 6 h using the urea combustion method. In this paper, nano-$CdFe_2O_4$ is prepared using the sol-gel process that is considered as one of the important methods used in nanoparticles' synthesis. This process is chosen because it gives enhanced homogeneity, better control for size, shape, and degree of agglomeration of the resulting nanocrystals, simple compositional control, and low processing temperature [2, 5, 8, 10–12]. In order to obtain sols and gels with desirable properties, the control of precursor reactivity may be achieved through the addition of chelating agents such as β-diketones, carboxylic acids, or other complex ligands [11, 13]. These characteristics, along with the chemical composition, are found to influence significantly the magnetic properties of nanoferrites [13]. The aim of this paper is to explain theoretically the reason of nano-$CdFe_2O_4$ formation when using the sol-gel process that gives a single phase at both lower sintering temperature and time than that used previously [8, 9]. The prepared nanomagnetic materials are characterized using X-ray diffraction (XRD), Fourier transform infrared spectra (FTIR), scanning electron microscopy (SEM), and Mössbauer effect spectroscopy techniques. Based on the achieved experimental results, a new model is proposed to explain the link between the use of the sol-gel process and the formation of nano-$CdFe_2O_4$ as a pure phase at low temperature.

2. Experimental

2.1. Materials and Method. α-Fe_2O_3 and $CdFe_2O_4$ sols are prepared using $Fe(NO_3)_3\cdot9H_2O$ (99% Merck, Germany), $Cd(NO_3)_2\cdot4H_2O$ (99% Merck, Germany), isopropyl alcohol (99.8% Scharlau, Spain), acetylacetone ($\geq$99% Merck, Germany), and distilled water. All the used reagents are of AR grade.

(i) α-Fe_2O_3 sol is prepared by a dropwise addition of an alcoholic mixture of isopropyl alcohol (5 mol) and distilled water (2 mol) to the stirred alcoholic solution of $Fe(NO_3)_3\cdot9H_2O$ and acetylacetone having molar ratio (isopropyl alcohol/$Fe(NO_3)_3\cdot9H_2O$/acetylacetone: $5:1:2$), respectively.

(ii) $CdFe_2O_4$ sol is prepared by a dropwise addition of an alcoholic mixture of isopropyl alcohol (5 mol) and distilled water (2 mol) to the alcoholic mixture of $Fe(NO_3)_3\cdot9H_2O$ and $Cd(NO_3)_2\cdot4H_2O$ as well as acetylacetone, where $Fe(NO_3)_3\cdot9H_2O$ and $Cd(NO_3)_2\cdot4H_2O$ are mixed in a stoichiometric ratio ($2:1$), respectively.

The prepared sols are aged for 168 h then dried at 80°C for 48 h and calcined at 400°C for 1 h in the air. The calcined powders are ground for 45 min then pressed uniaxially into tablets at 20 kN. The green compacts are sintered at 500° and 600°C for 3 h in the air. Table 1 shows the abbreviation of the prepared samples.

2.2. Material Characterization. Different compacts are characterized using X-ray diffraction (XRD) (Philips X'pert multipurpose diffractometer), where the used X-ray tube is a copper tube operating at 40 kV and 30 mA and the used wavelength is $K_{\alpha1}$ with wavelength 1.54056 Å. The scan is performed over the range of 2θ (10–70) degrees. The identification of the present crystalline phases is done using Joint Committee on Powder Diffraction Standards (JCPDS) database card numbers.

Fourier transform infrared spectroscopy (FTIR) analysis (Nicolet iS10 FTIR Spectrometer—Thermo Scientific) is carried out for the prepared powders using KBr disc technique. The scan is performed in the region of 400–4000 cm^{-1}.

Scanning electron microscopy (SEM) for the different samples is employed using JEOL JSM-5600 LV. The scan is performed at high vacuum mode using accelerating voltage 30 kV, working distance 20 mm, and magnification ×3000 and 4000.

Austin Science Mössbauer Spectrometer with constant acceleration laser-interferometer-controlled drive and data acquisition system is used in a standard transmission setup with a Personal Computer Analyzer (PCA II-card with 1024 channels). The radioactive source is ^{57}Co embedded in Rh matrix with initial activity of 50 mCi. Metallic iron spectrum is used for the calibration of both observed velocities and hyperfine magnetic fields. The absorber thickness is approximately 10 mg/cm^2 of natural iron. The Mössbauer effect (ME) spectra are analyzed with a computer program [14], where the

TABLE 1: Abbreviation of the prepared samples.

Sintering temperature (°C/3 h)	Sample	
	Fe_2O_3	$CdFe_2O_4$
500	F500	CF500
600	F600	CF600

analysis of ME spectra for these samples is done using the theories of N magnetic sextets and N quadrupole doublets.

2.3. Theoretical Background. The methods of synthesis based on either chemical or physical concepts allow the structural and microstructural properties of these materials to be highly controlled and to be well reproduced. It is generally agreed that the driving force for a phase transformation is the difference in free energy between the final and the initial states of the system [15]. The driving force for a certain reaction may be written as the derivative

$$-\left(\frac{\partial G}{\partial \xi}\right)_{P,T},\tag{1}$$

where G is the Gibbs free energy, P and T are pressure and temperature, and ξ is the extent of the reaction.

For the isobarothermal case and fixed composition, the integrated driving force is simply the difference in Gibbs energy between the final and the initial states of the system [15]. On the nanoscale, the phase transformation may involve several processes, for example, migration of coherent, semicoherent, or incoherent phase interface, solute diffusion both inside and ahead of the interface [15]. Each individual process needs some driving force, and the sum of these driving forces can exceed the integrated driving force [15].

For phase transformation with composition change, it is considered integrated force acting on the phase interface [15]. Hillert's analysis yields as the integrated driving force (counted per mole of growing phase)

$$\sum_{J=1}^{n} x_j \Delta\mu_j,\tag{2}$$

where n is the number of components, x_j is the mole fraction of component j in the material that is transferred over the phase interface, and $\Delta\mu_j$ is the difference in chemical potential for j across the phase interface [15].

If there is diffusion in both α and β phases, a different approach is required,

$$x_j = \frac{x_j^\alpha J_j^\beta - x_j^\beta J_j^\alpha}{J_j^\beta - J_j^\alpha},\tag{3}$$

where x and J, which denote the mole fraction and the diffusion flux, respectively, on each side of the phase interface, can be combined with (2) to give the driving force on the interface in the general case [15]. As concluded, the driving force is expressed not in terms of the total free energy change during crystallization, but rather as the change in the chemical potential of the crystallization species $\Delta\mu$ [16]. $\Delta\mu$ measures the free energy response to molecules transferring from one phase to the other [16].

It is clear that the determination of the driving force is connected with the evaluation of the thermodynamic properties as a function of composition, temperature, and pressure of a system [15]. Whether considering the nucleation or growth, the reason for the transformation from solution to solid is the same [16]. The free energy of the initial solution phase is greater than the sum of the free energies of the crystalline phase and the final solution phase [16]. As concluded from this brief survey, it is important to correlate the relation between the synthesis mode and the nanophase formation at low temperature based on thermodynamic features. We propose a new model derived from Sacanna et al.'s model [17] to explain the reason of the formation of nano-$CdFe_2O_4$ at low temperature.

3. Results and Discussion

In general, the sol-gel process involves several successive stages: (a) formation of sol which represents a colloidal suspension containing small particles with a diameter less than 1000 nm dispersed in a continuous liquid medium, (b) the gelation of the sol to give a three-dimensional M–O–M/M–OH–M network whose pores are filled with solvent molecules (wet gel), (c) the ageing of the resulting wet gel process known as syneresis, (d) the elimination of the solvent from the gel's pore (drying) [10].

In order to obtain narrow size distribution of nano-Fe_2O_3 and -$CdFe_2O_4$, agglomerations must be prevented during the processing step. So, to produce nanoparticles without agglomeration, a capping agent such as acetylacetone is used to control the particle size and shape, and also ageing time is increased in order to obtain a uniform morphology of the obtained structures [11, 12].

3.1. X-Ray Diffraction (XRD) Analysis. Figure 1 shows the X-ray diffraction pattern of both α-Fe_2O_3 and $CdFe_2O_4$ compacts thermally treated at 500° and 600°C for 3 h. α-Fe_2O_3 is formed at 500°C (JCPDS 79-1741), as shown in Figure 1(a). The obtained hematite phase has a rhombohedral structure with space group $R\bar{3}c$ (167). The broadening of the diffraction peaks suggests that the particle size is very small. The average particle size (D) is 61 nm as given from the Debye-Scherrer equation [18]

$$D = \frac{0.94\lambda}{\beta\cos\theta},$$ (4)

where λ is the wavelength of the X-ray radiation used and β is the full width at half-intensity maximum (FWHM) in radian. The increase in temperature to 600°C yields an increase of the obtained diffraction peaks sharpness due to particle size growth to 69 nm (Figure 1(b)). On the other hand, CF500 compact shows the presence of the cubic $CdFe_2O_4$ phase (JCPDS 79-1155) with minor α-Fe_2O_3 diffraction peaks (Figure 1(c)). The obtained spinel phase has a cubic structure with space group Fd3m (227). The average particle size of $CdFe_2O_4$

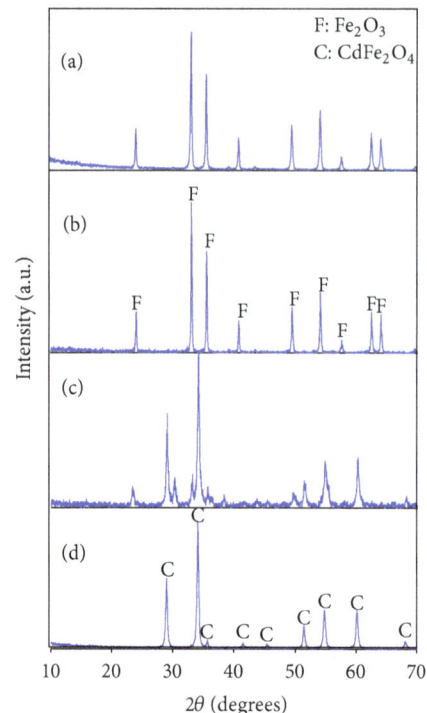

FIGURE 1: XRD patterns of Fe_2O_3 and $CdFe_2O_4$ compacts thermally treated at 500° and 600°C for 3 h: (a) F500, (b) F600, (c) CF500, and (d) CF600.

phase is 43 nm. CF600 compact shows the disappearance of Fe_2O_3 diffraction lines indicating the complete formation of $CdFe_2O_4$ phase having an average particle size of 48 nm (Figure 1(d)).

In general, the observed decrease in the average particle size of $CdFe_2O_4$ with respect to α-Fe_2O_3 is referred to the strong chemical affinity of Cd^{2+} to the tetrahedral A site and the metastable cation distribution in the nanoscale range of ferrite particles [5]. Based on this feature, one can explain the dependence of particle size on cation stoichiometry [5].

3.2. Fourier Transform Infrared Spectroscopy (FTIR) Analysis. Figure 2 shows the FTIR spectra of both α-Fe_2O_3 and $CdFe_2O_4$ powders thermally treated at 500° and 600°C for 3 h. Figure 2(a) shows the spectrum of F500 powder: a broad band at 3742 cm^{-1} that is assigned to the adsorbed H_2O from the atmosphere or OH group of isopropanol is observed [19, 20].

The very small band at 2300 cm^{-1} is due to adsorbed or atmospheric CO_2 [2, 21]. The characteristic small bands at 1716 and 1338 cm^{-1} are assigned to both asymmetric and symmetric C–O, respectively [2, 19, 21]. The C–O bands region is an indicative of the presence of organic residues due to acetylacetone species [21]. Two small peaks appeared at 533 and 440 cm^{-1} that are characteristic of poorly crystalline α-Fe_2O_3 [19].

The increase in temperature to 600°C shows the disappearance of the band at 3742 cm^{-1} and the increase of the characteristic α-Fe_2O_3 bands intensity (Figure 2(b)).

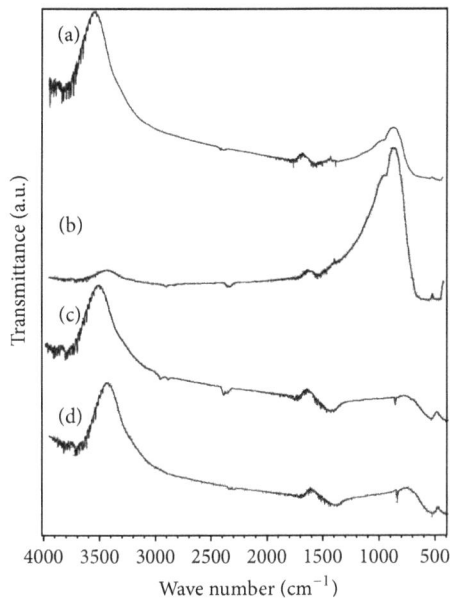

FIGURE 2: FTIR spectra of Fe_2O_3 and $CdFe_2O_4$ powders thermally treated at 500° and 600°C for 3 h: (a) F500, (b) F600, (c) CF500, and (d) CF600.

Figure 2(c) shows the spectrum of CF500 powder, where the bands at 3700, 2359, 1716, and 1450 cm^{-1} are assigned as in the case of the previous samples. Two major sharp bands at 858 and 545 cm^{-1} that are characteristic of spinel ferrite ($CdFe_2O_4$) are observed. The higher frequency band is caused by stretching vibration of the tetrahedral metal-oxygen band (Fe–O), and the lower frequency absorption band is caused by a metal-oxygen stretching vibration in octahedral sites (Cd–O) [21]. The sharpness of these bands is correlated to the high degree of crystallinity of the obtained phase. A very small band at 533 cm^{-1} is observed which is characteristic of poorly crystalline α-Fe_2O_3 as explained previously. This band is indicative of the presence of a minor α-Fe_2O_3 phase as confirmed by X-ray diffraction analysis (Figure 1(c)). A marked decrease in the height of 3700, 2359, 1716, and 1450 bands with increasing the temperature to 600°C is due to the decomposition of all the organic residues (Figure 2(d)) [2, 19]. The characteristic spinel ferrite bands at 858 and 545 cm^{-1} are still present. The Fe_2O_3 band disappears indicating the presence of $CdFe_2O_4$ as a pure phase which is also confirmed by the X-ray diffraction analysis given in Figure 1(d).

3.3. Scanning Electron Microscopy (SEM). SEM measurements are carried out in order to understand the morphology and the shape of the synthesized nanomaterials. SEM micrographs of both nano-α-Fe_2O_3 and -$CdFe_2O_4$ compacts thermally treated at 500° and 600°C for 3 h are shown in Figure 3. F500 is characterized by spherical particles with different sizes that are distributed in a homogeneous manner within fine and small granules (as seen in Figure 3(a)).

F600 is characterized by the presence of both enlarged spherical particles and tiny small spherical particles that are formed from the fine granules (as shown in Figure 3(b)). On the other hand, Figures 3(c) and 3(d) show the SEM of CF500 and CF600 at higher magnification (x = 4000) to show the microstructure of nano-$CdFe_2O_4$ particles in detail. The microstructures are characterized by the presence of uniform neck and the bonding of equiaxial grains. An increase in the interparticle spaces of the obtained nanoparticles of CF600 with respect to CF500 is observed (as seen in Figure 3(d)). Two marked features are obtained from SEM illustrations.

(i) The presence of nanoparticles in agglomerated form because in many cases of nanocrystalline ferrites, it is observed that there is a tendency of agglomeration among the nanoparticles [22], consequently, the particle sizes values from SEM do not represent the true particle sizes of the prepared nanoparticles. The aggregated nanoparticles are characterized by uniform shapes which reflect the tailor arrangements of the formed nanocrystallites.

(ii) A decrease in the porous structure of nano-α-Fe_2O_3 with respect to -$CdFe_2O_4$, which is correlated to the particle size increase [23], is the result of individual grains coming closer to each other increasing the effective area of grain to grain contact [23].

3.4. Mössbauer Studies. Figure 4(a) shows the Mössbauer effect (ME) spectrum recorded at room temperature (293 K) for the Fe_2O_3 sample. The figure displays a well-resolved spectrum consisting of one Zeeman sextet due to Fe^{3+}, where Figure 4(b) shows that the spectrum is consisting of one sextet due to the presence of Fe_2O_3 phase with 78% and one doublet of the $CdFe_2O_4$ phase with 22% as clearly shown in this figure. In contrast, the ME spectra at 293 K for $CdFe_2O_4$ ferrite heated at 600°C exhibit quadrupole doublet spectra as illustrated in Figure 5.

3.4.1. Quadrupole Interaction. The presence of chemical disorder in spinel structure produces an electric field gradient (EFG) of varying magnitude, direction, sign, and symmetry and a resulting distribution in the quadrupole splitting (QS). In other words, the EFG at ^{57}Fe nucleus arises from the asymmetrical charge distribution surrounding the ion. However, since an Fe^{3+} ion has a half-filled 3d shell ($3d^5$), the EFG in this case can arise only from an asymmetric charge distribution surrounding the iron ion. Cadmium is known to show strong preference for A sites in spinel ferrites. Consequently, $CdFe_2O_4$ is a normal spinel. This EFG may arise from the departure of the six nearest anion neighbours from their ideal octahedral symmetry and the nonspherical distribution of charges on the next nearest cation and anion neighbours of the B site. In the present work, the QS values at room temperature of the $CdFe_2O_4$ sample are in the range of 0.81 and 0.845 mms^{-1} as shown in Table 2. The values of QS obtained for the $CdFe_2O_4$ at 293 K are in good agreement with those reported earlier in [24].

3.4.2. Isomer Shift. The obtained isomer shift (IS) values for Fe_2O_3 sample and $CdFe_2O_4$ spinel ferrites are given in Table 2. The obtained result of IS is in agreement with other

(a)

(b)

(c)

(d)

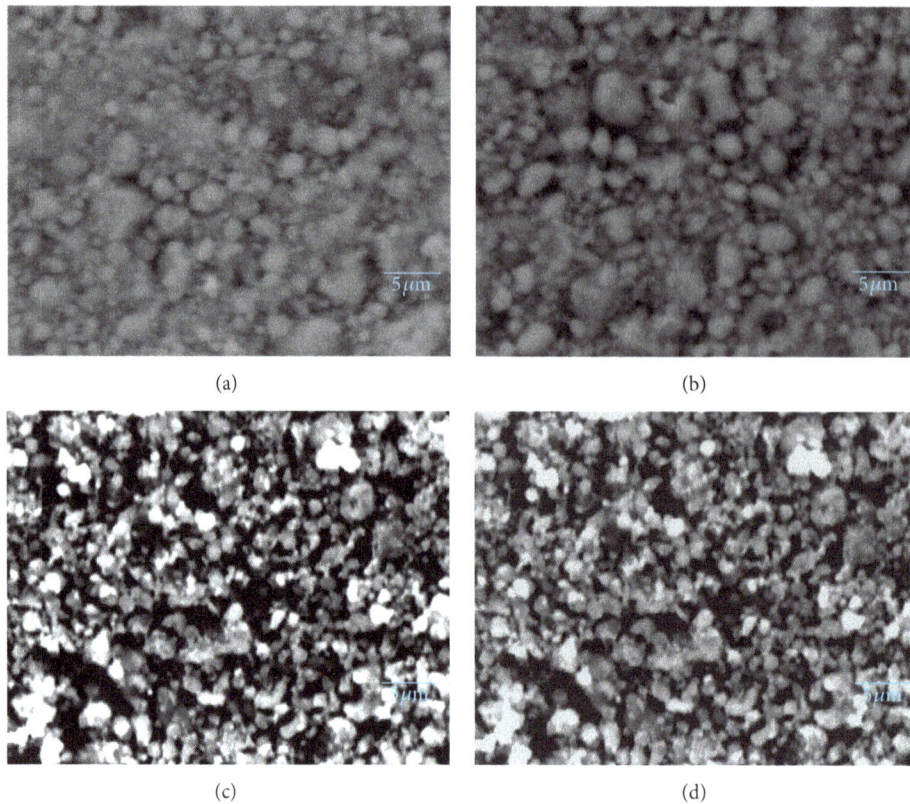

FIGURE 3: Scanning electron micrograph of Fe_2O_3 and $CdFe_2O_4$ compacts thermally treated at 500° and 600°C for 3 h: (a) F500, (b) F600, (c) CF500, and (d) CF600.

TABLE 2: Mössbauer parameters for the prepared samples at room temperature.

Sample	H_{hf} (kOe)	QS (mm s^{-1})	IS (mm s^{-1})	Line width Γ (mm s^{-1})	Area ratio (%)
a	515 ± 0.008	−0.215 ± 0.002	0.4964 ± 0.001	0.352 ± 0.004	100
	515.5 ± 0.008	−0.205 ± 0.002	0.514 ± 0.001	0.337 ± 0.004	22
b	—	0.81 ± 0.007	0.5122 ± 0.001	0.378 ± 0.01	78
c	—	0.845 ± 0.002	0.4388 ± 0.001	0.423 ± 0.003	100

previously reported data [24]. This can be interpreted as being due to the large band separation of Fe^{3+}–O^{2-} for the octahedral ions compared with that for the tetrahedral ions. As the orbitals of the Fe^{3+} and O^{2-} ions do not overlap, the covalency effect becomes smaller compared to the isomer shift at the octahedral site in the other spinel ferrite with inverse or partially inverse spinel.

3.4.3. Hyperfine Fields.

The hyperfine magnetic field at the iron nucleus is proportional to the spontaneous magnetization of the sublattice to which the particular nucleus belongs. The hyperfine field H_{hf} measured by the ME consists of three contributions: $H_{hf} = H_{core} + H_{dip} + H_{shift}$, where H_{core} results from the polarization of s electrons by the magnetic moments of the d electrons. This field is larger for free ions than for ions in a crystal because of covalency. H_{dip} represents the dipolar fields produced by the surrounding magnetic ions. At room temperature, the hyperfine field H_{hf} value for the

Fe_2O_3 sample "treated at 500°C" is 515 ± 0.02 kOe, which is in good agreement with previous work [24, 25]. While, for the $CdFe_2O_4$ sample "treated at 600°C," the intersublattice contributions h_{AB} and h_{BA} are not predominant. So, the paramagnetic doublet is observed as shown in Figure 5.

3.5. Theoretical Proposal.

The power of any mode of synthesis is the ability to form nanospinel structure, as a pure phase, at low temperature. A direct correlation exists between the decrease in the Gibbs free energy of the solution and the formation of the nanospinel phase at low temperature as shown in Figure 6.

So, we need to make a model for the synthesis method that explains the decrease of solution's Gibbs free energy and the formation of a nanospinel phase at low temperature.

In the present work, we benefit from the self-assembly model described by Sacanna et al. [17] and modify it to explain our purpose. The assembly model is based on a

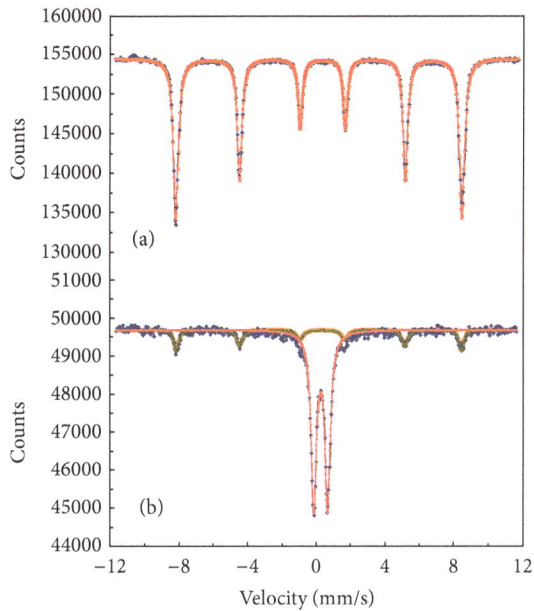

FIGURE 4: Mössbauer spectra at room temperature for Fe_2O_3 and $CdFe_2O_4$ thermally treated at 500°C for 3 h: (a) F500 and (b) CF500.

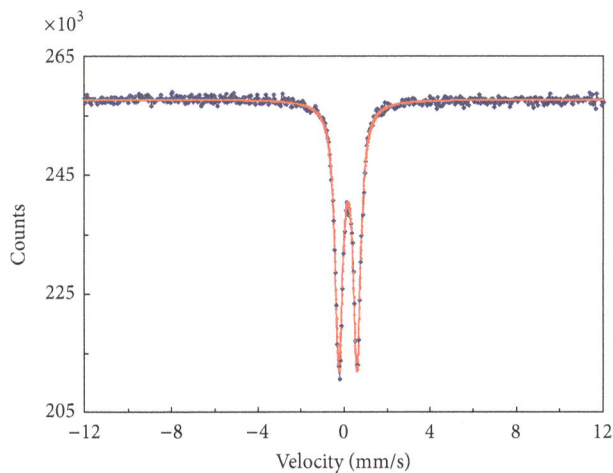

FIGURE 5: Mössbauer spectrum at room temperature for $CdFe_2O_4$ thermally treated at 600°C for 3 h: CF600.

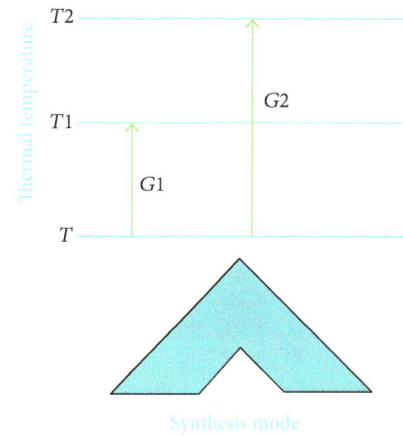

FIGURE 6: The relation between the decrease of Gibbs free energy and phase formation at low temperature.

binding mechanism between colloidal particles using a simple magnetostatic interaction [17]. There is an important reason to investigate this type of interaction, that is, unlike electrostatic and chemical interactions, magnetic forces are not screened in solutions and are independent of the changes in the experimental conditions, thus giving a significant experimental design freedom [17]. The magnetostatic interaction requires particles with localized and well-calibrated magnetic dipole moments [17]. The resulting magnetic forces should be strong enough to bind particles [17]. The model cited by Sacanna et al. describes the degree of dispersion or agglomeration of particles in a colloidal solution by assuming the presence of a magnet (Fe_2O_3) embedded underneath the

particle surface generating localized permanent magnetic dipoles [17].

In order to understand our model, we consider that the used colloidal solution gives the probability of obtaining three phases by sintering at the suitable temperatures: CdO, Fe_2O_3, and $CdFe_2O_4$ phases. In order to obtain nano-$CdFe_2O_4$ phase at low temperature as a pure phase, the Gibbs free energy of $CdFe_2O_4$ formation must be lower than that of CdO and Fe_2O_3 phases.

We consider the presence of both Cd(II) and Fe(III) ions linked together inside the chelating agent as the first nucleus of nanoparticle that breaks up to nanospinel phase by sintering. This nucleus is considered as a magnetic material, due to the presence of Fe(III) ions, that combines with another nucleus via a magnetostatic interaction leading to a more nucleation buildup of the nanospinel phase. By imagining that an oriented nucleus in a certain direction interacts with another one having a different orientation, consequently a building up of nuclei begins to grow sequentially. This imagination is based on considering that each nucleus is a magnet having positive and negative poles and so the unlike poles of different magnets attracts with each other. In this case, a self-magnetic field is generated around the buildup nuclei that are considered as magnets having dipole moments and connected together by a magnetostatic interaction. This self-induced magnetic field is responsible for the formation of the nanospinel phase at low temperature by lowering the Gibbs free energy of the $CdFe_2O_4$ phase formation with respect to that of the CdO and Fe_2O_3 phases. The reason is that many papers are talking about the decrease in the Gibbs free energy of a solution of magnetic particles when an external magnetic field is applied to it [26, 27]. So, the induced magnetic field has a tendency of decreasing the Gibbs free energy of the solution system [26] and so gives the reason of the nanospinel phase formation at low temperature.

The difference between the assembled model and our model is that the assembled model describes the formation of the agglomerated particles in the solution phase instead of the formation of dispersed particles. But here in our model, the assembled model is used in describing the buildup of nanospinel nuclei; that is, we describe nanophase formation (assembly of nuclei of the nanospinel phase).

4. Conclusion

The aim of this work is to explain theoretically the reason of nano-$CdFe_2O_4$ formation at low temperature when using the sol-gel process for synthesis.

So, nano-Fe_2O_3 and -$CdFe_2O_4$ are prepared using template sol-gel method. The XRD data investigate the formation of nano-Fe_2O_3 and -$CdFe_2O_4$ phases by sintering at 500° and 600°C for 3 h, respectively. Mössbauer studies confirm the results obtained by XRD. SEM illustrations reveal a uniform microstructure of agglomerated nano-Fe_2O_3 and -$CdFe_2O_4$ particles which reflect the tailor arrangements of nanocrystallites. Single-phase nano-$CdFe_2O_4$ (~48 nm) is formed at lower sintering temperature and time than that used in previous methods. A new model is proposed to explain the formation of nano-$CdFe_2O_4$ as a pure phase at low temperature. This model is based on the self-assembled model described by Sacanna et al. By assuming a simple magnetostatic interaction between the formed nuclei that are considered as magnets within the solution, a nanostable phase can be formed at low temperature. The presence of magnetostatic interaction between the nuclei creates a magnetic field, and consequently the Gibbs free energy of the solution is decreased. This explains the formation of $CdFe_2O_4$ as a pure phase at low temperature. Consequently, the assembled model is used in describing the buildup of nanospinel nuclei.

References

[1] Y. L. N. Murthy, I. V. K. Viswanath, T. K. Rao, and R. Singh, "Synthesis and characterization of nickel copper ferrite," *International Journal of ChemTech Research*, vol. 1, no. 4, pp. 1308–1311, 2009.

[2] S. Maensiri, C. Masingboon, B. Boonchom, and S. Seraphin, "A simple route to synthesize nickel ferrite ($NiFe_2O_4$) nanoparticles using egg white," *Scripta Materialia*, vol. 56, no. 9, pp. 797–800, 2007.

[3] S. A. Corr, Y. P. Rakovich, and Y. K. Gun'Ko, "Multifunctional magnetic-fluorescent nanocomposites for biomedical applications," *Nanoscale Research Letters*, vol. 3, no. 3, pp. 87–104, 2008.

[4] P. B. C. Rao and S. P. Setty, "Electrical properties of Ni-Zn nano ferrite particles," *International Journal of Engineering Science and Technology*, vol. 2, no. 8, pp. 3351–3354, 2010.

[5] R. Iyer, R. Desai, and R. V. Upadhyay, "Low temperature synthesis of nanosized $Mn_{1-x}Cd_xFe_2O_4$ ferrites," *Indian Journal of Pure and Applied Physics*, vol. 47, no. 3, pp. 180–185, 2009.

[6] S. P. Gubin, Y. A. Koksharov, G. B. Khomutov, and G. Y. Yurkov, "Magnetic nanoparticles: preparation, structure and properties," *Russian Chemical Reviews*, vol. 74, no. 6, pp. 489–520, 2005.

[7] P. K. Nayak, "Synthesis and characterization of cadmium ferrite," *Materials Chemistry and Physics*, vol. 112, no. 1, pp. 24–26, 2008.

[8] R. Desai, R. V. Mehta, R. V. Upadhyay, A. Gupta, A. Praneet, and K. V. Rao, "Bulk magnetic properties of $CdFe_2O_4$ in

nano-regime," *Bulletin of Materials Science*, vol. 30, no. 3, pp. 197–203, 2007.

[9] Y. Sharma, N. Sharma, G. V. S. Rao, and B. V. R. Chowdari, "Li-storage and cycling properties of spinel, $CdFe_2O_4$, as an anode for lithium ion batteries," *Bulletin of Materials Science*, vol. 32, no. 3, pp. 295–304, 2009.

[10] D. Caruntu, *Nanocrystalline transition metal ferrites: synthesis, characterization and functionalization [Ph.D. thesis]*, University of New Orleans, 2006.

[11] M. Cernea, "Sol-gel synthesis and characterization of $BaTiO_3$ powder," *Journal of Optoelectronics and Advanced Materials*, vol. 7, no. 6, pp. 3015–3022, 2005.

[12] M. A. Willard, L. K. Kurihara, E. E. Carpenter, S. Calvin, and V. G. Harris, "Chemically prepared magnetic nanoparticles," *International Materials Reviews*, vol. 49, no. 3-4, pp. 125–170, 2004.

[13] C. Sanchez, J. Livage, M. Henry, and F. Babonneau, "Chemical modification of alkoxide precursors," *Journal of Non-Crystalline Solids*, vol. 100, no. 1–3, pp. 65–76, 1988.

[14] S. S. Ata-Allah, M. K. Fayek, H. S. Refai, and M. F. Mostafa, "Mossbauer effect study of copper containing nickel-aluminate ferrite," *Journal of Solid State Chemistry*, vol. 149, no. 2, pp. 434–442, 2000.

[15] K. Thornton, J. Ågren, and P. W. Voorhees, "Modelling the evolution of phase boundaries in solids at the meso- and nano-scales," *Acta Materialia*, vol. 51, no. 19, pp. 5675–5710, 2003.

[16] P. Mathur, A. Thakur, and M. Singh, "Synthesis and characterization of $Mn_{0.4}Zn_{0.6}Al_{0.1}Fe_{1.9}O_4$ nano-ferrite for high frequency applications," *Indian Journal of Engineering and Materials Sciences*, vol. 15, no. 1, pp. 55–60, 2008.

[17] S. Sacanna, L. Rossi, and D. J. Pine, "Magnetic click colloidal assembly," *Journal of the American Chemical Society*, vol. 134, no. 14, pp. 6112–6115, 2012.

[18] B. D. Cullity, *Elements of X-Ray Diffraction*, Addison-Wesley, Reading, Mass, USA, 2nd edition, 1978.

[19] M. A. Gabal, A. A. El-Bellihi, and S. S. Ata-Allah, "Effect of calcination temperature on Co(II) oxalate dihydrate-iron(II) oxalate dihydrate mixture: DTA-TG, XRD, Mössbauer, FT-IR and SEM studies (part II)," *Materials Chemistry and Physics*, vol. 81, no. 1, pp. 84–92, 2003.

[20] S. W. Boland, S. C. Pillai, W.-D. Yang, and S. M. Haile, "Preparation of (Pb, Ba) TiO_3 powders and highly oriented thin films by a sol-gel process," *Journal of Materials Research*, vol. 19, no. 5, pp. 1492–1498, 2004.

[21] R. W. Schwartz, J. A. Voigt, B. A. Tuttle, D. A. Payne, T. L. Reichert, and R. S. DaSalla, "Comments on the effects of solution precursor characteristics and thermal processing conditions on the crystallization behavior of sol-gel derived lead zirconate titanate thin films," *Journal of Materials Research*, vol. 12, no. 2, pp. 444–456, 1997.

[22] S. A. Khorrami, G. Mahmoudzadeh, S. S. Madani, and F. Gharib, "Effect of calcination temperature on the particle sizes of zinc ferrite prepared by a combination of sol-gel auto combustion and ultrasonic irradiation techniques," *Journal of Ceramic Processing Research*, vol. 12, no. 5, pp. 504–508, 2011.

[23] M. A. Ahmed, N. Okasha, and S. I. El-Dek, "Preparation and characterization of nanometric Mn ferrite via different methods," *Nanotechnology*, vol. 19, no. 6, Article ID 065603, 6 pages, 2008.

[24] J. Plocek, A. Hutlová, D. Nižňanský, J. Buršík, J.-L. Rehspringer, and Z. Mička, "Preparation of $ZnFe_2O_4/SiO_2$ and $CdFe_2O_4/$

SiO$_2$ nanocomposites by sol-gel method," *Journal of Non-Crystalline Solids*, vol. 315, no. 1-2, pp. 70–76, 2003.

[25] A. Lančok, P. Bezdička, M. Klementová, K. Závěta, and C. Savii, "Fe$_2$O$_3$/SiO$_2$ hybrid nanocomposites studied mainly by Mössbauer spectroscopy," *Acta Physica Polonica A*, vol. 113, no. 1, pp. 577–581, 2008.

[26] X. J. Liu, Y. M. Fang, C. P. Wang, Y. Q. Ma, and D. L. Peng, "Effect of external magnetic field on thermodynamic properties and phase transitions in Fe-based alloys," *Journal of Alloys and Compounds*, vol. 459, no. 1-2, pp. 169–173, 2008.

[27] Y. Zhang, C. He, X. Zhao, L. Zuo, C. Esling, and J. He, "New microstructural features occurring during transformation from austenite to ferrite under the kinetic influence of magnetic field in a medium carbon steel," *Journal of Magnetism and Magnetic Materials*, vol. 284, no. 1–3, pp. 287–293, 2004.

24

Structural Elucidation of Some Borate Glass Specimen by Employing Ultrasonic and Spectroscopic Studies

S. Thirumaran and N. Karthikeyan

Department of Physics (DDE), Annamalai University, Annamalai Nagar 608 002, India

Correspondence should be addressed to S. Thirumaran; thirumaran64@gmail.com

Academic Editor: Joon-Hyung Lee

Quantitative analysis has been carried out in order to obtain more information about the structure of two glass systems, namely, $(B_2O_3\text{-}MnO_2\text{-}PbO)$ (BML glass system) and $(B_2O_3\text{-}Na_2CO_3\text{-}P_2O_5)$ (BSP glass system). Their structural elucidation has been carried out by studying the ultrasonic velocities (longitudinal velocities U_L and shear velocities U_S) and density of these glass samples. The present investigation has been interpreted by focusing more on elastic and mechanical properties of glass specimen through ultrasonic study and the elemental analysis study through spectroscopic studies. The scanning electron microscopic (SEM) study was also carried out with a view to throwing more light on their morphological aspects. The results are corroborated in the light of the role of borate (B_2O_3) glasses in the formation of glassy structural network.

1. Introduction

The acoustical properties are particularly suitable for describing glasses as a function of composition because they give some information about both the microstructure and the dynamics of the glasses. The elastic properties are related to microscopic properties through the behavior of the network and the modifier. Attempts have been made to measure and interpret the acoustical properties of borate glasses in terms of structural changes [1–3]. Bahatti and Singh [4] discussed the composition dependence of ultrasonic velocity. The ultrasonic nondestructive testing has been found to be one of the best techniques to study the microstructure, characterization, mechanical properties, and phase changes as well as to evaluate elastic constants. One can also characterize the materials such as semiconducting glasses, superconducting glasses, glass ceramics and bioactive glasses by this nondestructive testing technique. The propagation of ultrasonic wave in solids such as glass provides valuable information regarding the solid state motion in the material. Interest in glasses has rapidly increased in recent years because of diverse applications in electronic, nuclear, solar energy and acoustooptic devices. The acoustic wave propagation in bulk glasses has been of considerable interest to understand the mechanical

properties [5]. The velocity of sound is particularly suitable for characterising glasses as function of composition because it gives information about the microstructure and dynamics of glasses [6]. The study of elastic properties of the glasses has inspired many researches [7, 8] and significant information about the same has been obtained. The elastic moduli of glasses are influenced by the many physical parameters, which may in turn be studied by measuring the ultrasonic velocities. The dependence of ultrasonic velocity on the composition of glass indicates the various changes in the structural configuration between the network former and modifiers [9].

Glass structure is a basic issue to understand the behavior of the material. The velocity of ultrasonic waves and hence the elastic moduli are particularly suitable for characterizing glasses as a function of composition [10]. Ultrasonic investigation of solids is gaining much importance nowadays and interest in glasses has rapidly increased because of improving information technology. Elastic properties are very informative about the structure of solids and they are directly related to interatomic potentials. In recent years, attention has been focused more on glassy materials in few of their larger optical nonlinearity and high optical quality with fast response time [11]. Ultrasonic tools are very important for characterizing materials because they have many applications in physics, chemistry, engineering, biology, food industry, medicine, oceanography, seismology, and so forth.

Borate glass is one of the most characteristic glasses having unique superstructure (SS) of intermediate range order (IRO) such as boroxol ring and tetraborate. The conversion between threefold coordinated boron with the addition of modifier oxides is found in short-range order (SRO) and becomes one of the main factors bringing about the variety of SSs [12]. The ability of boron to exist in three- and four-oxygen coordinate environments and the high strength of covalent B–O bonds enable borate to form stable glasses. Coordination number and connectivity (Number of bridging bonds) usually determine the melting point and Poisson's ratio. Glasses having higher co-ordination number tend to lower the bond strength. A detailed investigation of coordination number is therefore needed to understand the structural properties of glasses. The thermal stability and the lattice vibrations within the glass systems have been related to the measurement of softening and Debye temperature. Boron oxide is one of the best glass formers and its structure consists of a sheet-like arrangement of boron-oxygen triangles connected at all corners to form a continuous network [13]. Physical properties of borate glasses can often be altered by the addition of a network modifier to the basic constituent. The commonly used network modifiers are the alkali and alkaline earth oxides. It was observed that the properties of borate glasses were modified with alkali and alkaline earth oxides. It was observed that the properties of borate glasses modified with alkali oxides showed nonlinear behavior when the alkali oxide was gradually increased [14]. The role of alkali oxide Na_2O in the B_2O_3 network is to modify the host structure through transformation of the structural units of the borate network from BO_3 to BO_4. The ability of boron atom to exist in three- and four-oxygen coordinated environments and the high strength of the (B–O) bond enable borates to form stable glasses.

The present work included preparation of two glass systems, namely, (i) MnO_2-B_2O_3-PbO (BML glasses system) and Na_2CO_3-B_2O_3-P_2O_5 (BSP glasses system) and its measurement of ultrasonic velocities (both longitudinal (U_L) and shear (U_S)) and density of the series. From these sample values, the elastic constants such as longitudinal modulus (L), shear modulus (S), bulk modulus (K), Young's modulus (E), Poisson's ratio (α), acoustic impedance $P(Z)$, microhardness (H), Debye temperature (θ_D), and thermal expansion coefficient (α_P) have been calculated, which give more vital information about the rigidity and structure of glasses. To add more support of our claims over the previous glass samples, we also carried out further studies like spectroscopic XRD and scanning electron microscope (SEM) to ascertain the validity of the structural elicitation of the prepared glass samples.

2. Materials and Methods

2.1. Sample Preparation. The chemicals used in the present research work were analytical reagent (AR) and spectroscopic reagent (SR) grade with minimum assay 99% being obtained

TABLE 1: Composition of glasses.

S. no.	Glass specimen	Composition in mol%
	System-I (B_2O_3-MnO_2-PbO)	
1	BML-1	60-10-30
2	BML-2	62-10-28
3	BML-3	64-10-26
4	BML-4	66-10-24
5	BML-5	68-10-22
6	BML-6	70-10-20
	System-II (B_2O_3-Na_2CO_3-P_2O_5)	
1	BSP-1	50-25-25
2	BSP-2	52-25-23
3	BSP-3	54-25-21
4	BSP-4	56-25-19
5	BSP-5	58-25-17
6	BSP-6	60-25-15

from Sd Fine Fhemicals, India, and E-Merck, Germany. The composition in mol% of (increase of B_2O_3 and decrease of PbO/P_2O_5 contents) glass specimen is listed in Table 1. The required amounts (approximately 20 g) in mol% of different chemicals in powder form were weighed using single pan digital balance (Model SHIMADZU AX 200) having an accuracy of 0.0001 g.

The homogenisation of the appropriate mixture of the component of chemicals was effected by repeating grinding using a pestle and mortar. The temperature controlled muffle furnace was gradually raised to a higher temperature at the rate of 100 K per hour and a glass structure was noticed for BML glass system at 1060 K and for BSP glass system 1080 K and eventually the molten glass melt was immediately poured on a heavy copper molding block having the dimension of 12 mm diameter and 6 mm length kept at room temperature. Then the glass samples were annealed at 400 K for two hours to avoid the mechanical strains developed during the quenching process. The two opposite faces of glass were highly polished to ensure a good parallelism. All glasses are cleaned with acetone to remove the presence of any foreign particles. The samples are prepared chemically stable and nonhygroscopic and such glass samples as system-I (BML) and system-II (BSP) are reported in plate-1 and plate-2.

2.2. Measurement of Ultrasonic Velocities in Glass Samples. The ultrasonic longitudinal and shear velocities of the glass specimen were determined by using the pulse-echo overlapping methods at room temperature by making use of 5 MHz X-cut and Y-cut transducers. These transducers act as both transmitters and receivers of the ultrasonic pulse. The transducers were brought into contact with each of the twelve samples by means of a couplant, in order to ensure that there was no air void between the transducer and the specimen.

By applying constant pressure on the probe the echo waveforms were obtained on the display unit and stored in the memory. Figures 2(a) and 2(b) show one of such echo waveforms obtained for longitudinal and shear waves (for BML and BSP systems).

TABLE 2: Values of density (ρ), longitudinal velocity (U_L), shear velocity (U_S), and elastic moduli of BML and BSP glass systems.

Name of the glass samples	Density ρ (Kg/m³)	Ultrasonic velocity (m/s)		Elastic moduli			
		Longitudinal velocity (U_L)	Shear velocity (U_S)	Longitudinal moduli (L) $\times 10^{-9}$ Nm^{-2}	Shear moduli (G) $\times 10^{-9}$ Nm^{-2}	Bulk moduli (K) $\times 10^{-9}$ Nm^{-2}	Young's moduli (E) $\times 10^{-9}$ Nm^{-2}
System-I BML glass systems (B₂O₃-MnO₂-PbO)							
BML-1	1304.19	4073.34	2463.94	21.63	7.91	11.08	19.17
BML-2	1324.64	4177.83	2472.42	23.12	8.08	12.32	19.92
BML-3	1353.70	4403.66	2533.10	26.25	8.68	14.66	21.76
BML-4	1359.31	4526.08	2610.90	27.84	9.27	15.49	23.17
BML-5	1372.54	4550.28	2661.44	28.42	9.72	15.55	24.44
BML-6	1384.01	4615.91	2693.84	29.48	10.00	16.12	25.35
System-II BSP glass systems (B₂O₃-Na₂CO₃-P₂O₅)							
BSP-1	1707.92	5145.35	2875.35	45.22	14.1	26.38	34.25
BSP-2	1743.67	5256.24	3055.06	48.17	16.2	26.47	40.52
BSP-3	1756.72	5291.56	3153.57	49.19	17.4	25.89	42.76
BSP-4	1762.19	5379.13	3186.69	50.98	17.7	27.62	43.39
BSP-5	1779.83	5631.24	3222.86	56.44	18.4	31.79	46.43
BSP-6	1792.60	6034.69	3296.34	65.28	1.95	39.31	50.13

2.3. Density of Glass Samples. The density of the glass samples was measured using relative measurement technique. Benzene was used as a buoyant liquid. The glass samples were weighed both in air and after immersing in benzene at 303 K. The weight of the glass samples was measured in a single pan with an accuracy of 0.0001 g. The density was calculated using the formula

$$\rho = \rho_B \frac{W_1}{W_1 - W_2}, \qquad (1)$$

where W_1 and W_2 are the weights of the glass samples in air and in benzene and ρ_B is the density of the benzene at 303 K.

3. Result and Discussion

3.1. Ultrasonic Study. The experimental values of density and ultrasonic velocity (both longitudinal (U_L) and shear (U_S)) of the different glass specimens with respect to change in mol% of PbO and P₂O₅ are listed in Table 2. The calculated elastic moduli such as longitudinal modulus (L), shear modulus (G), bulk modulus (K), and Young's modulus (E) are reported in Table 2. The perusal of Table 3 shows the value of Poisson's ratio (α), acoustic impedance (Z), microhardness (H), Debye's temperature (θ_D), and thermal expansion coefficient (α_P) for the two glass systems (BML and BSP). Figure 3 shows the variation of Debye temperature for the BML glasses with the composition of PbO (mol%). In a similar manner, the variation of the previous parameters with the composition of P₂O₅ (mol%) for BSP glasses is depicted in Figure 3.

Density is an effective tool to explore the degree of structural compactness, modification of the geometrical configuration of the glass network, change in coordination, and the variation of the dimension of the interstitial holes [15]. In our present study, the density of glasses (both BML and BSP glass systems) exhibits continuous increases with increase in

mol% B₂O₃. From Table 2, one can notice the monotonic increase of density with increase of B₂O₃ concentration which can be attributed to the structure changes occurring in the coordination of boron glass network. The structure of crystalline as well as amorphous B₂O₃ is made up of planar (BO₃) triangles units as reported by Bray [16]. In amorphous B₂O₃, most of these triangles are arranged into boroxol rings in which three oxygen are part of the rings and three oxygen are outside the ring. These rings are randomly interconnected through the loose (BO₃) units. Due to the addition of B₂O₃, the three coordinated triangle boron (BO₃) units are converted to four coordinated boron tetrahedra (BO₄) and thus the network dimensionality and connectivity increase. This would lead to efficient packing and compactness in the structure, which will be reflected in increase of density of the glass systems. In the present study, the higher values of density are found in BSP glass systems and hence the magnitudes of density values are in the order BSP glass system > BML glass system.

One can notice from Table 2 that the longitudinal (U_L) and shear velocities (U_S) increase linearly with increase in mol% B₂O₃ in both BML and BSP glass systems. It is observed that the rate of increase of U_L is greater than that of U_S. Bahatti and Singh [4] suggested that the addition of PbO and P₂O₅ to B₂O₃ changes the structure to a rigid and compact structure due to change in the coordination number of boron from 3 to 4, which causes increase in ultrasonic velocities. The variation of ultrasonic velocities and elastic moduli can be explained on the basis of the structural consideration of glassy network. From Table 2, the values of longitudinal ultrasonic velocity (U_L) increases from 4073.34 to 4615.91 ms^{-1} for BML glass and 5145.35 to 6034.69 ms^{-1} for BSP glass systems. Similarly, shear velocity (U_S) increases from 2463.94 to 2693.84 ms^{-1} for BML glass and 2875.35 to 3296.34 ms^{-1} for BSP glass systems on increasing the Borate content.

TABLE 3: Values of Poisson's ratio (σ), acoustic impedance (Z), microharness (H), Debye temperature (θ_D), and thermal expansion coefficient (α_p) of BML and BSP glass systems.

Name of the glass samples	Poisson's ratio (σ)	Acoustic impedance Z ($\times 10^7$ Kg·m²s⁻¹)	Microhardness H ($\times 10^{-9}$ Nm⁻²)	Debye temperature θ_D (K)	Thermal expansion coefficient α_P (K⁻¹)
System-I BML glass systems (B_2O_3- MnO_2-PbO)					
BML-1	0.211	0.5312	2.6390	144.92	94488.1
BML-2	0.230	0.5534	1.4100	146.49	96912.4
BML-3	0.252	0.5961	1.4300	151.54	102151.5
BML-4	0.250	0.6152	1.5370	156.37	104991.7
BML-5	0.257	0.6245	1.5750	159.72	105553.1
BML-6	0.262	0.6388	1.5935	162.12	107075.7
System-II BSP glass systems (B_2O_3-Na_2CO_3-P_2O_5)					
BSP-1	0.212	0.8782	2.7130	197.68	119358.7
BSP-2	0.244	0.9165	2.9830	210.31	121931.4
BSP-3	0.224	0.9295	3.0145	211.38	122750.8
BSP-4	0.238	0.9479	3.0600	223.36	124782.4
BSP-5	0.256	1.0022	3.0710	224.03	130631.4
BSP-6	0.287	1.0817	3.1765	230.16	139991.4

According to Higazy and Bridge [17] the longitudinal strain changes directly with bond stretching force constant. On the other hand, as reported by Reisfeld et al. [18], the shear strain changes with bond bending force constant. The increasing trend of both ultrasonic velocities may be attributed to the increase in rigidity of the glass network.

Glass is considered as an elastic substance and it can be characterized through a modulus of elasticity [19, 20]. This modulus increases as the lengthening of certain applied stress diminishes. This will be the case: if the glass structure is rigid and therefore contains the fewest possible nonbridging oxygen atoms with increase of B_2O_3 content, the structure becomes more rigid and so this leads to the increases in density as well as ultrasonic velocities [21–23]. The increase in velocities is attributed to the increases in rigidity of the glass network [21, 24]. Addition of B_2O_3 with PbO and P_2O_5 increases the elastic moduli such as the longitudinal modulus (L), shear modulus (G), bulk modulus (K), and Young's modulus (E) in both glass (BML and BSP) systems. The elevation of such elastic moduli in both glass systems can be interpreted on the basis of the structural consideration of borate network. Further, the addition of P_2O_5 and PbO with B_2O_3 network creates (BO_4) units and this leads to increase in the network dimensionality and connectivity of the network. The increasing trend of elastic moduli in all two glass systems indicates resistance to deformation and is most probably due to the presence of number of layers of covalent bonds. The continuous increase of Poisson's ratio as well as microhardness in the present study reveals the absence of nonbridging oxygen (NBO) and this causes the formation of glassy network. Rajendran et al. [25] observed that increase in elastic moduli with addition of B_2O_3 confirms the rigidity and hence the formation of stronger structural building units in this glassy network.

Poisson's ratio is an effective tool in exploring the degree of cross-link density of the glass network and its magnitude increases the cross-link density, and it is the ratio of transverse and linear strains for a linear stress. According to Bridge et al. [26] glass networks having a connectivity of two (zero cross-link density) have Poisson's ratio of = 0.4. Glass networks having connectivity of three (one cross-link density) have Poisson's ratio of = 0.3, while networks having a connectivity of four (two cross-link densities) have Poisson's ratio of = 0.15. It can be seen from the present study that Poisson's ratio increases from 0.211 to 0.262 for BML glass systems and from 0.212 to 0.287 for BSP glass systems (Figures 1 and 5; Tables 4 and 5). The perusal of Table 3 exhibits the values of Poisson's ratio obtained in this study suggesting that the network of these glasses has two-dimensional structure. The increase in Poisson's ratio with increasing content of B_2O_3 is attributed to increase in the average cross-link density of the glass as proposed by Higazy and Bridge [17]. The increasing values of Poisson's ratio are attached to strengthening of network linkage and hardening of the network structure. This has been attributed to the fact that the concentration of bonds resisting a transverse deformation decreases in that order [27].

Microhardness expresses the stress required to eliminate the free volume (deformation of the network) of the glass. The present study of increasing value of microhardness in all the two glass systems studied indicates the increase in rigidity of the glass. The softening point is temperature at which viscous flow changes to plastic flow. It determines the temperature stability of the glass. The higher the value of softening temperature, the greater is the stability of its elastic properties [28]. As seen from Table 3, the values of microhardness increase with B_2O_3 content. The increase of microhardness implies an increase in the rigidity of the glass system. The present investigation shows that the BSP

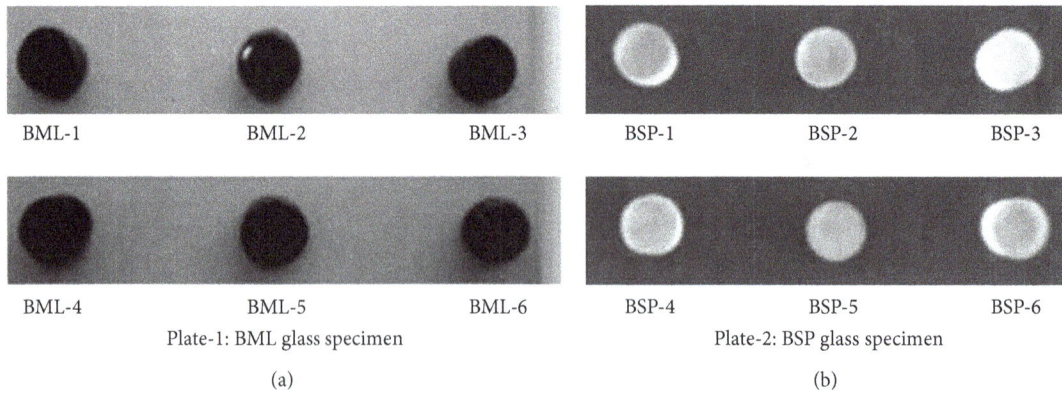

FIGURE 1: (a) and (b) Photograph for BML and BSP glass systems.

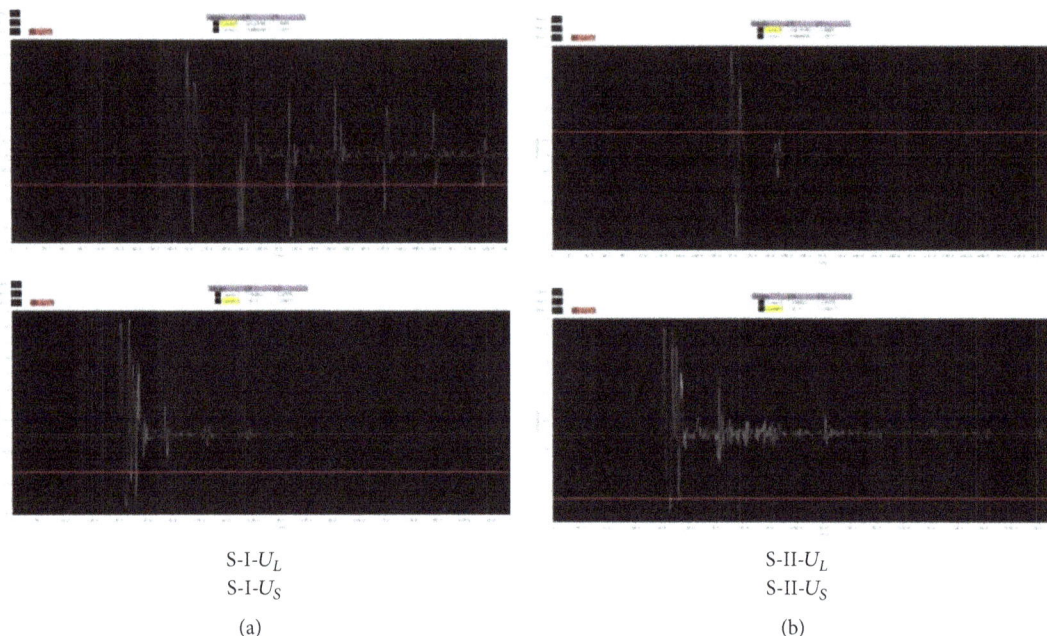

S-I-U_L
S-I-U_S

(a)

S-II-U_L
S-II-U_S

(b)

FIGURE 2: (a) Longitudinal and shear waveforms for BML-I glass system; (b) longitudinal and shear waveforms for BSP-I glass system.

glass systems possess the highest values of microhardness suggesting that BSP glasses are stronger than BML glass systems.

Debye's temperature represents the temperature at which nearly all modes of vibration in a solid are excited and its increasing trend implies an increase in the rigidity of the glass [28]. The perusal of Figure 3 describes the variation of Debye temperature with B_2O_3 content.

The gradual increase of Debye's temperature from 144.92 K to 162.12 K for BML glass systems and from 197.68 to 230.16 K for BSP glass systems indicates the increase in the rigidity of these glass systems. Such an enhancement of Debye's temperature is attributed to the increase in the number of atoms in the chemical formula of the glass and increase in the ultrasonic velocity [28]. This also further suggests the strengthening in the glass structure which is due to the creation of nonbridging oxygen ions [24]. In our present investigation, the higher values of Debye temperature

are reported in BSP glass system. The continuous increase of Debye temperature also suggests the compactness in the structure leading to increase in mean sound velocity [29].

The perusal of Table 3 exhibits the values of acoustic impedance (Z) and the thermal expansion coefficient of the two glass systems studied. The increasing trend of their values for both glass (BML and BSP) systems clearly support this and confirms the strengthening of the glass network in the system. This supports well our earlier conclusions derived from the previous parameters.

3.2. X-Ray Diffraction Study (XRD). X-ray diffraction studies were carried out on each glass sample to confirm the amorphous nature of prepared glasses. An X-ray diffractometer (PW1700: Philips Eindhoven, The New Netherlands) was used with CuK as a radiation source between 20° and 80°.

The prepared glass samples were washed gently in double-distilled water. The washed glass samples were dried at room

TABLE 4: System-I—B_2O_3-MnO_2-PbO (BML glass system).

Glass name	Peak's positions (cm^{-1})											
BMP-1	540.91	641.10	805.48	1021.55	1156.16	1194.52	1456.13	1646.37	2854.79	2926.03	3243.84	3426.50
BMP-2	541.39	697.29	876.71	1072.71	1156.16	—	1431.18	1645.70	2854.79	2915.07	3221.84	3428.01
BMP-3	541.84	641.10	—	1024.96	1156.16	1189.04	1448.39	1633.99	2854.32	2923.98	3243.84	3430.60
BMP-4	541.18	646.10	—	1021.48	—	1194.52	1451.49	1632.16	2854.79	2922.24	3232.88	3415.72
BMP-5	542.87	645.63	821.92	1059.75	—	1193.22	1457.52	1637.84	2854.79	2920.55	3227.57	3397.91
BMP-6	541.36	—	—	1062.96	—	—	1441.72	1631.96	2854.19	2923.87	—	3427.05

TABLE 5: System-II—B_2O_3-Na_2CO_3-P_2O_5 (BSL glass system).

Glass name	Peak's positions (cm^{-1})					
BSP-1	685.50	1025.04	1357.27	2849.32	2909.59	3408.38
BSP-2	679.45	—	1342.75	2854.79	2926.03	3441.50
BSP-3	676.52	1059.84	1355.92	2854.79	2923.52	3433.21
BSP-4	675.91	1062.56	1350.71	2854.79	2924.60	3431.07
BSP-5	686.37	1060.61	1351.76	2849.32	2920.55	3433.66
BSP-6	676.64	1028.97	1350.85	2843.84	2915.07	3431.14

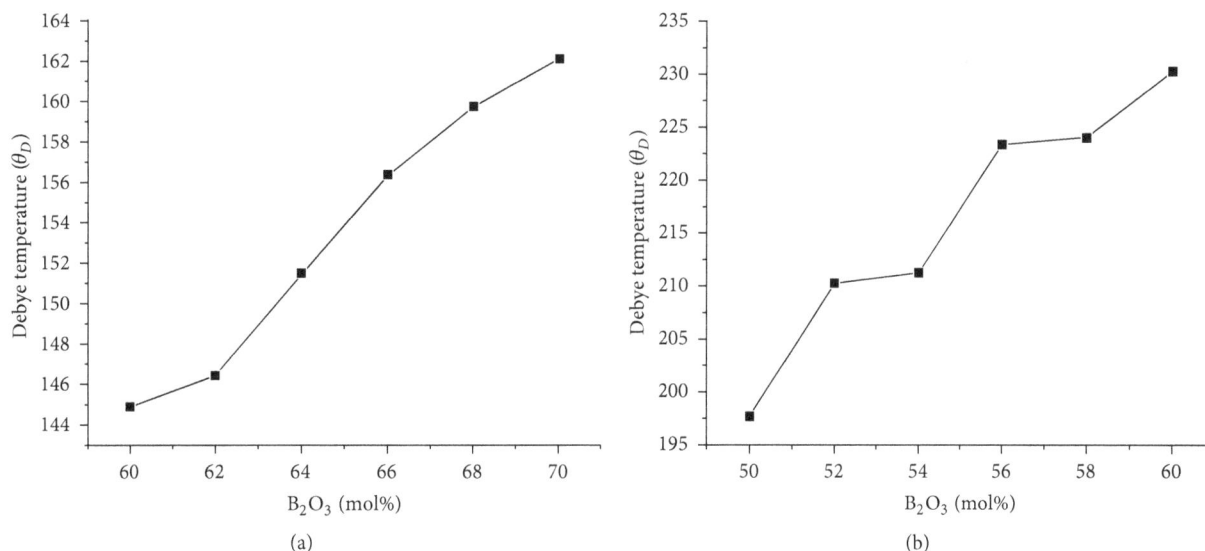

FIGURE 3: Debye temperature (θ_D) with variation of B_2O_3 (mol%) for system I and system II.

temperature, ground, and then used to obtain the XRD patterns as discussed previously. The XRD spectrum shows a number of lines of evidence of unmelted (W) crystalline particles in quarchal glasses. Such XRD spectrums of one of the BML-I and BSP-I glasses systems are shown in Figures 4(a) and 4(b).

3.3. Spectroscopic Study: FTIR Interpretation. The infrared transmission spectra of the glasses are measured at room temperature in the wavenumber range 400–4000 cm^{-1} by a Fourier transform computerised infrared spectrometer type (Perkin Elmer spectrometer). The prepared glass samples are mixed in the form of fine powder with KBr in the ratio of 1 : 100 mg glass powder : KBr, respectively. The weighed mixtures are then subjected to pressure of 150 kg/cm^2 to produce homogeneous pellets. The infrared transmission measurements are measured immediately after preparing the pellets. Normalized FTIR absorption spectra of B_2O_3-Na_2CO_3-P_2O_5 and B_2O_3-PbO-MnO_2 are shown in Figures. The goal of vibrational analysis to find vibrational modes is connected with content of glass samples. The observed infrared transmission peak around at 540–541.87 cm^{-1} is attributed to stretching vibrations of Pb–O [30, 31]. The composition of mole percentage of PbO is in decreasing trend giving rise to vibrational bands which are slightly shifted to lower wavenumber (Table 6). These vibrational bands do not appeared in the system II because of absence of PbO content. From FTIR and mole percentage results, the role of PbO in this glass structures has been explored. Infrared transmission spectra indicate that when composition of PbO is in the range

TABLE 6

Wavenumber (cm^{-1})	Band assignment
540.91–542.87	Deformation vibrations of the Pb–O.
641.10–697.29	B–O–B bonding vibration of BO$_3$ groups and stretching vibration of MnO.
805.48–876.71	Stretching vibration of B–O–B lineages in borate network.
1021.48–1072.71	B–O bond stretching of the tetrahedral BO$_4$ units.
1156.16	BO$_4$ stretching vibration.
1189.04–1646.37	B–O symmetric stretching vibration of BO$_3$ units in metaborate, paraborate, and orthoborate groups.
1025.04–1062.56	PO$_3$ symmetric stretching.
1342.75–1357.27	Symmetric stretching of P–O–P group.
1631–1646.3	O–H bending vibrations of water.
2854.32–3415.72	O–H stretching vibrations of water.

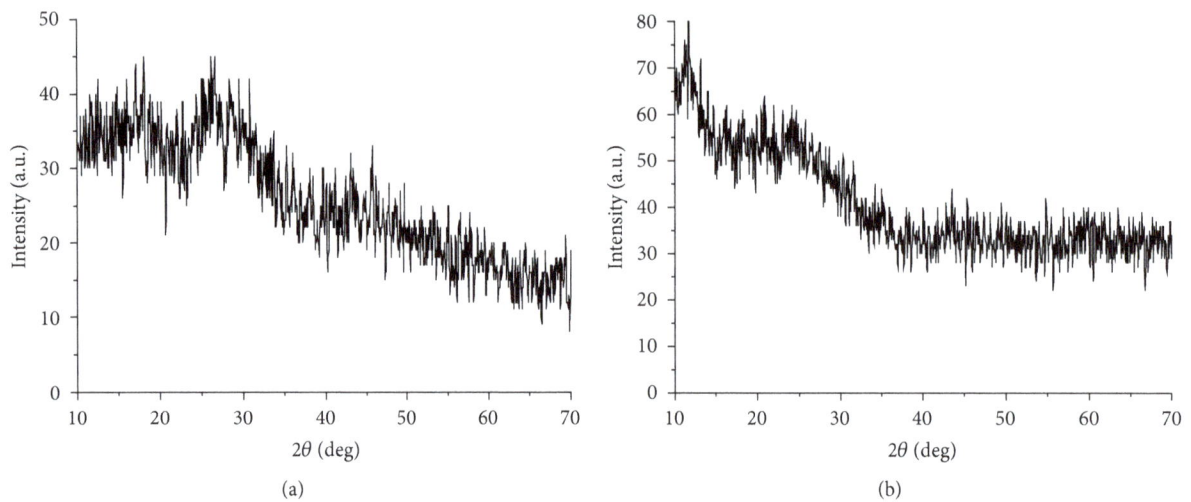

FIGURE 4: (a) X-ray diffractogram for BML-I glass system-I; (b) X-ray diffractogram for BSP-I glass system-II.

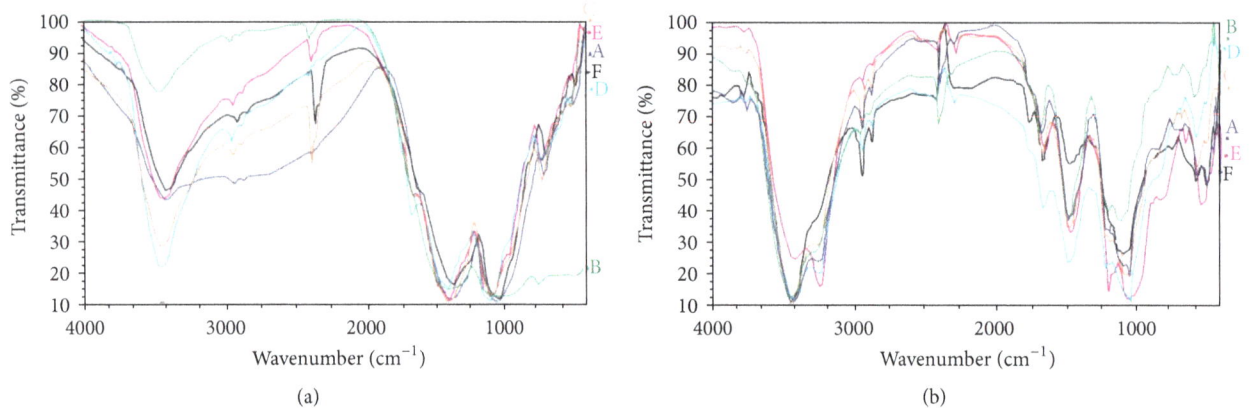

FIGURE 5: (a) FTIR spectrum for BML glass system; (b) FTIR spectrum for BSP glass system.

20% < x < 30%, the Pb^{2+} ions that are in the borate glass act as a network modifier [31, 32]. In general, the vibrational modes of the borate network are active mainly in three regions: the first region that lies between 600 and 800 cm^{-1} is due to bending vibration of various borate segments, the second region that lies between 850–1200 cm^{-1} is due to stretching vibrations of tetrahedral BO$_4$ units, and the third region that lies between 1200 and 1600 cm^{-1} is due to stretching vibrations of B–O in BO$_3$ triangles. In the present study, the peak observed at around the range 641.1–697.29 cm^{-1} is assigned as B–O–B bending vibration of borate groups which are composed of BO$_4$ and BO$_3$ units. The stretching normal

FIGURE 6: (a) SEM system-I for BML glass specimen; (b) SEM system-II for BSP glass specimen.

mode of MnO_2 is also observed in the previous region; these two vibrational modes overlap with each other [32–34]. The absorption peak identified at the region 805.48–876.71 cm^{-1} is due to stretching vibration of BO_4 groups [35, 36]. The infrared spectra of the existing structural borate groups and their rearrangement are observed to be slightly affected by the change of glass composition.

In system I, the bands that appeared in the region 1021.48–1072.71 cm^{-1} are assigned as symmetric stretching vibration of BO_4 tetrahedral units [34, 37]. But in system II, the absorption peaks at around the range 1025.04–1062.56 cm^{-1} are due to symmetric stretching vibration of PO_3 groups.

The previous two stretching vibration modes overlap with each other [38, 39]. The absorption peak identified at 1156.1 cm^{-1} is due to presence of BO_4 stretching vibration. The spectrum of borate glass exhibited vibrational bands at the region 1189.04–1646.37 cm^{-1} which is due to increasing percentage of B_2O_3 in both systems. The [40, 41] B–O asymmetric stretching vibrations of BO_3 units in metaborate, orthoborate, and paraborate appeared at the region 1431.18–1457.52 cm^{-1} [42, 43]. The vibrational bands observed at the region 1342.75–1357.27 cm^{-1} are assigned as antisymmetrical stretching vibrations of PO_2 groups. This region may also consist of bands due to P=O stretching vibrations.

This absorption is typically metaphosphate chains. Due to decreasing percentage of P_2O_5, in system II, the vibrational bands are slightly shifted to lower wavenumber [44, 45]. In particular, the atmospheric moisture is easily absorbed by the sample or by the pellet causing the appearance of IR band belonging to H_2O molecules. The peak at around the region $1631.96–1646.37$ cm^{-1} indicates a change from BO_3 triangles to BO_4 tetrahedra and this peak may also be assigned to the bending modes of molecular water. The IR bands from $2854.32–3415.72$ cm^{-1} are due to superposition of asymmetric and symmetric stretching vibration of OH. In system I, with the liberation of CO_2, the Na_2CO_3 is converted into Na_2O because of its glass transition temperature at $1050°$C. Thus the spectral absorption band of Na_2CO_3 is very feeble. IR absorption spectra of the studied samples show various vibrational bands which are characteristics of triangular and tetrahedral borate units together with small bands due to water OH and BOH. The FTIR studies confirm the presence of various contents of borate network. These vibrational assignments are in good agreement with literature values.

3.4. Scanning Electron Microscopic Study. Scanning electron microscope (SEM) is used in a wide range of fields as a tool for observing surfaces at nanometer level. The main advantage of SEM is its ability to study the heterogeneity of glass composites to visualize various mineral components and their relation in terms of overall micrographic and texture. SEM covers the observation from the fine structures of a surface of specimen to elemental analysis of a microarea without destroying the specimen. The SEM photographs of the samples are given in Figures 6(a) and 6(b). It is obtained that different sized grain particles are distributed. The particle size seems to vary in each micrograph. It consists of density packed grains free from holes.

The particles are angular or spherical in nature. From these figures, it is known that large particles may be monomineralic, but a majority is likely to be composite. Some sphere-like agglomerates found spreading are the glass surfaces, due to deposition of amorphous apatite. This suggests that during glass formation, the presence of clusters composed of fibers is formed. Also, the glass may consist of large particles, aggregates, agglomerates, and clusters, which clearly explain the surface of morphology of glass samples.

4. Conclusion

(i) The gradual increase in density with mol% B_2O_3 of the glass specimen indicates the dependence of density on weight of metal atom in the network modifier. The magnitude of the density is in the order BSP > BML.

(ii) The ultrasonic velocities (U_L and U_S) of BML and BSP glasses vary linearly with the addition of glass former (B_2O_3) and the magnitude is in the order BSP > BML.

(iii) The evaluated acoustical, elastic, and mechanical properties of BML and BSP glasses throw light on rigidity and compactness in structural network.

(iv) The observed increasing trend of microhardness and Poison's Ratio for both glass systems (BML and BSP) indicates that BSP glass system is stronger than BML system.

(v) The BSP glass possesses higher rigidity and compactness in structural network over the BML.

(vi) The functional groups present in the glass sample have been confirmed by FTIR spectral analysis. The topological aspects of the glass samples are reported from the SEM micrography.

References

[1] J. Philip, "Temperature dependence of elastic and dielectric properties of $(Bi_2O_3)1–X$, $(CUO)X$ oxide glasse," *Journal of Materials Science*, vol. 35, p. 229, 2000.

[2] F. A. Khalifa, H. A. El Batal, and V. Rajendran, "Behaviour of ultrasonic waves on chromium doped $XPbO(100X)B_2O_3$ glasses," *Egyptian Journal of Chemistry*, vol. 41, p. 329, 1998.

[3] S. P. Yawale, S. V. Pakade, and C. S. Adganokar, "New type of behavior in ultrasonic velocity and adiabatic compressibility of $Bi_2O_3–B_2O_3$ glass," *Acoustica*, vol. 81, p. 184, 1995.

[4] S. S. Bahatti and K. J. Singh, "Acoustical and mechanical properties of $xMnO.(0.9x)$ B_2O_3 $0.10Fe_2O_3$ glass system," *Journal of Pure and Applied Ultrasonics*, vol. 16, no. 3, p. 78, 1994.

[5] S. V. Pakade and S. P. Yawale, "Ultrasonic velocity and elastic constant measurement in some borate glasses," *Journal of Pure and Applied Ultrasonics*, vol. 18, p. 74, 1996.

[6] S. J. Jugan and R. Abraham, "Ultrasonic investigation on $Cao–B_2O_3–Al_2O_3–Na_2O$ and $Cao–B_2O_3–Al_2O_3–Fe_2O_3$ glass system," *J. Pure Appl.Ultrasonic*, vol. 19, p. 32, 1977.

[7] M. Kodamma, "Ultrasonic velocity in sodium borate glasses," *Journal of Materials Science*, vol. 26, no. 15, pp. 4048–4053, 1991.

[8] S. P. Yawale, S. V. Pakade, and C. S. Adgaonkar, "Ultrasonic velocity and absorption measurement in $xZno–(90–x)B_2O_3–10Bi_2O_3$ glasses," *Acoustica*, vol. 76, p. 103, 1992.

[9] A. P. Singh, A. Paul, and S. S. Bahatti, "Ultrasonic velocity and absorption measurement in bariumborate glasses and their elastic properties," *Indian Journal of Pure and Applied Physics*, vol. 483, 1990.

[10] M. S. Gaafar, Y. B. Saddeek, and L. Abd El-Latif, "Ultrasonic studies on alkali borate tungstate glasses," *Journal of Physics and Chemistry of Solids*, vol. 70, no. 1, pp. 173–179, 2009.

[11] A. Nishara Begum and V. Rajendran, "Structure investigation of $TeO_2–BaO$ glass employing ultrasonic study," *Materials Letters*, vol. 61, no. 11-12, pp. 2143–2146, 2007.

[12] T. Yano, N. Kunimine, S. Shibata, and M. Yamane, "Structural investigation of sodium borate glasses and melts by Raman spectroscopy. I. Quantitative evaluation of structural units," *Journal of Non-Crystalline Solids*, vol. 321, no. 3, pp. 137–146, 2003.

[13] G. D. Chryssikos, M. S. Bitsis, J. A. Kapoidsis, and E. I. Kamitsos, "Vibrational investigation of lithium metaborate-meta aluminate lasses and crystal," *Journal of Non-Crystalline Solids*, vol. 217, p. 278, 1997.

[14] A. H. Verhoef and H. W. den Hartog, "Infrared spectroscopy of network and cation dynamics in binary and mixed alkali borate glasses," *Journal of Non-Crystalline Solids*, vol. 182, no. 3, pp. 221–234, 1995.

[15] V. R. Roma, "Elastic properties of the lead containing bismuth tellurite glasses," in *Proceedings of the 15th World Conference on Non-Destructive Testing*, October 2000.

[16] P. J. Bray, *The Structure of Glass*, edited by E. Poraikoshits, Consulttant Bureau, New York, NY, USA, 1996.

[17] A. A. Higazy and B. Bridge, "Elastic constants and structure of the vitreous system Co_3O_4–P_2O_5 ," *Journal of Non-Crystalline Solids*, vol. 72, no. 1, pp. 81–108, 1985.

[18] R. Reisfeld, L. Boehm, Y. Eckstein, and N. Lieblich, "Multiphonon relaxation of rare earth ions in borate, phosphate, germanate and tellurite glasses," *Journal of Luminescence*, vol. 10, no. 3, pp. 193–204, 1975.

[19] H. B. Senin, H. A. A. Sidek, and G. A. Saunders, "Elastic behavior of terbium metaphosphate glasses under high pressures," *Australian Journal of Physics*, vol. 47, p. 795, 1994.

[20] H. A. A. Sidek, S. P. Chow, Z. A. Talib, and S. A. Halim, "Formation and electric behavior of lead-magnesium chlorophospate glasses," *Turkish Journal of Physics*, vol. 28, p. 67, 2004.

[21] Y. B. Saddex, "Ultrasonic study and physical properties of some borate glasses," *Materials Chemistry and Physics*, vol. 83, p. 222, 2004.

[22] Y. B. Saddex, "Efect of B_2O_3 on the structure and properties of tungsten-tellurite Glasses," *Philosophical Magazine*, vol. 89, p. 41, 2009.

[23] Y. B. Saddex, "Structure and acoustical studies of lead sodium borate glasses," *Journal of Alloys and Compounds*, vol. 467, p. 14, 2009.

[24] Y. B. Saddex, "Elastic properties of Gd_3-doped tellurovanadate glasses using pulse-echo technique," *Materials Chemistry and Physics*, vol. 91, p. 146, 2005.

[25] V. Rajendran, N. Palanivelu, H. A. El-Batal, F. A. Khalifa, and N. A. Shafi, "Effect of Al_2O_3 addition on the acoustical properties of lithium borate glasses," *Acoustics Letters*, vol. 23, no. 6, pp. 113–121, 1999.

[26] B. Bridge, N. D. Patel, and D. N. Waters, "On the elastic constant and structure of the pure inorganic oxide," *Physica Status Solidi A*, vol. 77, p. 655, 1983.

[27] K. V. Damodaran and K. J. Rao, "Elastic properties of phospho-tungstate glasses," *Journal of Materials Science*, vol. 24, p. 2380, 1989.

[28] M. A. Sidkey, A. Moneim, and L. Latif, "Ultrasonic studies on ternary Te_2O_2–V_2O_5–Sn_2O_3 glasses," *Materials Chemistry and Physics*, vol. 61, p. 103, 1999.

[29] K. V. Damodaran and K. J. Rao, "Elastic properties of Alkali phosphomolbybate glasses," *Journal of the American Ceramic Society*, vol. 72, no. 4, pp. 533–539, 1989.

[30] T. Satyanarayana, I. V. Kityk, M. Piasecki et al., "Structural investigations on PbO–SB_2O_3–B_2O_3: CoO glass ceramics by means of spectroscopic and dielectric studies," *Journal of Physics Condensed Matter*, vol. 21, no. 24, Article ID 245104, 2009.

[31] G. P. Singh, P. Kaur, S. Kaur, and D. P. Singh, "Role of V_2O_5 in structural properties of V_2O_5–MnO_2–Pbo–B_2O_3 glasses," *Materials Physics and Mechanics*, vol. 12, p. 58, 2011.

[32] G. P. Singh, P. Kaur, S. Kaur, D. Arora, P. Sing, and D. P. Singh, "Density and FTIR studies of multiple transition metal doped Borate Glass," *Materials Physics and Mechanics*, vol. 14, p. 31, 2012.

[33] G. D. Khattak and A. Mekki, "Structure and electrical properties of SrO-borovanadate $(V_2O_5)0.5(SrO)0.5$–$y(B_2O_3)y$ glassses," *Journal of Physics and Chemistry of Solids*, vol. 70, no. 10, pp. 1330–1336, 2009.

[34] H. Doweidar and Y. B. Saddeek, "FTIR and ultrasonic investigations on modified bismuth borate glasses," *Journal of Non-Crystalline Solids*, vol. 355, no. 6, pp. 348–354, 2009.

[35] P. Pascuta, R. Lungu, and I. Ardelean, "FTIR and Raman spectroscopic investigation of some strontium-borate glasses doped with iron ions," *Journal of Materials Science*, vol. 21, no. 6, pp. 548–553, 2010.

[36] I. Ardelean and S. Cora, "FT-IR, Raman and UV-VIS spectroscopic studies of copper doped Bi_2O_3. B_2O_3 glass matix," *Journal of Materials Science*, vol. 19, no. 6, pp. 584–588, 2008.

[37] C. N. Reddy, V. C. V. Gowda, and R. P. S. Chalaradhar, "Elastic properties and structural studies on lead-boro-vanadate glasses," *Journal of Non-Crystalline Solids*, vol. 354, p. 32, 2008.

[38] A. Shaim, M. Et-Tabirou, L. Montagne, and G. Palavit, "Role of bismuth and titanium in Na_2O–Bi_2O_3–TiO_2–P_2O_5 glasses and a model of structural units," *Materials Research Bulletin*, vol. 37, no. 15, pp. 2459–2466, 2002.

[39] A. H. Khafagy, "Infrared and ultrasonic investigations of some $[(MnO_2)x$–$(P_2O_5)100$–$x].1$ wt % Ncl_2O_3 glasses," *Physica Status Solidi A*, vol. 186, no. 1, p. 105, 2011.

[40] G. P. Singh, P. Kaur, S. P. Kaur, and D. P. Singh, "Role of V_2O_5 in structural properties of V_2O_5–MnO_2–Pbo–B_2O_3 glasses," *Materials Physics and Mechanics*, vol. 12, pp. 58–63, 2011.

[41] P. Y. Shih, S. W. Yung, and T. S. Chin, "FTIR and XPS studies of P_2O_5–Na_2O–CuO glasses," *Journal of Non-Crystalline Solids*, vol. 244, no. 2, pp. 211–222, 1999.

[42] E. Culea, L. Pop, M. Bosca, V. Dan, P. Pascuta, and S. Rada, "Structural and physical characteristics of $xGd_2O_3(100$–$x)[Bi_2O_3 B_2O_3]$ glasses," *Journal of Physics*, vol. 182, no. 1, Article ID 012062, 2009.

[43] P. Pasuta, M. Bosca, and S. Reda, "FTIR spectro scopic study of Gd_2O_3–Bi_2O_3–B_2O_3 glasses," *Journal of Optoelectronics and Advanced Materials*, vol. 10, pp. 2416–2419, 2008.

[44] G. D. Khattak and A. Mekki, "Structure and electrical properties of SrO-borovanadate $(V_2O_5)0.5(SrO)0.5$–$y(B_2O_3)y$ glassses," *Journal of Physics and Chemistry of Solids*, vol. 70, no. 10, pp. 1330–1336, 2009.

[45] D. Souri, "Dsc and FTIR spectra of Tellurite vanadate glasses containg Molebdenum," *Middle-East Journal of Science Research*, vol. 5, no. 1, p. 44, 2010.

25

Mean-Field Approach to Dielectric Relaxation in Giant Dielectric Constant Perovskite Ceramics

Shanming Ke,[1,2] Peng Lin,[1,2] Haitao Huang,[3] Huiqing Fan,[4] and Xierong Zeng[1,2]

[1] College of Materials Science and Engineering, Shenzhen University, Shenzhen 518060, China
[2] Shenzhen Key Laboratory of Special Functional Materials, Shenzhen 518060, China
[3] Department of Applied Physics and Materials Research Center, The Hong Kong Polytechnic University, Hong Kong
[4] School of Materials Science and Engineering, Northwestern Polytechnical University, Xi'an 710072, China

Correspondence should be addressed to Shanming Ke; smke@szu.edu.cn and Haitao Huang; aphhuang@polyu.edu.hk

Academic Editor: Matjaz Valant

The dielectric properties of $CaCu_3Ti_4O_3$ (CCTO) and A_2FeNbO_6 (AFN, A = Ba, Sr, and Ca) giant dielectric constant ceramics were investigated in the frequency range from 1 Hz to 10 MHz. The relaxation properties can be perfectly described by a polaron model, indicating that the dielectric relaxation is intimately related to the hopping motion caused by localized charge carriers.

1. Introduction

As driven by the impetus of smaller and smaller feature size of devices in microelectronics, researchers are looking for the so-called high-k materials. Perovskites such as lead zirconate titanate (PZT) usually possess high dielectric constant of a few hundreds at room temperature [1, 2] and are widely used as capacitive components. The perovskite-based oxide $CaCu_3Ti_4O_{12}$ (CCTO, space group $Im3$) was reported to have a colossal dielectric constant (CDC) in the order of 10^4 which is nearly constant from 400 K down to 100 K but drops rapidly to less than 100 below 100 K [3]. A huge amount of work [3–7] has thereafter been carried out in attempts to understand the origin of the remarkable dielectric properties. It has been found that the temperature at which the step-like decrease in dielectric constant takes place strongly depends on the measuring frequency and roughly follows an Arrhenius behavior. Similar phenomenon has also been reported in a number of materials, such as A_2FeBO_6 (A = Ba, Sr, and Ca; B = Nb, Ta, etc.) [8], $La_{1-x}Ca_xMnO_3$ (LCMO) [9], $Pr_{0.7}Ca_{0.3}MnO_3$ (PCMO) [10], $LaCuLiO_4$ [11], $LaSrNiO_4$ [12], $TbMnO_3$ [13], and Li/Ti doped NiO [14]. The remarkable low temperature dielectric relaxation in many manganites, cuprates, and nickelates [9–12, 15–17] has been attributed to localized hopping of polarons between lattice sites with a characteristic timescale. Furthermore, this relaxation behavior clearly displays a freezing temperature following a glass-like process, suggesting that an electronic glass state is realized. It seems plausible to think of the influence of the electric state on the dielectric response of a solid.

Polaronic relaxation usually involves a variable range hopping (VRH) or a nearest-neighbor hopping conduction process. Polarons as the main sources contributing to the conduction of CCTO ceramics at the low temperature relaxation region has been confirmed [18]. Tselév et al. [19] found a power-law frequency dependent dielectric response in CCTO epitaxial thin films suggesting that the dielectric contribution came from hopping carriers. Based on these results, the low temperature dielectric response in CCTO has been attributed to polaron relaxation related localized charge carriers [20]. It should be mentioned that Maxwell-Wagner (MW) origin, such as the internal barrier layer capacitance (IBLC) should not be excluded in the explanation of the colossal dielectric constants, especially at high temperatures [4, 6, 7, 21]. Although, the MW type mechanisms which were resulted from the surface layer [21], IBLC [4], and/or contact electrode [6] do have an influence on the dielectric constant values of CDC materials, they nearly have no influence on the relaxation process at low temperature.

In this paper, we present a systematic study on the low temperature dielectric relaxation process of CCTO and A_2FeNbO_6 (AFN, A = Ba, Sr, and Ca) systems. A polaron relaxation model was proposed as a possible explanation to the low temperature dielectric relaxation in CCTO and AFN ceramics. It should be mentioned that MW model and polaron model are both related to the moving of space charge carriers. The difference is that MW model emphasizes the interface, but polaron model considers the space electrons/ions exist in the whole bulk materials. Although the polaron concept has been used to explain the dipolar effects induced by charge carrier hopping motions inside the CCTO grains [20, 22], a clear theory formula to describe polaron relaxation is absence. Based on polaron theory, Jonscher's law (universal dielectric response (UDR)) could be used to fit the frequency dependence of the dielectric permittivity. By using a mean-field approach, we will present a simplicity expression on the temperature dependence of the dielectric permittivity driven from polaron relaxation.

2. Experimental

Single phase CCTO and AFN (A = Ba, Sr, and Ca) ceramics were prepared through a conventional mixed oxide route, and the detailed processing parameters can be found elsewhere [23, 24]. The Wolframite method, using $FeNbO_4$ as the B-site precursor, has been used to synthesize AFN with no secondary iron-oxide phases.

X-ray diffraction (XRD) was conducted on sintered ceramics samples of CCTO and AFN. Data were collected on an automated diffractometer (X'Pert PRO MPD, Philips) with $Cu\,K_{\alpha1}$ radiation. The XRD results confirm that the sintered ceramics are single phase. The fracture surfaces of the ceramic pellets were examined by scanning electron microscopy (SEM, JSM-6335F, and JEOL). Micro-Raman spectral measurements were performed on a JY HR800 Raman spectrometer under backscattering geometry. An Argon ion laser was used as the excitation source with an output power of 15 mw at 488 nm. For dielectric measurement, top and bottom electrodes were made by coating silver paint on both sides of the sintered disks and followed by a firing at 650°C for 20 minutes. The temperature and the frequency dependences of the dielectric permittivity of the samples were measured by a frequency response analyzer (Novocontrol Alpha-analyzer) over a broad frequency range (1 Hz–10 MHz).

3. Results and Discussion

3.1. Crystal Structure and Microstructure. Figure 1 shows the room temperature XRD pattern of sintered CCTO ceramics. All the diffraction peaks can be indexed according to a cubic cell of space group $Im3$ [3]. Standard group theory analysis predicts that the Raman active modes are distributed among the irreducible representations as $2A_g + 2E_g + 4F_g$ [25]. Inset of Figure 1 shows the Raman spectrum at room

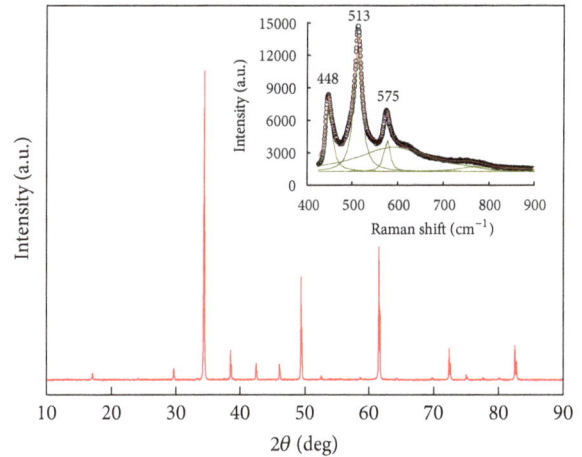

FIGURE 1: XRD pattern of CCTO ceramics. The inset shows the micro-Raman scattering spectra.

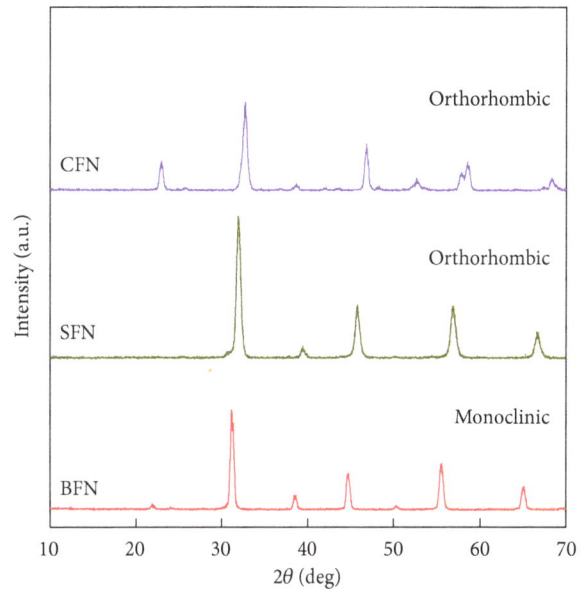

FIGURE 2: XRD patterns of BFN, SFN, and CFN ceramics.

temperature. The Raman lines at 448 and $512\,cm^{-1}$ have exact A_g symmetry, and a line of F_g symmetry is clearly pronounced at $575\,cm^{-1}$, which is assigned to the Ti–O–Ti antistretching mode of the oxygen octahedra [25]. The Raman spectrum also confirms a single cubic phase of our CCTO sample.

The XRD patterns of the Ba_2FeNbO_6 (BFN), Sr_2FeNbO_6 (SFN), and Ca_2FeNbO_6 (CFN) are given in Figure 2. The diffraction peaks of BFN samples can be indexed according to a monoclinic structure [26], while SFN and CFN samples show an orthorhombic structure with the space group *Pnma(62)* [27, 28]. Figure 3 displays typical SEM photographs of the fracture surface of CCTO and SFN ceramics. The average grain size of CCTO is found to be in the range of 2~8 μm, while SFN exhibits a morphology with the bimodal distribution of grain size. One has the grain size of several

(a)

(b)

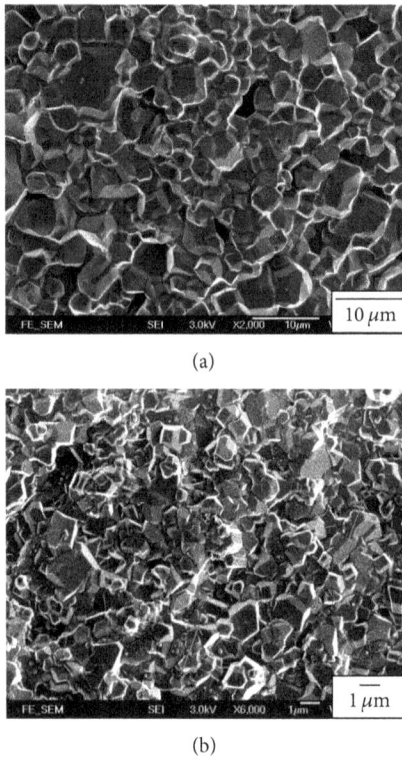

FIGURE 3: SEM images of the fracture surface of (a) CCTO and (b) SFN ceramics. The arrows denote nanograins in SFN.

micrometers, and the other has the smaller grain size in the order of several hundreds of nanometers. It should be noted that BFN and Ca_2FeNbO_6 (CFN) display a similar morphology to that of SFN, while BFN has a slightly larger average grain size than SFN and CFN.

3.2. Dielectric Behavior. Figure 4 shows the frequency dependences of the real $[\varepsilon'(f)]$ and imaginary $[\varepsilon''(f)]$ parts of the dielectric constant of CCTO, BFN, CFN, and SFN under different temperatures. The temperatures were chosen because they reveal the typical relaxation features of the mentioned ceramics. The high frequency region of CCTO and BFN shows the presence of a Debye-like dipolar relaxation process, and SFN and CFN show a generalized dipolar loss peak possessing a more serious departure from Debye, as summarized by Jonscher [29]. The peak frequency of $\varepsilon''(f)$ decreases with decreasing temperatures for all the samples, indicating the freezing in dipolar moment. The width of the $\varepsilon''(f)$ peak of AFN (A = Ba, Sr, Ca) is much larger than that of CCTO (as shown in Figure 4), which implies a broader distribution of relaxation time. Above the $\varepsilon''(f)$ peak frequencies, the high-frequency branches of CCTO and AFN follow the fractional power law [29]:

$$\varepsilon''(\omega) = \cot\left(\frac{n\pi}{2}\right)\varepsilon'(\omega) \propto (i\omega)^{n-1}, \qquad (1)$$

where ω is the angular frequency ($2\pi f$) and n is the exponential constant. The calculated exponent n from Figure 4 is 0.11, 0.12, 0.25, and 0.38 for CCTO, BFN, CFN, and SFN,

respectively. It should be noted that n is nearly independent of temperature for all these CCTO and AFN samples. Larger n means a much lower loss ($\varepsilon''(\omega)/\varepsilon'(\omega) = \cot(n\pi/2)$) and flatter frequency dependence than that in a Debye relaxation.

The peak frequency of the above relaxation process obeys the following Arrhenius law:

$$f = f_0 \exp\left(-\frac{E_a}{kT}\right), \qquad (2)$$

where f_0 is a prefactor and E_a is the activation energy. The Arrhenius fits for CCTO, BFN, SFN, and CFN ceramics are summarized in Figure 5, where E_a = 0.1, 0.18, 0.23, and 0.34 eV for CCTO, BFN, SFN, and CFN, respectively. The obtained values are very close to the previous reported ones for the same materials [4–7, 30].

Although, the Maxwell-Wagner type mechanism is always responsible for the dielectric permittivity of ceramics as an extrinsic source due to the existence of grain boundaries. This model cannot explain why the one with very similar microstructure and electrical resistivity (BFN, CFN, and SFN) shows so different relaxation regions and activation energies. In addition, in terms of the Maxwell-Wagner model, it is very difficult to explain the similar dielectric relaxation behavior appearing in the same temperature region for single crystals, polycrystalline ceramics, and epitaxial thin films whose microstructure is certainly different. In Figure 6, the imaginary part of dielectric permittivity ε'' of CCTO and BFN is shown as a function of the real part ε', namely, the Cole-Cole plot. From Figure 6, all the data at different temperatures collapse on an arc described by the Cole-Cole equation, showing universal scaling behavior in the dielectric response. We note that in the lower frequency region there is a relaxation relating to the extrinsic effect. This relaxation is probably due to a Schottky-type barrier existing in this ceramic [31].

3.3. Mean-Field Approach. In a previous study, It has been confirmed that localized charge carriers are the main sources contributing to the conduction of CCTO [18] and AFN [32] ceramics at low temperature relaxation region. Under an external electric field, the hopping motion of localized carriers like polarons gives rise to dipolar effect and sizable polarization contribution to the dielectric permittivity. Therefore a quantitative description of the polaron contribution to the dielectric response is highly demanding.

Ramirez et al. [33] have proposed a defect model as a possible explanation of the low temperature dielectric response of CCTO. In this model, isolated defects such as Cu vacancies produce a local disruption of the ideal cubic structure and then the defective regions relax between alternate equivalent configurations. However, high resolution experiments [33] have revealed that Cu vacancies most probably exist in the grain boundary regions, and thus they cannot be shown in the bulk response. Therefore, polarons are mainly responsible for the bulk dielectric relaxation. Essentially, it is reasonable to presume that polarons are isolated; then the cooperative effects are absent at low densities of polarons. The relaxation of polarons occurs at the rate γ determined by the energy

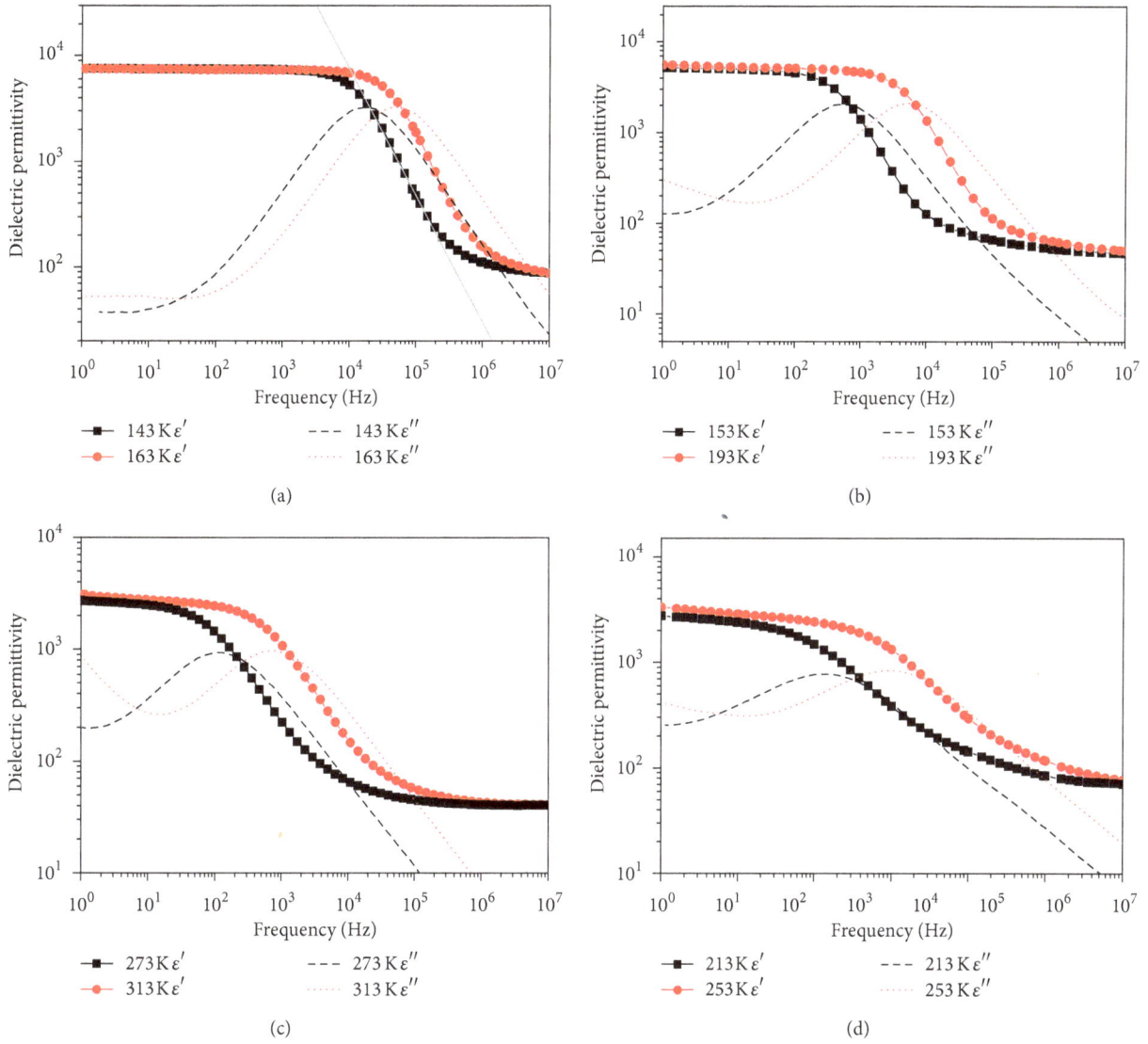

FIGURE 4: Frequency dependence of the real and imaginary parts of dielectric permittivity for (a) CCTO, (b) BFN, (c) CFN, and (d) SFN. The straight gray line in (a) is a guide to the eye.

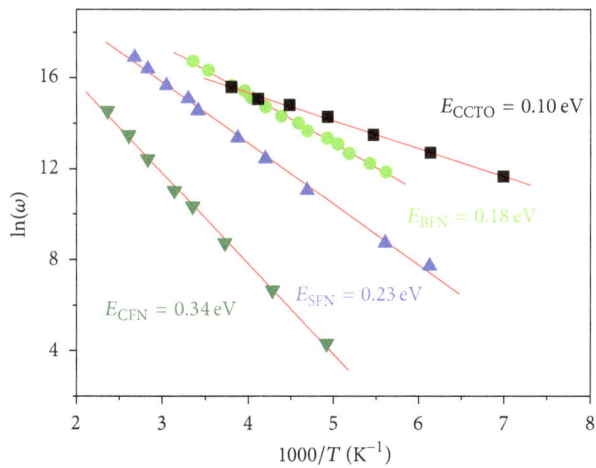

FIGURE 5: Arrhenius fit of the frequency dependence of the dynamic transition temperatures T_m for the dielectric relaxation in CCTO, BFN, SFN, and CFN.

barrier between alternate equivalent configurations and the temperature:

$$\gamma = \gamma_0 \exp\left(\frac{-\Delta}{T}\right), \quad \left(\Delta = \frac{E}{k_B}\right), \tag{3}$$

where γ_0 depends on the effective mass of the defect and is in the order of 10^{10}–10^{12} Hz. The local polarizability of a polaron is

$$\chi_p(\omega, T) = \frac{\mu_p \gamma}{-i\omega + \gamma}, \tag{4}$$

where μ_p is the static polarizability of a polaron. It should be noted that (3) and (4) could be obtained directly from [33]. As suggested by Ramirez et al. [33], a mean-field approximation

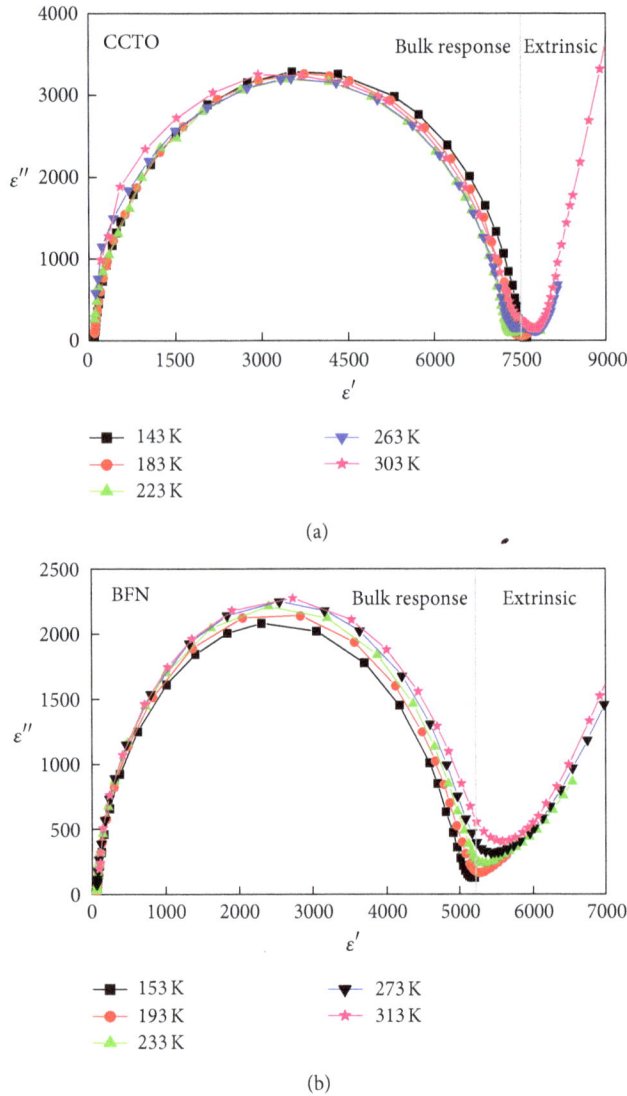

FIGURE 6: The imaginary part of dielectric permittivity ε'' of CCTO and BFN is plotted against the real part ε'.

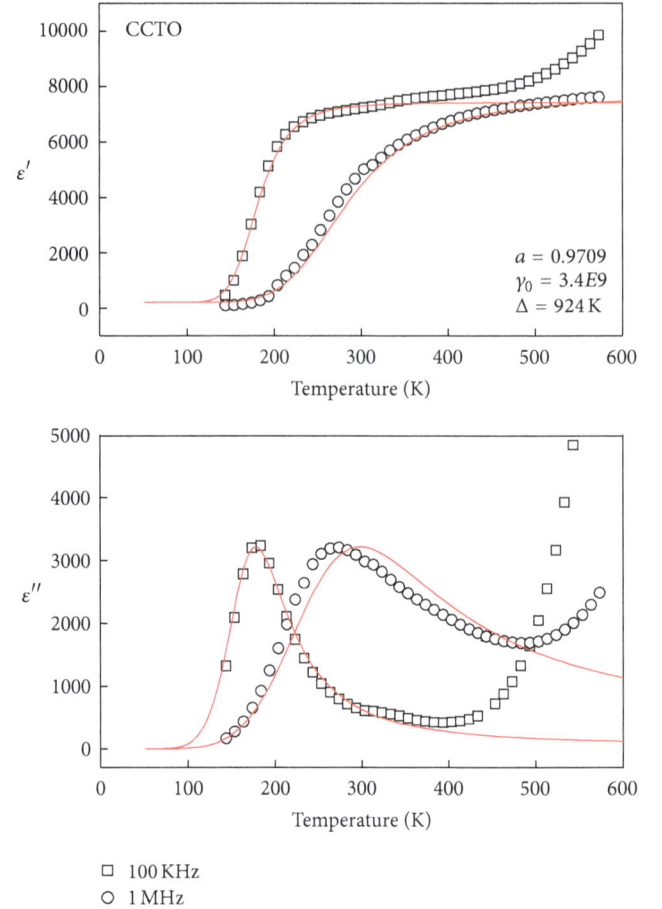

FIGURE 7: The temperature dependence of dielectric constant of CCTO. The solid lines are the best fits from polaron model (see text).

can be adopted and the dielectric response can be calculated by the Clausius-Mossotti equation:

$$\varepsilon(\omega, T) \approx \frac{\varepsilon_0}{1 - (4\pi/3)\rho\varepsilon_0\chi_p}$$
$$= \frac{\varepsilon_0}{1 - (4\pi/3)\rho\varepsilon_0\left(\mu_p\gamma/(-i\omega + \gamma)\right)}, \quad (5)$$

where ρ is the density of the polarons, and the details can be found in [34]. It is worth noting that the Clausius-Mossotti equation works best for gases and is only approximately true for liquids or solids, particularly for polar materials, or the dielectric constant is large. In addition, the limitations of the approximation in (5), which does not consider the variations around each polaron properly, should also be kept in mind.

Let $a = (4\pi/3)\rho\varepsilon_0\mu_p$; then we get

$$\varepsilon(\omega, T) \approx \frac{\varepsilon_0}{1 - a\gamma/(-i\omega + \gamma)}. \quad (6)$$

The real and imaginary parts of the dielectric permittivity are

$$\varepsilon'(\omega, T) = \frac{\varepsilon_0\left[(1-a)\gamma^2 + \omega^2\right]}{(1-a)^2\gamma^2 + \omega^2},$$
$$\varepsilon''(\omega, T) = \frac{\varepsilon_0 a\gamma\omega}{(1-a)^2\gamma^2 + \omega^2}. \quad (7)$$

The measured temperature dependent $\varepsilon'(T)$ and $\varepsilon''(T)$ (see Figures 7, 8, and 9) for CCTO, BFN, and CFN are similar to those reported in previous work [3–7, 30]. The real part dielectric permittivity ε' reveals a pronounced step-like increase accompanied by a loss peak. The dielectric permittivity of CCTO, BFN, and CFN is fitted according to (7) by a least-square method. For CCTO measured at 100 kHz, the best fitted result was obtained with $a = 0.971$, $\gamma_0 = 3.4 \times 10^9$, and $\Delta = 924$ K. It is very interesting to note that the obtained barrier energy E (obtained from $\Delta = E/k_B$) is 0.08 eV, a value quite close to the characteristic

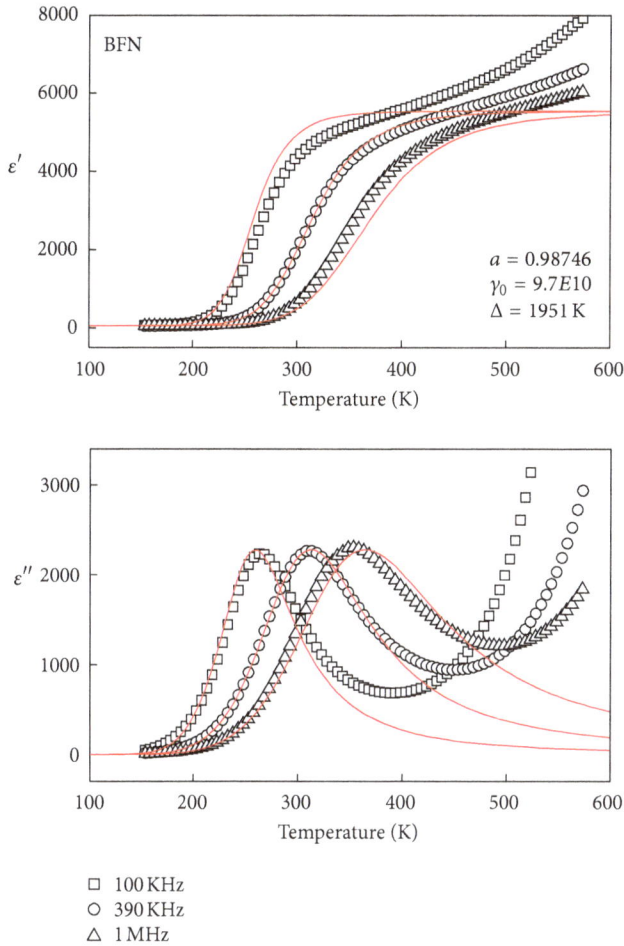

FIGURE 8: The dielectric constant of BFN is shown against temperature. The solid lines are the beat fits of polaron model.

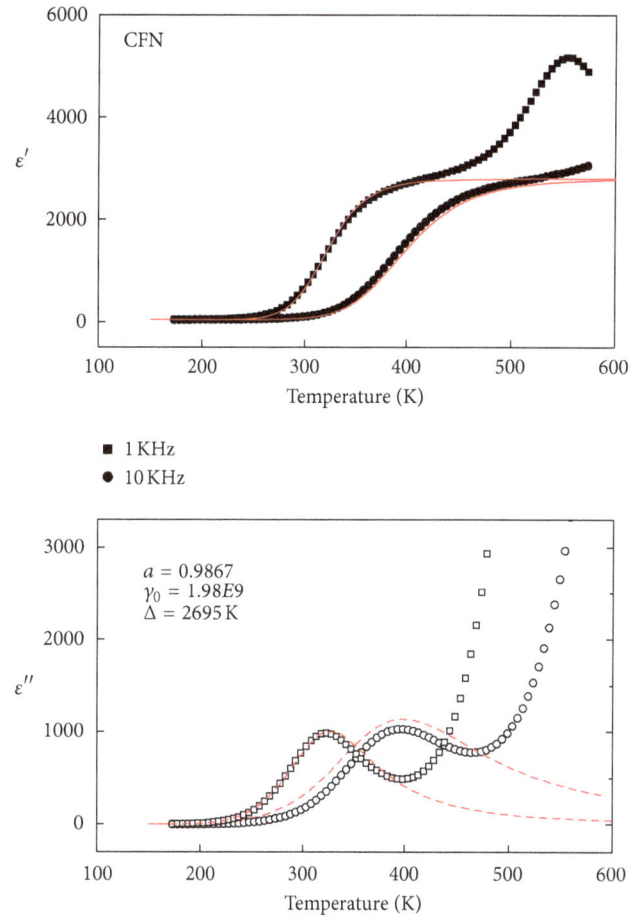

FIGURE 9: The dielectric constant of CFN is shown as a function of temperature. The solid lines are the beat fits of polaron model.

activation energy obtained from (2) (Section 3.2, E_a = 0.1 eV for CCTO). The same fitting parameters were used to calculate the temperature dependence of ε' and ε'' at different frequencies. The solid lines in Figure 7 are calculated from the previous model, which reproduce the experimental data quite well. We have also performed the least square fitting process on BFN and CFN. The best result was obtained with a = 0.987, γ_0 = 9.7E10, and Δ = 1951 K for BFN, and a = 0.987, γ_0 = 1.98E9, and Δ = 2695 K for CFN, respectively. The calculated barrier energy E is 0.17 eV for BFN, which is also quite close to the result obtained from (2) (see Figure 5). However, the calculated barrier energy for CFN is only 0.24 eV, which is much smaller than the one obtained from (2) (0.34 eV). Presumably, this may be attributed to an overlap between the polaron relaxation and a high temperature relaxation process as shown in Figure 9.

4. Conclusions

In summary, we have prepared single phase CCTO and AFN (BFN, CFN, and SFN) ceramics by conventional solid state reaction. Based on the dielectric study, it can be concluded that the low temperature step-like dielectric relaxation in CCTO and AFN can be well explained by a polaron model involving a dipolar-type relaxation. By using a mean-field approach, we get a dielectric function which describes the dielectric relaxation in CCTO and BFN quite well.

Acknowledgments

This work was supported by the Hong Kong Polytechnic University (Projects nos. A-PL54 and 1-BD08), the National Natural Science Foundations of China (nos. 51302172 and 51272161), and Shenzhen Innovation and Technology Commission under the strategic emerging industries development project (Contract no. ZDSY20120612094418467).

References

[1] L. Kerkache, A. Layadi, E. Dogheche, and D. Remiens, "Structural, ferroelectric and dielectric properties of In$_2$O$_3$:Sn (ITO) on PbZr$_{0.53}$Ti$_{0.47}$O$_3$ (PZT)/Pt and annealing effect," *Journal of Alloys and Compounds*, vol. 509, no. 20, pp. 6072–6076, 2011.

[2] N. Vittayakorn, G. Rujijanagul, and D. P. Cann, "Investigation of the influence of thermal treatment on the morphologies, dielectric and ferroelectric properties of PZT-based ceramics,"

Journal of Alloys and Compounds, vol. 440, no. 1-2, pp. 259–264, 2007.

[3] M. A. Subramanian, D. Li, N. Duan, B. A. Reisner, and A. W. Sleight, "High dielectric constant in $ACu_3Ti_4O_{12}$ and $ACu_3Ti_3FeO_{12}$ phases," *Journal of Solid State Chemistry*, vol. 151, no. 2, pp. 323–325, 2000.

[4] D. C. Sinclair, T. B. Adams, F. D. Morrison, and A. R. West, "$CaCu_3Ti_4O_{12}$: one-step internal barrier layer capacitor," *Applied Physics Letters*, vol. 80, no. 12, pp. 2153–2155, 2002.

[5] A. P. Ramirez, M. A. Subramanian, M. Gardel et al., "Giant dielectric constant response in a copper-titanate," *Solid State Communications*, vol. 115, no. 5, pp. 217–220, 2000.

[6] P. Lunkenheimer, R. Fichtl, S. G. Ebbinghaus, and A. Loidl, "Nonintrinsic origin of the colossal dielectric constants in $CaCu_3Ti_4O_{12}$," *Physical Review B*, vol. 70, no. 17, Article ID 172102, 4 pages, 2004.

[7] C. C. Homes, T. Vogt, S. M. Shapiro, S. Wakimoto, and A. P. Ramirez, "Optical response of high-dielectric-constant perovskite-related oxide," *Science*, vol. 293, no. 5530, pp. 673–676, 2001.

[8] I. P. Raevski, S. A. Prosandeev, A. S. Bogatin, M. A. Malitskaya, and L. Jastrabik, "High dielectric permittivity in $AFe_{1/2}B_{1/2}O_3$ nonferroelectric perovskite ceramics (A = Ba, Sr, Ca; B = Nb, Ta, Sb)," *Journal of Applied Physics*, vol. 93, no. 7, pp. 4130–4136, 2003.

[9] K. P. Neupane, J. L. Cohn, H. Terashita, and J. J. Neumeier, "Doping dependence of polaron hopping energies in $La_{1-x}Ca_xMnO_3$ ($0 \leq x \leq 0.15$)," *Physical Review B*, vol. 74, no. 14, Article ID 144428, 2006.

[10] R. S. Freitas, J. F. Mitchell, and P. Schiffer, "Magnetodielectric consequences of phase separation in the colossal magnetoresistance manganite $Pr_{0.7}Ca_{0.3}MnO_3$," *Physical Review B*, vol. 72, no. 14, Article ID 144429, 2005.

[11] T. Park, Z. Nussinov, K. R. A. Hazzard et al., "Novel dielectric anomaly in the hole-doped $La_2Cu_{1-x}Li_xO_4$ and $La_{2-x}Sr_xNiO_4$ insulators: signature of an electronic glassy state," *Physical Review Letters*, vol. 94, no. 1, Article ID 017002, 2005.

[12] J. Rivas, B. Rivas-murias, A. Fondado, J. Mira, and M. A. Señarís-Rodríguez, "Dielectric response of the charge-ordered two-dimensional nickelate $La_{1.5}Sr_{0.5}NiO_4$," *Applied Physics Letters*, vol. 85, no. 25, Article ID 6224, 3 pages, 2004.

[13] C. C. Wang, Y. M. Cui, and L. W. Zhang, "Dielectric properties of $TbMnO_3$ ceramics," *Applied Physics Letters*, vol. 90, no. 1, Article ID 012904, 3 pages, 2007.

[14] J. B. Wu, C. W. Nan, Y. H. Lin, and Y. Deng, "Giant dielectric permittivity observed in Li and Ti doped NiO," *Physical Review Letters*, vol. 89, no. 21, Article ID 217601, 2002.

[15] J. L. Cohn, M. Peterca, and J. J. Neumeier, "Low-temperature permittivity of insulating perovskite manganites," *Physical Review B*, vol. 70, no. 21, Article ID 214433, pp. 1–6, 2004.

[16] A. Seeger, P. Lunkenheimer, J. Hemberger et al., "Charge carrier localization in $La_{1-x}Sr_xMnO_3$ investigated by ac conductivity measurements," *Journal of Physics Condensed Matter*, vol. 11, no. 16, pp. 3273–3290, 1999.

[17] J. L. Cohn, M. Peterca, and J. J. Neumeier, "Giant dielectric permittivity of electron-doped manganite thin films, $Ca_{1-x}La_xO_3$ ($0 \leq x \leq 0.03$)," *Journal of Applied Physics*, vol. 97, no. 3, Article ID 034102, 2005.

[18] L. Zhang and Z. J. Tang, "Polaron relaxation and variable-range-hopping conductivity in the giant-dielectric-constant material $CaCu_3Ti_4O_{12}$," *Physical Review B*, vol. 70, no. 17, Article ID 174306, 6 pages, 2004.

[19] A. Tselév, C. M. Brooks, S. M. Anlage et al., "Evidence for power-law frequency dependence of intrinsic dielectric response in the $CaCu_3Ti_4O_{12}$," *Physical Review B*, vol. 70, no. 14, Article ID 144101, 2004.

[20] C. C. Wang and L. W. Zhang, "Polaron relaxation related to localized charge carriers in $CaCu_3Ti_4O_{12}$," *Applied Physics Letters*, vol. 90, no. 14, Article ID 142905, 3 pages, 2007.

[21] L. Liu, C. C. Wang, X. H. Sun, G. J. Wang, C. M. Lei, and T. Li, "Oxygen-vacancy-related relaxations of $Sr_3CuNb_2O_9$ at high temperatures," *Journal of Alloys and Compounds*, vol. 552, pp. 279–282, 2013.

[22] L. Fang, M. R. Shen, F. Zheng, Z. Li, and J. Yang, "Dielectric responses and multirelaxation behaviors of pure and doped $CaCu_3Ti_4O_{12}$ ceramics," *Journal of Applied Physics 104*, vol. 104, no. 6, Article ID 064110, 8 pages, 2008.

[23] S. M. Ke, H. T. Huang, and H. Q. Fan, "Relaxor behavior in $CaCu_3Ti_4O_{12}$ ceramics," *Applied Physics Letters*, vol. 89, no. 18, Article ID 182904, 3 pages, 2006.

[24] S. Ke, H. Huang, H. Fan, H. L. W. Chan, and L. M. Zhou, "Colossal dielectric response in barium iron niobate ceramics obtained by different precursors," *Ceramics International*, vol. 34, no. 4, pp. 1059–1062, 2008.

[25] N. Kolev, C. L. Chen, M. Gospodinov et al., "Raman spectroscopy of $CaRuO_3$," *Physical Review B*, vol. 66, no. 1, Article ID 014101, pp. 141011–141014, 2002.

[26] S. Saha and T. P. Sinha, "Low-temperature scaling behavior of $BaFe_{0.5}Nb_{0.5}O_3$," *Physical Review B*, vol. 65, no. 13, Article ID 134103, 7 pages, 2002.

[27] K. Tezuka, K. Henmi, Y. Hinatsu, and N. M. Masaki, "Magnetic susceptibilities and Mossbauer spectra of perovskites A_2FeNbO_6 (A = Sr, Ba)," *Journal of Solid State Chemistry*, vol. 154, no. 2, pp. 591–597, 2000.

[28] Y. Y. Liu, X. M. Chen, X. Q. Liu, and L. Li, "Dielectric relaxations in $Ca(Fe_{1/2}Nb_{1/2})O_3$ complex perovskite ceramics," *Applied Physics Letters*, vol. 90, no. 26, Article ID 262904, 3 pages, 2007.

[29] A. K. Jonscher, "The universal dielectric response and its physical significance," *IEEE Transactions on Electrical Insulation*, vol. 27, no. 2, pp. 407–423, 1992.

[30] Z. Wang, X. M. Chen, L. Ni, and X. Q. Liu, "Dielectric abnormities of complex perovskite $Ba(Fe_{1/2}Nb_{1/2})O_3$ ceramics over broad temperature and frequency range," *Applied Physics Letters*, vol. 90, no. 2, Article ID 022904, 3 pages, 2007.

[31] P. R. Bueno, M. A. Ramirez, J. A. Varela, and E. Longo, "Dielectric spectroscopy analysis of $CaCu_3Ti_4O_{12}$ polycrystalline systems," *Applied Physics Letters*, vol. 89, no. 19, Article ID 191117, 3 pages, 2006.

[32] S. Ke, H. Fan, and H. Huang, "Dielectric relaxation in A_2FeNbO_6 (A = Ba, Sr, and Ca) perovskite ceramics," *Journal of Electroceramics*, vol. 22, no. 1–3, pp. 252–256, 2009.

[33] A. P. Ramirez, G. Lawes, V. Butko, M. A. Subramanian, and C. M. Varma, "Colossal dielectric constants in braced latticeswith defects," http://arxiv.org/abs/cond-mat/0209498 .

[34] B. I. Halperin and C. M. Varma, "Defects and the central peak near structural phase transitions," *Physical Review B*, vol. 14, no. 9, pp. 4030–4044, 1976.

26

The Lattice Compatibility Theory: Arguments for Recorded I-III-O$_2$ Ternary Oxide Ceramics Instability at Low Temperatures beside Ternary Telluride and Sulphide Ceramics

K. Boubaker

École Supérieure de Sciences et Techniques de Tunis (ESSTT), Université de Tunis, 63 Rue Sidi Jabeur, 5100 Mahdia, Tunisia

Correspondence should be addressed to K. Boubaker; mmbb11112000@yahoo.fr

Academic Editor: Joon-Hyung Lee

Some recorded behaviours differences between chalcopyrite ternary oxide ceramics and telluride and sulphides are investigated in the framework of the recently proposed Lattice Compatibility Theory (LCT). Alterations have been evaluated in terms of Urbach tailing and atomic valence shell electrons orbital eigenvalues, which were calculated through several approximations. The aim of the study was mainly an attempt to explain the intriguing problem of difficulties of elaborating chalcopyrite ternary oxide ceramics (I-III-O$_2$) at relatively low temperatures under conditions which allowed crystallization of ternary telluride and sulphides.

1. Introduction

I-III-VI$_2$ ternary ceramics are attractive materials in possible photovoltaic and optoelectronic applications [1–9] like coating films for IR reflection, smart windows, functional glasses, transparent electrodes in flat panel displays, and solar cells [10, 11]. During the last two decades, p-type ternary transparent conducting oxide ceramics, like CuAlO$_2$, CuInO$_2$, AgAlO$_2$, and ZnAlO$_2$, have attracted more and more attention for their high absorption coefficient as well as their band gap energy (1.8–3.5 eV). Currently, many experimental efforts are underway to search for new p-type TCOs. In 1997, Kawazoe et al. [12] elaborated p-type CuAlO$_2$ films resistivity of 10.5 Ω cm, which is ten times less than that of typical ZnO [13], the most available candidate for similar applications.

p-type Cu-Al-O-like ceramics have been prepared by several techniques. Pulsed laser deposition [14], sputtering [15], more information of these two techniques will be introduced in the following sections.

In our laboratories, we tried to elaborate Cu-Al-O-like ceramics using wet techniques (Spray, vaporization, etc.) at relatively low temperatures (300–450°C). After hundreds of attempts, we came to the conclusion that such temperature ranges are not suitable but for elaborating ternary oxides. Dlala et al. [16], succeeded to stabilize Ag$_2$S thin films; Amlouk et al. [17] elaborated SnO$_2$:F and CdS airless sprayed layers Khélia et al. [18], Kamoun et al. [19], and Amlouk et al. [20] fabricated and characterized β-SnS$_2$ and β-In$_{2-x}$Al$_x$S$_3$ compounds at low temperatures. Binary sulphide compounds such as Ag$_2$S, Cu$_2$S, SnS$_2$, and In$_2$S$_3$ have been obtained in our laboratory in 200–320°C domain of temperature. Later, ternary sulphide compounds based on copper and silver which generally crystallized in chalcopyrite structure I-III-VI$_2$ (CuInS$_2$, AgInS$_2$ AgInS$_{2-x}$Se$_x$) were prepared, by Guezmir et al. [21], Kamoun et al. [22, 23], Aissa [24, 25], and Lazzez et al. [26], in substrate temperature lying in 340–420°C. Finally, Bouaziz et al. [27, 28] obtained Cu$_2$SnS$_3$ and Cu$_3$SnS$_4$ ternary compounds in 300–400°C domain.

The subject of this paper is an attempt to explain, starting from experimental facts, why was it impossible to stabilize ternary oxides under humid condition below 450°C, while both binary and ternary sulphides were easily obtained. This paper is arranged as follows. In Section 2, we present some experimental details concerning ternary oxide ceramics fabrication. In Section 3, the fundamentals of the lattice Compatibility Theory (LCT) are presented as a plausible explanation to ternary oxide ceramics unexpected instability at low temperatures. Conclusion will be appeared in Section 4.

2. Experimental Details

CuAlS$_2$ thin films ceramics have been deposited on glass substrates using chemical bath deposition CBD technique with CuSO$_4$, (NH$_2$)$_2$SC and hydrated Al$_2$(SO$_4$)$_3$ as precursors. Commercial solutions of 0.10 M CuSO$_4$(NH$_2$)$_2$, and 1.0 M (NH$_2$)$_2$SC, 0.1 M Al$_2$(SO$_4$)$_3$·14H$_2$O were used. A primal solution of CuSO$_4$(NH$_2$)$_2$ with an excess triethanolamine was stirred for few minutes until colour changed, before adding a stirred Al$_2$(SO$_4$)$_3$ solution. Substrate slides were finally immerged in the obtained dark orange black solution for 22 hours. Removal of the slides was carried out at room temperature and was followed by a few rinsing-drying stages. CuAlTe$_2$ layers have been synthesized by annealing, under vacuum in an argon-rich medium, a multilayer structure of thin Cu, Al, and Te layers which were deposited by evaporation. A mixture of pure copper, aluminium, and tellurium weighed in stoichiometric ratios was sealed in a vacuum quartz tube. The tube was subjected to several 35 h annealing stages. CuAlTe$_2$ layers were consecutively grown by the evaporation of the obtained powder under vacuum. More details about experimental procedure can be found elsewhere [21, 23, 24, 26–28]. Similar techniques and disposals have been used in order to synthesize CuAlO$_2$. A wide variety of precursors, including mineral CuFeO$_2$, high-purity CuO, Al$_2$O$_3$, and metallic Aluminium have been tried unsuccessfully in our laboratories, particularly at temperatures beyond 450°C.

3. Oxygen Beside Sulfur-Like Structural Disparity

3.1. *Atomic Scale Patterns.* In order to explain the mentioned paradox (Section 2), a quick comparative review of oxygen-sulfur properties is needed. While both elements have the same valence, oxygen electronegativity is 3.44, less than that of sulfur (2.58). Some other differences can be outlined:

  (i) O=O double bonds are stronger than S=S.

  (ii) S–S single bonds are almost twice as strong as O–O.

  (iii) Sulfur can expand its valence shell to hold more eight to twelve electrons.

The last difference has been illustrated by Carelli et al. [29] who stated that in combined media, "sulfur can expand its valence shell to produce a soft nucleophilic center" which refers to the so-called S-nucleophilicity. This property, accounts, in the case of I-III-O$_2$ and I-III-S$_2$ compounds for the ability of sulfur to expand its valence shell from 8 to 10 or 12 electrons using its available 3d orbitals, allowing oxidation states not available to its oxygen analogs (Figure 1).

Moreover, opposite to Cu–O bonding patterns, which obey a simple oxidation state formalism, Cu–S bonding systems have a high degree of delocalization resulting in additional electronic band structures (Figure 1) and are hence covalent rather than ionic.

Moreover, calculation of atomic valence shell electrons orbital eigenvalues for sulfur, oxygen, and tellurium in delafossite structures confirms this delocalization plausibility. Orbital eigenvalues were calculated through local density (LDA) [30] and scalar-relativistic local density (ScRLD) [31] approximations using OPIUM code [32, 33]. In the first approach, the many-electron configuration is approximated by a set of single-particle equations while the second takes into account some of the effects of relativity, such as Darwin shift and mass-velocity term.

Table 1 gathers atomic orbital eigenvalues (in eV) for oxygen, sulfur, and tellurium according to (LDA) and (ScRLD) approximations.

According to (LDA) and (ScRLD) calculations, most of Tellurium bonding in the valence region is accomplished by the 5p orbitals. The in-plane locally π-type 5p orbitals (Figure 2) form a narrowband between −0.2 and −0.3 eV, which is far below the Fermi energy. This range confirms the diffuse character of tellurium p-orbitals, which also confers to the lattice a covalent rather than ionic aspect.

(Figure 3) shows the two-dimensional arrangement with Cu$^+$ ions located between layers of B^{III}O$_2$ octahedra. The structure is characterized by alternating layers of slightly distorted edge-shared B^{III}O$_2$ (or B^{III}S$_2$) octahedra sandwiching planes of close-packed A$^+$ cations (here Cu$^+$) in linear or "dumbbell" coordination to oxygen anions in the adjacent octahedra.

The difference in length between Cu–O and Cu–S bonds (24%) has been ascribed to the electrons being shifted to the orbital 3d. All these observations plea on favor of the presence of a delocalized valence "hole" favorizing natural bonding between two opposed octahedra planes (Figure 3) at low temperatures. If we consider the results of Liang et al. [34, 35], which stated that the Cu–Cu bonding character was very weak in such structures, we can conclude that the crystalline behavior of A^{I}B^{III}O$_2$ (Figure 3) and A^{I}B^{III}S$_2$ lattices depends mainly on is the status of A$^+$ ion (here Copper). According to Landolt-Börnstein [36–38] and Greenwood and Earnshaw [39], any formulation of copper monosulphide as Cu^{II}S with no meaningful S–S bonds is incompatible both with the crystal structure, and the observed diamagnetism as a Cu(II) compound would have a d^9 configuration and be expected to be paramagnetic. An alternative formulation as (Cu$^+$)$_3$(S^{2-})(S$_2$)$^-$ was proposed and supported by other studies.

3.2. *Urbach Energy and Lattice Compatibility Theory (LCT) Consideration.* An additional argument can also be formulated through the recently proposed Urbach energy-related Lattice Compatibility Theory (LCT) [40–42]. Urbach energy E_u has been determined, for the ternary oxides and telluride and sulphides samples through the equations:

$$\text{Ln}\left(\alpha\left(h\nu\right)\right) = \text{Ln}\left(\alpha_0\right) + \frac{h\nu}{E_u},$$

$$E_u = \alpha\left(h\nu\right)\left(\frac{d\left[\alpha\left(h\nu\right)\right]}{d\left[h\nu\right]}\right)^{-1} = h\left[\frac{d}{d\nu}\left(\text{Ln}\alpha\left(\nu\right)\right)\right]^{-1}, \tag{1}$$

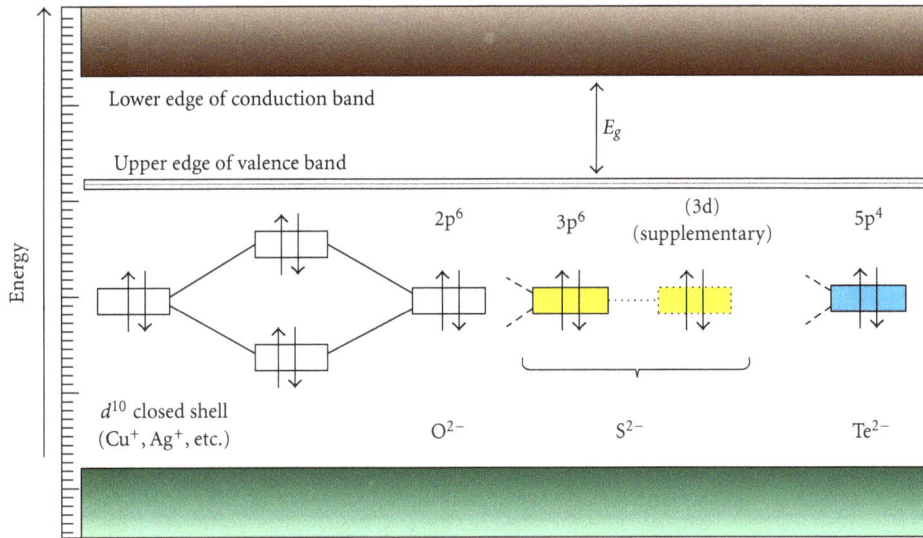

FIGURE 1: Sulfur-oxygen-tellurium different outer shell configuration inside a Cu-like combination.

TABLE 1: Atomic orbital eigenvalues (in eV) for oxygen, sulfur, and tellurium.

| Oxygen | | | Sulfur | | | Tellurium | | |
| Structure: [He] $2s^2\ 2p^4$ | | | Structure: [Ne] $3s^2\ 3p^4$ | | | Structure: [Kr] $4d^{10}\ 5s^2\ 5p^4$ | | |
Orbital	LDA	ScRLD	Orbital	LDA	ScRLD	Orbital	LDA	ScRLD
1s	−18.7582	−18.7585	1s	−87.78993	−87.94937	4s	−5.57284	−6.00853
2s	−0.87136	−0.87246	2s	−7.69994	−7.737587	4p	−4.10008	−4.19056
2p	−0.33838	−0.33816	2p	−5.75125	−5.749477	4d	−1.60838	−1.53477
			3s	−0.63091	−0.634502	5s	−0.52099	−0.56288
			3p	−0.261676	−0.261260	5p	−0.22659	−0.22459

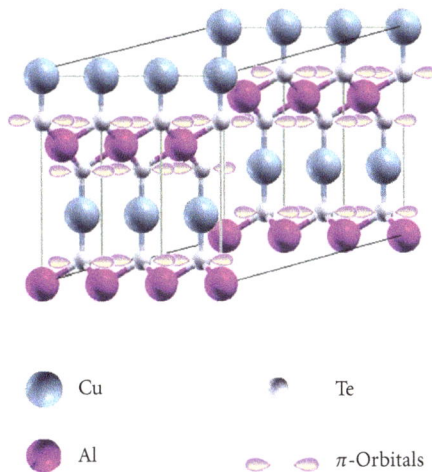

Cu

Te

Al

π-Orbitals

FIGURE 2: CuAlTe$_2$ ceramics crystal structure.

where $\alpha(h\nu)$ represents, for each sample, the experimentally deduced optical absorption profile.

Urbach energy E_u has been defined as an aggregated for the evaluating of the intrinsic disorder and atomic scale dispersion inside structures as it indicates the width of the band tails of the localized states in presence of defects. The width of the localized states (band tail energy or Urbach energy E_u) has been estimated from the slopes of the plots of $(\text{Ln}\ \alpha(\nu))$ versus energy $h\nu$ (Figure 4).

It can be noticed that the atomic-scale intrinsic disorder is higher in CuAlS$_2$ and CuAlTe$_2$ than in films, and disposals have been used in order to synthesize CuAlO$_2$ matrices. This result is in good agreement with the precedent analyses and with the fundaments of the Lattice Compatibility Theory (LCT) [40–42].

Finally, additional investigation should be carried out in order to explain the need of calcination (at 1200°C [43], 800°C [44], and 900°C [45]), as un avoidable step for establishing A^{I}B^{III}O$_2$ stable structures, while A^{I}B^{III}S$_2$ are autostabilized at low temperatures apparently, thanks to the above-described valence "hole."

4. Conclusion

In this study, we tried to give some explanation to the intriguing problem of the impossibility of elaborating chalcopyrite ternary oxide ceramics (I-III-O$_2$) at relatively low temperatures in wet media. Contrast has been highlighted with regard to ternary telluride and sulphide ceramics.

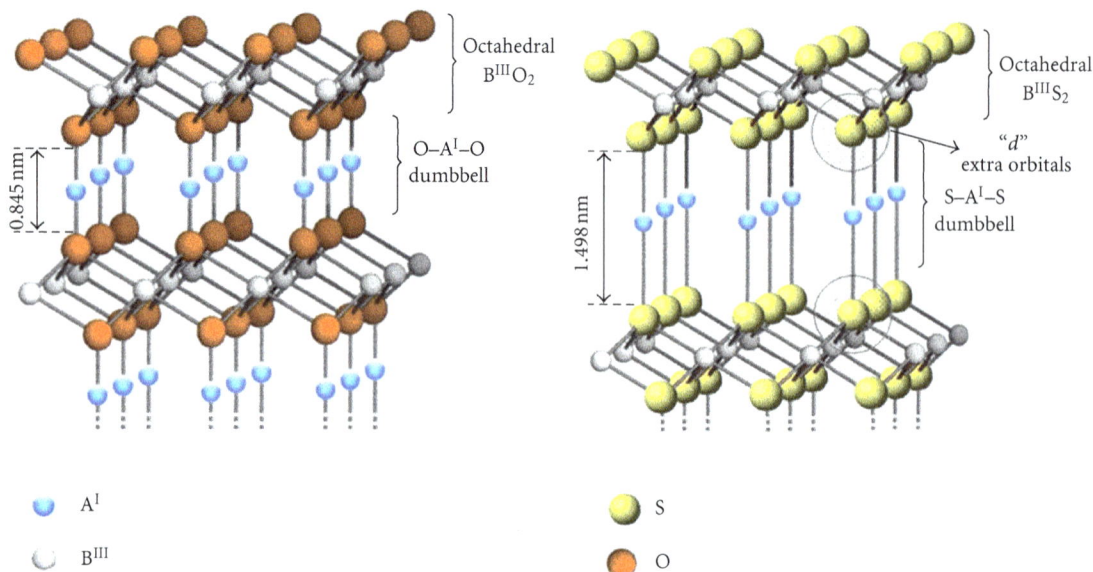

FIGURE 3: $A^I B^{III} S_2$ and $A^I B^{III} O_2$ ceramics Delafossite type structure (with additional sulfur orbitals).

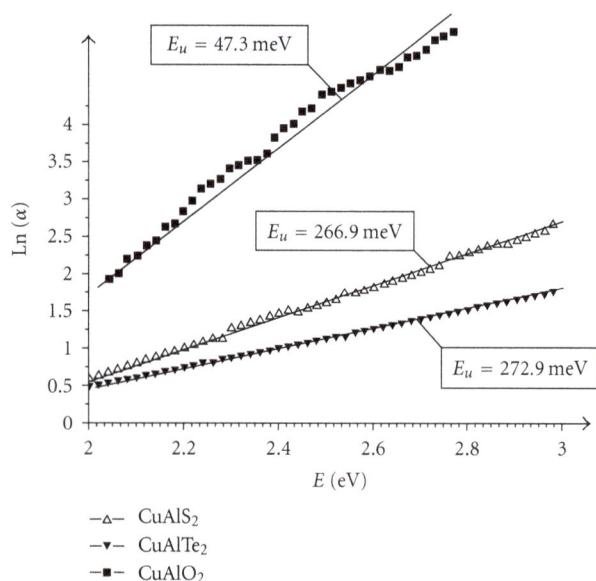

FIGURE 4: Plots of $(Ln \alpha(\nu))$ versus energy $h\nu$ (as guides for evaluating E_u).

An attempt to explain this paradox has been formulated in terms of particularities of $A^I B^{III} O_2$ and $A^I B^{III} S_2$ ceramics delafossite structure and bonding interaction.

References

[1] M. Singh, A. R. Rao, and V. Dutta, "Effect of pH on structural and morphological properties of spray deposited p-type transparent conducting oxide CuAlO₂ thin films," *Materials Letters*, vol. 62, no. 21-22, pp. 3613–3616, 2008.

[2] T. Mei, J. Zhang, L. Wang, Z. Xing, Y. Zhu, and Y. Qian, "Preparation of mixed oxides Ca₉Co₁₂O₂₈ and their electrochemical properties," *Materials Letters*, vol. 82, pp. 1–3, 2012.

[3] Z. F. Zhang, Z. M. Sun, and H. Hashimoto, "Deformation and fracture behavior of ternary compound Ti₃SiC₂ at 25–1300°C," *Materials Letters*, vol. 57, no. 7, pp. 1295–1299, 2003.

[4] Y. F. Zhang, J. X. Zhang, Q. M. Lu, and Q. Y. Zhang, "Synthesis and characterization of Ca₃Co₄O₉ nanoparticles by citrate sol-gel method," *Materials Letters*, vol. 60, no. 20, pp. 2443–2446, 2006.

[5] S. Matar, M. Lelogeas, D. Michau, and G. Demazeau, "Investigations on the high-pressure varieties of GaAsO₄," *Materials Letters*, vol. 10, no. 1-2, pp. 45–48, 1990.

[6] F. M. Túnez, J. A. Gamboa, J. A. González, and M. Esquivel, "A new polymorph of GaAsO₄," *Materials Letters*, vol. 79, pp. 202–204, 2012.

[7] A. N. Banerjee, R. Maity, and K. K. Chattopadhyay, "Preparation of p-type transparent conducting CuAlO₂ thin films by reactive DC sputtering," *Materials Letters*, vol. 58, no. 1-2, pp. 10–13, 2003.

[8] S. Gao, Y. Zhao, P. Gou, N. Chen, and Y. Xie, "Preparation of CuAlO₂ nanocrystalline transparent thin films with high conductivity," *Nanotechnology*, vol. 14, no. 5, pp. 538–541, 2003.

[9] Y. Q. Chen, J. Jiang, B. Wang, and J. G. Hou, "Synthesis of tin-doped indium oxide nanowires by self-catalytic VLS growth," *Journal of Physics D*, vol. 37, pp. 3319–3322, 2004.

[10] S. Y. Li, P. Lin, C. Y. Lee, T. Y. Tseng, and C. J. Huang, "Effect of Sn dopant on the properties of ZnO nanowires," *Journal of Physics D*, vol. 37, pp. 2274–2282, 2004.

[11] M. Liu, J. Yang, S. Feng et al., "Composite photoanodes of Zn₂SnO₄ nanoparticles modified SnO₂ hierarchical microspheres for dye-sensitized solar cells," *Materials Letters*, vol. 76, pp. 215–218, 2012.

[12] H. Kawazoe, M. Yasukawa, H. Hyodo, M. Kurita, H. Yanagi, and H. Hosono, "P-type electrical conduction in transparent

thin films of CuAlO$_2$," *Nature*, vol. 389, no. 6654, pp. 939–942, 1997.

[13] K. Minegishi, Y. Koiwai, Y. Kikuchi, YanoK, and A. Shimizu, "Characterization of interface electronic properties of low-temperature ultrathin oxides and oxynitrides formed on Si(111) surfaces by contactless capacitance-voltage and photoluminescence methods," *Journal of Applied Physics*, vol. 36, pp. 1453–1460, 1997.

[14] D. M. Hoffman, B. Singh, and J. H. Thomas, *Handbook of Vacuum Science and Technology*, Academic Press, San Diego, Calif, USA, 1998.

[15] R. A. Powell and S. M. Rossnagel, *PVD for Microelectronics: Sputter Deposition Applied to Semiconduct or Manufacturing*, Academic Press, San Diego, Calif, USA, 1999.

[16] H. Dlala, M. Amlouk, S. Belgacem, P. Girard, and D. Barjon, "Structural and optical properties of Ag$_2$S thin films prepared by spray pyrolysis," *EPJ Applied Physics*, vol. 2, no. 1, pp. 13–16, 1998.

[17] M. Amlouk, M. Dachraoui, S. Belgacem, and R. Bennaceur, "Structural, optical and electrical properties of SnO$_2$:F and CdS airless sprayed layers," *Solar Energy Materials*, vol. 15, no. 6, pp. 453–461, 1987.

[18] C. Khélia, K. Boubaker, T. Ben Nasrallah et al., "Morphological and thermal properties of β-SnS$_2$ crystals grown by spray pyrolysis technique," *Journal of Crystal Growth*, vol. 311, no. 4, pp. 1032–1035, 2009.

[19] N. Kamoun, S. Belgacem, M. Amlouk et al., "Structure, surface composition, and electronic properties of β-In$_2$S$_3$ and β-In$_{2-x}$AlxS$_3$," *Journal of Applied Physics*, vol. 89, no. 5, pp. 2766–2771, 2001.

[20] M. Amlouk, M. A. Ben Saïd, N. Kamoun, S. Belgacem, N. Brunet, and D. Barjon, "Acoustic properties of β-In$_2$S$_3$ thin films prepared by spray," *Japanese Journal of Applied Physics, Part 1*, vol. 38, no. 1, pp. 26–30, 1999.

[21] N. Guezmir, T. B. Nasrallah, K. Boubaker, M. Amlouk, and S. Belgacem, "Optical modeling of compound CuInS$_2$ using relative dielectric function approach and Boubaker polynomials expansion scheme BPES," *Journal of Alloys and Compounds*, vol. 481, no. 1-2, pp. 543–548, 2009.

[22] N. Kamoun, R. Bennaceur, M. Amlouk et al., "Optical properties of InS layers deposited using an airless spray technique," *Physica Status Solidi A*, vol. 169, no. 1, pp. 97–104, 1998.

[23] N. Kamoun, S. Belgacem, M. Amlouk, R. Bennaceur, K. Abdelmoula, and A. Belhadj Amara, "Structural and morphological characterizations of airless spray CuInS$_2$ and InS thin films," *Journal de Physique*, vol. 4, no. 3, pp. 473–491, 1994.

[24] Z. Aissa, M. Amlouk, T. Ben Nasrallah, J. C. Bernède, and S. Belgacem, "Effect of S/In concentration ratio on the physical properties of AgInS$_2$-sprayed thin films," *Solar Energy Materials and Solar Cells*, vol. 91, no. 6, pp. 489–494, 2007.

[25] Z. Aissa, T. Ben Nasrallah, M. Amlouk, J. C. Bernède, and S. Belgacem, "Some physical investigations on AgInS$_2$ sprayed thin films," *Solar Energy Materials and Solar Cells*, vol. 90, no. 7-8, pp. 1136–1146, 2006.

[26] S. Lazzez, K. Boubaker, T. B. Nasrallah et al., "Structural and optoelectronic properties of In-Zn-S sprayed layers," *Acta Physica Polonica A*, vol. 114, no. 4, pp. 869–880, 2008.

[27] M. Bouaziz, K. Boubaker, M. Amlouk, and S. Belgacem, "Effect of Cu/Sn concentration ratio on the phase equilibrium-related properties of Cu-Sn-S sprayed materials," *Journal of Phase Equilibria and Diffusion*, vol. 31, no. 6, pp. 498–503, 2010.

[28] M. Bouaziz, M. Amlouk, and S. Belgacem, "Structural and optical properties of Cu$_2$SnS$_3$ sprayed thin films," *Thin Solid Films*, vol. 517, no. 7, pp. 2527–2530, 2009.

[29] V. Carelli, F. Liberatore, L. Scipione, R. Musio, and O. Sciacovelli, "On the structure of intermediate adducts arising from dithionite reduction of pyridinium salts: a novel class of derivatives of the parent sulfinic acid," *Tetrahedron Letters*, vol. 41, no. 8, pp. 1235–1240, 2000.

[30] S. H. Vosko, L. Wilk, and M. Nusair, "Accurate spin-dependent electron liquid correlation energies for local spin density calculations: a critical analysis," *Canadian Journal of Physics*, vol. 58, p. 1200, 1980.

[31] D. D. Koelling and B. N. Harmon, "A technique for relativistic spin-polarised calculations," *Journal of Physics C*, vol. 10, no. 16, pp. 3107–3114, 1977.

[32] H. Dixit, R. Saniz, D. Lamoen, and B. Partoens, "The quasiparticle band structure of zincblende and rocksalt ZnO," *Journal of Physics: Condensed Matter*, vol. 22, no. 12, Article ID 125505, 2010.

[33] S. H. Vosko and L. Wilk, "Influence of an improved local-spin-density correlation-energy functional on the cohesive energy of alkali metals," *Physical Review B*, vol. 22, no. 8, pp. 3812–3815, 1980.

[34] W. Liang and M. H. Whangbo, "Conductivity anisotropy and structural phase transition in Covellite CuS," *Solid State Communications*, vol. 85, no. 5, pp. 405–408, 1993.

[35] H. Nozaki, K. Shibata, and N. Ohhashi, "Metallic hole conduction in CuS," *Journal of Solid State Chemistry*, vol. 91, no. 2, pp. 306–311, 1991.

[36] *Landolt-Börnstein: Numerical Data and Functional Relationships in Science and Tech*, New Series, II/16, Diamagnetic Susceptibility, Springer, Heidelberg, Germany, 1986.

[37] *Landolt-Börnstein: Numerical Data and Functional Relationships in Science and Technology*, New Series, III/19, Subvolumes a to i2, Magnetic Properties of Metals, Springer, Heidelberg, Germany, 1986–1992.

[38] *Landolt-Börnstein: Numerical Data and Functional Relationships in Science and Technology*, New Series, II/2, II/8, II/10, II/11,and II/12a,Coordination and Organometallic Transition Metal Compounds, Springer, Heidelberg, Germany, 1966-1984.

[39] N. N. Greenwood and A. Earnshaw, *Chemistry of the Elements*, Butterworth-Heinemann, Oxford, UK, 2nd edition, 1997.

[40] P. Petkova and K. Boubaker, "The Lattice Compatibility Theory (LCT): an attempt to explain Urbach tailing patterns in copper-doped bismuth sillenites (BSO) and germanates (BGO)," *Journal of Alloys and Compounds*, vol. 546, pp. 176–179, 2013.

[41] K. Boubaker, "Preludes to the Lattice Compatibility Theory LCT: urbach tailing controversial behavior in some nanocompounds," *ISRN Nanomaterials*, vol. 2012, Article ID 173198, 4 pages, 2012.

[42] K. Boubaker, M. Amlouk, Y. Louartassi, and H. Labiadh, "About unexpected crystallization behaviors of some ternary oxide and sulfide ceramics within lattice compatibility theory LCT framework," *Journal of the AustralianCeramic Society*, vol. 49, no. 1, pp. 115–117, 2013.

[43] A. Nattestad, X. Zhang, U. Bach, and Y. Cheng, "Dye-sensitized CuAlO$_2$ photocathodes for tandem solar cell applications," *Journal of Photonics for Energy*, vol. 1, no. 1, Article ID 011103, 2011.

[44] J. E. Clayton, D. P. Cann, and N. Ashmore, "Synthesis and processing of AgInO₂ delafossite compounds by cation exchange
reactions," *Thin Solid Films*, vol. 411, pp. 140–146, 2002.

[45] K. Y. Jung, S. B. Park, and S. K. Ihm, "Local structure and
photocatalytic activity of B_2O_3-SiO_2/TiO_2 ternary mixed oxides
prepared by sol-gel method," *Applied Catalysis B*, vol. 51, no. 4,
pp. 239–245, 2004.

27

Preparation of Stable ZrB_2-SiC-B_4C Aqueous Suspension for Composite Based Coating: Effect of Solid Content and Dispersant on Stability

Mehri Mashadi,[1] Mohsen Mohammadijoo,[2] Alireza Honarkar,[3] and Zeinab Naderi Khorshidi[2]

[1] *Department of Materials Engineering, Faculty of Engineering, Tarbiat Modares University, P.O. Box 1411713116, Tehran, Iran*
[2] *Department of Chemical and Materials Engineering, University of Alberta, Edmonton, AB, Canada T6G 2V4*
[3] *School of Engineering, Shahid Rajaee University, P.O. Box 16785-136, Tehran, Iran*

Correspondence should be addressed to Mohsen Mohammadijoo; mo1@ualberta.ca

Academic Editor: Keizo Uematsu

ZrB_2-SiC-B_4C aqueous suspension has been prepared using poly(ethyleneimine) as a dispersant. Since increasing the solid content of suspension leads to high compaction and consequently low porosities through final coat, the effect of solid content has been studied. The dispersant and solid content were changed in the range of 0.3–1.5 wt.% and 45–55 vol.%, respectively, to assess the optimal conditions effect on stability and characteristics of suspension. Results of zeta potential measurements and rheological analysis at pH 7.8 showed that the composite suspension including 45 vol.% solid content and 1.5 wt.% dispersant was in stable state.

1. Introduction

The interest in ultrahigh temperature ceramics (UHTCs) has increased significantly in recent years [1–3] because of their remarkable properties, such as high melting point, high thermal conductivities, excellent corrosion resistance, and good oxidation resistance [4, 5], which make them promising candidates for high temperature structural applications. Among the UHTCs, zirconium diboride (ZrB_2) is a material of particular interest owing to the excellent and unique combination of high melting point, high electrical and thermal conductivity, good thermal shock and wear resistance, and chemical inertness [6]. These properties make it an attractive candidate for ultrahigh temperature applications where corrosion-wear-oxidation resistance is demanded [6, 7].

In spite of the excellent high temperature and mechanical properties of ZrB_2-based ceramics, their rather low fracture toughness (3-4 MPa m$^{1/2}$) has limited the sample size because of the low reliability and, thus, has reduced the chances for application of boride ceramics [8].

Several studies have demonstrated that the addition of SiC could improve the oxidation resistance and mechanical properties of ZrB_2 ceramics, so a lot of works have been carried out on ZrB_2-SiC ceramics [9–16]. Nowadays, colloidal processing such as tape casting, slip casting [17, 18], and dip coating method, which can produce a more homogeneous green microstructure, is becoming more and more important in the fabrication of advanced ceramics because it offers the potential to produce reliable ceramic films and bulk forms through careful control of initial suspension "structure" and its evolution during fabrication [19].

For the fabrication of highly dense ceramic composites, the preparation of well-dispersed and stable ceramic suspensions is one of the most important issues in order to guarantee a homogeneous filling of the interstices among ceramic powders [8]. However, dispersion behaviors of aqueous ZrB_2-SiC-B_4C slurries have been rarely analyzed [17, 18] and reports on the successful preparation of highly concentrated aqueous suspensions have not been available. To maintain the stability through an aqueous suspension, it is needed to prevent particles to (i) stick when they collide and (ii) sediment when they are introduced in a colloidal system. This can be achieved by enhancing the charge associated with the particles, that is, zeta potential [20]. Also, the proper viscosity of suspension causes a slower settling velocity and therefore better stability of suspension.

In the present research, the dispersion behaviors of ZrB_2-SiC-B_4C (ZSB) composite in aqueous medium are investigated with the application of a dispersant (poly(ethylenimine), PEI) for the preparation of highly concentrated aqueous ZSB suspensions.

2. Experimental Procedure

2.1. Aqueous Suspension Preparation. Commercially available ZrB_2 powder (initial particle size ~6 μm), SiC powder (average particle size 1.5 μm), and B_4C powder (average particle size as 3 μm) were used as raw materials. Deionized water and PEI (molecular weight (M_w): 2,000; 50 wt.% in H_2O; Sigma-Aldrich, Belgium) were used for the preparation of ZSB aqueous suspension.

The as-received ZrB_2 powder was milled using a planetary mill with ethanol, a WC ball (diameter: 10 mm), and a stainless steel (coated by WC) jar at 150 revolutions per minute (rpm) for 2 h. The average size of ZrB_2 powders was reduced to 3–5 μm after milling (Figure 1). A mixture of 80 vol.% zirconium diboride and 20 vol.% silicon carbide powders was selected as starting materials. 3 wt.% boron carbide was also used as sintering aid which causes a better sintering of the ceramic coating through next step of preparation.

To produce slurries, powders were mixed with various amounts of PEI (0.3–1.5 wt.%) as a dispersant. The pH of slurries was also set 7.8. The initial ratio of solid to liquid was set as 45 vol.% according to previous researches [17–19]. To investigate the effect of solid content on ZSB suspension properties, slurries with 45, 46, 47.5, 50, 52.5, and 55 vol.% solid contents were prepared. Table 1 shows the composition of slurries.

2.2. Characterization. Size distribution of milled ZrB_2 powders was measured by zeta size analyzer (ZEN3600, England). Rheological behavior and viscosity of slurries were characterized by rheometer apparatus (Physica CPR300, Japan). Ultrasonic vibrator (Hielscher-UP200H, Germany) was used to disperse suspensions prior to zeta potential analyzing. The zeta potential analyzer (ZEN3600, England) was also used to measure the zeta potential of slurries.

3. Results and Discussion

Prior to any investigation of a colloidal system, it is necessary to bring a proper understanding of a stable suspension to the system, for example, how could a colloidal stability be achieved? The force balance associated with the particles in the suspension is simply demonstrated in two terms [21] (1):

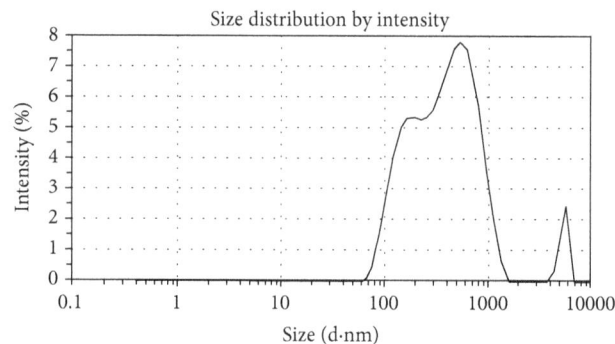

FIGURE 1: Size distribution of ZrB_2 milled powders.

TABLE 1: Characteristics of slurries.

Solid content	PEI
Investigation of effect of PEI on slurry properties	
	0.3 wt.%
	0.5 wt.%
45 vol.%	0.7 wt.%
	1 wt.%
	1.5 wt.%
Investigation of effect of solid content on slurry properties	
45 vol.%	
46 vol.%	
47.5 vol.%	
50 vol.%	1.5 wt.%
52.5 vol.%	
55 vol.%	
45 vol.%	1 wt.%
46 vol.%	

(i) gravitational forces (numerator in (1)) and (ii) Brownian forces (denominator in (1)). Consider

$$\Delta F \approx \frac{a^4 \Delta \rho g}{k_B T}, \qquad (1)$$

where ΔF, a, and $\Delta \rho$ are the force balance, particles size, and density difference of particles and continuous phase, respectively. k_B and T are Boltzmann constant and temperature, respectively. In submicron colloidal systems, the Brownian motion is usually significant to overcome the effect of gravity. To maintain stability through Brownian motion, it is necessary to prevent particles sticking when they collide. This can be achieved by increasing the charge associated with the particles, that is, zeta potential of particles. Figure 2 indicates the zeta potential of a sphere particle within an aqueous medium.

By increasing the zeta potential (over ±30 mV), the significance of long range electrostatic double layer surrounding particles increases which leads to a repulsion between particles in suspension [22]. Figure 3 depicts the schematic view of the effect of zeta potential on dispersion and coagulation of particles in a suspension.

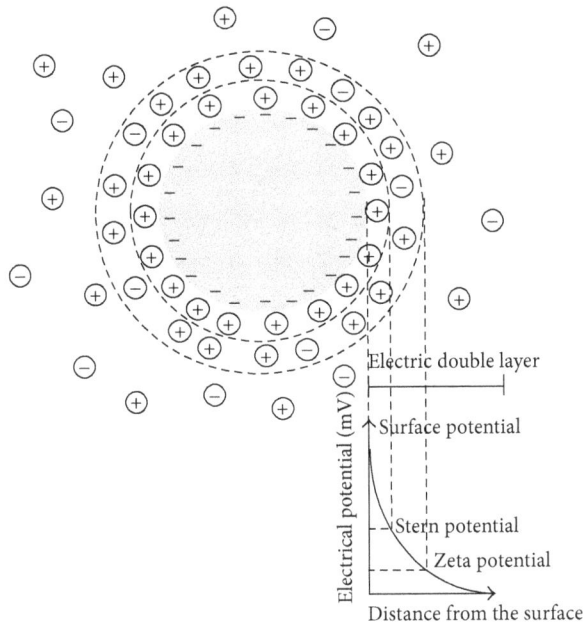

FIGURE 2: Electrostatic double layer surrounding sphere particles in an aqueous medium.

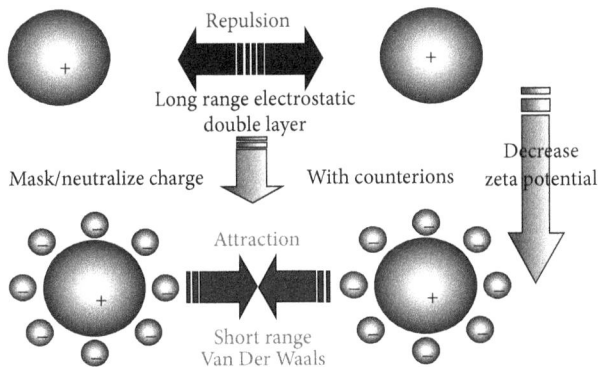

FIGURE 3: Schematic view of the particles interactions in an aqueous suspension [22].

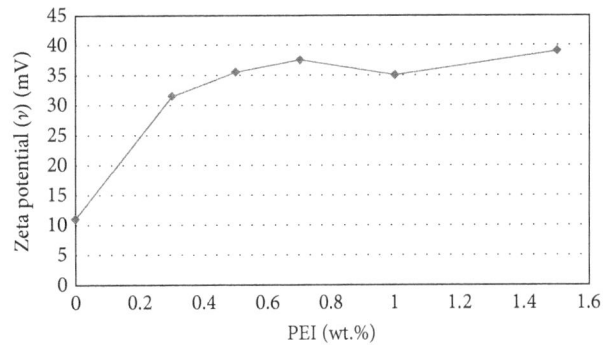

FIGURE 4: Zeta potential of ZSB slurry with various amounts of PEI dispersant.

Zeta potential (ν) of slurries including 45 vol.% solid content and different amounts of dispersant is shown in Figure 4. pH of slurries was set as 7.8. As it is indicated, the zeta potential of slurry without PEI is in unstable area ($-30 < \nu < 30$). Increasing PEI amount, 0.3 wt.%, led to the zeta potential falling in stable area ($\nu > 30$ and $\nu < -30$). A slight increase occurred in the zeta potential while PEI amount increased from 0.3 to 1.5 wt.%. This increase could be explained in terms of adsorption of PEI molecules on surface of particles [23]. Although the data concerning the reaction between ZrB$_2$-SiC-B$_4$C particles and PEI are currently unavailable, Wang and Gao [24] have reported that the adsorption of PEI on ZrB$_2$ is of a high affinity type, and hydrogen bonding was proposed to be the predominant mechanism between PEI and ZrB$_2$ under both acidic and basic conditions.

Protons are adsorbed on PEI molecules when PEI is dissolved in a neutral solution which results in the protonation of the amine group in the molecule [23]. So, the adsorption of positively charged PEI on the surface of the powder is increased. Therefore, as stated earlier, the higher the charge density on particles is, the higher the zeta potential and repulsive force on particles would be [22]. Consequently, it eventually leads to stability of suspension. On the other hand, viscosity of suspension plays a significant role in the stability of suspension. The settling velocity of suspension could be explained by Stokes equation (2) for concentrated colloidal systems. Consider

$$V = \frac{2\Delta\rho g a^2 (1 - \varphi)^{5\pm0.25}}{9\eta}, \qquad (2)$$

where $\Delta\rho$, a, and η are the density difference of particles and continuous phase, particles size, and continuous phase viscosity, respectively. φ is the phase volume percent which adversely affects settling velocity. Although higher solid phase volume percent decreases the velocity of sedimentation and increases viscosity, it causes a higher chance of coagulation of particles. Therefore, a proper value of solid content is necessary to be estimated in a colloidal system.

Rheological behavior of slurries including different amounts of PEI is showed in Figure 5(a). The viscosity of the concentrated slurry increased with the addition of PEI amount up to 1.5 wt.% which was due to very high charge absorbed by PEI molecules on particles surface. Thus, high charge density helps slow down sedimentation of suspension. According to the rheological behavior and viscosity of slurries (Figure 5(b)), as well as zeta potential of slurries, it seems slurries with 1 and 1.5 wt.% PEI would have appropriate stability and viscosity depending on coating process. To fabricate composite based coating by plasma spray, slip casting, and tape casting methods [17–19], the stable slurry with low viscosity is suitable. On the other hand, the stable slurry with absolute proper viscosity (almost in the range of 0.8–1.0 Pa·s) is definitely necessary to fabricate the coat by dip coating.

Previous researches [17–19, 25, 26] show that the optimum slurry which was used for coating includes 40–45 vol.% solid content and different amounts of 1–1.5 wt.% of various

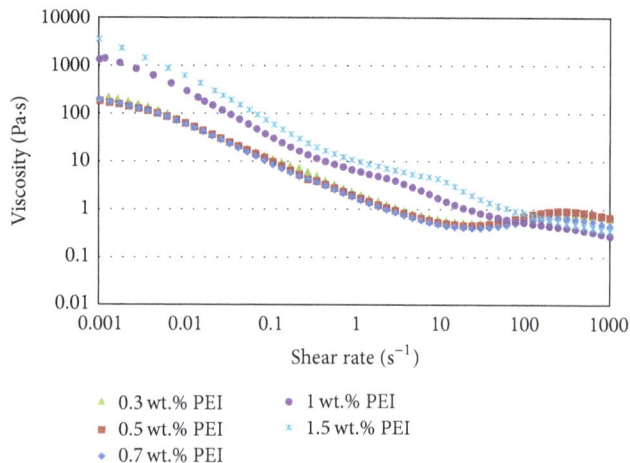

(a)

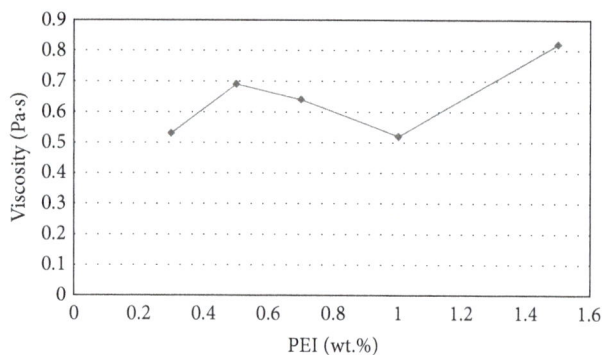

(b)

FIGURE 5: Effect of dispersant content on (a) rheological behavior of 45 vol.% ZSB slurries and (b) the viscosity at 96.2 s^{-1} (pH = 7.8).

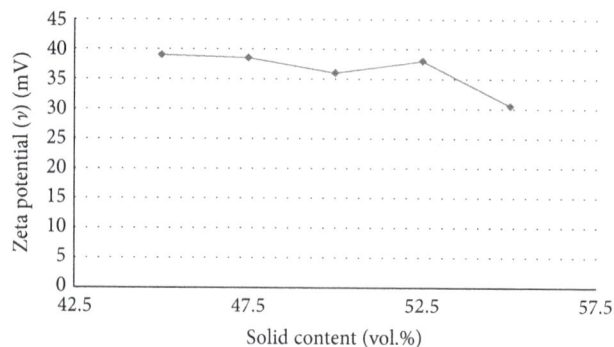

FIGURE 6: Zeta potential of ZSB slurries with 1.5 wt.% PEI and different amounts of solid content.

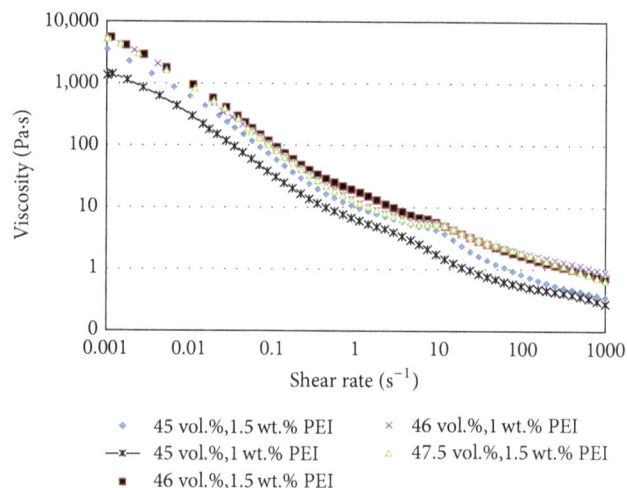

FIGURE 7: Rheological behavior of ZSB slurries including different amounts of PEI and solid content (pH = 7.8).

dispersants (e.g., poly(acrylamine), Dolapix, and Duramax). It is obviously supposed that increasing the solid content of suspension leads to high homogeneity and compaction and subsequently low porosities of coat. Accordingly, the effect of solid content has been studied to determine proper suspension to fabricate ZSB composite based coating. The changes of zeta potential of slurries with various amounts of solid content are depicted in Figure 6.

According to the zeta potential and viscosity of slurries with different amounts of PEI, 1.5 wt.% PEI was a proper amount to prepare the stable slurry for dip coating. Hence, solid content of slurries with 1.5 wt.% PEI was changed to determine the proper slurry for subsequent step, fabrication of ZSB based coating by dipping method. By increasing the solid content from 45 to 52.5 vol.%, the zeta potential of slurries was in stable area; however, increasing to 55 vol.% caused slight diminution of zeta potential falling into the unstable state.

Figure 7 depicts the rheological behavior of ZSB slurries. It indicates that increasing the solid content of slurry, up to

47.5 vol.%, leads to the high viscosity; therefore, slurry with solid content higher than 47.5 vol.% is not suitable for coating step by dip coating method. On the other hand, the high viscose suspension is not suitable for coating by slip casting or tape casting. Table 2 shows the viscosity of slurries. As it seems, the viscosity of slurries with 46 and 47.5 vol.% solid content and 1.5 wt.% PEI is very high. The rheological test was used to estimate the viscosity of slurry including 46 vol.% solid content and 1 wt.% PEI to improve solid content from 45 to 46 vol.%, but unfortunately the viscosity was high again. Indeed, increasing the solid content is very critical since it enhances the chance of the particles agglomeration and consequently increases the particle size. According to (1) and (2), increasing the particle size leads to increasing the gravitation forces and also higher settling velocity. Although increasing φ in (2) causes lower settling velocity (V) and higher viscosity (η), its effect on agglomeration and increasing particle size is more dominant than that on V and η. Therefore, the slurry with optimal condition was obtained including 45 vol.% solid content along with 1 and 1.5 wt.% PEI to be used as composite based coating by slip or tape casting and dip coating method.

TABLE 2: Specification of slurries.

Solid content/vol.%	PEI/wt.%	Viscosity/Pa·s (at 96.2 s^{-1})
45	1.5	0.82
45	1	0.53
46	1.5	1.64
46	1	1.84
47.5	1.5	1.82

4. Conclusion

ZSB composite aqueous suspension was prepared at pH 7.8, adjacent to the neutral state, using PEI as a dispersant. Different values of PEI (0.3–1.5 wt.%) and different amounts of solid content (45–55 vol.%) were used to investigate their effect on stability of aqueous suspensions. A slight increase occurred in the zeta potential while PEI amount increased from 0.3 to 1.5 wt.%. Increasing the solid content from 45 to 52.5 vol.% led to being the zeta potential of slurries in stable state ($\nu > 30$); however, increasing till 55 vol.% caused the sudden decrease to the unstable state. Viscosity of slurry would be low or in a proper amount depending on coating method. Incorporating zeta potential and viscosity showed that the stable composite suspension including 45 vol.% solid content as well as 1 and 1.5 wt.% PEI is an appropriate slurry to produce ZSB composite coat by slip and tape casting and dip coating methods, respectively.

Conflict of Interests

The authors declare that there is no conflict of interests regarding the publication of this paper.

References

[1] S. R. Levine, E. J. Opila, M. C. Halbig, J. D. Kiser, M. Singh, and J. A. Salem, "Evaluation of ultra-high temperature ceramics for aeropropulsion use," *Journal of the European Ceramic Society*, vol. 22, no. 14-15, pp. 2757–2767, 2002.

[2] M. M. Opeka, I. G. Talmy, and J. A. Zaykoski, "Oxidation-based materials selection for 2000°C + hypersonic aerosurfaces: Theoretical considerations and historical experience," *Journal of Materials Science*, vol. 39, no. 19, pp. 5887–5904, 2004.

[3] D. M. van Wie, D. G. Drewry Jr., D. E. King, and C. M. Hudson, "The hypersonic environment: required operating conditions and design challenges," *Journal of Materials Science*, vol. 39, no. 19, pp. 5915–5924, 2004.

[4] S. Kumar, "Self-propagating high temperature synthesis of refractory nitrides, carbides and borides," *Key Engineering Materials*, vol. 56, pp. 183–188, 1991.

[5] K. Upadhya, J.-M. Yang, and W. P. Hoffman, "Materials for ultrahigh temperature structural applications," *The American Ceramic Society Bulletin*, vol. 76, no. 12, pp. 51–56, 1997.

[6] M. M. Opeka, I. G. Talmy, E. J. Wuchina, J. A. Zaykoski, and S. J. Causey, "Mechanical, thermal, and oxidation properties of refractory hafnium and zirconium compounds," *Journal of the European Ceramic Society*, vol. 19, no. 13-14, pp. 2405–2414, 1999.

[7] Y. Yan, H. Zhang, Z. Huang, J. Liu, and D. Jiang, "In situ synthesis of ultrafine ZrB2-SiC composite powders and the pressureless sintering behaviors," *Journal of the American Ceramic Society*, vol. 91, no. 4, pp. 1372–1376, 2008.

[8] S.-H. Lee, Y. Sakka, and Y. Kagawa, "Dispersion behavior of ZrB2 powder in aqueous solution," *Journal of the American Ceramic Society*, vol. 90, no. 11, pp. 3455–3459, 2007.

[9] H. Zhang, Y. J. Yan, Z. Huang, X. Liu, and D. Jiang, "Pressureless sintering of ZrB$_2$-SiC ceramics incorporating sol-gel synthesized ultra-fine ceramic powders," *Key Engineering Materials*, vol. 434-435, pp. 193–196, 2010.

[10] D. W. Ni, G. J. Zhang, Y. M. Kan, and Y. Sakka, "Highly textured ZrB2-based ultrahigh temperature ceramics via strong magnetic field alignment," *Scripta Materialia*, vol. 60, no. 8, pp. 615–618, 2009.

[11] F. Monteverde and A. Bellosi, "Development and characterization of metal-diboride-based composites toughened with ultra-fine SiC particulates," *Solid State Sciences*, vol. 7, no. 5, pp. 622–630, 2005.

[12] Y. Yan, Z. Huang, S. Dong, and D. Jiang, "Pressureless sintering of high-density ZrB2-SiC ceramic composites," *Journal of the American Ceramic Society*, vol. 89, no. 11, pp. 3589–3592, 2006.

[13] S. Zhu, W. G. Fahrenholtz, and G. E. Hilmas, "Influence of silicon carbide particle size on the microstructure and mechanical properties of zirconium diboride-silicon carbide ceramics," *Journal of the European Ceramic Society*, vol. 27, no. 4, pp. 2077–2083, 2007.

[14] S. S. Hwang, A. L. Vasiliev, and N. P. Padture, "Improved processing and oxidation-resistance of ZrB2 ultra-high temperature ceramics containing SiC nanodispersoids," *Materials Science and Engineering A*, vol. 464, no. 1-2, pp. 216–224, 2007.

[15] J. Han, P. Hu, X. Zhang, and S. Meng, "Oxidation behavior of zirconium diboride-silicon carbide at 1800°C," *Scripta Materialia*, vol. 57, no. 9, pp. 825–828, 2007.

[16] F. Monteverde, "Beneficial effects of an ultra-fine α-SiC incorporation on the sinterability and mechanical properties of ZrB2," *Applied Physics A: Materials Science and Processing*, vol. 82, no. 2, pp. 329–337, 2006.

[17] Z. Lü, D. Jiang, J. Zhang, and Q. Lin, "Processing and properties of ZrB$_2$-SiC composites obtained by aqueous tape casting and hot pressing," *Ceramics International*, vol. 37, no. 1, pp. 293–301, 2011.

[18] X. G. Wang, J. X. Liu, Y. M. Kan, G. J. Zhang, and P.-L. Wang, "Slip casting and pressureless sintering of ZrB2-SiC ceramics," *Journal of Inorganic Materials*, vol. 24, no. 4, pp. 831–835, 2009.

[19] F. F. Lange, "Powder processing science and technology for increased reliability," *Journal of the American Ceramic Society*, vol. 72, no. 1, pp. 3–15, 1989.

[20] J. N. Israelachvili, *Intermolecular and Surface Forces*, Academic Press, New York, NY, USA, 2009.

[21] E. J. W. Verwey and J. T. G. Overbeek, *Theory of the Stability of Lyophobic Colloids*, Elsevier, Amsterdam, The Netherlands, 1948.

[22] R. J. Hunter, *Zeta Potential in Colloid Science: Principles and Applications*, Academic Press, London, UK, 1988.

[23] X. Zhu, F. Tang, T. S. Suzuki, and Y. Sakka, "Role of the initial degree of ionization of polyethylenimine in the dispersion of silicon carbide nanoparticles," *Journal of the American Ceramic Society*, vol. 86, no. 1, pp. 189–191, 2003.

[24] J. Wang and L. Gao, "Surface properties of polymer adsorbed zirconia nanoparticles," *Journal of Nanostructured Materials*, vol. 11, no. 4, pp. 451–457, 1999.

[25] Z. Lü, D. Jiang, J. Zhang, and Q. Lin, "Aqueous tape casting of zirconium diboride," *Journal of the American Ceramic Society*, vol. 92, no. 10, pp. 2212–2217, 2009.

[26] Z. Lü, D. Jiang, J. Zhang, and Q. Lin, "Microstructure and mechanical properties of zirconium diboride obtained by aqueous tape casting process and hot pressing," *Journal of the American Ceramic Society*, vol. 93, no. 12, pp. 4153–4157, 2010.

28

A Comprehensive Study on Gamma-Ray Exposure Build-Up Factors and Fast Neutron Removal Cross Sections of Fly-Ash Bricks

Vishwanath P. Singh[1,2] and N. M. Badiger[1]

[1] *Department of Physics, Karnatak University, Dharwad 580003, India*
[2] *Health Physics Section, Kaiga Atomic Power Station-3&4, NPCIL, Karwar 581400, India*

Correspondence should be addressed to Vishwanath P. Singh; kudphyvps@rediffmail.com

Academic Editor: Shaomin Liu

Geometric progression (GP) method was utilized to investigate gamma-ray exposure build-up factors of fly-ash bricks for energies from 0.015 to 15 MeV up to 40 mfp penetration depth. The EBFs of the fly-ash bricks are dependent upon the photon energy, penetration depths, and the chemical compositions of the elements. Appreciable variations in exposure build-up factor (EBF) are noted for the fly-ash bricks. The EBFs were found to be small in low and high photon energy regions whereas very large in medium energy region. EBF of the bricks is inversely proportional to equivalent atomic number below 10 mfp for entire energy region of interest 0.015 to 15 MeV. The EBFs of fly-ash, brick of mud, and common brick were similar at 1.5 MeV photon energy. The EBF of the fly-ash bricks was found to be higher than that of the brick of mud, and common brick. The fast neutron removal cross sections of the fly-ash bricks, brick of mud, and common bricks were also calculated which were found to be in the same order. It is expected that this study should be very directly useful for shielding effectiveness of fly-ash brick materials and dose estimation.

1. Introduction

Safety inside residential and nonresidential building against the radiation is evaluated by the shielding properties by the parameters such as mass attenuation coefficients, energy absorption coefficients, and half-value layer. Gamma-ray interacts with material by photoelectric absorption, compton scattering, and pair production, which depends upon photon energy and element compositions. The intensity of a gamma-ray beam through a medium follows the Lambert Beer law under three conditions namely, (i) monochromatic rays, (ii) thin absorbing material, and (iii) narrow beam geometry. In case, any of the above conditions is not being met, this law is no longer applicable. The law can be applicable by using a correction factor, called as "build-up factor."

The concept of build-up factor was introduced in late 1950 [2] by obtaining experimentally the build-up factor at 1.25 MeV gamma-ray energy for water up to 16 mfp. The importance of build-up factor in attenuation studies was further recognized for multienergy gamma-rays with poor geometry [3]. Since 1950, due to the availability of reasonably accurate values of attenuation coefficients and cross section of the various mediums, a great progress has been made in the computation of build-up factor in different types of materials such as medical, dosimetric, shielding, and radiation protection.

The build-up factor is a dimensionless multiplication factor which corrects the response of uncollided photon beam. The build-up is defined as the ratio of total value of specified radiation quantity at any point to the contribution to that value from radiation reaching the point without having undergone a collision. There are two types of build-up factors which are the quantity of interest: (a) the absorbed or deposited energy in the interacting materials and detector response function is that of absorption in the interacting medium; (b) the exposure build-up factor in which quality of interest is the exposure and detector response function is that of absorption in air [4]. The build-up factors have been computed by various codes such as PALLAS [5], ADJMON-I [6, 7], ASFIT [8], and EGS4 [9]. These codes are using

TABLE 1: Elemental compositions of fly-ash bricks, brick of mud, and common brick [1].

Element	Elemental composition (% by weight)									
Density (g cm^{-3})	FAB1	FAB2	FAB3	FAB4	FAB5	FAB6	FAB7	FAB8	BOM	COM
	1.87	1.81	1.76	1.72	1.85	1.79	1.74	1.70	1.82	1.87
H	0.09	0.09	0.09	0.09	0.09	0.09	0.09	0.09	—	0.24
C	3.69	3.41	3.13	2.84	1.70	1.70	1.70	1.70	—	0.06
N	—	—	—	—	—	—	—	—	—	0.01
O	48.71	48.70	48.69	48.68	49.96	49.77	49.58	49.39	54.6	45.35
Na	0.01	0.01	0.01	0.01	0.11	0.09	0.08	0.07	0.11	0.01
Mg	1.07	1.03	0.99	0.94	0.77	0.77	0.77	0.77	0.11	0.04
Al	11.77	11.13	10.5	9.86	10.78	10.28	9.79	9.29	4.95	7.88
Si	23.22	23.09	22.96	22.83	25.08	24.68	24.29	23.9	5.96	20.78
P	—	—	—	—	0.07	0.06	0.05	0.04	—	—
S	0.04	0.04	0.03	0.03	0.02	0.02	0.02	0.02	—	0.02
K	0.42	0.80	1.18	1.56	0.44	0.82	1.20	1.57	—	2.66
Ca	6.92	6.77	6.61	6.45	7.32	7.10	6.88	6.67	34.01	4.75
Ti	0.01	0.01	0.01	0.01	0.57	0.49	0.41	0.33	—	0.09
Mn	0.01	0.01	0.01	0.01	0.01	0.01	0.01	0.01	—	0.17
Fe	4.07	4.95	5.84	6.73	3.12	4.15	5.17	6.19	0.30	17.96

TABLE 2: Equivalent atomic number of fly-ash bricks and other bricks.

Energy (MeV)	Equivalent atomic number (Z_{eq})									
	FAB1	FAB2	FAB3	FAB4	FAB5	FAB6	FAB7	FAB8	BOM	COB
0.015	13.16	13.39	13.61	13.81	13.12	13.35	13.58	13.79	14.61	15.97
0.02	13.32	13.56	13.78	14.00	13.27	13.51	13.75	13.97	14.78	16.22
0.03	13.50	13.74	13.98	14.20	13.43	13.69	13.93	14.16	14.95	16.50
0.04	13.61	13.87	14.11	14.34	13.53	13.80	14.06	14.30	15.06	16.67
0.05	13.69	13.95	14.20	14.44	13.62	13.89	14.15	14.40	15.14	16.79
0.06	13.76	14.03	14.27	14.51	13.69	13.96	14.21	14.47	15.20	16.88
0.08	13.85	14.13	14.37	14.61	13.77	14.04	14.31	14.56	15.28	17.02
0.1	13.91	14.21	14.43	14.68	13.83	14.10	14.37	14.63	15.33	17.12
0.15	14.02	14.33	14.54	14.80	13.91	14.19	14.47	14.74	15.41	17.24
0.2	14.07	14.33	14.60	14.87	13.96	14.25	14.53	14.81	15.46	17.31
0.3	14.13	14.40	14.67	14.95	14.02	14.31	14.60	14.89	15.51	17.40
0.4	14.16	14.44	14.72	15.00	14.05	14.34	14.64	14.93	15.54	17.45
0.5	14.18	14.46	14.74	15.02	14.07	14.36	14.66	14.96	15.55	17.48
0.6	14.19	14.47	14.76	15.04	14.08	14.38	14.67	14.97	15.56	17.50
0.8	14.20	14.48	14.77	15.05	14.08	14.39	14.68	14.98	15.57	17.52
1	14.20	14.49	14.77	15.05	14.09	14.39	14.69	14.99	15.57	17.52
1.5	12.22	12.44	12.65	12.86	12.23	12.44	12.66	12.87	13.66	15.15
2	11.69	11.85	12.01	12.16	11.74	11.89	12.04	12.19	13.00	14.01
3	11.56	11.71	11.85	11.99	11.62	11.75	11.89	12.02	12.82	13.70
4	11.52	11.66	11.80	11.94	11.57	11.71	11.84	11.97	12.76	13.63
5	11.51	11.64	11.78	11.92	11.56	11.70	11.83	11.96	12.75	13.59
6	11.50	11.63	11.77	11.91	11.56	11.69	11.81	11.94	12.73	13.57
8	11.49	11.62	11.76	11.90	11.54	11.67	11.80	11.93	12.72	13.54
10	11.48	11.61	11.75	11.89	11.54	11.67	11.79	11.92	12.71	13.53
15	11.47	11.61	11.74	11.88	11.53	11.66	11.79	11.91	12.70	13.51

an accurate algorithmic for the Klein-Nishina cross section which eliminated other sources of errors.

The compilation for build-up factors by various codes was reported in ANSI/ANS-6.4.3-1991 by American Nuclear Society [10]. The data in the report covers energy range from 0.015 to 15 MeV up to penetration depth of 40 mean free path (mfp). The build-up factors in the ANS-6.4.3 are for 23 elements of atomic number, $Z = 4$ to 92. The build-up factors of ANS-6.4.3 can be calculated by invariant embedding [11, 12]. Harima et al. [13] developed a fitting formula, called geometric progression (GP), which gave build-up factors of the good agreement with the ANS-6.4.3. The GP fitting is more accurate than three exponential fit in the water medium. The GP fitting formula is known to be accurate within the

TABLE 3: GP fitting parameters for FAB1 and FAB2.

Energy (MeV)	b	c	a	X_K	d
			FAB1		
0.015	1.028	0.386	0.215	14.919	−0.146
0.02	1.062	0.398	0.206	15.078	−0.116
0.03	1.201	0.408	0.207	14.534	−0.110
0.04	1.421	0.476	0.180	14.500	−0.100
0.05	1.677	0.605	0.124	15.483	−0.065
0.06	1.963	0.701	0.095	14.727	−0.053
0.08	2.435	0.851	0.055	14.413	−0.048
0.1	2.646	1.026	0.014	13.593	−0.029
0.15	2.731	1.272	−0.042	10.413	−0.006
0.2	2.670	1.371	−0.057	7.813	−0.005
0.3	2.476	1.464	−0.077	17.191	0.013
0.4	2.350	1.467	−0.080	16.046	0.016
0.5	2.247	1.451	−0.079	16.302	0.018
0.6	2.171	1.422	−0.076	18.225	0.021
0.8	2.049	1.383	−0.073	15.355	0.021
1	1.976	1.323	−0.063	16.064	0.019
1.5	1.864	1.232	−0.048	14.866	0.017
2	1.789	1.155	−0.033	15.608	0.011
3	1.678	1.061	−0.012	13.922	−0.001
4	1.602	0.995	0.005	12.945	−0.009
5	1.537	0.948	0.020	12.601	−0.021
6	1.474	0.924	0.027	11.700	−0.022
8	1.406	0.902	0.033	13.689	−0.027
10	1.345	0.881	0.042	13.141	−0.033
15	1.258	0.835	0.061	14.297	−0.052
			FAB2		
0.015	1.027	0.376	0.228	14.103	−0.155
0.02	1.058	0.408	0.197	15.983	−0.113
0.03	1.190	0.402	0.212	14.267	−0.113
0.04	1.397	0.466	0.185	14.389	−0.104
0.05	1.646	0.583	0.133	15.174	−0.071
0.06	1.890	0.700	0.093	15.276	−0.052
0.08	2.374	0.827	0.061	14.429	−0.051
0.1	2.586	1.001	0.016	13.739	−0.032
0.15	2.692	1.248	−0.037	10.823	−0.009
0.2	2.647	1.356	−0.055	8.168	−0.006
0.3	2.462	1.454	−0.076	17.822	0.013
0.4	2.342	1.458	−0.078	16.424	0.015
0.5	2.239	1.445	−0.078	16.305	0.017
0.6	2.162	1.420	−0.076	17.592	0.020
0.8	2.044	1.380	−0.072	15.458	0.021
1	1.970	1.323	−0.063	16.408	0.019
1.5	1.862	1.232	−0.048	15.186	0.017
2	1.788	1.155	−0.033	15.865	0.011
3	1.677	1.062	−0.012	14.771	−0.001
4	1.601	0.995	0.005	12.967	−0.009
5	1.537	0.946	0.021	11.908	−0.021
6	1.477	0.927	0.026	11.786	−0.021
8	1.405	0.902	0.034	13.732	−0.028
10	1.344	0.882	0.042	13.128	−0.033
15	1.257	0.838	0.060	14.282	−0.052

TABLE 4: GP fitting parameters for FAB3 and FAB4.

Energy (MeV)	b	c	a	X_K	d
			FAB3		
0.015	1.025	0.366	0.240	13.337	−0.164
0.02	1.054	0.417	0.188	16.831	−0.110
0.03	1.180	0.396	0.216	14.016	−0.116
0.04	1.377	0.458	0.189	14.341	−0.108
0.05	1.618	0.567	0.141	15.031	−0.086
0.06	1.854	0.683	0.099	15.149	−0.055
0.08	2.327	0.808	0.068	14.016	−0.051
0.1	2.542	0.984	0.020	13.835	−0.034
0.15	2.667	1.232	−0.034	11.093	−0.010
0.2	2.624	1.341	−0.052	8.518	−0.007
0.3	2.448	1.444	−0.074	18.446	0.013
0.4	2.333	1.449	−0.077	16.798	0.015
0.5	2.231	1.439	−0.077	16.307	0.016
0.6	2.154	1.418	−0.075	16.966	0.019
0.8	2.039	1.377	−0.071	15.559	0.021
1	1.964	1.323	−0.063	16.748	0.020
1.5	1.859	1.231	−0.048	15.503	0.017
2	1.787	1.155	−0.033	16.105	0.011
3	1.675	1.063	−0.013	15.611	−0.002
4	1.600	0.996	0.005	12.989	−0.009
5	1.538	0.944	0.022	11.222	−0.021
6	1.480	0.930	0.026	11.870	−0.020
8	1.405	0.902	0.034	13.775	−0.028
10	1.343	0.884	0.041	13.114	−0.033
15	1.255	0.841	0.059	14.267	−0.052
			FAB4		
0.015	1.024	0.357	0.252	12.616	−0.172
0.02	1.051	0.426	0.179	17.628	−0.108
0.03	1.171	0.395	0.215	14.076	−0.033
0.04	1.360	0.453	0.190	14.365	−0.107
0.05	1.592	0.553	0.147	14.928	−0.080
0.06	1.820	0.666	0.105	15.025	−0.058
0.08	2.278	0.788	0.074	13.592	−0.051
0.1	2.494	0.964	0.025	13.941	−0.037
0.15	2.635	1.213	−0.030	11.426	−0.012
0.2	2.601	1.326	−0.049	8.863	−0.008
0.3	2.434	1.435	−0.072	19.060	0.013
0.4	2.325	1.440	−0.075	17.165	0.014
0.5	2.223	1.433	−0.076	16.309	0.016
0.6	2.146	1.415	−0.075	16.510	0.018
0.8	2.035	1.374	−0.071	15.691	0.020
1	1.959	1.322	−0.063	16.955	0.020
1.5	1.857	1.230	−0.048	15.815	0.017
2	1.786	1.155	−0.033	15.976	0.011
3	1.674	1.064	−0.013	16.447	−0.002
4	1.598	0.997	0.005	13.010	−0.009
5	1.538	0.942	0.023	10.541	−0.022
6	1.482	0.932	0.025	11.955	−0.020
8	1.405	0.902	0.034	13.818	−0.028
10	1.343	0.885	0.041	13.101	−0.033
15	1.254	0.844	0.058	14.253	−0.051

estimated uncertainty (<5%) and Harima [4] had extensive historical review and reported the current gamma photon

build-up factors and applications [14]. Various researchers have investigated gamma-ray build-up factors in different

TABLE 5: GP fitting parameters for FAB5 and FAB6.

Energy (MeV)	b	c	a	X_K	d
			FAB5		
0.015	1.028	0.389	0.213	15.089	−0.144
0.02	1.062	0.395	0.209	14.855	−0.117
0.03	1.204	0.410	0.206	14.612	−0.110
0.04	1.429	0.480	0.178	14.539	−0.099
0.05	1.687	0.611	0.122	15.574	−0.063
0.06	1.984	0.700	0.096	14.558	−0.056
0.08	2.453	0.857	0.053	14.293	−0.038
0.1	2.663	1.033	0.009	13.540	−0.021
0.15	2.745	1.280	−0.043	11.334	0.006
0.2	2.679	1.378	−0.059	8.048	0.011
0.3	2.482	1.468	−0.078	16.930	0.014
0.4	2.354	1.471	−0.081	15.887	0.016
0.5	2.250	1.454	−0.080	16.301	0.018
0.6	2.175	1.423	−0.076	18.496	0.021
0.8	2.051	1.384	−0.073	15.311	0.021
1	1.978	1.323	−0.063	15.916	0.019
1.5	1.864	1.232	−0.048	14.873	0.017
2	1.789	1.155	−0.033	15.683	0.011
3	1.678	1.061	−0.012	14.248	−0.001
4	1.601	0.995	0.005	12.954	−0.009
5	1.537	0.947	0.021	12.308	−0.021
6	1.475	0.925	0.027	11.737	−0.022
8	1.405	0.902	0.034	13.708	−0.027
10	1.344	0.881	0.042	13.135	−0.033
15	1.258	0.836	0.060	14.290	−0.052
			FAB6		
0.015	1.027	0.378	0.226	14.230	−0.153
0.02	1.058	0.406	0.199	15.816	−0.114
0.03	1.192	0.403	0.211	14.326	−0.113
0.04	1.403	0.468	0.184	14.417	−0.103
0.05	1.653	0.588	0.131	15.251	−0.069
0.06	1.907	0.702	0.093	15.187	−0.052
0.08	2.391	0.834	0.059	14.583	−0.051
0.1	2.607	1.010	0.014	13.694	−0.031
0.15	2.709	1.258	−0.039	10.645	−0.008
0.2	2.654	1.361	−0.055	8.052	−0.006
0.3	2.467	1.458	−0.076	17.611	0.013
0.4	2.344	1.461	−0.079	16.295	0.015
0.5	2.242	1.447	−0.079	16.304	0.017
0.6	2.165	1.421	−0.076	17.812	0.020
0.8	2.046	1.381	−0.072	15.422	0.021
1	1.972	1.323	−0.063	16.288	0.019
1.5	1.862	1.232	−0.048	15.195	0.017
2	1.788	1.155	−0.033	15.931	0.011
3	1.676	1.062	−0.013	15.053	−0.001
4	1.600	0.996	0.005	12.975	−0.009
5	1.537	0.945	0.022	11.655	−0.021
6	1.478	0.928	0.026	11.817	−0.021
8	1.405	0.902	0.034	13.749	−0.028
10	1.344	0.883	0.042	13.122	−0.033
15	1.256	0.839	0.059	14.276	−0.052

TABLE 6: GP fitting parameters for FAB7 and FAB8.

Energy (MeV)	b	c	a	X_K	d
			FAB7		
0.015	1.025	0.368	0.239	13.437	−0.162
0.02	1.055	0.416	0.189	16.700	−0.111
0.03	1.182	0.397	0.215	14.062	−0.116
0.04	1.381	0.459	0.188	14.336	−0.107
0.05	1.624	0.570	0.139	15.055	−0.074
0.06	1.863	0.687	0.098	15.178	−0.054
0.08	2.339	0.813	0.066	14.122	−0.051
0.1	2.554	0.989	0.019	13.808	−0.038
0.15	2.675	1.237	−0.035	11.003	−0.010
0.2	2.630	1.345	−0.053	8.423	−0.007
0.3	2.452	1.447	−0.074	18.271	0.013
0.4	2.336	1.452	−0.077	16.691	0.015
0.5	2.233	1.440	−0.077	16.307	0.017
0.6	2.157	1.419	−0.075	17.148	0.019
0.8	2.041	1.378	−0.072	15.529	0.021
1	1.965	1.323	−0.063	16.649	0.020
1.5	1.859	1.231	−0.048	15.509	0.017
2	1.787	1.155	−0.033	16.068	0.011
3	1.675	1.063	−0.013	15.844	−0.002
4	1.599	0.996	0.005	12.995	−0.009
5	1.538	0.944	0.023	11.014	−0.021
6	1.480	0.930	0.025	11.897	−0.020
8	1.405	0.902	0.034	13.789	−0.028
10	1.343	0.884	0.041	13.110	−0.033
15	1.255	0.842	0.058	14.263	−0.052
			FAB8		
0.015	1.024	0.358	0.251	12.694	−0.171
0.02	1.051	0.425	0.180	17.526	−0.108
0.03	1.173	0.395	0.215	14.061	−0.116
0.04	1.363	0.453	0.190	14.361	−0.107
0.05	1.597	0.555	0.146	14.946	−0.079
0.06	1.826	0.669	0.104	15.048	−0.057
0.08	2.071	0.792	0.073	13.676	−0.051
0.1	2.422	0.968	0.024	13.919	−0.036
0.15	2.642	1.217	−0.031	11.354	−0.012
0.2	2.606	1.329	−0.050	8.786	−0.008
0.3	2.437	1.437	−0.073	18.919	0.013
0.4	2.327	1.442	−0.075	17.080	0.014
0.5	2.225	1.434	−0.076	16.310	0.016
0.6	2.148	1.416	−0.075	16.496	0.018
0.8	2.035	1.375	−0.071	15.634	0.021
1	1.959	1.323	−0.063	17.003	0.020
1.5	1.857	1.230	−0.048	15.819	0.017
2	1.786	1.155	−0.033	15.955	0.010
3	1.674	1.064	−0.013	16.350	−0.002
4	1.598	0.997	0.005	13.015	−0.009
5	1.538	0.942	0.024	10.380	−0.022
6	1.483	0.933	0.024	11.975	−0.020
8	1.405	0.902	0.034	13.828	−0.028
10	1.342	0.885	0.041	13.098	−0.033
15	1.254	0.845	0.058	14.249	−0.051

materials such as concretes [14–16], gaseous mixture [17], human tissues [18], soils and ceramic [19, 20] which showed that the GP fitting is a very useful method for estimation of exposure and energy absorption build-up factors. Recently

TABLE 7: GP fitting parameters for BOM and COB.

Energy (MeV)	b	c	a	X_K	d
			BOM		
0.015	1.020	0.379	0.231	11.998	−0.148
0.02	1.043	0.431	0.176	14.650	−0.084
0.03	1.143	0.394	0.213	14.390	−0.115
0.04	1.310	0.438	0.196	14.441	−0.109
0.05	1.519	0.515	0.164	14.678	−0.091
0.06	1.727	0.621	0.122	14.749	−0.067
0.08	2.135	0.759	0.081	13.507	−0.050
0.1	2.380	0.916	0.038	13.703	−0.042
0.15	2.561	1.171	−0.022	12.064	−0.016
0.2	2.548	1.296	−0.044	9.929	−0.010
0.3	2.419	1.397	−0.064	13.335	0.001
0.4	2.303	1.427	−0.073	18.953	0.015
0.5	2.211	1.423	−0.074	16.293	0.015
0.6	2.140	1.402	−0.072	17.362	0.016
0.8	2.032	1.362	−0.068	16.187	0.018
1	1.954	1.317	−0.062	16.346	0.019
1.5	1.846	1.230	−0.048	15.414	0.016
2	1.781	1.153	−0.032	15.316	0.009
3	1.673	1.058	−0.011	11.803	−0.002
4	1.605	0.979	0.011	12.405	−0.015
5	1.531	0.953	0.019	10.723	−0.017
6	1.482	0.933	0.025	12.246	−0.023
8	1.398	0.913	0.031	13.872	−0.027
10	1.336	0.897	0.039	13.047	−0.033
15	1.247	0.863	0.052	14.713	−0.049
			COB		
0.015	1.017	0.296	0.321	10.490	−0.248
0.02	1.038	0.337	0.231	26.819	−0.340
0.03	1.106	0.376	0.225	13.775	−0.048
0.04	1.230	0.409	0.208	14.579	−0.116
0.05	1.388	0.465	0.187	14.455	−0.105
0.06	1.553	0.545	0.153	14.502	−0.085
0.08	1.846	0.706	0.093	14.638	−0.054
0.1	2.089	0.837	0.056	13.860	−0.042
0.15	2.342	1.074	−0.003	13.216	−0.023
0.2	2.398	1.206	−0.027	11.700	−0.017
0.3	2.330	1.327	−0.050	8.675	−0.010
0.4	2.247	1.361	−0.058	11.468	−0.003
0.5	2.159	1.387	−0.068	20.771	0.017
0.6	2.099	1.374	−0.067	18.708	0.015
0.8	2.002	1.344	−0.064	16.552	0.017
1	1.931	1.305	−0.059	15.991	0.014
1.5	1.834	1.224	−0.047	15.476	0.015
2	1.771	1.155	−0.033	14.576	0.010
3	1.670	1.058	−0.010	10.534	−0.003
4	1.593	0.997	0.006	14.562	−0.014
5	1.525	0.963	0.017	10.978	−0.016
6	1.475	0.945	0.022	13.346	−0.021
8	1.392	0.924	0.028	13.653	−0.025
10	1.332	0.905	0.037	13.208	−0.032
15	1.240	0.878	0.048	14.114	−0.045

the radiation shielding by fly-ash concretes [14] and building materials [21] has been reported.

The innovative bricks using the residual fly-ash are considered high-quality building materials by the manufacturers which will potentially decrease some of the negative environmental impact of coal-fired power generation while meeting increasing demands for greener building materials [22, 23]. Fly-ash brick (FAB), an environment-friendly cost-saving building product, is an alternative to burnt clay bricks. The FAB is approximately stronger than common bricks with consistent strength. The FABs are ideally suited for internal, external, load bearing, and nonload bearing walls. FABs are durable, economical, and eco-friendly and have low water absorption (8–12%), less mortar consumption, and low energy consumption with the lowest green house impact. These bricks are not affected by environmental conditions and remain static thus ensuring longer life of the building. These bricks are economical/cost-effective and nil wastage while transporting and handling. The houses and buildings in which people are living are constructed by the bricks made up of soil and environmental-friendly fly-ashes. The potential applications of fly-ash are shielding materials [24], glasses [25, 26], X-ray shielding [27], electromagnetic radiation in X-band and Ku-band shielding [28], and houses and building construction [22, 23].

In view of radiation safety inside the houses or buildings constructed by FAB, a theoretical gamma-ray exposure build-up factor (EBF) and fast neutron removal cross section have been calculated. We have calculated the EBF of eight types of FABs, brick of mud (BOM), and common brick (COB) by GP fitting in the photon energy range from 0.015 to 15 MeV up to 40 mfp. Comparative analysis shows that higher build-up factors exist for FABs and lower fast neutron removal cross sections. High EBF proves that FABs are poor gamma shielding for the construction of houses and buildings. The study reveals that brick of mud and common brick are low-cost safe building materials against radiation. It should be noted that this study is valuable in shielding analysis and estimation of emergency dose.

2. Computation Method

The elemental compositions of the fly-ash bricks are given in Table 1 [1]. These fly-ash bricks samples are prepared by a formula (Lime)$_{0.15}$ (Gypsum)$_{0.05}$ (Fly Ash) × (Soil)$_{0.8-x}$, where the values of x range from 0.4 to 0.7. Two other bricks of mud (BOM) and common brick (COM) are also analyzed for comparison of shielding properties for gamma-ray and neutron. The build-up calculation by GP fitting method and fast neutron removal cross section is explained below.

2.1. Exposure Build-Up Factor. The EBF of the bricks and the GP fitting parameters are calculated by method of interpolation from the equivalent atomic number, Z_{eq}, of the bricks. The computational work of these parameters is done in three steps as follows:

(1) calculation of equivalent atomic number (Table 2),

(2) calculation of GP fitting parameters (Tables 3, 4, 5, 6, and 7),

(3) calculation of build-up factors.

TABLE 8: Fast neutron removal cross section of fly-ash brick and other bricks.

(a)

Ele.	FAB1 $\rho = 1.87\,\mathrm{g\,cm^{-3}}$		FAB2 $\rho = 1.81\,\mathrm{g\,cm^{-3}}$		FAB $\rho = 1.76\,\mathrm{g\,cm^{-3}}$		FAB4 $\rho = 1.72\,\mathrm{g\,cm^{-3}}$		FAB5 $\rho = 1.85\,\mathrm{g\,cm^{-3}}$	
	Partial density	Σ_R (cm^{-1})	Partial density	Σ_R (cm^{-1})	Partial density	Σ_R (cm^{-1})	Partial density	Σ_R (cm^{-1})	Partial density	Σ_R (cm^{-1})
H	1.68E − 03	1.01E − 03	1.63E − 03	9.74E − 04	1.58E − 03	9.47E − 04	1.55E − 03	9.26E − 04	1.67E − 03	9.96E − 04
C	6.90E − 02	3.46E − 03	6.17E − 02	3.10E − 03	5.51E − 02	2.77E − 03	4.88E − 02	2.45E − 03	3.15E − 02	1.58E − 03
N	0.00E + 00	0.00E + 00	0.00E + 00	0.00E + 00	0.00E + 00	0.00E + 00	0.00E + 00	0.00E + 00	0.00E + 00	0.00E + 00
O	9.11E − 01	3.69E − 02	8.81E − 01	3.57E − 02	8.57E − 01	3.47E − 02	8.37E − 01	3.39E − 02	9.24E − 01	3.74E − 02
Na	1.87E − 04	6.38E − 06	1.81E − 04	6.17E − 06	1.76E − 04	6.00E − 06	1.72E − 04	5.87E − 06	2.04E − 03	6.94E − 05
Mg	2.00E − 02	6.66E − 04	1.86E − 02	6.21E − 04	1.74E − 02	5.80E − 04	1.62E − 02	5.38E − 04	1.42E − 02	4.74E − 04
Al	2.20E − 01	6.45E − 03	2.01E − 01	5.90E − 03	1.85E − 01	5.42E − 03	1.70E − 01	4.97E − 03	1.99E − 01	5.84E − 03
Si	4.34E − 01	1.28E − 02	4.18E − 01	1.23E − 02	4.04E − 01	1.19E − 02	3.93E − 01	1.16E − 02	4.64E − 01	1.37E − 02
P	0.00E + 00	0.00E + 00	0.00E + 00	0.00E + 00	0.00E + 00	0.00E + 00	0.00E + 00	0.00E + 00	1.30E − 03	3.66E − 05
S	7.48E − 04	2.07E − 05	7.24E − 04	2.01E − 05	5.28E − 04	1.46E − 05	5.16E − 04	1.43E − 05	3.70E − 04	1.02E − 05
K	7.85E − 03	1.94E − 04	1.45E − 02	3.58E − 04	2.08E − 02	5.13E − 04	2.68E − 02	6.63E − 04	8.14E − 03	2.01E − 04
Ca	1.29E − 01	3.15E − 03	1.23E − 01	2.98E − 03	1.16E − 01	2.83E − 03	1.11E − 01	2.70E − 03	1.35E − 01	3.29E − 03
Ti	1.87E − 04	3.83E − 06	1.81E − 04	3.71E − 06	1.76E − 04	3.61E − 06	1.72E − 04	3.53E − 06	1.05E − 02	2.16E − 04
Mn	1.87E − 04	3.80E − 06	1.81E − 04	3.67E − 06	1.76E − 04	3.57E − 06	1.72E − 04	3.49E − 06	1.85E − 04	3.76E − 06
Fe	7.61E − 02	1.63E − 03	8.96E − 02	1.92E − 03	1.03E − 01	2.20E − 03	1.16E − 01	2.48E − 03	5.77E − 02	1.24E − 03
Total	1.87	0.0663	1.81	0.0639	1.76	0.0619	1.72	0.0602	1.85	0.0651

(b)

Ele.	FAB6 $\rho = 1.79\,\mathrm{g\,cm^{-3}}$		FAB7 $\rho = 1.74\,\mathrm{g\,cm^{-3}}$		FAB8 $\rho = 1.70\,\mathrm{g\,cm^{-3}}$		BOM $\rho = 1.82\,\mathrm{g\,cm^{-3}}$		COB $\rho = 1.87\,\mathrm{g\,cm^{-3}}$	
	Partial density	Σ_R (cm^{-1})	Partial density	Σ_R (cm^{-1})	Partial density	Σ_R (cm^{-1})	Partial density	Σ_R (cm^{-1})	Partial density	Σ_R (cm^{-1})
H	1.61E − 03	9.63E − 04	1.57E − 03	9.36E − 04	1.53E − 03	9.15E − 04	0.00E + 00	0.00E + 00	4.49E − 03	2.68E − 03
C	3.04E − 02	1.53E − 03	2.96E − 02	1.48E − 03	2.89E − 02	1.45E − 03	0.00E + 00	0.00E + 00	1.12E − 03	5.63E − 05
N	0.00E + 00	0.00E + 00	0.00E + 00	0.00E + 00	0.00E + 00	0.00E + 00	0.00E + 00	0.00E + 00	1.87E − 04	8.38E − 06
O	8.91E − 01	3.61E − 02	8.62E − 01	3.49E − 02	8.39E − 01	3.40E − 02	9.93E − 01	4.02E − 02	8.48E − 01	3.43E − 02
Na	1.61E − 03	5.49E − 05	1.39E − 03	4.74E − 05	1.19E − 03	4.06E − 05	2.00E − 03	6.82E − 05	1.87E − 04	6.38E − 06
Mg	1.38E − 02	4.59E − 04	1.34E − 02	4.46E − 04	1.31E − 02	4.36E − 04	2.00E − 03	6.66E − 05	7.48E − 04	2.49E − 05
Al	1.84E − 01	5.39E − 03	1.70E − 01	4.99E − 03	1.58E − 01	4.63E − 03	9.01E − 02	2.64E − 03	1.47E − 01	4.32E − 03
Si	4.42E − 01	1.30E − 02	4.22E − 01	1.25E − 02	4.06E − 01	1.20E − 02	1.08E − 01	3.20E − 03	3.89E − 01	1.15E − 02
P	1.07E − 03	3.04E − 05	8.70E − 04	2.46E − 05	6.80E − 04	1.92E − 05	0.00E + 00	0.00E + 00	0.00E + 00	0.00E + 00
S	3.58E − 04	9.91E − 06	3.48E − 04	9.64E − 06	3.40E − 04	9.41E − 06	0.00E + 00	0.00E + 00	3.74E − 04	1.04E − 05
K	1.47E − 02	3.62E − 04	2.09E − 02	5.16E − 04	2.67E − 02	6.59E − 04	0.00E + 00	0.00E + 00	4.97E − 02	1.23E − 03
Ca	1.27E − 01	3.09E − 03	1.20E − 01	2.91E − 03	1.13E − 01	2.75E − 03	6.19E − 01	1.50E − 02	8.88E − 02	2.16E − 03
Ti	8.77E − 03	1.80E − 04	7.13E − 03	1.46E − 04	5.61E − 03	1.15E − 04	0.00E + 00	0.00E + 00	1.68E − 03	3.45E − 05
Mn	1.79E − 04	3.63E − 06	1.74E − 04	3.53E − 06	1.70E − 04	3.45E − 06	0.00E + 00	0.00E + 00	3.18E − 03	6.45E − 05
Fe	7.43E − 02	1.59E − 03	8.99E − 02	1.92E − 03	1.05E − 01	2.25E − 03	5.46E − 03	1.17E − 04	3.36E − 01	7.19E − 03
Total	1.79	0.0628	1.74	0.0608	1.70	0.0593	1.82	0.0614	1.87	0.0636

Z_{eq} is a parameter which describes the material properties in terms of equivalent elements similar to atomic number for a single element. Since interaction processes of gamma-ray photon with matter, photo-electric absorption, Compton scattering, and pair-production are energy dependent, therefore Z_{eq} for each interaction varies according to the photon energy. However the build-up of photons in the medium is mainly due to multiple scattering events by Compton scattering, so that Z_{eq} is derived from the Compton scattering interaction process.

The Z_{eq}, for individual brick, is estimated by the ratio of $(\mu/\rho)_{Compton}/(\mu/\rho)_{Total}$, at a specific energy with the

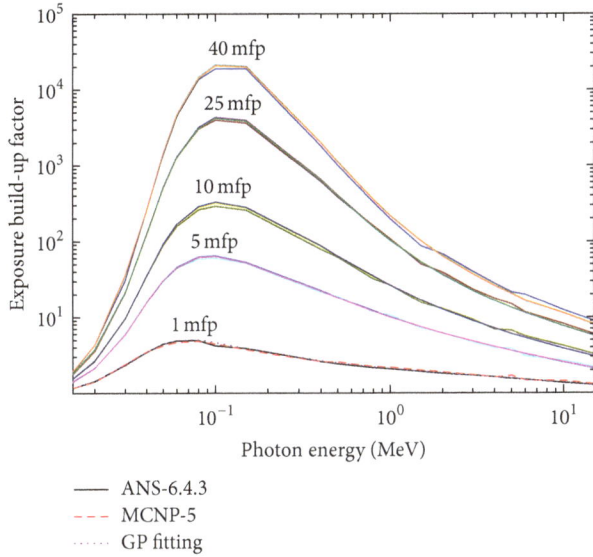

FIGURE 1: Exposure build-up factors for water by ANS-6.4.3, MCNP-5, and GP fitting.

correspondance of an element at the same energy. Thus first the Compton partial mass attenuation coefficient, $(\mu/\rho)_{Compton}$, and the total mass attenuation coefficients, $(\mu/\rho)_{Total}$, are obtained for elements $Z = 4$ to 40 for the selected bricks in the energy region from 0.0015 to 15 MeV using WinXCom [29, 30].

The interpolation of Z_{eq} is employed by the following formula [31, 32]:

$$Z_{eq} = \frac{Z_1 (\log R_2 - \log R_1) + Z_2 (\log R - \log R_1)}{\log R_2 - \log R_1}, \quad (1)$$

where Z_1 and Z_2 are the atomic numbers of the elements corresponding to the ratios R_1 and R_2, respectively. R is the ratio, $(\mu/\rho)_{Compton}/(\mu/\rho)_{Total}$, at specific energy and the ratio $(\mu/\rho)_{Compton}/(\mu/\rho)_{Total}$ for Z_{eq} lies between two successive ratios of the elements.

The GP fitting parameters are calculated in a similar fashion of interpolation procedure for Z_{eq}. The GP fitting parameters for the elements were taken from the ANS-6.4.3 standard reference database which provides the GP fitting parameters for twenty three elements ($Z = 4$ to 92) in the energy region from 0.015 to 15 MeV up to 40 mfp penetration depth. The GP fitting parameters for the bricks were interpolated using a similar formula:

$$C = \frac{C_1 (\log Z_2 - \log Z_{eq}) + C_2 (\log Z_{eq} - \log Z_1)}{\log Z_2 - \log Z_1}, \quad (2)$$

where C_1 and C_2 are the values of the GP fitting parameters corresponding to the atomic numbers of Z_1 and Z_2, respectively, at a given energy.

The third and final step is build-up factors estimation by GP fitting parameters (b, c, a, X_K, and d) in the photon

energy range of 0.015–15 MeV up to a 40 mfp by the following equations [13]:

$$B(E, X) = 1 + \frac{b-1}{K-1}(K^x - 1) \quad \text{for } K \neq 1,$$

$$B(E, X) = 1 + (b-1)X \quad \text{for } K = 1,$$

$$\quad (3)$$

$$K(E, X) = CX^a + d \frac{\tanh(X/X_K - 2) - \tanh(-2)}{1 - \tanh(-2)}$$

$$\text{for penetration depth } (X) \leq 40 \text{ mfp},$$

where X is the source-detector distance for the medium in terms of mfp and b the value of the exposure build-up factor at 1 mfp, $K(E, X)$ is the dose multiplicative factor, and b, c, a, X_K, and d are computed GP fitting parameters which depend on the attenuating medium and source energy.

2.2. Fast Neutron Removal Cross Section. An approximate method for calculation of the attenuation of fast neutrons by use of an effective removal cross section has been developed to allow for scattering or build-up. The effective removal crosssection for compounds and homogenous mixtures may be calculated from the value of Σ_R (cm^{-1}) or Σ_R/ρ (cm^2/g) for various elements in the compounds by mixture rule. Difference in application of mixture for neutron interaction differs as weight fraction is replaced by partial density and mass attenuation coefficient by neutron removal cross section:

$$\frac{\Sigma_R}{\rho} = \sum_i w_i \left(\frac{\Sigma_R}{\rho}\right)_i,$$

$$\frac{\Sigma_R}{\rho} = \sum_i \rho_i \left(\frac{\Sigma_R}{\rho}\right)_i.$$

$$\quad (4)$$

The values obtained for effective neutron removal cross section by the above equations are accurate within 10% of the experimental values investigated for aluminium, beryllium, graphite, hydrogen, iron, lead, oxygen, boron carbide, and so forth [33]. The Σ_R/ρ values of elements of the bricks have been taken from Kaplan and Chilton et al. [34, 35].

3. Uncertainties

The uncertainties in the build-up factor estimated by GP fitting are comparable with ANS-6.4.3 standard and MCNP-5 [36] for air and water. Figure 1 shows that the EBFs in water by GP fitting, ANS-6.4.3 standard, and MCNP-5 at different photon energies are comparable. The MCNP-5 results vary from those ANS-6.4.3 standards with greatest 13.83% due to difference in cross section libraries, method of solution for codes, calculation methods, standard deviation, and physics assumptions for bremsstrahlung and coherent scattering [36]. It is found that the invariant embedding, GP fitting, and MCNP simulation are in good agreement with 18 low-Z materials with small discrepancies [12]. This quantitative and comparative approach shows that our results for gamma-ray EBF will be of small uncertainties for easy establishment of the data and analysis by GP fitting methodology.

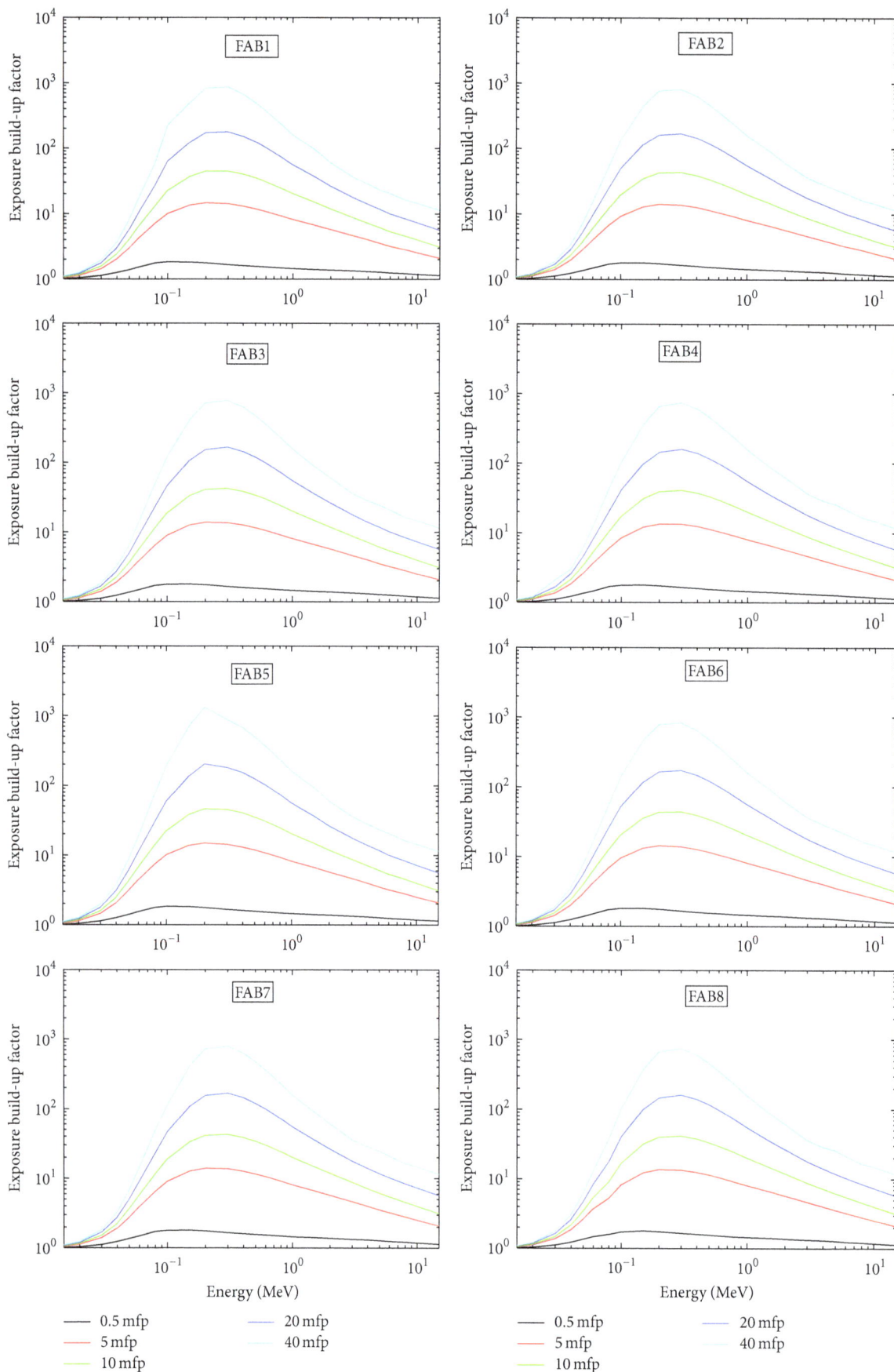

FIGURE 2: Continued.

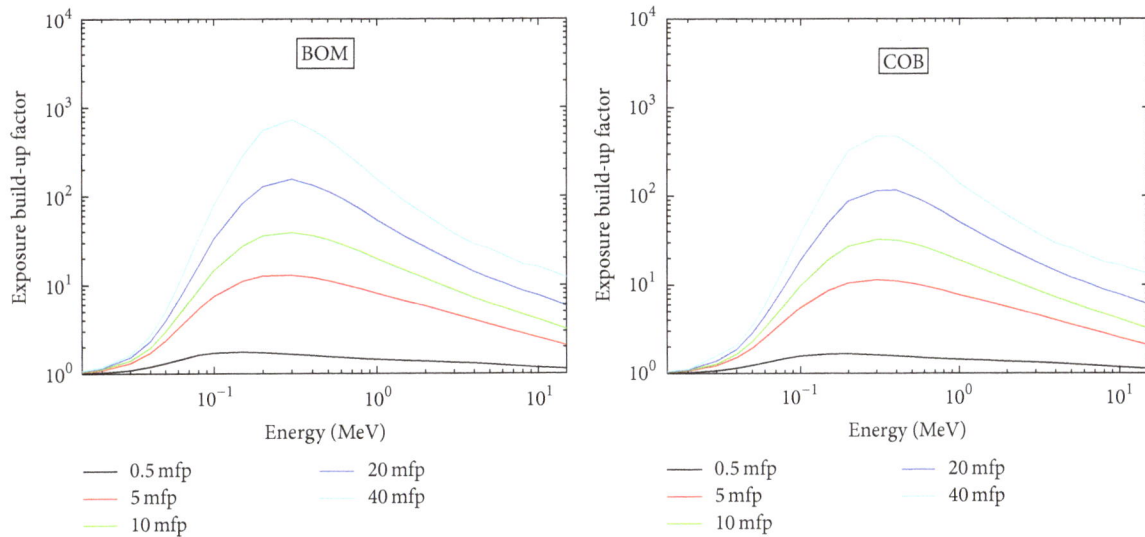

FIGURE 2: Variation of exposure build-up factor of fly-ash bricks, brick of mud, and common brick with incident photon energy.

4. Result and Discussion

The photon energy dependency of the EBFs is shown in Figure 2 at various penetration depths (0.5, 5, 10, 20, 30, and 40 mfp) for each FAB, BOM, and COB. Variation of EBF with penetration depths is shown in Figure 3 at photon energies 0.015, 0.15, 1.5, and 15 MeV. The chemical compositions of the selected materials are investigated at penetration depths of 0.5, 5, 10, 20, 30, and 40 mfp and shown in Figure 4. The fast neutron removal cross sections of the studies bricks are given in Tables 8(a) and 8(b). The variation of EBF with photon energy, penetration depth, and chemical compositions is explained in next coming sections.

4.1. Photon Energy Dependency of Exposure Build-Up Factor. The variation of EBF values of the FAB, BOM, and COB with photon energy is shown in Figure 2. It is noted that the EBF values of the bricks are minimum in low-energy and high-energy photons whereas they are higher in the intermediate-energy photons. The build-up factor in low-energy is small because the photons are completely absorbed by photo-electric absorption, gradually increases with the energy due to multiple scattering by the Compton scattering in the intermediate-energy region and finally again reduces in high-energy region due to pair-production. The analysis shows that the brick having, lowest EBF should be considered for gamma-ray shielding. The gamma build-up factors are the lowest for 0.5 mfp, and highest for 40 mfp, due to multiple scattering events for large penetration depth of 40 mfp. The EBFs of the FAB, BOM, and COB are found higher than the ordinary concretes [16].

4.2. Penetration Depth Dependency of Exposure Build-Up Factor. The variation of EBFs of the selected FABs, BOM, and COB with penetration depths at photon energies 0.015, 1.5, 5, and 15 MeV is shown in Figure 3. It is observed that the EBF values increase with the penetration depth initially

and afterwards become stable. The EBFs for the BOM and COB are found lesser than FABs at photon energies 0.15, 0.15, and 1.5 MeV photon energy. It can be seen that the values of EBF of the bricks increase with increase in Z_{eq} at 15 MeV as penetration depth increases. In the low photon energy region (~1.5 MeV), exposure build-up factor becomes independent of penetration depth above 25 mfp. The reason may be that beyond 3 MeV, the pair-production dominates the Compton scattering and produces the electron-positron pairs. These particles may escape from the medium of lower penetration depth whereas they will scatter in large penetration depths as well as originate the secondary gamma photon by annihilation to increase the gamma photon intensities. The smallest value of EBFs of COB for low energy (<1.22 MeV) gradually increases with increase in photon energy due pair-production.

4.3. Chemical Composition Dependency of Exposure Build-Up Factor. The EBFs are dependent on the atomic number of elements; hence EBF values of the materials are dependent on chemical compositions [16]. The chemical element composition dependency on the EBF is analyzed by for the photon energy at constant penetration depths as shown in Figure 4. It is observed that the EBF values are small at low- as well as high-energy photons whereas maximum in the intermediate-energy photons. The reason for peaked EBF values is that the photo-electric and pair-production are dominant interaction processes in low- and high-energy, respectively, which completely remove the photons. The intermediate-energy photon takes over Compton scattering where photon goes under multiple scattering events to build-up the exposure. The brick materials having large weight fraction of low-Z elements are showing high EBF values due to less removal of the photons.

The fly-ash brick (low Z_{eq}) is having larger EBF compared with other bricks (higher Z_{eq}) for entire energy region from 0.015 to 15 MeV in low penetration depths. The EBF of

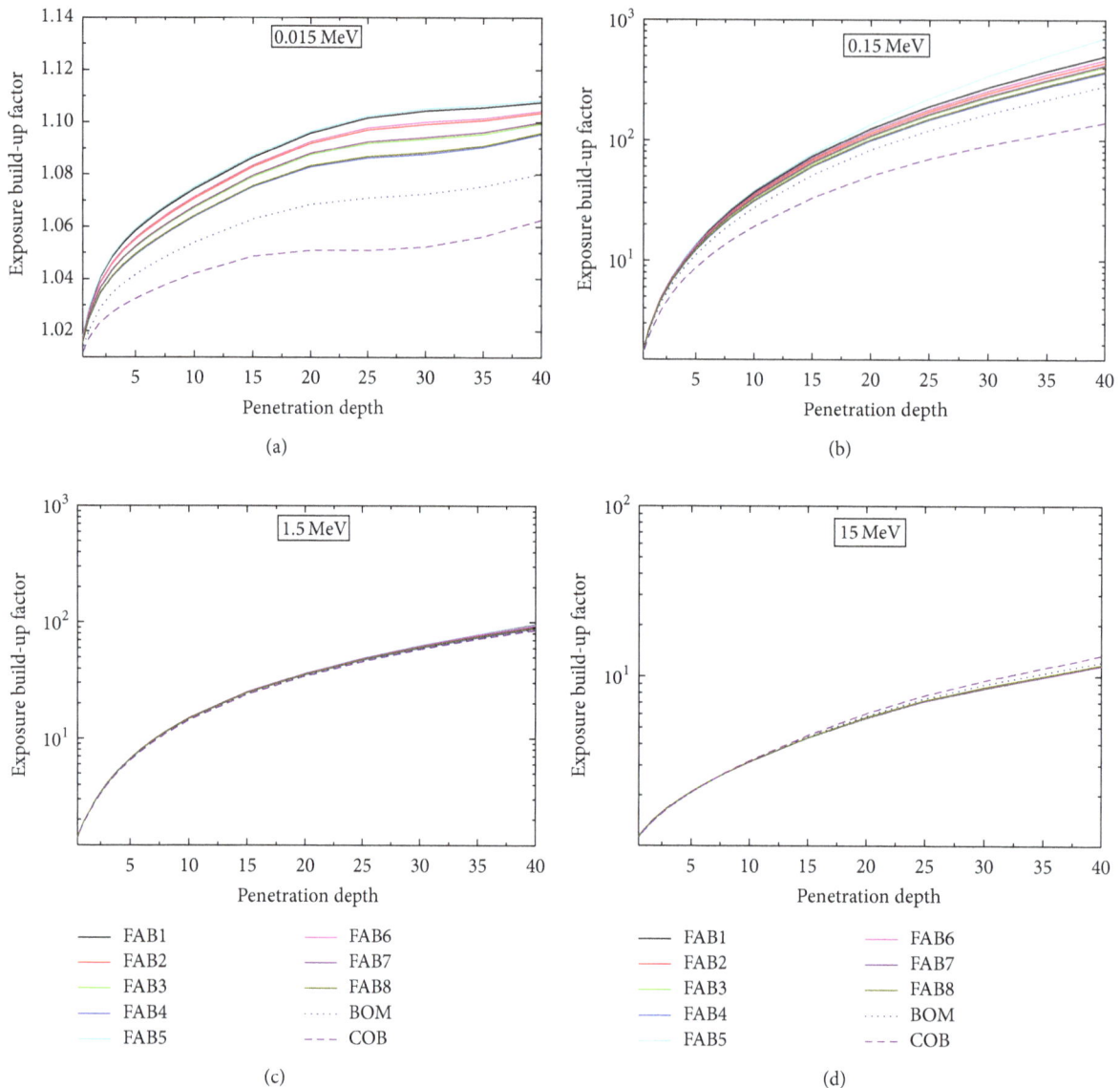

(a)

(b)

(c)

(d)

FIGURE 3: Variation of exposure build-up factor of fly-ash bricks, brick of mud, and common brick with penetration depth for photon energies 0.015, 0.15, 1.5, and 15 MeV.

BOM and COB shows behavior in reverse order at 3 MeV for large penetration depth. Above 3 MeV photon energy and for large penetration depths (above 10 mfp), exposure build-up factor becomes directly proportional to Z_{eq}. It may be due to dominance of pair-production above 1.22 MeV which increases with increase in photon energy and results in production of electron and positron. The positron annihilates at zero kinetic energy and produces new two photons of energy 0.511 MeV which build-up in the medium due to multiple scattering events. The pair-production dependent on Z^2 signifies that the COB and BOM should have large photon intensities of energies 0.511 MeV. The EBF is found large in the higher penetration depths due to possibilities of multiple scattering whereas the positron may escape though the low penetration depth.

4.4. *Fast Neutron Removal Cross Section.* In case of neutron attenuation, the effective fast neutron removal (FNR) cross sections of the different FAB, BOM, and COB are given in Tables 8(a) and 8(b). The effective removal cross section is approximately constant for neutron energy from 2 to 12 MeV [34]. In Tables 8(a) and 8(b), the calculations are being listed for various elemental compositions of the FAB and other bricks. The effective FNR cross section Σ_R (cm^{-1}) of the selected brick materials is calculated by partial density of each element and FNR cross section. It is observed that the removal cross section is higher for FAB1, and this leads to the conclusion that the FAB1 is more effective for neutron attenuation than the FABs, BOM, and COB. Also, from point of neutron shielding, fly-ash and common brick materials are more suitable replacement of brick of mud. The low

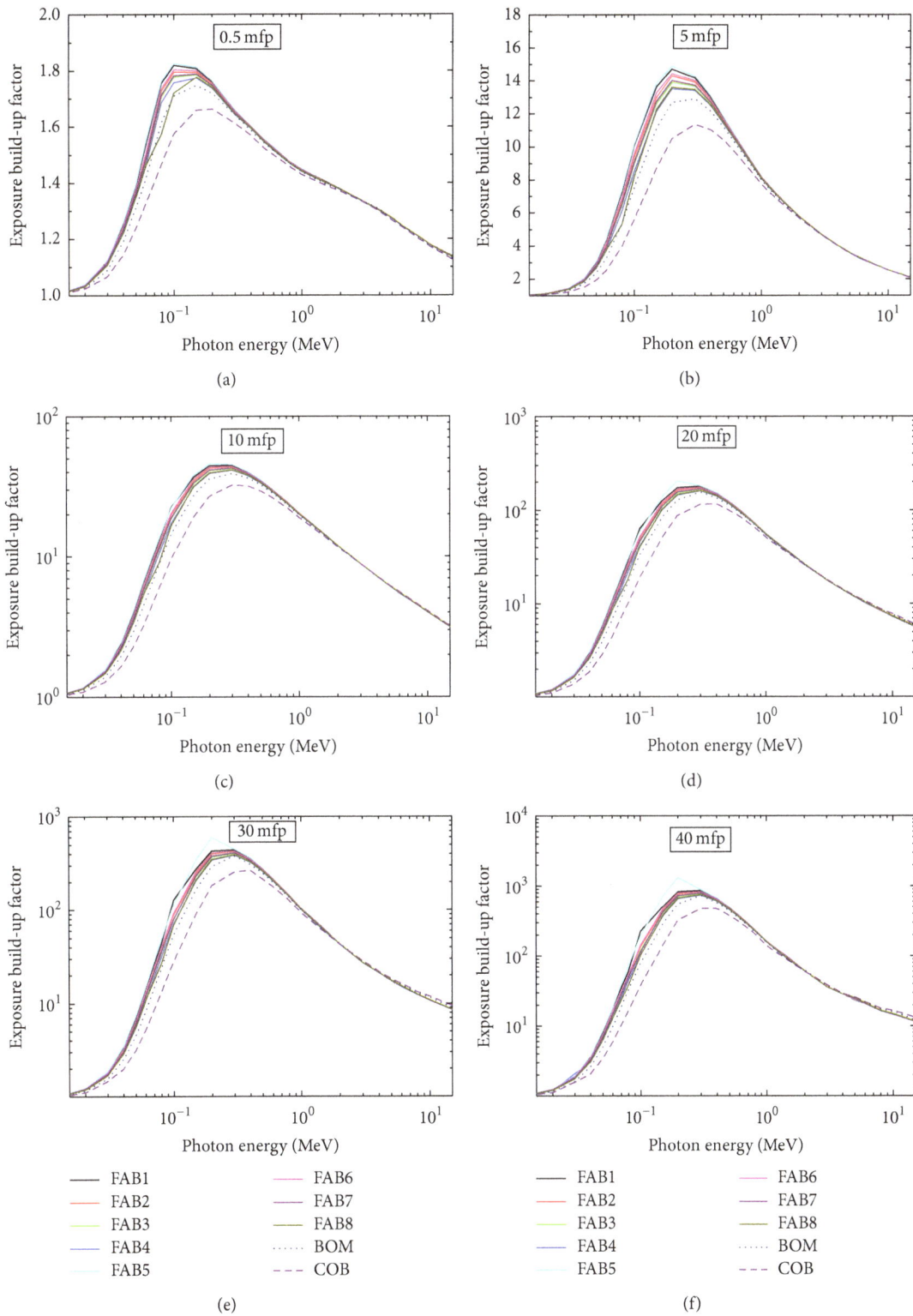

FIGURE 4: Variation of exposure build-up factor of fly-ash bricks, brick of mud, and common brick with photon energy for penetration depths 0.5, 5, 10, 20, 30, and 40 mfp.

variation in removal cross section is being noticed for all types of the bricks. Elevated values of removal cross section of FAB1 and COB are attributed due to high density and large contribution of low atomic number weight fractions of H and C. With constant H weight fraction and variable densities, the removal cross section varies. In cases of COB, the removal cross section is slightly higher due to large C weight fraction with equal value of density of FAB1. Therefore it is concluded that low atomic number elemental composition and density contribute a vital role in gamma-ray and neutron shielding properties.

5. Conclusions

In the present studies, we have calculated the exposure build-up factors of fly-ash bricks for photon energies from 0.015 to 15 MeV up to 40 mfp penetration depth by geometrical progression (GP) method. Exposure build-up factor increases with the increase in penetration depth for all the fly-ash bricks. For the entire energy region (0.015–15.0 MeV), in case of small penetration depths (below 10 mfp), exposure build-up factor is inversely proportional to the Z_{eq}. In the low photon energy region ($\approx$1.5 MeV), exposure build-up factor becomes dependent upon penetration depth above 25 mfp. Above 3 MeV photon energy and for large penetration depths (above 10 mfp), exposure build-up factor becomes directly proportional to Z_{eq}. Bricks containing low atomic number elements show high fast neutron removal cross. Low atomic number elemental composition and density contribute a vital role in gamma-ray and neutron shielding properties. The study should be very useful for shielding effectiveness of fly-ash bricks and dose estimation.

Acknowledgments

The authors are grateful to respected Dr. M. J. Berger and J. H. Hubbell and L. Gerward for providing the state-of-the-art and user-friendly computer programs XCOM/WinXCom.

References

[1] S. M. Kulwinder, K. Baljit, S. S. Gurdeep, and K. Ajay, "Investigations of some building materials for γ-rays shielding effectiveness," *Radiation Physics and Chemistry*, vol. 87, pp. 16–25, 2013.

[2] G. R. White, "The penetration and diffusion of Co^{60} gamma-rays in water using spherical geometry," *Physical Review*, vol. 80, no. 2, pp. 154–156, 1950.

[3] U. Fano, "Gamma-ray attenuation—part II: analysis of penetration," *Nucleonics*, vol. 11, pp. 55–61, 1953.

[4] Y. Harima, "An historical review and current status of buildup factor calculations and applications," *Radiation Physics and Chemistry*, vol. 41, no. 4-5, pp. 631–672, 1993.

[5] K. Takeuchi and S. Tanaka, "PALLAS-ID (VII). A code for direct integration of transport equation in one-dimensional plane and spherical geometries," Tech. Rep. 84, Japan Atomic Energy Research Institute, 1984.

[6] G. L. Simmons, "An adjoint gamma-ray moments computer code, ADJMOM-I," NBS Technical Note 748, National Bureau of Standards, 1973.

[7] A. B. Chilton, C. M. Eisenhauer, and G. L. Simmons, "Photon point source buildup factors for air, water and iron," *Nuclear Science and Engineering*, vol. 73, pp. 97–107, 1980.

[8] D. V. Gopinath and K. Samthanam, "Radiation transport in one dimensional finite system—part I: development in anisotropic source flux technique," *Nuclear Science and Engineering*, vol. 43, pp. 186–196, 1971.

[9] W. R. Nelson, H. Hirayama, and D. W. O. Rogers, *EGS4 Code System, SLAC-265*, Stanford Linear Accelerator Centre, Stanford, Calif, USA, 1985.

[10] ANSI/ANS-6.4.3, "Gamma ray attenuation coefficient and buildup factors for engineering materials," 1991.

[11] A. Shimizu, "Calculation of gamma-ray buildup factors up to depths of 100 mfp by the method of invariant embedding, (I): analysis of accuracy and comparison with other data," *Journal of Nuclear Science and Technology*, vol. 39, no. 5, pp. 477–486, 2002.

[12] A. Shimizu, T. Onda, and Y. Sakamoto, "Calculation of gamma-ray buildup factors up to depths of 100 mfp by the method of invariant embedding, (III) generation of an improved data set," *Journal of Nuclear Science and Technology*, vol. 41, no. 4, pp. 413–424, 2004.

[13] Y. Harima, Y. Sakamoto, S. Tanaka, and M. Kawai, "Validity of the geometric-progression formula in approximating gamma-ray buildup factors," *Nuclear Science and Engineering*, vol. 94, no. 1, pp. 24–35, 1986.

[14] S. Singh, S. S. Ghumman, C. Singh, K. S. Thind, and G. S. Mudahar, "Buildup of gamma ray photons in flyash concretes: a study," *Annals of Nuclear Energy*, vol. 37, no. 5, pp. 681–684, 2010.

[15] G. S. Brar, K. Singh, M. Singh, and G. S. Mudahar, "Energy absorption buildup factor studies in water, air and concrete up to 100 mfp using G-P fitting formula," *Radiation Physics and Chemistry*, vol. 43, no. 6, pp. 623–627, 1994.

[16] V. P. Singh and N. M. Badiger, "Comprehensive study of energy absorption and exposure buildup factor for concrete shielding in photon energy range 0.015–15 MeV upto 40 mfp penetration depth: dependency of density, chemical element, photon energy," *International Journal of Nuclear Energy Science and Technology*, vol. 7, no. 1, pp. 75–99, 2012.

[17] V. P. Singh and N. M. Badiger, "Photon energy absorption buildup factor of gaseous mixtures used in radiation detectors," *Radioprotection*, vol. 48, no. 1, pp. 63–78, 2013.

[18] M. Kurudirek, B. Doğan, M. Ingeç, N. Ekinci, and Y. Özdemir, "Gamma-ray energy absorption and exposure buildup factor studies in some human tissues with endometriosis," *Applied Radiation and Isotopes*, vol. 69, no. 2, pp. 381–388, 2011.

[19] G. Sandeep and S. S. Gurdeep, "A comprehensive study on energy absorption and exposure buildup factors for some soils and ceramic materials," *Journal of Applied Physics*, vol. 2, no. 3, pp. 24–30, 2012.

[20] S. Tejbir, K. Gurpreet, and S. S. Parjit, "Study of gamma ray exposure buildup factor for some ceramics with photon energy, penetration depth and chemical composition," *Journal of Ceramics*, vol. 2013, Article ID 721606, 6 pages, 2013.

[21] E. Yilmaz, H. Baltas, E. Kiris, I. Ustabas, U. Cevik, and A. M. El-Khayatt, "Gamma ray and neutron shielding properties of some concrete materials," *Annals of Nuclear Energy*, vol. 38, no. 10, pp. 2204–2212, 2011.

[22] *Building Materials in India: 50 Years: A Commemorative Volume*, 1998.

[23] Eco-Friendly building materials and technologies, Ecohousing assessment criteria-Version II, Annexure-4, August 2009.

[24] P. David and B. Dalibor, "Barite mortar with fluid fly ash as shielding material," *Intersections*, vol. 6, no. 2, article 3, pp. 28–34, 2009.

[25] S. Singh, A. Kumar, D. Singh, K. S. Thind, and G. S. Mudahar, "Barium-borate-flyash glasses: as radiation shielding materials," *Nuclear Instruments and Methods in Physics Research B*, vol. 266, no. 1, pp. 140–146, 2008.

[26] T. Suparat, K. Jakrapong, L. Pichet, and C. Weerapong, "Development of $BaO:B_2O_3$: flyash glass system for gamma-rays shielding materials," *Progress in Nuclear Science and Technology*, vol. 1, pp. 110–113, 2011.

[27] S. S. Amritphale, A. Anshul, N. Chandra, and N. Ramakrishnan, "Development of celsian ceramics from fly ash useful for X-ray radiation-shielding application," *Journal of the European Ceramic Society*, vol. 27, no. 16, pp. 4639–4647, 2007.

[28] A. P. Singh, A. S. Kumar, A. Chandra, and S. K. Dhawan, "Conduction mechanism in Polyaniline-flyash composite material for shielding against electromagnetic radiation in X-band Ku band," *AIP Advances*, vol. 1, no. 2, Article ID 022147, 2011.

[29] L. Gerward, N. Guilbert, K. B. Jensen, and H. Levring, "X-ray absorption in matter. Reengineering XCOM," *Radiation Physics and Chemistry*, vol. 60, no. 1-2, pp. 23–24, 2001.

[30] L. Gerward, N. Guilbert, K. B. Jensen, and H. Levring, "WinXCom—a program for calculating X-ray attenuation coefficients," *Radiation Physics and Chemistry*, vol. 71, no. 3-4, pp. 653–654, 2004.

[31] Y. Harima, "An approximation of gamma-ray buildup factors by modified geometrical progression," *Nuclear Science and Engineering*, vol. 83, no. 2, pp. 299–309, 1983.

[32] M. J. Maron, *Numerical Analysis: A Practical Approach*, Macmillan, New York, NY, USA, 2007.

[33] G. Samuel and S. Alexander, *Nuclear Reactor Engineering*, vol. 1, Chapman and Hall, 4th edition, 2004.

[34] M. F. Kaplan, *Concrete Radiation Shielding*, Longman Scientific and Technology, Essex, UK, 1989.

[35] A. B. Chilten, J. K. Shultis, and R. E. Faw, *Principle of Radiation Shielding*, Prentice-Hall, Englewood Cliffs, NJ, USA, 1984.

[36] D. Luis, *Update to ANSI/ANS-6.4.3-1991 for Low-Z Materials and Compound Materials and Review of Particle Transport Theory*, UNLV, Las Vegas, Nev, USA, 2009.

29

Possibility of NiCuZn Ferrites Composition for Stress Sensor Applications

M. Penchal Reddy,[1] **W. Madhuri,**[2] **M. Venkata Ramana,**[3] **I. G. Kim,**[1] **D. S. Yoo,**[1]
N. Ramamanohar Reddy,[4] **K. V. Siva Kumar,**[5] **D. V. Subbaiah,**[5] **and R. Ramakrishna Reddy**[5]

[1] *Department of Physics, Changwon National University, Changwon 641773, Republic of Korea*
[2] *School of Advanced Sciences, VIT University, Vellore 632014, India*
[3] *School of Materials Science and Engineering, Harbin Institute of Technology, Shenzhen 518055, China*
[4] *Department of Materials Science and Nanotechnology, Yogi Vemana University, Kadapa 516227, India*
[5] *Department of Physics, Sri Krishnadevaraya University, Anantapur 515055, India*

Correspondence should be addressed to I. G. Kim; igkim@changwon.ac.kr

Academic Editor: Zhenxing Yue

NiCuZn ferrite with composition of ($Ni_{0.42+x}Cu_{0.10}Zn_{0.60}Fe_{1.76-2x}O_{3.76-2x}$) (where $x = 0.00, 0.02, 0.04, 0.06, 0.08$, and 0.10) was prepared by the conventional ceramic double sintering technique. The formation of single phase was confirmed by X-ray diffraction. The microstructural features were also studied by electronic microscopy and are reported. Initial permeability measurements on these samples were carried out in the temperature range of 30 to 300°C. The effect of external applied stress on the open magnetic circuit type coil with these ferrite cores was studied by applying uniaxial compressive stress parallel to the magnetizing direction and the change in the inductance was measured. The variation of inductance ($\Delta L/L$)% increases up to certain applied compressive stress and there after it decreases, showing different stress sensitivities for different compositions of ferrites studied in the present work. The variation of ratio of inductance ($\Delta L/L$)% with external applied compressive stress was examined. These results show that the $Ni_{0.42}Cu_{0.10}Zn_{0.60}Fe_{1.76}O_{3.76}$ and $Ni_{0.44}Cu_{0.10}Zn_{0.60}Fe_{1.72}O_{3.72}$ samples are found to be suitable for inductive stress sensor applications.

1. Introduction

It is well known that the effect of stress is very important in the magnetization process of ferrites [1]. The magnetic properties of ferrites such as permeability, coercivity, and hysteresis loop change with the application of stress [2, 3]. The effect of external applied stress can be observed in ferrites as a change of inductance ($\Delta L/L$)% when compressive stresses are applied. These variations of permeability with applied stress in these samples can be attributed to the magnetostrictive contributions of varied amounts of nickel and iron compositions. For small compressive stresses, the stress raises the initial permeability with negative magnetostriction and for large tensile stresses the permeability decreases [3]. The magnetoelastic properties of ferrites are interesting subjects

for investigations to determine the possibilities of their use in the construction of measuring sensors [4, 5]. Over the last decade, manganese substituted cobalt ferrites have been the dominant ferrite materials for stress sensor applications due to their large magnetomechanical effect and high sensitivity to stress [6–11]. Magnetic sensors play a significant role in physical measurements used in all kinds of applications [12, 13]. The most often used magnetic phenomena in today's magnetic sensor technology are the magnetoresistance [14, 15], the magnetoimpedance [16, 17], the magnetostriction [18, 19], the electromagnetic induction [20], and the Hall effect [21]. There also exist other effects usable in sensing applications, both macroscopic and microscopic [22]. Due to superior mechanical and magnetic properties in ferrites they seem to be suitable for construction of force and stress sensors [23, 24].

This led to systematic investigation of studies on the stress sensitivity of series of NiCuZn [25], MgCuZn, and NiMgCuZn [26] ferrites. Incidentally it has been noticed from these studies that certain NiCuZn ferrite compositions can be used for stress sensor applications. In view of this the details of synthesis of those NiCuZn ferrite compositions suitable for stress sensor applications and their stress dependence studies are reported in this paper.

2. Experimental Procedure

A series of six nonstoichiometric compositions of NiCuZn ferrite with compositions shown in Table 1 have been prepared by the conventional double sintering method using analytical grade NiO, CuO, ZnO, and Fe_2O_3 in their respective proportions. These oxides were weighed and intimately mixed in stoichimetric proportions. These constituents were ball milled (RETSCH PM - 200, Germany) in agate bowls with agate balls in acetone medium for 20 h. The slurry was dried and the dried powders were loosely packed in the form of cakes. These cakes were presintered in closed alumina crucibles at 800°C for 2 h. The presintered cakes removed from the furnace were crushed and ball milled in an acetone medium in agate bowls with agate balls for another 24 hours to obtain fine particle size. These slurries after drying were sieved to obtain uniform particle size.

The presintered green powders were mixed with 2% polyvinyl alcohol as a binder and were compacted in the form of disks of diameter 10 mm and 2 mm height, toroids of 12 mm outer diameter (OD), 8 mm inner diameter (ID) and 4 mm height and cylinders of diameter 10 mm and length nearly 20 mm at 200 MPa with a suitable die. The compacted bodies were conventionally sintered in a programmable furnace (V.B. Ceramic Consultants, Chennai, India) at a temperature of 1250°C for 2 hours and were cooled to room temperature at 80°C/hr. The sintering schedule includes half an hour dwelling time at 110°C to remove moisture from the samples and one-hour binding burning time at 600°C. Proper care was taken to avoid zinc evaporation by providing $ZnFe_2O_4$. The density of the sintered specimens was measured by Archimedes's method. All the samples were structurally characterized using Philip high resolution X-ray diffraction system (PM 1730, Germany) with CuK$_\alpha$ radiation. Microstructures of sintered samples were investigated using JOEL (JSM Model 6360, Japan) scanning electron microscope. The initial permeability (μ_i)of these ferrite toroids was evaluated using the standard formulae from the inductance measurements carried out at a frequency of 10 kHz using computer controlled impedance analyzer (Hioki Model 3532–50 LCR Hi-Tester, Japan). These measurements were carried out in the temperature range 30 to 300°C. In order to study the effect of external stress, uniaxial compressive stress parallel to the magnetizing direction was applied to the cylindrical shaped ferrite cores using uniaxial press system. The stress magnitudes were varied from 0 to 10 MPa.

TABLE 1: Chemical composition of the NiCuZn ferrites.

S. No.	Sample	NiO	CuO	ZnO	Fe_2O_3
1	a	21	5	30	44
2	b	22	5	30	43
3	c	23	5	30	42
4	d	24	5	30	41
5	e	25	5	30	40
6	f	26	5	30	39

a: $Ni_{0.42}Cu_{0.10}Zn_{0.60}Fe_{1.76}O_{3.76}$, b: $Ni_{0.44}Cu_{0.10}Zn_{0.60}Fe_{1.72}O_{3.72}$, c: $Ni_{0.46}Cu_{0.10}Zn_{0.60}Fe_{1.68}O_{3.68}$, d: $Ni_{0.48}Cu_{0.10}Zn_{0.60}Fe_{1.64}O_{3.64}$, e: $Ni_{0.50}Cu_{0.10}Zn_{0.60}Fe_{1.60}O_{3.60}$, f: $Ni_{0.52}Cu_{0.10}Zn_{0.60}Fe_{1.56}O_{3.56}$.

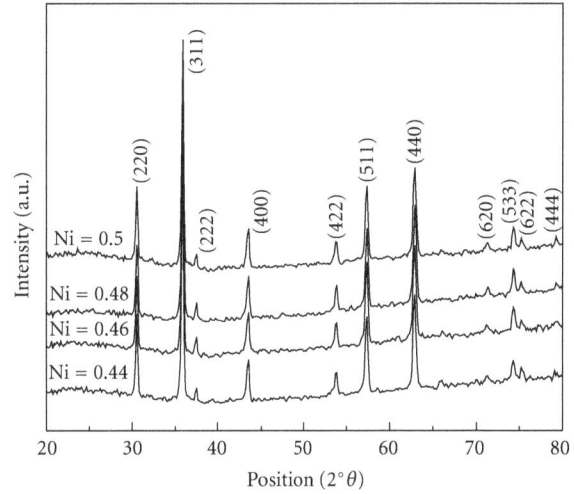

FIGURE 1: Typical X-ray diffractograms of x = 0.2, 0.4, 0.6, and 0.8.

The change in inductance was measured using the abovementioned LCR Hi-Tester by employing 100–120 turns coil on each cylinder. The initial permeability was calculated using the relation

$$L = 0.0046N^2\mu_i h \log_{10}\left(\frac{D_o}{D_i}\right), \tag{1}$$

where L is the series inductance; N is the number of turns of winding; h is the height of the toroid in inches; D_o is the outer diameter of toroid; and D_i is the inner diameter of toroid, respectively.

3. Results and Discussion

Figure 1 represents the typical X-ray diffractograms of nonstoichiometric NiCuZn ferrite samples, respectively. The X-ray diffraction analysis of the ferrite samples shows the formation of single phase spinel structure. The X-ray lines show considerable broadening, indicating the fine particle nature of the ferrite powder.

The variation of the lattice parameter "a" as a function of nickel ion content is depicted in Figure 2. It is noticed that the lattice parameter increases with the nickel ion content in the lattice. This variation can be explained on the basis of an ionic size difference of the component ions. It may be

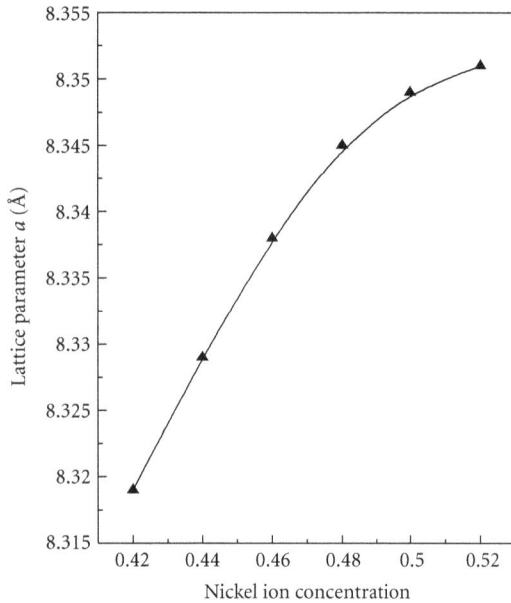

FIGURE 2: Lattice constant (a) as a function of nickel ion concentration of the NiCuZn system.

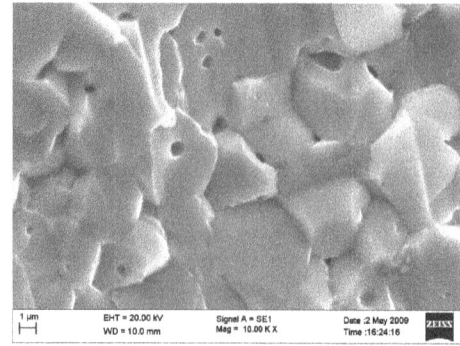

(a)

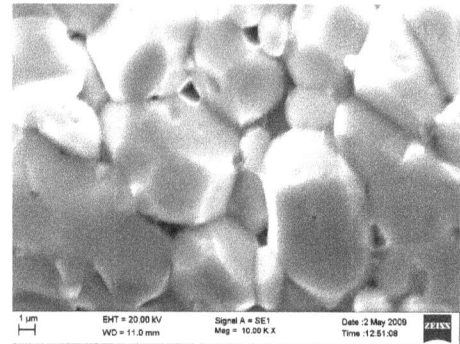

(b)

FIGURE 3: SEM photographs of sintered ($Ni_{0.42+x}Cu_{0.10}Zn_{0.60}Fe_{1.76-2x}O_{3.76-2x}$) ferrites (a) $x = 0.2$ and (b) 0.6.

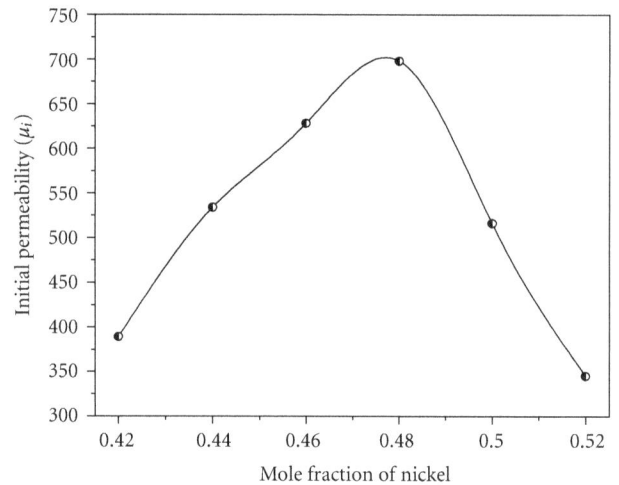

FIGURE 4: Variation of initial permeability with mole fraction of nickel at room temperature.

pointed out here that the ionic radii of Ni^{2+}, Cu^{2+}, and Zn^{2+} are almost same with the experimental error that is, 0.72 Å, 0.74 Å, and 0.74 Å, respectively [27]. However the lattice parameter is found to increase systematically with increase in nickel content.

The SEM photographs of the sintered samples were recorded to understand the microstructure of the NiCuZn ferrites. Scanning electron microscopy (SEM) of powder sintered at 1250°C/2 h shows an increase in particle size increasing the nickel concentration [28]. Figures 3(a) and 3(b) show the microstructures of fracture surfaces of the sample b (Ni 0.44) and sample d (Ni 0.48) (Table 1). It is well known that the initial permeability characteristics depend not only on chemical composition but also on the microstructure of the sintered body. Desired magnetic properties of ferrite can be achieved by the control of microstructures [29]. Small and uniform grain size is favourable to obtain low power loss, but large grain size is favourable to get high permeability [30].

Figure 4 shows the variation of the initial permeability (μ_i) versus the nickel content for samples at room temperature (27°C). It can be noticed from the figure that the initial permeability shows a maximum at 0.48 nickel content in these nonstoichiometric NiCuZn ferrospinels. This may be attributed to the fact that the magnetocrystalline anisotropy constant (K_1) becoming zero at this composition of nickel [31, 32].

The initial permeability (μ_i) as the function of temperature (at constant frequency, 10 kHz) from room temperature to Curie point was also studied. The temperature dependence of magnetic anisotropy can be inferred from the temperature dependence of initial permeability as shown in Figure 5. It can be noted from Figure 5 that as the temperature increases the initial permeability (μ_i) remains constant up to a certain temperature and increases to a peak value and

then abruptly falls to a minimum value. The temperature at which this abrupt fall takes place is the magnetic Curie transition temperature T_c. The magnetic initial permeability for the material is expected to strongly depend on the microstructure, as the initial permeability represents the mobility of magnetic domain wall in response to the small applied field [33]. It is proved by Figures 3(a) and 3(b) that

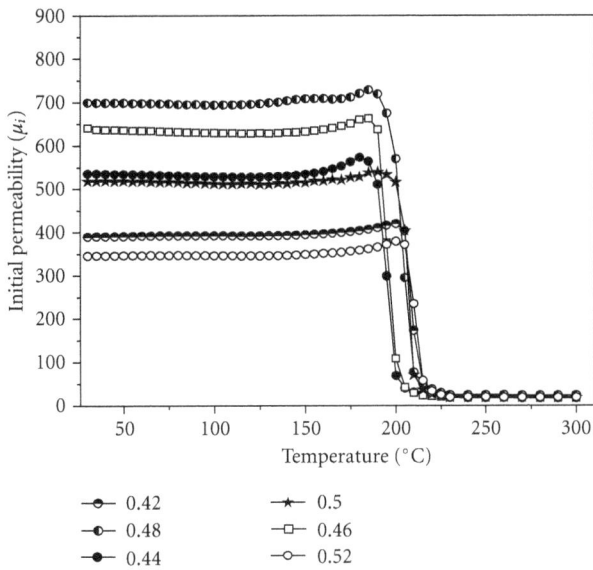

FIGURE 5: Temperature dependence of the initial permeability of $(Ni_{0.42+x}Cu_{0.10}Zn_{0.60}Fe_{1.76-2x}O_{3.76-2x})$ ferrites sintered at $1250°C/2$ h.

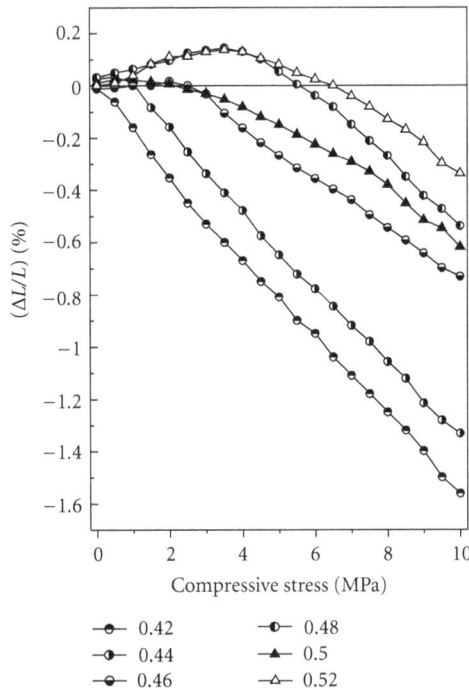

FIGURE 6: Variation of ratio of inductance change $(\Delta L/L)\%$ as a function of applied compressive stress in non-stoichiometric NiCuZn ferrite series.

the magnetic permeability of the large-grain samples is larger than the fine-grain ones.

The initial permeability is an important magnetic property to study the quality of soft ferrites. Generally, the initial permeability (μ_i) depends on two factors [34] namely, (1) contribution from spin rotation and (2) contribution from domain wall motion. But the contribution from spin rotation

is smaller than domain wall motion [35]. The permeability due to domain wall motion is given by [36]

$$(\mu_i - 1) = \frac{3\pi M_s^2 D}{4\gamma}, \tag{2}$$

where M_s is saturation magnetization, D is mean grain size, and γ is magnetic domain wall energy, which is proportional to magnetocrystalline anisotropy constant K_1 [37, 38].

Thus

$$(\mu_i)_w = \frac{M_s^2 D}{K_1}, \tag{3}$$

where $(\mu_i)_w$ is the initial permeability of domain wall motion.

The magnitudes of initial permeability (μ_i) increased with increasing content of nickel up to 0.48 after decreasing with increasing nickel content and interestingly the Curie transition temperature remained fairly constant around $200°C$. The figure reveals that there is a flat temperature response of (μ_i) in all these samples. The Curie transition temperature and permeability behaviour as a function of nickel content in these ferrites were published in earlier communication [25].

The variation of ratio of inductance change $(\Delta L/L)\%$ as a function of applied compressive stress is shown in Figure 6. From an examination of Figure 6 it is clear that in samples a and b the $(\Delta L/L)\%$ values decrease with increasing external stress. In the case of samples d and f, there is an increase of $(\Delta L/L)\%$ up to certain applied stress and thereafter it decreases. At higher concentrations of nickel the $(\Delta L/L)\%$ increases up to a certain stress and decreases with further raise in external compressive stress showing a peak value. Exactly, similar behaviour was noticed by Kanada et al., [39] in NiMgCuZn ferrites at 4 MPa. For example these peak values are 3.5 MPa, 2 MPa, and 3.5 MPa for nickel contents 0.48, 0.50, and 0.52, respectively. In the present work, we have studied variation of inductance at low concentration of nickel. This type of behaviour dependency of $(\Delta L/L)\%$ on stress can be utilized for the development of a sensor for detection of stress variations.

4. Conclusions

The initial permeability measurements were carried out in the temperature range 30–300°C in NiCuZn ferrites. In the present study it has been noticed that there is a linear decrease of $(\Delta L/L)\%$ with increasing external applied stress in ferrite compositions with nickel 0.42, and 0.44 concentration. The compositions $Ni_{0.42}Cu_{0.10}Zn_{0.60}Fe_{1.76}O_{3.76}$, and $Ni_{0.44}Cu_{0.10}Zn_{0.60}Fe_{1.72}O_{3.72}$ showed best magnetic properties among the five compositions and would be prominent materials for inductive stress sensor applications.

References

[1] G. Mian and T. Yamaguchi, "Stress effect on the magnetization process in Ni-Cu-Zn ferrite in a weak field," *Journal of Magnetism and Magnetic Materials*, vol. 68, pp. 351–357, 1987.

[2] A. Biehkowski, "Ferrite," in *Proceedings of the ICF3*, Center for Academic Publishers, Tokyo, Japan, 1981.

[3] J. Smit and H. P. Wijn, *Ferrites*, Philips Technical Library, Eindhoven, The Netherlands, 1959.

[4] A. Bierikowski, *Acta IMEKO*, p. 591, 1979.

[5] A. Bienkowski and J. Kulikowski, *Prace ITR*, vol. 97, p. 10, 1984.

[6] S. S. Shinde and K. M. Jadhav, "Bulk magnetic properties of cobalt ferrite doped with Si4+ ions," *Journal of Materials Science Letters*, vol. 17, no. 10, pp. 849–851, 1998.

[7] J. A. Paulsen, C. C. H. Lo, J. E. Snyder, A. P. Ring, L. L. Jones, and D. C. Jiles, "Study of the curie temperature of cobalt ferrite based composites for stress sensor applications," *IEEE Transactions on Magnetics*, vol. 39, no. 5, pp. 3316–3318, 2003.

[8] J. A. Paulsen, A. P. Ring, C. C. H. Lo, J. E. Snyder, and D. C. Jiles, "Manganese-substituted cobalt ferrite magnetostrictive materials for magnetic stress sensor applications," *Journal of Applied Physics*, vol. 97, no. 4, Article ID 044502, 3 pages, 2005.

[9] O. Caltun, H. Chiriac, N. Lupu, I. Dumitru, and B. P. Rao, "High magnetostrictive doped cobalt ferrite," *Journal of Optoelectronics and Advanced Materials*, vol. 9, no. 4, pp. 1158–1160, 2007.

[10] O. F. Caltun, G. S. N. Rao, K. H. Rao et al., "High magnetostrictive cobalt ferrite for sensor applications," *Sensor Letters*, vol. 5, no. 1, pp. 45–47, 2007.

[11] O. Caltun, I. Dumitru, M. Feder, N. Lupu, and H. Chiriac, "Substituted cobalt ferrites for sensors applications," *Journal of Magnetism and Magnetic Materials*, vol. 320, no. 20, pp. e869–e873, 2008.

[12] G. Brumfiel, "Magnetic effect sends physicists into a spin," *Nature*, vol. 426, no. 6963, p. 110, 2003.

[13] E. Hristoforou, "Magnetic effects in physical sensor design and development," *Journal of Optoelectronics and Advanced Materials*, vol. 4, no. 2, pp. 245–260, 2002.

[14] I. Bakonyi, B. L. Peter, V. Weihnacht, J. Toth, L. F. Kiss, and C. M. Schneider, "Giant magnetoresistance in electrodeposited multilayer films. The influence of superparamagnetic regions," *Journal of Optoelectronics and Advanced Materials*, vol. 7, no. 2, pp. 589–598, 2005.

[15] V. Georgescu and M. Daub, "Magnetism and magnetoresistance in electrodeposited ($L1_0$) CoPt superlattices," *Journal of Optoelectronics and Advanced Materials*, vol. 7, no. 2, pp. 853–858, 2005.

[16] J. M. Barandiarán, M. L. Fdez-Gubieda, J. Gutiérrez, I. Orúe, A. G. Arribas, and G. V. Kurlyandskaya, "Magnetic films of technical interest prepared by pulsed laser deposition," *Journal of Optoelectronics and Advanced Materials*, vol. 6, no. 2, pp. 565–574, 2004.

[17] L. V. Panina, D. P. Makhnovskiy, and K. Mohri, "Magneto-impedance in amorphous wires and multifunctional applications: from sensors to tunable artificial microwave materials," *Journal of Magnetism and Magnetic Materials*, vol. 272, part 2, pp. 1452–1459, 2004.

[18] H. C. Jiang, W. L. Zhang, W. X. Zhang, S. Q. Yang, and H. W. Zhang, "Influences of sputtering angles and annealing temperatures on the magnetic and magnet ostrictive performances of TbFe films," *Journal of Materials Science & Technology*, vol. 21, no. 3, p. 315, 2005.

[19] P. Ciureanu, G. Rudkowska, L. Clime, A. Sklyuyev, and A. Yelon, "Anisotropy optimization of giant magnetoimpedance sensors," *Journal of Optoelectronics and Advanced Materials*, vol. 6, no. 3, pp. 905–910, 2004.

[20] D. De Cos, A. García-Arribas, and J. M. Barandiarán, "Simplified electronic interfaces for sensors based on inductance changes," *Sensors and Actuators A*, vol. 112, no. 2-3, pp. 302–307, 2004.

[21] G. Boero, I. Utke, T. Bret et al., "Submicrometer Hall devices fabricated by focused electron-beam-induced deposition," *Applied Physics Letters*, vol. 86, no. 4, Article ID 042503, 3 pages, 2005.

[22] H. Chiriac, M. Tibu, V. Dobrea, and I. Murgulescu, "Thin magnetic amorphous wires for GMI sensor," *Journal of Optoelectronics and Advanced Materials*, vol. 6, no. 2, pp. 647–650, 2004.

[23] A. Bienkowski and R. Szewczky, "new possibility of utilizing amorphous ring cores as stress sensor," *Physica Status Solidi A*, vol. 189, no. 3, pp. 787–790, 2002.

[24] A. Bienkowski and J. Kulikowski, "The magneto-elastic Villari effect in ferrites," *Journal of Magnetism and Magnetic Materials*, vol. 19, no. 1–3, pp. 120–122, 1980.

[25] N. R. Reddy, M. V. Ramana, G. Rajitha, E. Rajagopal, K. V. Sivakumar, and V. R. K. Murthy, "Stress sensitivity of inductance in NiCuZn ferrites," *Journal of Magnetism and Magnetic Materials*, vol. 292, pp. 159–163, 2005.

[26] N. Varalaxmi, N. R. Reddy, M. V. Ramana et al., "Stress sensitivity of inductance in NiMgCuZn ferrites and development of a stress insensitive ferrite composition for microinductors," *Journal of Materials Science: Materials in Electronics*, vol. 19, pp. 399–405, 2008.

[27] N. Varalaxmi and D. Ph, *[Ph.D. thesis]*, Sri Krishnadevaraya University, Anantapur, India, 2007.

[28] U. R. Lima, M. C. Nasar, R. S. Nasar, M. C. Rezende, J. H. Araujo, and J. F. Olivera, "Synthesis of NiCuZn ferrite nanoparticles and microwave absorption characterization," *Materials Science and Engineering B*, vol. 151, no. 3, pp. 238–242, 2008.

[29] S. I. Pyun and J. T. Baek, "Microstructural dependence of permeability and permeability spectra in Ni-Zn ferrites," *American Ceramic Society Bulletin*, vol. 64, no. 4, pp. 602–605, 1985.

[30] T. Y. Byun, S. C. Byeon, K. S. Hong, and C. K. Kim, "Factors affecting initial permeability of Co-substituted Ni-Zn-Cu ferrites," *IEEE Transactions on Magnetics*, vol. 35, no. 5, pp. 3445–3447, 1999.

[31] U. B. Deshmukh, S. M. Kabbur, N. D. Chaudhari, S. R. Sawant, and S. Suryavansi, in *Proceedings of the National Conference on Electronic Materials Devices and Systems (NCEMDS '99)*, p. 231, Gulbarga, India, 1999.

[32] J. Loaec, "Thermal hysteresis of the initial permeability of soft ferrites at transition temperatures," *Journal of Physics D*, vol. 26, no. 6, p. 963.

[33] C. Y. Tsay, K. S. Liu, and I. N. Lin, "Microwave sintering of $(Bi_{0.75}Ca_{1.2}Y_{1.05})(V_{0.6}Fe_{4.4})O_{12}$ microwave magnetic materials," *Journal of the European Ceramic Society*, vol. 24, no. 6, pp. 1057–1061, 2004.

[34] A. Globus, P. Duplex, and M. Guyot, "Determination of initial magnetization curve from crystallites size and effective anisotropy field," *IEEE Transactions on Magnetics*, vol. 7, no. 3, p. 617, 1971.

[35] O. F. Caltun, L. Spinub, A. L. Stancua, L. D. Thungb, and W. Zhou, "Study of the microstructure and of the permeability spectra of Ni-Zn-Cu ferrites," *Journal of Magnetism and Magnetic Materials*, vol. 242–245, pp. 160–162, 2002.

[36] G. Baca, R. Valenzuela, M. A. Escobar, and L. F. Magaña, "Temperature dependence of the critical magnetic field in polycrystalline ferrites," *Journal of Applied Physics*, vol. 57, no. 8, pp. 4183–4185, 1985.

[37] T. Y. Byun, S. C. Bycon, and K. S. Hong, "Factors affecting initial permeability of Co-substituted Ni-Zn-Cu ferrites," *IEEE Transactions on Magnetics*, vol. 35, no. 5, pp. 3445–3447, 1999.

[38] M. Guyot and A. Globus, "wall displacement and bulging in magnetization mechanisms of the hysteresis loop," *Physica Status Solidi B*, vol. 52, no. 2, pp. 427–431, 1972.

[39] I. Kanada, T. Murse, and T. Nomura, "Effect of chemical composition and microstructure on stress sensitivity of ferrite," *Journal of the Japan Society of Powder and Powder Metallurgy*, vol. 48, pp. 135–139, 2001.

30

Durability Modeling of Environmental Barrier Coating (EBC) Using Finite Element Based Progressive Failure Analysis

Ali Abdul-Aziz,[1] **Frank Abdi,**[2] **Ramakrishna T. Bhatt,**[1] **and Joseph E. Grady**[1]

[1] *NASA Glenn Research Center, Cleveland, OH 44135, USA*
[2] *AlphaSTAR Corporation, Long Beach, CA 90804, USA*

Correspondence should be addressed to Ali Abdul-Aziz; ali.abdul-aziz-1@nasa.gov

Academic Editor: Guillaume Bernard-Granger

The necessity for a protecting guard for the popular ceramic matrix composites (CMCs) is getting a lot of attention from engine manufacturers and aerospace companies. The CMC has a weight advantage over standard metallic materials and more performance benefits. However, these materials undergo degradation that typically includes coating interface oxidation as opposed to moisture induced matrix which is generally seen at a higher temperature. Additionally, other factors such as residual stresses, coating process related flaws, and casting conditions may influence the degradation of their mechanical properties. These durability considerations are being addressed by introducing highly specialized form of environmental barrier coating (EBC) that is being developed and explored in particular for high temperature applications greater than 1100°C. As a result, a novel computational simulation approach is presented to predict life for EBC/CMC specimen using the finite element method augmented with progressive failure analysis (PFA) that included durability, damage tracking, and material degradation model. The life assessment is carried out using both micromechanics and macromechanics properties. The macromechanics properties yielded a more conservative life for the CMC specimen as compared to that obtained from the micromechanics with fiber and matrix properties as input.

1. Introduction

Durability and damage related issues concerning fiber reinforced ceramic matrix composites (FRCMC), specifically SiC fiber reinforced SiC matrix composites (SiC/SiC), are of significance for low maintenance, dependability, and cost efficiency. Typically, most of damage and failure are caused by environmental conditions. These conditions are confined to moisture, thermal-mechanical load, creep, and fatigue. Lab, burner rig, and field tests have been performed to capture the service environment, induced damage, and resulting strength/stiffness reduction for several classes of CMCs being considered as components in aeroengines [1]. The CMCs are lightweight materials and operate at higher temperature than metals of at least 200°C. In dry air conditions, these materials form a protective layer on the surface called silica which makes them stable at a temperature up to 1300°C for long-term applications. However, in combustion environment containing moisture, the silica layer disintegrates causing surface recession [2]. Therefore, in order for these CMC materials to be useful in aeroengine applications, their surface must be protected. Such protection is being considered by applying environmental barrier coating (EBC) that has a range of operating temperature between 1200 and 1500°C depending on the composition [2–6].

There are three classes of EBC currently being evaluated for SiC/SiC turbine components. They are barium aluminum strontium (BSAS), rare earth di- and monosilicates (REMS and REDS), and hafnia/zirconia based systems [7–9]. The rare earth series include elements from lanthanum to lutetium. In general, an EBC system consists of two or more layers of coating, in which each layer serves a specific purpose. The total thickness of the EBC applied depends on the components and the coating can be applied by different processing methods depending on the intended microstructure and durability. Static components such as combustor liners, turbine vanes, and shrouds are subjected to thermal and gas pressure loads only. As a result, these components can accommodate coating thickness as much as 525 μm. On the other hand, rotating components such as blades are subjected to a combination of thermal and mechanical loads. To reduce overall weight of the rotating component, the thickness of EBC is limited to ~125 μm [10].

The coating system can be applied via variety of application systems. Among the most common ones are techniques such as air plasma spray (APS), physical vapor deposition (PVD), and slurry depending on the components, manufacturing cost, and intended durability. These systems in general have different material properties than the substrate since the sublayers of the coating are applied at different temperatures and as a result residual stresses develop. Depending on the magnitude and nature of these stresses, damage can occur in the coating after deposition and after exposing the coated substrate to turbine operating conditions. The damage has to be minimized or controlled; otherwise, the coating will spall which will reduce and limit the life of the component [7–9].

Therefore, damage drivers such residual stresses are of concern to EBC development, durability, and application. These stresses can be determined or measured by non-destructive techniques such as X-ray diffraction, Raman spectroscopy, and neutron diffraction. However, because of the complex crystal structure of some of the EBC compositions, it is cumbersome to use these techniques for these measurements. A means to tackling these factors is to control the constituent properties and thickness of the coating and develop physics based models that enable prediction of the durability and service life of the EBC under typical environmental conditions such as moisture, creep, fatigue, and crack propagation at the coating-CMC interface. PFA is used to determine the residual stresses in the specimen and to evaluate their role in damage initiation and propagation.

This paper is an extension of a prior work [1] where the focus of the research was based on examining an analytical methodology to model the durability of the EBC using a multiscale progressive failure analysis [14, 15] approach. Prior work detected damage initiation events using lamina fatigue properties for the CMC as input to the PFA analysis. However, in CMC composites damage initiates at the microscale level of the material. The use of fiber and matrix constituent properties enables the evaluation of damage events at their inception source. With macromechanics, the lamina properties are degraded at the onset of damage. But, with micromechanics, the properties of the constituent that is damaged are degraded, while the other constituent retains its properties.

The analysis used an updated material model for the EBC as compared to the one used in the prior work. Also, it used strength-time exposure degradation model from literature with improved accuracy for the SiC/SiC CMC material. The life prediction was performed once using reverse engineered micromechanics properties as input and once more using macromechanics properties as input. The lamina properties consisted of stiffness, strength, and fatigue properties. Similarly, derived fiber matrix properties consisted of stiffness, strength, and fatigue properties for each constituent. Strength based failure criteria based on maximum stress were employed to determine material damage. Stiffness of damaged elements is reduced once a specific failure criterion is invoked. Damaged elements are not removed from the

TABLE 1: Physical, thermal, and mechanical properties of uncoated SiC/SiC substrate at 21°C [12].

SiC/SiC property	Value	SiC/SiC property	Value
E_{11} (MPa)	$2.85E + 05$	S_{11T} (MPa)	$3.21E + 02$
E_{22} (MPa)	$2.85E + 05$	S_{11C} (MPa)	$3.21E + 02$
E_{33} (MPa)	$1.57E + 05$	S_{22T} (MPa)	$3.21E + 02$
G_{12} (MPa)	$1.13E + 05$	S_{22C} (MPa)	$3.21E + 02$
G_{23} (MPa)	$9.90E + 04$	S_{33T} (MPa)	$4.00E + 01$
G_{13} (MPa)	$9.90E + 04$	S_{33C} (MPa)	$1.00E + 02$
NU12	$1.30E - 01$	S_{12S} (MPa)	$2.10E + 02$
NU23	$1.70E - 01$	S_{23S} (MPa)	$1.05E + 02$
NU13	$1.70E - 01$	S_{13S} (MPa)	$1.05E + 02$
ALPHA11 (1/°C)	$2.71E - 06$	ALPHA22 = 33 (1/°C)	$2.71E - 06$

finite element model. Future work can use fracture mechanics principle using the damage path predicted by PFA to assess fracture growth in the EBC coated CMC specimen. Results from the analytical effort are discussed next.

2. Description of Analytical Approach

The analyses utilized the finite element method to model the combined EBC/CMC bar specimen sample which included the three layers of EBC and the coated substrate or the CMC part. Finite element model was developed for estimating the stress response based on the known processing conditions of the coating, the specimen geometry of the coated substrate, and the thermomechanical properties of the coated layers and the substrate.

Coating was applied via the plasma spray technique on SiC/SiC composite substrates [9]. It is assumed that the substrate is maintained at 1300°C during deposition of the plasma spray coating and then cooled to room temperature ~25°C. Also, thermomechanical properties of standalone individual layers of the EBC system required for the model were obtained from [13].

The analytical calculations covered modeling the beam specimen with defined EBC layers on top of a SiC/SiC substrate. Thermomechanical properties, including thermal expansion coefficient, stiffness, Poisson's ratio, and strength for all four materials constituting the EBC/CMC specimen, are used as input to the durability and life prediction analysis; see Tables 1 and 2. The SiC/SiC CMC material lamina properties of the fabric were obtained from [12]. Plastic deformation and microcracking that may occur in the plasma-sprayed coating were not considered in the model. The specimen dimensions were 2 by 3 by 45 mm in addition to the EBC thickness on the top, Figure 1. The thermal boundary conditions associated with the coating application methodology were all incorporated into the thermal model.

2.1. Multiscale Progressive Failure Analysis (PFA). Micromechanics and macromechanics composite analysis are integrated with finite element analysis and damage and fracture tracking to perform progressive failure analysis, Figure 2. The capability is integrated in the GENOA [14, 15] software.

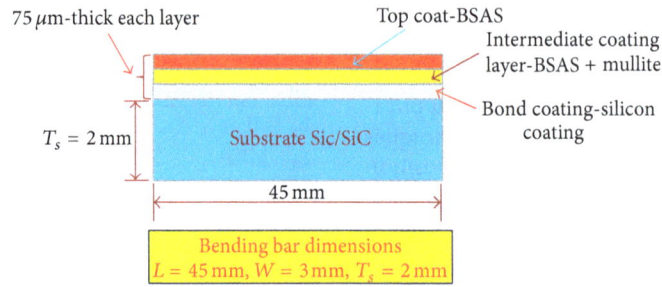

FIGURE 1: Two-dimensional section of the beam bar specimen showing dimension.

TABLE 2: Material properties of top and intermediate coats and bond [13].

(a)

Bond coat properties	Value	Units
Young's modulus	97	(GPa)
Poisson's ratio	0.21	(—)
CTE	$4.50E - 06$	(1/°C)
Tension strength	40	(MPa)
Compression strength	40	(MPa)
Shear strength	40	(MPa)

(b)

Intermediate coat properties	Value	Units
Young's modulus	37.4	(GPa)
Poisson's ratio	0.179	(—)
CTE	$5.70E - 06$	(1/°C)
Tension strength	28	(MPa)
Compression strength	28	(MPa)
Shear strength	28	(MPa)

(c)

Top coat properties	Value	Units
Young's modulus	32	(GPa)
Poisson's ratio	0.19	(—)
CTE	$5.60E - 06$	(1/°C)
Tension strength	28	(MPa)
Compression strength	28	(MPa)
Shear strength	28	(MPa)

Traditionally, failure is assessed at the macroscale using lamina or laminate properties. The software enables assessment of failure and damage in composites at a lower scale, that is, the fiber, matrix, and interface level. The methodology augments finite element analysis (FEA), with a full-hierarchical modeling capability that goes down to the microscale of subdivided unit cells composed of fiber bundles and their surrounding matrix [16]. The life prediction strategy uses a PFA-FEA based approach shown in Figure 2 [14, 15].

Damage is tracked at the micro- or macroscale levels leading to local material degradation and recalculation of stiffness. This is done by evaluating series of physics based

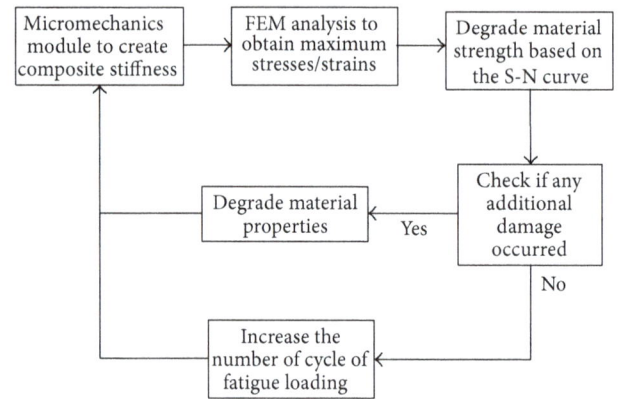

FIGURE 2: General flow of the progressive failure analysis methodology for life prediction [1].

failure criteria (shown in Table 3) at increased load increments or fatigue cycles. In addition to degradation, stress damage, strength-cycle, or strength-time curves are used as input to the analysis at each fatigue cycle block to degrade the strength. Damage is accumulated as life cycles are increased until the ultimate life of the structure is reached.

The life prediction analysis uses PFA to determine how many cycles of temperature ramping the specimen can sustain before the SiC/SiC and the coating materials are damaged; see Figure 2. Ideally, strength-time curves would be required as input to the analysis for all the materials constituting the EBC/CMC specimen. Such data are typically obtained from physical testing or from literature. The analysis assumed that the EBC coating materials do not degrade as function of exposure time to temperature. Only the SiC/SiC CMC is degraded using test data obtained from literature for strength degradation as function of time [11].

2.2. Life Prediction with PFA. The mathematical approach used in applying the PFA includes the integration of composite mechanics and damage/fracture mechanics with finite element analysis. The damage mechanics account for matrix cracking under transverse, compressive, and shear loading. The ply fracture mechanisms include fiber failure under tension, compression (crushing, microbuckling, and debonding), and delamination. This is invoked via the GENOA code [16] by allowing a sequence of analytical steps that includes

TABLE 3: Failure criteria used in life prediction of composite specimens.

Mode number	Fiber failure criteria	Event description
1	Longitudinal tensile (S_{11T})	Failure of ply controlled by fiber tensile strength and fiber volume ratio
2	Longitudinal compressive (S_{11C})	(1) Fiber/matrix delamination under compression loading (2) Fiber microbuckling (3) Fiber crushing
3	Transverse tensile	Matrix cracking under tensile loading, event controlled by matrix tensile strength, matrix modulus, and fiber volume ratio
	Transverse compressive	Matrix cracking under compressive loading, event controlled by matrix compressive strength, matrix modulus, and fiber volume ratio
	Normal tensile (S_{22T})	Plies are separating due to normal tension
4	Normal compressive (S_{22C})	Due to very high surface pressure that is crushing of laminate
5	In-plane shear (S_{33C})	Failure in-plane shear relative to laminate
6	Transverse normal shear (S_{12S})	Shear failure acting on transverse cross-oriented in a normal direction of the ply
7	Longitudinal normal shear (S_{13S})	Shear failure on longitudinal cross section that is oriented in a normal direction of ply
	Normal tensile (S_{23T})	Combined stress failure criteria used for isotropic materials
	Relative rotation criterion (RROT)	Considers failure if the adjacent plies rotate excessively with one another
8	Transverse normal shear (S_{23S})	Considers invariant through-the-thickness
9	Linear elastic fracture	Virtual crack closure technique (VCCT), discrete cohesive zone model (DCZM)

(1) the use of a finite element stress solver, (2) user selection of 2D or 3D architectural details (through-the-thickness fibers, resin rich interphase layer between weave plies, fiber volume ratio, void shape, size and location, cure condition, etc.), (3) assigning static (thermomechanical) or spectrum loading, (4) automatic update of the finite element model prior to executing FEA stress solver for accurate lamina and laminate properties, and (5) degradation of material properties at increased loading (including number of cycles) based on detected damage. Additional details can be found in [14].

All stages of damage evolution within the composite structure are identified. They are damage initiation, damage propagation, fracture initiation, and fracture propagation. The damage events include matrix cracking, delamination, and fiber failure. Displacements, stresses, and strains derived from the structural scale FEA solution at a node or element of the finite element model are passed to the laminate and lamina scales using laminate theory. Stresses and strains at the microscale are derived from the lamina scale using microstress theory. The analytical capability offers microscale modeling and damage assessment capability for composite materials such as ceramic, metal, and polymer matrix composites. For the failure criteria shown in Table 3, the code automatically distinguishes between the criteria that are applicable to laminated composites versus those that are applicable to isotropic material. For example, for longitudinal compression, the code evaluates three failure potentials under longitudinal compression. They are fiber and matrix delamination, fiber microbuckling, and fiber crushing. For isotropic materials, the compression stress or strain is compared to the allowable to determine whether or not failure had occurred.

In addition to maximum stress failure criteria, the code evaluates failure due to maximum strain and interactive stress criteria. More details can be obtained from [14, 15].

PFA stress based evaluation is accurate up to fracture initiation [14, 15]. Due to stress singularity often experienced in finite element analysis, fracture mechanics based approach is used to grow the crack. In linear fracture mechanics approach, it is required to have fracture toughness for static crack growth and da/dn versus ΔK for fatigue crack growth; da/dn is the change of crack length with loading cycles, while ΔK is the stress intensity factor change. Since such data are not available for the analysis, the focus in this paper is to identify cycles that caused damage to initiate and propagate to the substrate SiC/SiC material. It should be noted that the PFA strength based approach has key advantages as compared to fracture mechanics methods. For example, in terms of advantages, PFA does not require prior knowledge of crack path. Crack growth will be the subject of future work once the data required for the prediction becomes available.

Additionally, for the analyses used in this paper, linear elastic fracture mechanics approach (item no. 9 in Table 3) is not used. Fracture mechanics application would require knowledge of fracture path, toughness, and fracture energy. This type of analysis will be considered in the future.

3. Specimen Geometry and FE Model

The finite element analyses to generate the thermal profile were conducted using the commercial finite element code

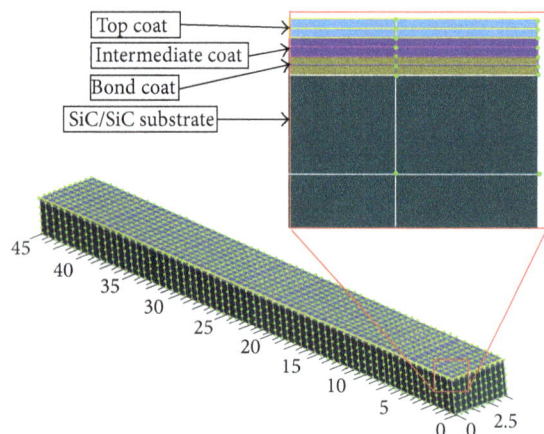

FIGURE 3: Representative finite element model of the thermal barrier coating. Work plane rulers shown are in units of mm.

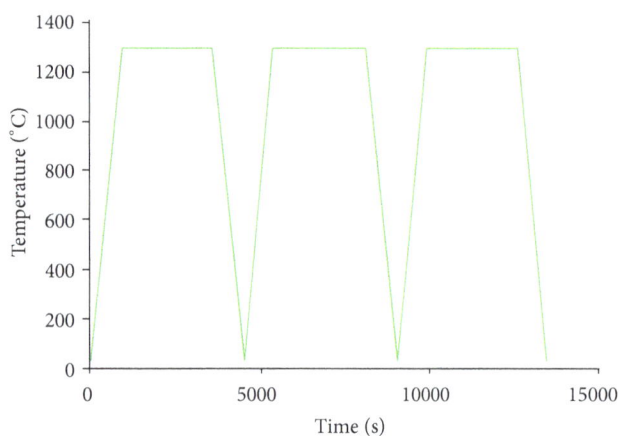

FIGURE 4: Representative thermal loading profile applied in the analyses.

Abaqus [17]. The finite element model dimensions and sections are shown in Figure 3. The thermal profile was predicted under transient loading conditions as noted in Figure 4. The thermal cycle assumes that the bar specimen is initially at 21°C and, within 15 minutes, it heats up to 1300°C and remains constant for a duration of 45 minutes until shutdown, where it cools off back to 21°C. One complete cycle constitutes exposing the specimen to these thermal conditions for a total time of one hour. Material properties of both the coating and the substrate were input into the model under linear isotropic condition for the coating systems and linear orthotropic condition for the SiC/SiC substrate. Temperature dependency of all the materials was accounted for in the analyses. The mesh included a 3D model of the bar specimen with high density mesh along the substrate and the coatings interfaces. Eight-node brick element was employed.

For the durability solution, GENOA-PFA augmented the FEA solver (Abaqus) for life prediction to determine life cycles that caused damage to initiate and propagate. PFA used Abaqus iteratively at increased number of cycles to evaluate damage after each FEA run. Damage stabilization is attained

to ensure material and structural equilibrium before the number of cycles is increased again. The process is repeated until ultimate life is obtained. It was assumed that the top coat BSAS, intermediate coating layer (BSAS+Mullite), and the bond silicon coating do not degrade as a function of time and temperature and the results presented pertained only to determining the number of cycles that it would take for damage to initiate and for damage to propagate. Furthermore, data obtained from literature [11] shows that some degradation of the SiC/SiC substrate at elevated temperature after exposure time is expected.

4. Analytical Description

To evaluate the effects of thermal fatigue on the EBC SiC/SiC using micromechanics based approach, in situ (effective) fiber and matrix properties for the SiC/SiC system were generated. To generate the effective SiC/SiC constituent properties, a $[0, 90]s$ laminate with 50% fiber volume content was modeled using materials characterization and qualifications (MCQ) composite software [18] and an iterative process (Figure 5) was implemented. The code used an optimization algorithm to derive a unique set of fiber/matrix properties (mainly stiffness and strength) that are capable of reproducing test data of the lamina or laminate. The process iterates until values of the predicted lamina or laminate in-plane and out-of-plane properties are in good agreement with data from test. The fiber and matrix properties for the SiC/SiC derived through the elaborate reverse engineering process shown in Tables 4(a) and 4(b) are used as input to PFA to determine the life cycle that would cause damage to initiate and propagate in the CMC specimen. This makes it possible to run the comparative assessment of the behavior of the CMC using both microscale (fiber and matrix) and macroscale (lamina level). PFA treats the top, intermediate, and bond materials as isotropic during the life prediction analysis.

The fiber and matrix properties obtained from reverse engineering are listed in Table 4(a). The fiber and matrix properties are then used as input to MCQ [18] to evaluate the mechanical properties of the CMC (0/90)s laminate. As indicated in Table 4(b), the MCQ predictions starting from microscale properties yielded an accurate representation of the laminate properties from test [12]. The out-of-plane predictions for the laminate could improve as the properties were taken from a plain weave system where the effects of the fiber weave on the out-of-plane strength and stiffness was significant. If more details were available on the architecture, the out-of-plane predictions with MCQ would improve as the architecture details would be included in material characterization.

4.1. SiC/SiC Stress-Cycle Curve as Function of Exposure Time. The stress cycle (S-N) criterion adopted for these analyses utilized a set of S-N curve for each of the laminate, fiber, and matrix as shown in Figure 6. The first set used the typical or the standard S-N curve that represented the degradation of the SiC/SiC composite under thermal loading conditions [11]. This S-N curve started to degrade after 27.78 hours

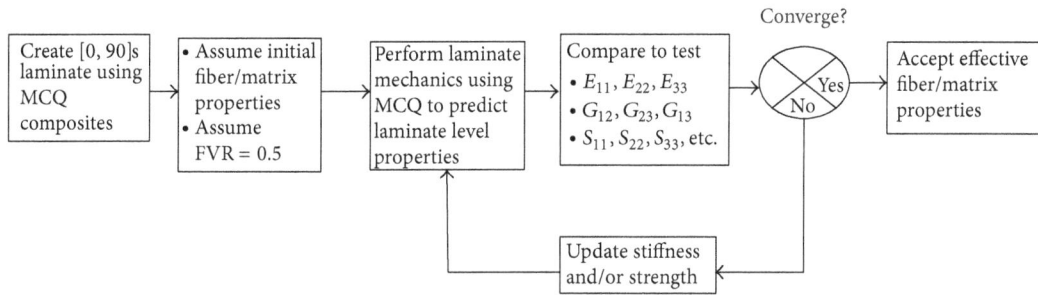

FIGURE 5: Iterative process used to determine effective constituent properties of SiC/SiC laminate.

TABLE 4: (a) Effective fiber and matrix properties. (b) Effective ply correlating to effective fiber/matrix properties.

(a)

Effective fiber and matrix properties (FVR = 0.5 [0, 90]s)					
SiC fiber			SiC matrix		
Property	Units	Value	Property	Units	Value
E_{11}	MPa	380000	E	MPa	380000
E_{22}	MPa	156000	NU		0.19
G_{12}	MPa	100000	ST	MPa	250
G_{23}	MPa	70000	SC	MPa	250
NU12		0.19	SS	MPa	190
NU23		0.17	ALPHA	1/degC	$2.71E-06$
S_{11T}	MPa	600			
S_{11C}	MPa	600			
ALPHA11	1/degC	$2.71E-06$			
ALPHA22	1/degC	$2.71E-06$			

(b)

Properties of [0, 90]s using effective fiber/matrix properties				
Property	Units	Target values	Calibrated	Discrepancy (%)
E_{11}	MPa	284900	285000	0.0
E_{22}	MPa	284900	285000	0.0
E_{33}	MPa	190600	157000	17.6
G_{12}	MPa	112300	113000	0.6
G_{23}	MPa	104800	99000	5.5
G_{13}	MPa	104800	99000	5.5
NU12	MPa	0.126	0.13	3.2
NU23	MPa	0.2132	0.17	20.3
NU13	MPa	0.2132	0.17	20.3
S_{11T}	MPa	318.5	321	0.8
S_{11C}	MPa	318.5	321	0.8
S_{22T}	MPa	318.5	321	0.8
S_{22C}	MPa	318.5	321	0.8
S_{12S}	MPa	213.5	210	1.6

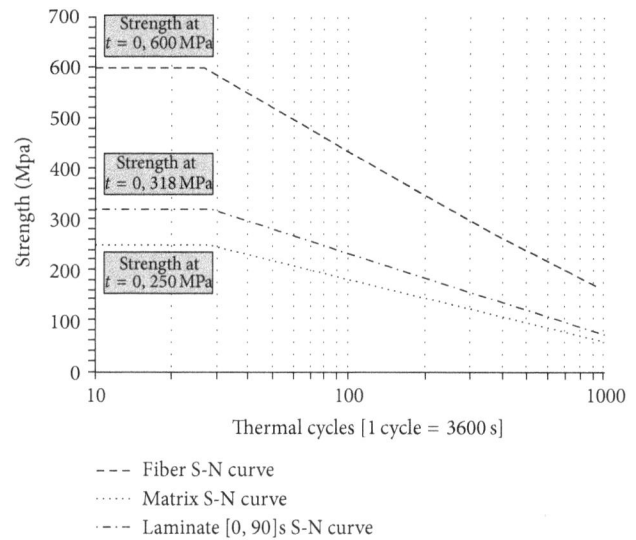

FIGURE 6: Strength as a function of thermal cycles for laminate, fiber, and matrix of SiC-SiC material [11].

level. To achieve these trends, the laminate level trend was scaled accordingly using the fiber and matrix ultimate tensile strength at $t = 0$.

Characteristics and thickness dimensions of the coating system used are shown in Table 5 and, as noted, the 3 layers of coating had the same magnitude of thickness which is 75 μm and the substrate had a 2 mm thickness. Photographs of the top surface and cross section of a typical APS trilayered environmental barrier coated SiC/SiC composite are shown in Figure 7. The sublayers of the coating are inhomogeneous and contain microcracks and significant levels of nonuniform pores.

5. Results and Discussion

The durability analysis performed indicated that the material damage initiated in the top EBC coat and then propagated down to the intermediate layer then to the bond. This took place during the first one hour of thermal loading (in one cycle). For each cycle, the PFA analysis assumed that the specimen reached the 1300°C temperature, which means that a gradient of 1279°C is applied instantly to simulate each cycle.

from 318 MPa to 159 MPa (50% of the original ultimate tensile strength at $t = 0$) at 1000 hr. The logarithmic degradation continued until failure. Since S-N curve was not available for the constituent materials, it was assumed that the same degradation trends occurred for the fiber and matrix

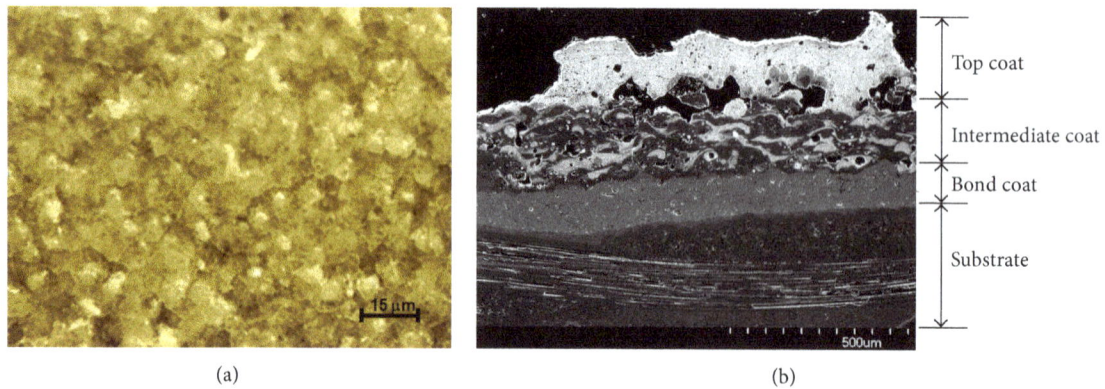

(a)   (b)

FIGURE 7: Typical microstructure of plasma sprayed EBC on SiC/SiC composites: (a) top view (optical micrograph) and (b) cross-sectional view (scanning electron micrograph).

TABLE 5: Coating systems and thicknesses considered.

Coating system	Coating thickness
Top coat-BSAS	75 μm
Intermediate coat Mullite + barium strontium aluminum silicate (BSAS)	75 μm
Bond coat-silicon	75 μm
Sic-SiC substrate	2 mm

TABLE 6: Summary of cycles to damage and associated damage modes for each EBC and substrate layer.

Layer	Cycle of damage initiation	Damage mode
Top layer	1	Tension
Intermediate layer	1	Tension and shear
Bond layer	1	Tension and shear
SiC/SiC	Macro: 1070 Micro: 1090	Tension

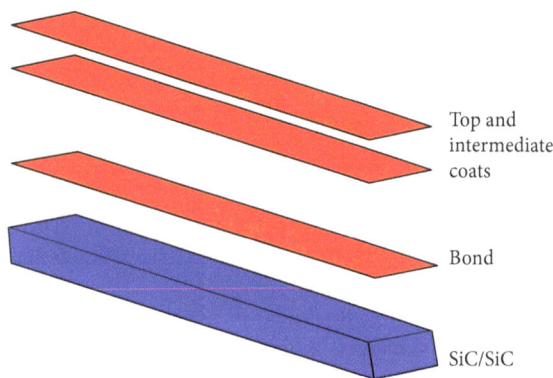

FIGURE 8: Damage in the EBC layers due to tension stress (red color indicates damage; blue color represents undamaged elements).

As mentioned earlier in the paper, the maximum stress criteria listed in Table 3 are used in the durability evaluation performed by the PFA. When an element stress exceeds the allowable value, the stiffness of the element was degraded accordingly in the direction where the damage occurred. No elements were removed when damaged, which meant that the mesh size remained unchanged throughout the analysis. The analysis was repeated twice, once with macrolevel properties for the CMC material and once more with microlevel properties using the reverse engineered properties of the fiber and matrix.

Figure 8 shows the damaged elements in red color in the three EBC layers due to tension stress. As noted, all elements in the three EBC layers are damaged at the end of cycle 1, that is, after 3600 seconds of exposure to 1300°C. The damage started in the top layer and propagated down to the bond. The SiC/SiC substrate was undamaged until the cycle 1070 was reached for macrotype input and until the cycle 1090 for micro input.

The damage volume is computed to keep track of total number of elements that are damaged during a given loading cycle. It provides useful inspection criterion of critical parts. For example, a sudden increase in damage volume indicates the onset of major damage event in the structure. Since only degradation in the SiC/SiC substrate was considered, no additional damage was detected until cycle 1070 when the SiC/SiC substrate's tensile failure criterion was detected. A summary of the life cycles for the EBC-SiC/SiC system is shown in Table 6. A macrobased simulation seemed to offer more conservative life cycle compared with the microbased approach showing greater life by approximately 20 hours. This is assuming the same in-plane ply properties and when using effective fiber/matrix properties rather than macrobased laminate properties.

The results indicated that macromechanics or ply mechanics approach is more conservative when it comes to assessing damage initiation in the substrate as compared to microscale simulation. This was expected as the postdamage degradation was more severe at the macrolevel as compared to the one at the microscale. With micromechanics, if one

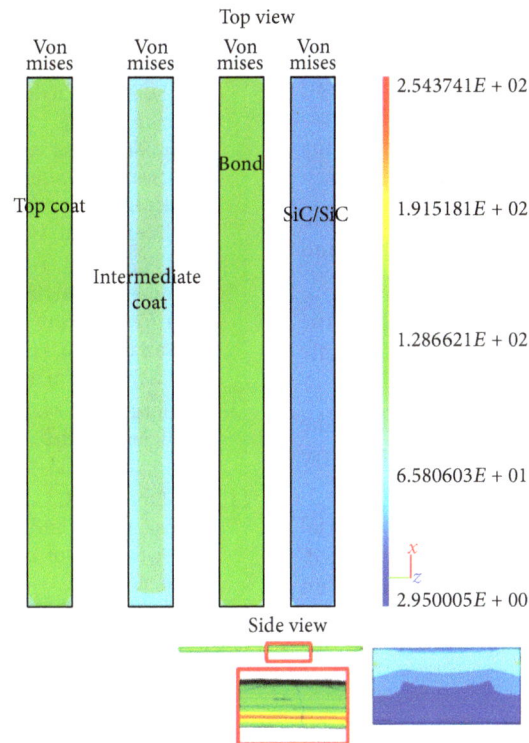

FIGURE 9: Von Mises residual stresses in MPa as a result of cool down from 1300°C to room temperature after the first cycle of exposure to elevated temperature (top and side views).

constituent is damaged, the other constituent retains the stiffness. In the case of macromechanics approach, the whole lamina stiffness in the direction of damage is reduced to a small value.

Figure 9 shows the von Mises residual stresses in each material obtained during cool down from 1300°C to room temperature. The residual stresses are a good indicator of where damage is likely to start. It supports the findings presented in Figure 9, whereby the top, intermediate, and bond layers were damaged first before propagating after several hundred cycles to the CMC substrate. The PFA analysis is accounting for the residuals' stresses during the life prediction as the residual stresses are translated into damage indices when damage is introduced. The damage indices are then stored for use as input in the subsequent cycle analysis.

It should be further noted that the von Mises stresses calculated during cool down show that the bond material experienced the highest stresses. However, this does not mean that the bond is failing more than the top or intermediate coats. Failure is driven by the allowable material stress or strain. In the case of the EBC specimen, the maximum stress criteria were used to guide the assessment of the damage evolution. Comparing stress to strength for the top three materials, the residual stresses do indicate that material damage is experienced by all three materials, whereby the top coat experienced the most damage because of the ratio of stress to strength. Future analysis will include material nonlinearity as well as coupled structural thermal analysis to

determine the effect of nonuniform heating on the specimen's life prediction.

6. Conclusions and Future Work

A novel computational simulation approach is presented to predict life for EBC/CMC specimen using the finite element method augmented with progressive failure analysis (PFA) that included durability, damage tracking, and material degradation model. The following conclusions and recommendations can be drawn from the work presented in the paper.

(1) Damage initiates predominantly in the top coat due to tensile strength failure and in the intermediate/bond coat due to delamination.

(2) Damage propagates into the SiC/SiC substrate due to tensile failure, eventually redistributing the stress into the EBC causing further damage propagation.

(3) Multiscale progressive failure analysis allowed a systematic prediction of the life cycles for damage initiation and propagation in EBC SiC/SiC specimens. The technical approach applied combined composite mechanics and damage tracking and fracture.

(4) Use of micromechanics properties as input to PFA resulted in life prediction that is approximately 20 hours greater as compared to that obtained from the

use of lamina properties indicating that macromechanics is more conservative than micromechanics.

(5) Accurate life prediction requires strength-time exposure behavior for all the materials used in the specimen. This will allow reliable assessment of any structural component made of the same materials.

(6) Defects such as flaws and initial cracks in coating will add more accuracy to the life prediction analysis and it all must be accounted for in any future work.

(7) Material characterization can help optimization of the laminate thickness which in return can increase life and delay damage.

(8) Material architecture should be considered in the material characterization to yield an accurate reverse engineering of constituent properties.

(9) Future work should include nonlinear material behavior in the analysis and simulations performed. This will require data from ATSM tests at different temperatures.

Nomenclature

E_{11}: Lamina modulus in fiber direction
E_{22}: Lamina modulus perpendicular to fiber direction
E_{33}: Lamina modulus perpendicular to fiber direction
G_{12}: Lamina in-plane shear modulus
S_{11C}: Lamina compressive strength in fiber direction
S_{11T}: Lamina tensile strength in fiber direction
S_{12S}: Lamina in-plane shear strength
S_{22C}: Lamina compressive strength perpendicular to fiber direction
S_{22T}: Lamina tensile strength perpendicular to fiber direction
S_{13}: Lamina strength in longitudinal shear direction
S_{23}: Lamina strength in transverse shear direction
S_{33T}: Lamina tensile strength in normal out-of-plane direction
S_{33C}: Lamina compressive strength in normal out-of-plane direction
Alpha11: Thermal expansion coefficient in fiber direction
Alpha22: Thermal expansion coefficient transverse to fiber direction
Alpha33: Thermal expansion coefficient normal to fiber direction
da/dn: Change of crack length with loading cycles
ΔK: Stress intensity factor change.

Conflict of Interests

The authors declare that there is no conflict of interests regarding the publication of this paper.

References

[1] A. Abdul-Aziz, G. Abumeri, W. Troha, R. T. Bhatt, J. E. Grady, and D. Zhu, "Environmental barrier coating (EBC) durability modeling using a progressive failure analysis approach," in *Smart Structures and Materials & Nondestructive Evaluation and Health Monitoring, Behavior and Mechanics of Multifunctional and Composite Materials*, vol. 8346 of *Proceedings of SPIE*, San Diego, Calif, USA, March 2012.

[2] K. N. Lee, D. S. Fox, R. C. Robinson, and N. P. Bansal, "Environmental barrier coatings for silicon-based ceramics," in *High Temperature Ceramic Matrix Composites*, W. Krenkel, R. Naslain, and H. Schneider, Eds., pp. 224–229, Wiley-Vch, Weinheim, Germany, 2001.

[3] P. J. Jorgensen, M. E. Wadsworth, and I. B. Cutler, "Oxidation of silicon carbide," *Journal of the American Ceramic Society*, vol. 42, no. 12, pp. 613–616, 1959.

[4] J. L. Smialek, R. C. Robinson, E. J. Opila, D. S. Fox, and N. S. Jacobson, "SiC and Si_3N_4 recession due to SiO_2 scale volatility under combustor conditions," *Advanced Composite Materials*, vol. 8, no. 1, pp. 33–45, 1999.

[5] K. L. More, P. F. Tortorelli, and L. R. Walker, "Effects of high water vapor pressures on the oxidation of SiC-based fiber-reinforced composites," *Materials Science Forum*, vol. 369—372, pp. 385–394, 2001.

[6] K. L. More, P. F. Tortorelli, L. R. Walker, N. Miriyala, J. R. Price, and M. Van Roode, "High-temperature stability of SiC-based composites in high-water-vapor-pressure environments," *Journal of the American Ceramic Society*, vol. 86, no. 8, pp. 1272–1281, 2003.

[7] K. N. Lee, "Current status of environmental barrier coatings for Si-based ceramics," *Surface and Coatings Technology*, vol. 133-134, pp. 1–7, 2000.

[8] D. M. Zhu, R. A. Miller, and D. S. Fox, "Thermal and environmental barrier coating development for advanced propulsion engine systems," NASA TM-2008-215040, 2008.

[9] D. M. Zhu, N. P. Bansal, and R. A. Miller, "Thermal conductivity and stability of HfO_2-Y_2O_3 and $La_2Zr_2O_7$ evaluated for $1650^\circ C$," in *Advances in Ceramic Matrix Composites*, N. P. Bansal, J. P. Singh, W. M. Kriven, and H. Schnneider, Eds., John Wiley & Sons, Hoboken, NJ, USA.

[10] D. Zhu and R. A. Miller, "Thermal conductivity and elastic modulus evolution of thermal barrier coatings under high heat flux conditions," *Journal of Thermal Spray Technology*, vol. 9, no. 2, pp. 175–180, 2000.

[11] S. Ochiai, S. Kimura, H. Tanaka et al., "Degradation of SiC/SiC composite due to exposure at high temperatures in vacuum in comparison with that in air," *Composites A: Applied Science and Manufacturing*, vol. 35, no. 1, pp. 33–40, 2004.

[12] M. van Roode, A. K. Bhattacharya, M. K. Ferber, and F. Abdi, "Creep resistance and water vapor degradation of sic/sic ceramic matrix composite gas turbine hot section components," in *ASME Turbo Expo 2010: Power for Land, Sea, and Air*, pp. 455–469, June 2010.

[13] A. Abdul-Aziz and R. T. Bhatt, "Modeling of thermal residual stress in environmental barrier coated fiber reinforced ceramic matrix composites," *Journal of Composite Materials*, 2011.

[14] M. Garg, G. H. Abumeri, and D. Huang, "Predicting failure design envelop for composite material system using finite element and progressive failure analysis approach," in *Proceedings of the 52nd International SAMPE Symposium: Material and Process Innovations: Changing our World (SAMPE '08)*, May 2008.

[15] F. Abdi, Z. Qian, and M. Lee, *The Premature Failure of 3D Woven Composites*, ACMA Composites, Columbus, Ohio, USA, 2005.

[16] "GENOA durability and damage tolerance and life prediction software," AlphaSTAR Corporation, Long Beach, Calif, USA, http://www.ascgenoa.com.

[17] "Abaqus commercial finite element code," Providence, RI, USA, 2909—2499.

[18] "MCQ software," AlphaSTAR Corp, Long Beach, Calif, USA, 2012.

31

Experimental Study on LTCC Glass-Ceramic Based Dual Segment Cylindrical Dielectric Resonator Antenna

Ravi Kumar Gangwar,[1] S. P. Singh,[2] Meenakshi Choudhary,[3] D. Kumar,[3] G. Lakshmi Narayana Rao,[4] and K. C. James Raju[4]

[1] Department of Electronics Engineering, Indian School of Mines, Dhanbad 826004, India
[2] Department of Electronics Engineering, Indian Institute of Technology, Banaras Hindu University, Varanasi 221005, India
[3] Department of Ceramic Engineering, Indian Institute of Technology, Banaras Hindu University, Varanasi 221005, India
[4] School of Physics, University of Hyderabad, Hyderabad 500 046, India

Correspondence should be addressed to Ravi Kumar Gangwar; ravi8331@gmail.com

Academic Editor: Baolin Wang

The measured characteristics in C/X bands, including material properties of a dual segment cylindrical dielectric resonator antenna (CDRA) fabricated from glass-ceramic material based on B_2O_3–La_2O_3–MgO glass and $La(Mg_{0.5}Ti_{0.5})O_3$ ceramic, are reported. The sintering characteristic of the ceramic in presence of glass is determined from contact angle measurement and DTA. The return loss and input impedance versus frequency characteristics and radiation patterns of CDRA at its resonant frequency of 6.31 GHz are studied. The measured results for resonant frequency and return loss bandwidth of the CDRA are also compared with corresponding theoretical ones.

1. Introduction

Several investigators have focused attention on dielectric resonator antennas (DRAs) due to their simple geometry, small size, low cost, high radiation efficiency, flexible feed arrangement, wide range of material dielectric constants, ease of excitation, and easily controlled characteristics [1–5]. DRAs are available in various basic classical shapes such as rectangular, cylindrical, spherical, and hemispherical geometries. The techniques used for bandwidth enhancement of the DRAs include changing the aspect ratio of DRA, employing multisegments and stacked DRAs, and varying the dielectric constant of DRA material. DRAs having lower dielectric constant values are preferred for wideband applications. This results in weak coupling. Multisegment DRAs can be used to surmount this problem [6, 7].

Due to the incessantly increasing demands of miniaturizing the important passive microwave components, ceramic dielectrics are used as dielectric resonators and filters and antennas in many small microwave systems including wireless portable devices. However, microwave dielectric materials that possess high dielectric constant and quality factor usually need a very high sintering temperature and long soaking time to achieve enough density. There are three approaches to reducing the sintering temperature of the dielectric ceramics: (i) addition of low-melting glass [8] and sintering aids including carbonates and oxides [9], (ii) chemical processing [10], and (iii) smaller particle sizes for the starting materials [11]. Liquid-phase sintering with glass additives is the least expensive among them. However, if the amount of frits is large, the network formers contained in the remaining glass materials such as B_2O_3 and SiO_2 can profoundly absorb the microwave power at high frequencies, degrading the quality factor of the final materials [12]. Compared to the Liquid-phase sintered ceramics with low-melting glass frits in which the glass phase remains, another approach of using low temperature cofired ceramics (LTCCs), "glass-ceramic" where glass frits crystallize during the sintering stage produces a better result. The advantages offered by this glass-ceramic approach include shape stability after sintering, improved dielectric properties, mechanical strength, and controlled thermal expansion. Therefore, low temperature cofired ceramics (LTCCs) have a great importance to the electronic industry for building smaller RF modules and for fulfilling the need to miniaturize the devices in the wireless communication industry [13]. La based complex perovskite ceramic $La(Mg_{0.5}Ti_{0.5})O_3$ has negative temperature coefficient of resonant frequency (τ_f) and high Q value. It exhibits dielectric constant around 25–28; $Q \times f$ is 63100–73000 GHz depending upon processing conditions. $La(Mg_{0.5}Ti_{0.5})O_3$ ceramics have suitable microwave dielectric properties for the application of dielectric resonators, antenna, and filters [14, 15]. Sintering temperatures of these materials are high, that is, in the range of 1500°–1600°C. Lin et al. [16] prepared LTCC material with $La(Mg_{0.5}Ti_{0.5})O_3$ ceramics and B_2O_3–La_2O_3–MgO glass and had shown that these LTCC glass-ceramics can be sintered at 850°C and had excellent microwave dielectric properties ε_r = 11.8, $Q \times f$ = 14700 GHz and τ_f = 7.4 ppm°C.

In this paper, experimental investigation on a wide-band dual segment cylindrical dielectric resonator antenna (CDRA) fabricated from glass-ceramic materials based on B_2O_3–La_2O_3–MgO (BLM) glass with $La(Mg_{0.5}Ti_{0.5})O_3$ (LMT) ceramic along with its material properties is carried out. The crystallization behavior of BLM glass is investigated by differential thermal analysis (DTA). The sintering characteristic of the ceramic in presence of glass was determined from contact angle measurement. The dielectric properties of glass-ceramic material at a microwave frequency in X-band is measured. The variations of return loss and input impedance with frequency in C/X bands and radiation characteristics of the proposed glass-ceramic antenna at the antenna resonant frequency obtained through measurement are reported. The experimental results for resonant frequency and return loss bandwidth of CDRA are also compared with corresponding theoretical values.

2. Material Synthesis and Characterization

2.1. Preparation of Glass-Ceramic Material Based on B_2O_3–La_2O_3–MgO Glass with $La(Mg_{0.5}Ti_{0.5})O_3$ Ceramic.
Microwave ceramic $La(Mg_{0.5}Ti_{0.5})O_3$ (LMT) was prepared by the conventional mixed oxide method. High purity powders, La_2O_3, MgO, and TiO_2, with a molar ratio of nominal composition LMT were mixed and calcined at 1450°C for 4 h. The glass $60B_2O_3$–$12La_2O_3$–28MgO (BLM) (in mol %) was prepared by normal melting in a platinum crucible at 1200°C for 1 h. Differential thermal analysis (DTA) was carried out for powdered glass specimens with NETZSCH STA429C analyzer at 10°C/min heating rate. Contact angle between ceramic and glass was measured at different temperatures using DSA1 v 1.9-03, Kruss Gmbh, Hamburg. Sessile drop method was used for measuring contact angle. The sessile drop technique is a method used for the characterization of solid surface energies and, in some cases, aspects of liquid surface energies. The mixture of 50 wt. % calcined ceramic

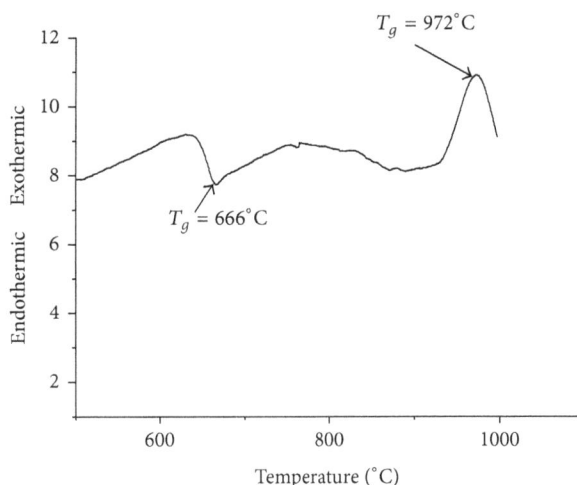

FIGURE 1: DTA pattern of glass (BLM).

LMT $(La(Mg_{0.5}Ti_{0.5})O_3)$ with 50 wt% reactive glass BLM $(60B_2O_3$–$12La_2O_3$–28MgO) was first ball-milled for 24 h. The milled powder was then dried at 120°C and granulated with PVA. Pellets were prepared by hydraulic press using cylindrical mold of 12.5 mm diameter at a pressure of 60 kN. The pellets of glass-ceramic were sintered at 850–950°C for 2 h in air with a heating rate of 5°C/min. The bulk densities of glass-ceramic pellets were measured from their dimension and mass. The sintered pellets were ground and powder X-ray diffraction patterns were recorded using X-ray Diffractometer (Seifert) employing Cu – $K\alpha_I$ radiation using an Ni filter to identify different crystalline phases present in the sintered samples. Microstructure observations of the natural surface of sintered glass-ceramic were studied by using Scanning Electron Microscope (JSM-840 LV, JEOL, Japan).

2.2. Differential Thermal Analysis (DTA).
The differential thermal analysis (DTA) trace for BLM glass is shown in Figure 1. The convention for determining the glass transition temperature is to extend the straight-line portions of the baseline and the linear portion of the upward slope, marking their intersection. This shift in the base line shows a change in specific heat of the glass, which is attributed to the glass transition temperature T_g. The glass transition temperature has been found to be 666°C. DTA pattern of glass samples shows one exothermic peak. It represents the temperature at which the rate of crystallization of different phases is maximum. Glass crystallization temperature has been found to be 972°C.

2.3. Contact Angle.
We took a small drop of glass (BLM) and placed it upon the sintered ceramic pellet (LMT). The pellet and the glass were inserted inside the instrument furnace. The temperature of the furnace was increased at the rate of 10°C/min. Shape of the glass droplet was recorded at some fixed interval of time. The recorded photographs are shown in Figures 2(a) and 2(b). From Figures 2(a) and 2(b) it can be observed that at 800°C the glass drop

(a)   (b)

(c)

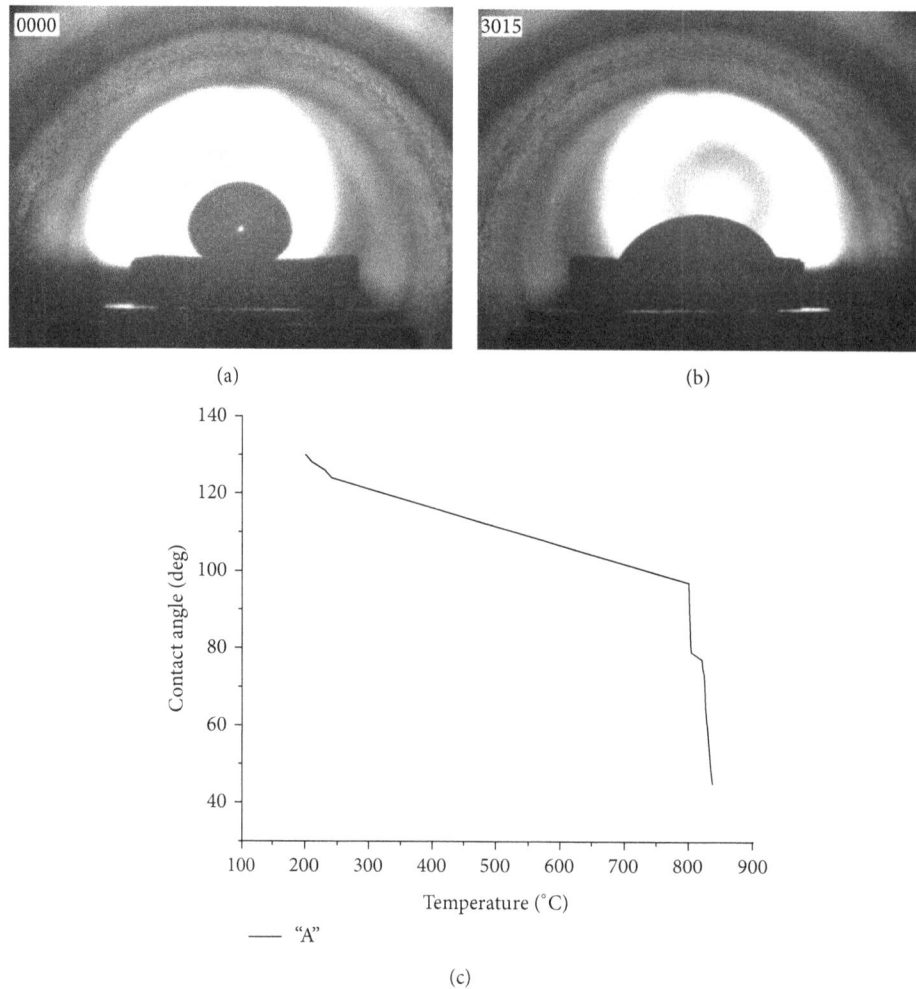

FIGURE 2: Contact angle variation with respect to temperature (a) at 200°C, (b) at 850°C, (c) from 200°C to 850°C.

became softer and the contact angle reduces. At 850°C glass drop reaches the condition of melting. Figure 2(c) shows the variation of contact angle with temperature for glass. It is clear from Figure 2(c) that 850°C is the minimum temperature for sintering the LTCC glass-ceramic sample. In this way glass material lowers the sintering temperature of microwave glass-ceramic and makes it LTCC glass-ceramic.

2.4. Density Measurement. The density of glass-ceramic as a function of sintering temperature is shown in Figure 3. Densities of the pressed pellets were measured from the dimensions and weight of the samples.

The measured density was 2.546 gm/cc. With the increase of sintering temperature, density of glass-ceramic decreased. The reaction between BLM glass and LMT ceramic was enhanced with sintering temperature, which is evident from XRD analysis.

2.5. X-Ray Diffraction. X-ray diffraction (XRD) patterns for LTCC glass-ceramic samples (LMT-BLM) sintered at 850,

900, and 950°C for 4 hr are shown in Figure 4. All the peaks in XRD patterns were matched with JCPDS files of various compound oxides of the constituent elements. LTCC sample sintered at 850°C contains ceramic phases LMT and $LaBO_3$. With increasing sintering temperature LMT phase disappears, and $LaBO_3$ and TiO_2 form as reported by Lin et al. [16].

2.6. Fourier Transform Spectroscopy. The infrared (IR) spectra for the glass-ceramic sample are shown in Figure 5. The IR spectra of these glass-ceramics generally consist of broad and sharp bands in different regions (500 to 5000 cm^{-1}). IR spectra of each sample show a number of absorption bands. The IR absorption spectra of these samples show two peaks for glass-ceramic sample. The peaks are sharp and broad in nature. All the absorption peaks are numbered 1 and 2 starting from low wave number side. In Figure 5 the first absorption peak (1) lies between 492 and 750 cm^{-1}. This observed broad peak is due to stretching of BO_3^- bond inside the glassy network. The second absorption peak (2) lies at 1280 cm^{-1}. It is due to TiO_4^- bond.

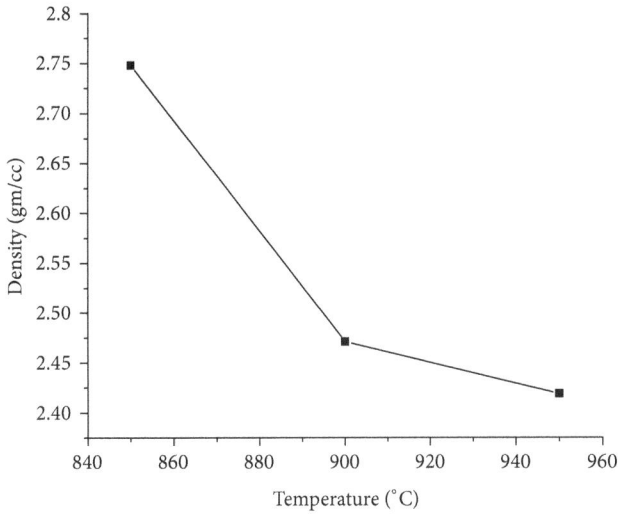

FIGURE 3: Variation of density with temperature.

2.7. Measurement of Dielectric Constant of Glass-Ceramic Material. Microwave dielectric properties of glass-ceramic materials in the frequency range of 2–10 GHz were measured using a vector network analyzer (Agilent 8722ES). The ε_r was obtained from the TE_{011} resonance mode of the end-shorted sample placed between two conducting plates, using the method of Hakki and Coleman [17] and modified by Courtney (Figure 6(a)). The reflection method was used for measuring the Q-factor, in which the sample was placed at the center of a cylindrical resonant cavity having dimensions three times greater than the sample dimensions (Figure 6(b)). The measurement was done with the TE_{011} mode considering weakly-coupled case and the correction for the coupling coefficient was applied [18]. The obtained dielectric constant and loss tangent of the glass-ceramic material are 7.735 and 0.00422, respectively, at 8 GHz.

3. Antenna Configuration, Theoretical Expressions for Resonant Frequency, Q-Factor, and Return Loss Bandwidth

The dual segment CDRA consists of lower segment made from Teflon sheet with dielectric constant $\varepsilon_{r1} = 2.08$ and an upper segment of glass-ceramic (B_2O_3–La_2O_3–MgO glass with $La(Mg_{0.5}Ti_{0.5})O_3$ ceramic) block with $\varepsilon_{r2} = 7.735$ as shown in Figure 7. The dual segment CDRA is placed on a ground plane of size $60 \times 60 \times 4 \, mm^3$. The lower and upper segments of the CDRA have dimensions of $D \times l = 14.15 \times 10 \, mm^2$ and $D \times l_1 = 14.15 \times 6 \, mm^2$, respectively. The CDRA is assumed to be excited by a 50 Ω coaxial probe of outer radius 2 mm and inner radius 0.6 mm. The probe height above the surface of ground plane is found by trial and error to be 9.5 mm to provide lowest return loss at resonant frequency.

The resonant frequency of single segment CDRA excited in $HEM_{11\delta}$ mode can be written as, [1, 19],

$$f = \frac{6.324c}{2\pi a \sqrt{2 + \varepsilon_r}} \left\{ 0.27 + 0.36\frac{a}{2h} + 0.02\left(\frac{a}{2h}\right)^2 \right\}, \quad (1)$$

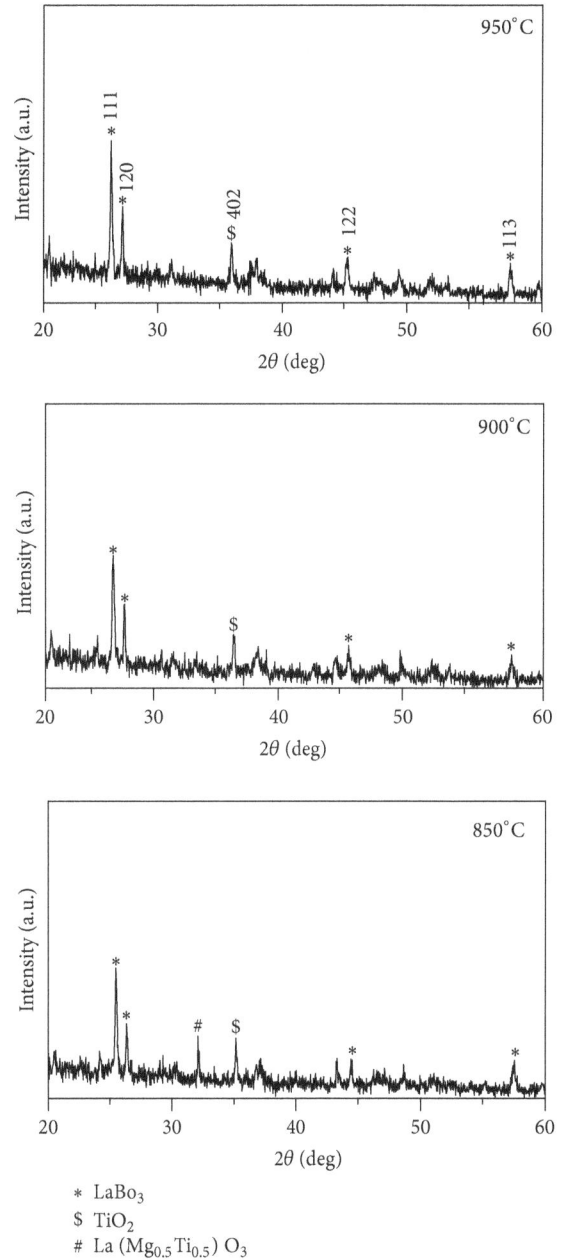

FIGURE 4: X-ray diffraction pattern of glass-ceramic.

where $a = D/2$, D is the diameter of CDRA, h is the height of the CDRA above ground plane, c is the velocity of microwave in free space ($= 3 \times 10^8$ m/sec), and ε_r is the relative permittivity of CDRA material.

Equation (1) has been obtained through curve fitting and numerical simulations based on the method of moments [1, 20].

Radiation Q-factor of isolated CDRA can be written as, [1, 19],

$$Q = 0.01007 \, (\varepsilon_r)^{1.3} \left(\frac{a}{h}\right) \left[1 + 100 \, e^{-[2.05 \, (a/2h)-(1/80(a/h)^2)]} \right]. \quad (2)$$

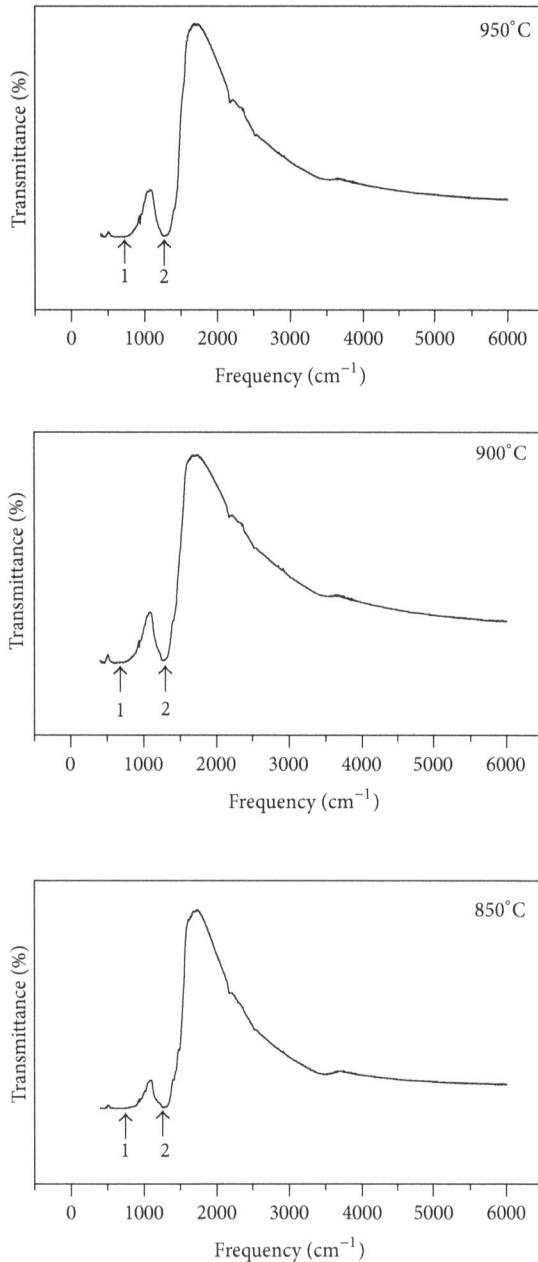

FIGURE 5: Infrared spectra of glass-ceramic sample.

The percentage bandwidth of the isolated CDRA is given by, [1, 19],

$$\% \; BW = \frac{S - 1}{\sqrt{SQ}} \times 100, \qquad (3)$$

where S and Q are the VSWR and the radiation Q-factor of isolated CDRA.

Equations (1)–(3) can be used to compute the resonant frequency, radiation Q-factor, and bandwidth of the dual segment CDRA by replacing CDRA material dielectric constants ε_r and h and ε_{eff} with h_{eff}, respectively. Adopting a simple static capacitance model, the following expressions

for effective permittivity (ε_{eff}) and effective height (h_{eff}) are obtained:

$$\varepsilon_{\text{eff}} = \frac{h_{\text{eff}}}{(1/\varepsilon_{r1}) + (1/\varepsilon_{r2})}, \qquad (4)$$

$$h_{\text{eff}} = l + l_1.$$

See [6], where l and l_1 are the length of lower and upper segments of CDRA, respectively.

The resonant frequency, Q-factor, and bandwidth of the dual segment CDRA computed using (1), (2), and (3) are found to be 6.225 GHz, 1.6644, and 42.4842%, respectively.

4. Experimental Results and Discussion

4.1. Variations of Return Loss and Input Impedance versus Frequency. Measurement of return loss and input impedance of dual segment CDRA as a function of frequency in C/X bands was done using Agilent PNA series vector network analyzer (model number E8364 B). The measured variations of return loss and input impedance with frequency for the CDRA are shown in Figures 8 and 9, respectively. From Figure 8 the resonant frequency, operating frequency range, and the percentage bandwidth of the proposed CDRA are extracted. The resonant frequency, operating frequency range, and the percentage bandwidth of the CDRA in presence of coaxial feed are found to be 6.31 GHz, 6.01–7.69 GHz, and 26.62%, respectively.

The measured resonant frequency of the antenna is nearly in agreement with the theoretical value of 6.225 GHz. The measured bandwidth of the antenna is lower than the theoretical value. The deviation in the results may be due to fabrication tolerances, the possibility of misalignment in the placement of two DRA segments, the effect of finite ground plane not considered in theoretical computation, and the effect of glue used to bind the two DRA segments during fabrication of antenna. In fact using LTCC material in combination with Teflon reduces the effective dielectric constant of the antenna structure as a whole, thereby reducing the radiation Q-factor of DRA which enhances the bandwidth of the antenna. Some other ceramic materials have been used earlier to design probe fed multisegment DRA/DRAs with air gap in which wide bandwidth has been obtained with the help of lower dielectric constant inserts [21–23].

The input resistance at resonant frequency of the CDRA is found to be 50.3 Ω (Figure 9) providing very good impedance match to 50 Ω coaxial feeder. It is also worth noting from Figure 9 that the frequency at which input resistance becomes maximum is slightly lower than that at which input reactance becomes zero. This can be due to the effect of coaxial feed inductance.

4.2. Far Field Performance. The copolar and cross-polar radiation patterns of the CDRA for $\Phi = 0°$ plane (x-z plane) and $\Phi = 90°$ plane (y-z plane) were measured in anechoic chamber at the measured resonant frequency of 6.31 GHz. The experimental setup is not shown here for brevity. The measured E- and H-plane patterns of the CDRA are shown in Figures 10 and 11, respectively.

FIGURE 6: Setup for (a) ε_r measurement and (b) Q-factor measurement.

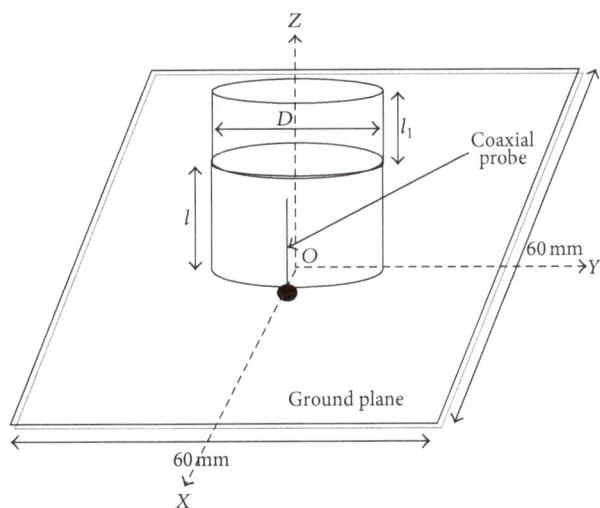

FIGURE 7: Geometry of dual segment CDRA.

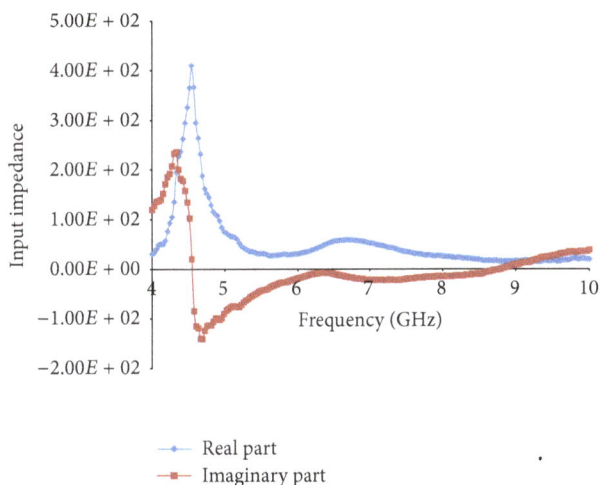

FIGURE 9: Variation of input impedance with frequency for dual segment CDRA fabricated from glass-ceramic material.

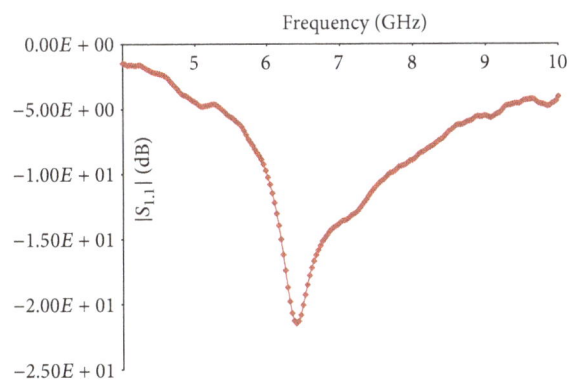

FIGURE 8: Variation of return loss versus frequency of dual segment cylindrical DRAs fabricated from glass-ceramic material.

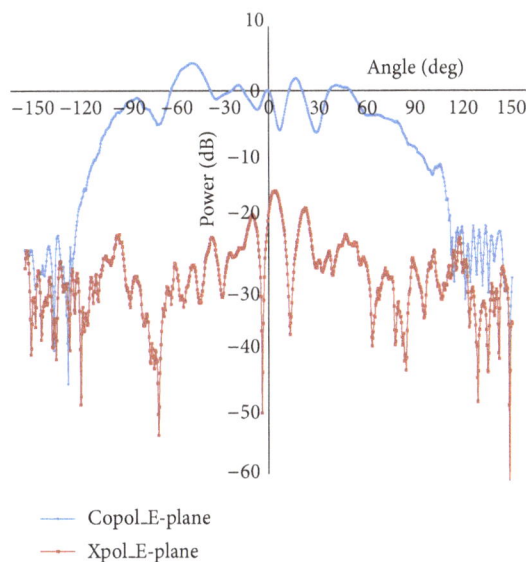

FIGURE 10: E-plane radiation pattern of dual segment CDRA.

The half power beam width (HPBW), first side lobe level and maximum cross-polarized lobe level, of the CDRA are extracted from Figures 10 and 11. The HPBWs of the antenna in E- and H-planes are found to be 51.2° and 59.2°,

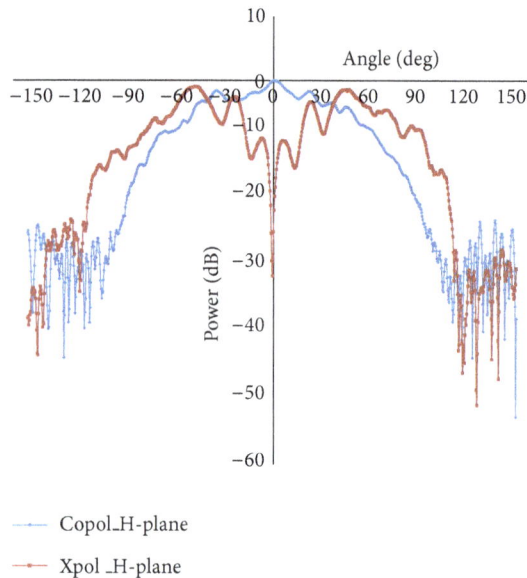

Copol_H-plane

Xpol _H-plane

FIGURE 11: H-plane radiation pattern of dual segment CDRA.

respectively. The first side lobe levels of CDRA in E- and H-planes are 1.904 dB and −1.629 dB loose, respectively, whereas cross-polarized lobe levels of the antenna in these planes are found to be −15.72 dB and −1.335 dB, respectively. Asymmetry in E-plane copolarization pattern can be seen in Figure 10. This may be due to the use of antenna material having very low effective dielectric constant value, which degrades the radiation pattern of the antenna [6]. But at the same time performance improvement of antenna in terms of bandwidth is also obtained due to reduction in its quality factor. Higher cross-polarized lobe level in H-plane ($\Phi = 0°$) is obtained due to significant height (above ground plane) of coaxial probe coupling energy into the antenna which may result in spurious radiation.

The measured gain of the CDRA is found to be 2.18 dB at the resonant frequency of 6.31 GHz.

5. Conclusion

An experimental study on wideband dual segment CDRA fabricated from glass-ceramic material, which is based on $(B_2O_3–La_2O_3–MgCO_3)$ glass and $La(Mg_{0.5}Ti_{0.5})O_3$ ceramic, along with its material characteristics has been carried out. Low temperature cofired glass-ceramic material based on $(B_2O_3–La_2O_3–MgCO_3)$ glass and $La(Mg_{0.5}Ti_{0.5})O_3$ ceramic were fabricated at sintering temperatures of less than 1000°C. From DTA and contact angle studies the minimum sintering temperature for LTCC glass-ceramic material through Liquid-phase sintering procedure is found to be 850°C. With increase in temperature, density of material increases. From the experimental study on proposed dual segment CDRA it is inferred that the proposed antenna provides wide bandwidth and reasonable gain though at the cost of degradation in radiation pattern.

From these results it is concluded that LTCC dielectric resonator antenna may find potential application in wireless communication field for designing a multisegment wideband antenna through proper selection and optimization of segment materials and material composition.

References

[1] R. K. Mongia and P. Bhartia, "Dielectric resonator antennas: a review and general design relations for resonant frequency and bandwidth," *International Journal of Microwave and Millimeter-Wave Computer-Aided Engineering*, vol. 4, no. 3, pp. 230–247, 1994.

[2] R. K. Mongia and A. Ittipiboon, "Theoretical and experimental investigations on rectangular dielectric resonator antennas," *IEEE Transactions on Antennas and Propagation*, vol. 45, no. 9, pp. 1348–1356, 1997.

[3] P. Rezaei, M. Hakkak, and K. Forooraghi, "Design of wide-band dielectric resonator antenna with a two-segment structure," *Progress in Electromagnetics Research*, vol. 66, pp. 111–124, 2006.

[4] D. Kajfez and A. A. Kishk, "Dielectric resonator antenna-possible candidate for adaptive antenna arrays," in *Proceedings of the VITEL, International Symposium on Telecommunications, Next Generation Networks and Beyond*, Portoroz, Slovenia, May 2002.

[5] M. Saed and R. Yadla, "Microstrip-fed low profile and compact dielectric resonator antennas," *Progress in Electromagnetics Research*, vol. 56, pp. 151–162, 2006.

[6] A. Petosa, *Dielectric Resonator Antennas Handbook*, Artech House, London, UK, 2007.

[7] Y. Ge and K. P. Esselle, "A dielectric resonator antenna for UWB applications," in *Proceedings of the IEEE International Symposium on Antennas and Propagation and USNC/URSI National Radio Science Meeting (APSURSI '09)*, Charleston, SC, USA, June 2009.

[8] D. W. Kim, D. G. Lee, and K. S. Hong, "Low-temperature firing and microwave dielectric properties of $BaTi_4O_9$ with Zn-B-O glass system," *Materials Research Bulletin*, vol. 36, no. 3-4, pp. 585–595, 2001.

[9] T. Mailadil Sebastian, "Dielectric materials for wireless communication," in *Low Temperature Cofired Ceramics*, Chapter 12, pp. 445–511, Elsevier publication, 2008.

[10] C. C. Cheng, T. E. Hsieh, and I. N. Lin, "The effect of composition on Ba-Nd-Sm-Ti-O microwave dielectric materials for LTCC application," *Journal of European Ceramic Society*, vol. 79, no. 2-3, pp. 119–123, 2003.

[11] J. M. Wu and H. L. Huang, "Microwave properties of zinc, barium and lead borosilicate glasses," *Journal of Non-Crystalline Solids*, vol. 260, no. 1-2, pp. 116–124, 1999.

[12] M. Valant and D. Suvorov, "Microstructural phenomena in low-firing ceramics," *Materials Chemistry and Physics*, vol. 79, no. 2-3, pp. 104–110, 2003.

[13] R. R. Tummala, "Ceramic and glass-ceramic packaging in the 1990s," *Journal of American Ceramic Society*, vol. 74, no. 5, pp. 895–908, 1991.

[14] D.-Y. Lee, S.-J. Yoon, J. H. Yeo et al., "Crystal structure and microwave dielectric properties of $La(Mg_{1/2}Ti_{1/2})O_3$ ceramics," *Journal of Materials Science Letters*, vol. 19, no. 2, pp. 131–134, 2000.

[15] M. P. Seabra and V. M. Ferreira, "Synthesis of $La(Mg_{0.5}Ti_{0.5})O_3$ ceramics for microwave applications," *Materials Research Bulletin*, vol. 37, no. 2, pp. 255–262, 2002.

[16] H. Lin, L. Luo, and W. Chen, "Microwave dielectric properties of low temperature co-fired glass-ceramic based on B_2O_3-La_2O_3-MgO glass with $La(Mg_{0.5}Ti_{0.5})O_3$ ceramics," *Materials Letters*, vol. 62, no. 4-5, pp. 611–614, 2008.

[17] B. W. Hakki and P. D. Coleman, "A dielectric resonator method of measuring inductive capacities in the millimeter range," *IEEE Transactions on Microwave Theory and Techniques*, vol. 8, no. 4, pp. 402–410, 1960.

[18] D. Kajfez and E. J. Hwan, "Q-factor measurement with network analyzer," *IEEE Transactions on Microwave Theory and Techniques*, vol. 32, no. 7, pp. 666–670, 1984.

[19] A. A. Kishk, A. W. Glisson, and G. P. Junker, "Study of broadband dielectric resonator antennas," in *Proceedings of the Antenna Application Symposium*, pp. 45–68, 1999.

[20] A. A. Kishk, A. W. Glisson, and D. Kajfez, "Computed resonant frequency and far fields of isolated disks," in *Proceedings of the IEEE Antennas and Propagation International Symposium*, pp. 408–411, Ann Arbor, Mich, USA, 1993.

[21] M. Cooper, A. Petosa, A. Ittipiboon, and J. S. Wight, "Investigation of dielectric resonator antennas for L-band communications," in *Antenna Technology and Applied Electromagnetics Symposium (ANTEM '96)*, pp. 167–170, Ottawa, Canada, August 1996.

[22] G. P. Junker, A. A. Kishk, A. W. Glisson, and D. Kajfez, "Effect of fabrication imperfections for ground-plane-backed dielectric-resonator antennas," *IEEE Antennas and Propagation Magazine*, vol. 37, no. 1, pp. 40–47, 1995.

[23] M. Cooper, *Investigation of current and novel rectangular dielectric resonator antennas for broadband applications at L-Band [M.S. thesis]*, Carleton University, 1997.

Permissions

List of Contributors

Kleber Franke Portella and Alex Joukoski
Instituto de Tecnologia para o Desenvolvimento, CP 19067, 81531-980 Curitiba, PR, Brazil

João Bosco Moreira do Carmo and Camila Freitas
PIPE, UFPR, Centro Politécnico, Jardim das Américas, 81531-980 Curitiba, PR, Brazil

Carlos Vicente Gomes Filho
PRODETEC, Instituto de Tecnologia para o Desenvolvimento, CP 19067, 81531-980 Curitiba, PR, Brazil

Cinthya Hoppen
PGERHA, UFPR, Centro Politécnico, Jardim das Américas, 81531-980 Curitiba, PR, Brazil

R. Papitha, M. Buchi Suresh and Roy Johnson
Center for Ceramic Processing, International Advanced Research Centre for Powder Metallurgy and New Materials (ARCI), Balapur, Hyderabad 500005, India

Dibakar Das
School of Engineering Sciences and Technology, University of Hyderabad, Hyderabad 500046, India

Mohannad M. S. Al Bosta
Ph.D. Program in Engineering Science, College of Engineering, Chung Hua University, Hsinchu 30012, Taiwan

Keng-JengMa and Hsi-Hsin Chien
College of Engineering, Chung Hua University, Hsinchu 30012, Taiwan

Robert Lugolole and Sam Kinyera Obwoya
Department of Physics, Kyambogo University, P.O. Box 1, Kyambogo, Kampala, Uganda

V. Moreno and R. M. Bernardino
Graduate Program on Materials Science and Engineering (PGMAT), Federal University of Santa Catarina, 88040-900 Florianópolis, SC, Brazil
Department of Mechanical Engineering, Federal University of Santa Catarina, 88040-900 Florianópolis, SC, Brazil

D. Hotza
Graduate Program on Materials Science and Engineering (PGMAT), Federal University of Santa Catarina, 88040-900 Florianópolis, SC, Brazil
Department of Chemical Engineering, Federal University of Santa Catarina, 88040-900 Florianópolis, SC, Brazil

M. V. Silva, H. A. Al-Qureshi and D. Hotza
Núcleo de Pesquisa emMateriais Cerâmicos eVítreos (CERMAT), Programa de Pós-Graduação emCiência e Engenharia deMateriais (PGMAT), Universidade Federal de Santa Catarina (UFSC), 88040-900 Florianópolis, SC, Brazil

D. Stainer
CMC Tecnologia, Avenida Roberto Galli 1220, 88845-000 Cocal do Sul, SC, Brazil

O. R. K. Montedo
Programa de Pós-Graduação em Ciência e Engenharia de Materiais (PPGCEM), Laboratório de Cerâmica Técnica (CerTec), Universidade do Extremo Sul Catarinense (UNESC), Avenida Universitária, 1105, 88806-000 Criciúma, SC, Brazil

Maria TeresaMalachevsky
Centro Atómico Bariloche, CNEA, Avenida Bustillo 9500, 8400 San Carlos de Bariloche, Argentina
CONICET, Argentina

Diego Rodríguez Salvador, Sergio Leiva and Claudio Alberto D'Ovidio
Centro Atómico Bariloche, CNEA, Avenida Bustillo 9500, 8400 San Carlos de Bariloche, Argentina

J. Wang, Di Yang and H. Conrad
Materials Science and Engineering Department, North Carolina State University, Raleigh, NC 27695-7907, USA

A. Du
Materials and Metallurgy Department, Kunming University of Science and Technology, Kunming, Yunnan 650093, China

R. Raj
Department of Mechanical Engineering, Engineering Center, ECME 114, University of Colorado, Boulder, CO 80309-0427, USA

Tejbir Singh
Department of Physics, Sri Guru Granth Sahib World University, Fatehgarh Sahib, Punjab 140407, India

Gurpreet Kaur
Department of Physics, Maharishi Markandeshwar University, Mullana, Haryana 133207, India

Parjit S. Singh
Department of Physics, Punjabi University, Patiala, Punjab 147002, India

Elias Hanna Bakraji, Rana Abboud and Haissm Issa
Atomic Energy Commission, Chemistry Department, P.O. Box 609, Damascus, Syria

Jon C. Goldsby
Glenn Research Center, National Aeronautics and Space Administration, 21000 Brookpark Road, Cleveland, OH 44135, USA

Ashwini Kumar, Poorva Sharma and Dinesh Varshney
Materials Science Laboratory, School of Physics, Vigyan Bhawan, Devi Ahilya University, Khandwa Road Campus, Indore 452001, India

A. Fernández Solarte, N. Pellegri, O. de Sanctis and M. G. Stachiotti
Laboratorio de Materiales Cerámicos, Universidad Nacional de Rosario, IFIR, CONICET, Avenida Pellegrini 250, S2000BTP Rosario, Argentina

Rashmi Gupta, Seema Verma, Vishal Singh and K. K. Bamzai
Crystal Growth & Materials Research Laboratory, Department of Physics and Electronics, University of Jammu, Jammu 180006, India

E. Bwayo and S. K. Obwoya
Department of Physics, Kyambogo University, P.O. Box 1, Kyambogo, Kampala, Uganda

JunWei, Lokeswarappa R. Dharani and K. Chandrashekhara
Department of Mechanical and Aerospace Engineering, Missouri University of Science and Technology, Rolla, MO 65409-0050, USA

Gregory E. Hilmas and William G. Fahrenholtz
Department of Materials Science and Engineering, Missouri University of Science and Technology, Rolla, MO 65409-0340, USA

Alfonso Salinas and Karen Lozano
Mechanical Engineering Department, The University of Texas-Pan American, Edinburg, TX 78539, USA

Maricela Lizcano
National Aeronautics and Space Administration, Materials and Structures Division, Glenn Research Center at Lewis Field, Cleveland, OH 44135, USA

Omar Aguilar-García, Rafael Lara-Hernández, José L. Gil-Vázquez and Jaime Aguilar-García
Departamento de Ingeniería Industrial, Instituto Tecnológico de Morelia, Avenida Tecnológico 1500, Col. Lomas de Santiaguito, 58120 Morelia, MICH, Mexico

Azucena Arellano-Lara
Departamento de Cerámica, Instituto de Investigaciones Metalúrgicas, Universidad Michoacana de San Nicolás de Hidalgo, Apdo. Postal 888, 58000 Morelia, MICH, Mexico

S. R. Murthy
Department of Physics, Osmania University, Hyderabad 500 007, India

Vijayeta Pal and R. K. Dwivedi
Department of Physics and Material Science & Engineering, Jaypee Institute of Information Technology, Noida 201307, India

O. P. Thakur
Electroceramics Group, Solid State Physics Laboratory, Defence Research and Development Organization (DRDO), Timarpur, Delhi 110054, India

AnaMaría Herrera, Amir AntônioMartins de Oliveira Jr., Antonio Pedro Novaes de Oliveira and Dachamir Hotza
Graduate Program in Materials Science and Engineering (PGMAT), Departments of Chemical (EQA) and Mechanical Engineering (EMC), Federal University of Santa Catarina (UFSC), 88040-900 Florian´opolis, SC, Brazil

Galina Volkova, Oleksandr Doroshkevych, Artem Shylo, Tetyana Zelenyak, Valeriy Burkhovetskiy, Igor Danilenko and Tetyana Konstantinova
Donetsk Institute for Physics and Engineering named after O.O. Galkin, NAS of Ukraine, R. Luxembourg Street, 72, Donetsk 83114, Ukraine

S. M. Ismail and S. S. Attallah
Reactor Physics Department, Nuclear Research Center, Egyptian Atomic Energy Authority, P.O. Box 13759, Cairo, Egypt

Sh. Labib
Nuclear Chemistry Department, Hot Laboratories Center, Egyptian Atomic Energy Authority, P.O. Box 13759, Cairo, Egypt

S. Thirumaran and N. Karthikeyan
Department of Physics (DDE), Annamalai University, Annamalai Nagar 608 002, India

Shanming Ke, Peng Lin and Xierong Zeng
College of Materials Science and Engineering, Shenzhen University, Shenzhen 518060, China
Shenzhen Key Laboratory of Special Functional Materials, Shenzhen 518060, China

Haitao Huang
Department of Applied Physics and Materials Research Center, The Hong Kong Polytechnic University, Hong Kong

Huiqing Fan
School of Materials Science and Engineering, Northwestern Polytechnical University, Xi'an 710072, China

K. Boubaker
École Supérieure de Sciences et Techniques de Tunis (ESSTT), Université de Tunis, 63 Rue Sidi Jabeur, 5100 Mahdia, Tunisia

MehriMashadi
Department ofMaterials Engineering, Faculty of Engineering, Tarbiat ModaresUniversity, P.O. Box 1411713116, Tehran, Iran

Mohsen Mohammadijoo and Zeinab Naderi Khorshidi
Department of Chemical and Materials Engineering, University of Alberta, Edmonton, AB, Canada T6G 2V4

Alireza Honarkar
School of Engineering, Shahid Rajaee University, P.O. Box 16785-136, Tehran, Iran

Vishwanath P. Singh
Department of Physics, Karnatak University, Dharwad 580003, India
Health Physics Section, Kaiga Atomic Power Station-3&4, NPCIL, Karwar 581400, India

N.M. Badiger
Department of Physics, Karnatak University, Dharwad 580003, India

M. Penchal Reddy, I. G. Kim and D. S. Yoo
Department of Physics, Changwon National University, Changwon 641773, Republic of Korea

W. Madhuri
School of Advanced Sciences, VIT University, Vellore 632014, India

M. Venkata Ramana
School of Materials Science and Engineering, Harbin Institute of Technology, Shenzhen 518055, China

N. Ramamanohar Reddy
Department of Materials Science and Nanotechnology, Yogi Vemana University, Kadapa 516227, India

K. V. Siva Kumar, D. V. Subbaiah and R. Ramakrishna Reddy
Department of Physics, Sri Krishnadevaraya University, Anantapur 515055, India

Ali Abdul-Aziz, Ramakrishna T. Bhatt and Joseph E. Grady
NASA Glenn Research Center, Cleveland, OH 44135, USA

Frank Abdi
AlphaSTAR Corporation, Long Beach, CA 90804, USA

Ravi Kumar Gangwar
Department of Electronics Engineering, Indian School of Mines, Dhanbad 826004, India

S. P. Singh
Department of Electronics Engineering, Indian Institute of Technology, Banaras Hindu University, Varanasi 221005, India

Meenakshi Choudhary and D. Kumar
Department of Ceramic Engineering, Indian Institute of Technology, Banaras Hindu University, Varanasi 221005, India

G. Lakshmi Narayana Rao and K. C. James Raju
School of Physics, University of Hyderabad, Hyderabad 500 046, India

www.ingramcontent.com/pod-product-compliance
Lightning Source LLC
Chambersburg PA
CBHW070152240326

41458CB00126B/4262